Proceedings of the Conference on Transformation Groups

New Orleans, 1967

Edited by Paul S. Mostert

Springer-Verlag New York Inc. 1968

Paul S. Mostert
Tulane University
Department of Mathematics
New Orleans, Louisiana/USA

ISBN-13: 978-3-642-46143-9 e-ISBN-13: 978-3-642-46141-5
DOI: 10.1007/978-3-642-46141-5

 Library of Congress Catalog Card Number 68-27313. Title No. 1514

Softcover reprint of the hardcover 1st edition 1968

Preface

These Proceedings contain articles based on the lectures and informal discussions at the Conference on Transformation Groups held at Tulane University, May 8 to June 2, 1967 under the sponsorship of the Advanced Science Seminar Projects of the National Science Foundation (Contract No. GZ 400). They differ, however, from many such Conference proceedings in that particular emphasis has been given to the review and exposition of the state of the theory in its various manifestations, and the suggestion of direction to further research, rather than purely on the publication of research papers. That is not to say that there is no new material contained herein. On the contrary, there is an abundance of new material, many new ideas, new questions, and new conjectures—carefully incorporated within the framework of the theory as the various authors see it.

An original objective of the Conference and of this report was to supply a much needed review of and supplement to the theory since the publication of the three standard works, MONTGOMERY and ZIPPIN, *Topological Transformation Groups*, Interscience Publishers, 1955, BOREL et al., *Seminar on Transformation Groups*, Annals of Math. Surveys, 1960, and CONNER and FLOYD, *Differential Periodic Maps*, Springer-Verlag, 1964. Considering this objective ambitious enough, it was decided to limit the survey to that part of Transformation Group Theory derived from the Montgomery School. Thus, that part of the theory generally referred to as *Topological Dynamics* has been purposely excluded, in order to concentrate on the narrower objektive.

The survey articles contain few proofs. However, they do give adequate bibliographical data to enable the reader to check with the original sources. Since the objective of the Conference (as reflected in these Proceedings) was to review, consolidate, and give new direction to the research in Transformation Groups, such omission is not inappropriate. However, not all articles are of a primarily expository nature. But, in the ordering of the papers, precedence is given to the expository papers, so that the first three or four papers in each part are of this nature, and indeed these papers form the nucleus around which the part is organized, and give it its character. With very few exceptions, all papers, whether expository or not, were either reported on during the Conference or were inspired by work during the Conference.

The most dynamic movement lately in the Theory of Transformation Groups, and at the same time the subject almost totally lacking in coverage by the standard works referred to above, is the theory of compact connected Lie groups acting differentiably on a manifold. Main tools in this branch of the theory have been representation theory and differential topology—particularly the Browder-Novikov theory. We have prevailed upon W. BROWDER to write an exposition of this theory as it applies or is applicable to the Theory of Transformation Groups. This article immediately follows the Introductory Remarks by MONTGOMERY and precedes the division of the material into parts. The four parts are then:

Part I Differentiable Transformation Groups
Part II Algebraic Topological and Other Techniques
Part III Compact Non-Lie Transformation Groups
Part IV Non-Compact Transformation Groups.

These divisions must at some point become arbitrary. Thus, for example, Parts II and III have considerable overlap, with BREDON's review of cohomological methods in II rather than in III, though much of the work there applies equally well to compact non-Lie group actions.

Each part contains a section for short notes and observations, examples, etc., and a section on problems. It should be pointed out, however, that many problems are contained within the various papers and are not repeated in the sections on problems. There is one paper by the HSIANG brothers devoted entirely to problems.

To the contributors of this volume, I wish to extend my sincere thanks for the enthusiasm and energy which they have given to this task, and for the cooperation they have shown in meeting the deadlines. Although it is perhaps unjust to single out any of the individual authors, still the enormous contributions of G. BREDON and W.-Y. HSIANG deserve special mention. Also, special thanks are due H.-T. and M.-C. KU for their help in reading some of the manuscripts for mathematical and typographical errors. To C.-T. YANG for suggesting the Conference, to K. H. HOFMANN for his help in organizing the Conference, and to the Mathematics Department of Tulane University for making available unusually comfortable facilities at the expense of some invonvenience to faculty and graduate students, there is further thanks due. For comments and suggestions as to the proper ordering of problems and about other matters, I wish to express my appreciation to PETER ORLIK.

Finally, to Dr. KLAUS PETERS and Springer-Verlag for the skillful and expeditious handling of the printing and publication, the entire community of transformation group theorists owes a debt of gratitude.

Princeton, May 15, 1968 PAUL S. MOSTERT

Contents

PART II *Algebraic Topological and Other Techniques*

SHORT NOTES

PART III *Compact (Non-Lie) Transformation Groups*

SHORT NOTES

PART IV *Non-Compact Transformation Groups*

List of Participants at the Conference

Bredon, G. E.
Browder, W.
Connell, E.
Fáry, I.
Feldman, L. A.
Floyd, E. E.
Goto, M.
Hofmann, K. H.
Hoo, C.-S.
Hsiang, W.-C.
Hsiang, W.-Y.
Jänich, K.
Kinoshita, S.
Ku, H.-T.
Ku, M.-C.
Kwun, K. W.
Landweber, P. S.
Lee, C. N.
Lee, R.
Lininger, L.
Livesay, G. R.
López de Medrano, S.
Mann, L. N.
Montgomery, D.
Mostert, P. S.
Novikov, S. P.
Orlik, P.
Pak, J.
Raymond, F.
Richardson, R. W.
Roseman, D.
Samelson, H.
Su, J. C.
Thomas, C. B.
Wang, H.-C.
Wasserman, A.
West, J. E.
Williams, R. F.
Yang, C. T.

List of Contributors

BREDON, G. E., Rutgers University, Department of Mathematics, New Brunswick, New Jersey/USA.

BROWDER, W., Princeton University, Department of Mathematics, Princeton, New Jersey/USA.

CALABI, E., Institut des Hautes Études Scientifiques, Bures-sur-Yvette/France.

FÁRY, I., University of California at Berkeley, Department of Mathematics, Berkeley, California/USA.

HIRZEBRUCH, F., Mathematisches Institut der Universität, Bonn/Germany.

HOFMANN, K. H., Tulane University, Department of Mathematics, New Orleans, Lousiana/USA.

HSIANG, W.-C., Yale University, Department of Mathematics, New Haven, Connecticut/USA.

HSIANG, W.-Y., University of Chicago, Department of Mathematics, Chicago, Illinois/USA.

JÄNICH, K., Mathematisches Institut der Universität, Bonn/Germany.

KINOSHITA, S., Florida State University, Department of Mathematics, Tallahassee, Florida/USA.

KU, H.-T., University of Massachusetts, Department of Mathematics, Amherst, Massachusetts/USA.

KU, M.-C., University of Massachusetts, Department of Mathematics, Amherst, Massachusetts/USA.

KWUN, K. W., Michigan State University, Department of Mathematics, East Lansing, Michigan/USA.

LEE, C. N., University of Michigan, Department of Mathematics, Ann Arbor, Michigan/USA.

LEE, R., University of Michigan, Department of Mathematics, Ann Arbor, Michigan/USA.

LIVESAY, G. R., Cornell University, Department of Mathematics, Ithaca, New York/USA.

LÓPEZ DE MEDRANO, S., Princeton University, Department of Mathematics, Princeton, New Jersey/USA, and Universidad National Autónoma de Mexico, INIC, Mexico City/Mexico.

MANN, D. N., University of Massachusetts, Department of Mathematics, Amherst, Massachusetts/USA.

MONTGOMERY, D., The Institute for Advanced Study, Princeton, New Jersey/USA.

MOSTERT, P. S., Tulane University, Department of Mathematics, New Orleans, Lousiana/USA.

NEUMANN, W. D., Mathematisches Institut der Universität, Bonn/Germany.

ORLIK, P., The Institute for Advanced Study, Princeton, New Jersey/USA.

RAYMOND, F., University of Michigan, Department of Mathematics, Ann Arbor, Michigan/USA.

RICHARDSON, R., University of Washington, Department of Mathematics, Seattle, Washington/USA.

SCHAFER, J. A., University of Michigan, Department of Mathematics, Ann Arbor, Michigan/USA.

SU, J. C., University of Massachusetts, Department of Mathematics, Amherst, Massachusetts/USA.

THOMAS, C. B., The University of Hull, Department of Pure Mathematics, Hull/England.

WEST, J. E., University of Kentucky, Department of Mathematics, Lexington, Kentucky/USA.

WILLIAMS, R. F., Northwestern University, Department of Mathematics, Evanston, Illinois/USA.

YANG, C. T., University of Pennsylvania, Department of Mathematics, Philadelphia, Pennsylvania/USA.

Introductory Remarks

DEANE MONTGOMERY

The study of transformation groups originated in the use of groups in geometry, in the work of Lie on continuous groups, and in the work of POINCARÉ on dynamical systems. L. E. J. BROUWER's work on periodic transformations and on the foundations of continuous groups provided further stimulus. Around 1930 topological groups were given a setting in general spaces and there was a renewed interest in foundations of continuous groups. A subgroup of a topological group acts on the group by left or right translation, which provides a relation between topological groups and transformation groups. In considering this relation it was often felt that a "small" compact group could not act effectively on a manifold. This question in full generality remains open although the part of it relating to continuous groups has been settled, that is, it is known that a locally euclidean group is a Lie group so that it can not contain "small" non-trivial subgroups. In 1932 NEWMAN established some basic results on periodic transformations of manifolds. Beginning a little later P. A. SMITH showed that a periodic transformation of prime power period on a sphere or euclidean space resembles rather closely a linear transformation in the sense of homology modulo the prime. Later examples showed that the same result was not true with integer coefficients. In the period beginning at this time most of the work made use of the methods of point set or algebraic topology. However in recent years there has been considerable interest in using the methods of differential topology. This has not only permitted progress on some of the older problems but has led to several new kinds of problems on many of which there is now a considerable body of results.

It is good to meet to talk about these new results and the further problems that suggest themselves. Mathematicians need to meet for this professional reason, and an added pleasure in meeting is that they understand each other better than they understand others and than others understand them. It may not be true mathematics is as thick as blood but it *is* thicker than water.

Introductory Remarks

Surgery and the Theory of Differentiable Transformation Groups

William Browder*

Recent progress in differential topology has developed many powerful techniques for the study of differentiable manifolds, for classification of manifolds, embeddings, diffeomorphisms, etc. These methods have been applied to the study of differentiable transformation groups with some notable successes. I will try to outline in this paper some results of the theory of surgery (or spherical modification) which have given many good results about manifolds, and show how these results may be applied to study actions of groups, in particular actions of S^1 and S^3.

For the most part, the results of Chapters I and II are either well known, or in print in various places, or both, and include the work of many persons. Many of the results of Chapter III on semifree actions of S^1 are new however.

Chapter I is an exposition of the theory of surgery on 1-connected manifolds as developed by J. Milnor, M. Kervaire, S. P. Novikov, the author, and others, and gives some general applications in differential topology. The point of view follows that of [7] in general.

Chapter II discusses free actions of S^1 and S^3 on homotopy spheres, ways of constructing them, and questions about invariant and characteristic spheres. The results in this area are due for the most part to the Hsiang brothers, Montgomery and Yang, and M. Rothenberg (unpublished in the latter case).

In Chapter III, we discuss semifree actions of S^1 on homotopy spheres, that is, actions which are free outside the fixed point set F, under the additional assumption that F is a homotopy sphere. The approach here is new and yields many new results as well as new proofs of known results. In particular, we construct infinitely many inequivalent semifree S^1 actions on homotopy spheres with fixed point set the standard sphere.

We have omitted many things. In Chapter I we have not discussed Sullivan's point of view on the surgery problem [47], or surgery on non-simply connected manifolds, such as the theory of [51]. In Chap-

* The author was partially supported by an NSF grant.

ters II and III we have emphasized the construction of examples, rather than classification. In Chapter III we have discussed only S^1 actions, although a similar theory may be constructed for S^3 actions. Also the results may be extended to construct infinitely many actions with S^1 as fixed point set, by extending some of the results of Chapter I in a simple way. Throughout, we have not discussed $\mathbb{Z}_2$-actions, though many results are known.

Many questions remain open which seem amenable to attack in this direction:

1. What are the homotopy spheres which are being operated on in our constructions, and what is the knot type of the fixed point set?

2. In Chapter III, in the cases where our theorem does not construct an infinite number of semifree S^1 actions, are there in reality only a finite number? It seems likely when the dimension of the fixed point set F^q (a homotopy sphere) is even.

3. How can one study the problem of invariant spheres for non-free actions, in particular, for semifree actions?

This work on transformation groups has benefited greatly from contact with many people. In particular my collaboration with G. R. LIVESAY began my interest in this direction, and it was further stimulated and educated by conversations and lectures by W.-C. HSIANG, W.-Y. HSIANG, D. MONTGOMERY, C. T. YANG, G. BREDON, C. GIFFEN, and others.

I. The Surgery Problem and the Fundamental Results

All manifolds we shall deal with will be compact oriented with boundary, equipped with a differential structure, and diffeomorphisms will preserve orientation. In fact all the theorems of this section are valid in the piecewise linear (p. l.) category also (see [15]), using p. l. microbundles instead of linear bundles (see [32]) for stable normal bundles, and using p. l. "block bundles" (see [43, 39]), in place of normal bundles of embeddings. However we shall restrict our attention for the most part to the differentiable case.

In this chapter we shall describe the results of the theory of surgery for simply connected manifolds. The account here follows that of [7] closely, and may be considered a globalization of the paper of KERVAIRE-MILNOR [29] due to S. P. NOVIKOV ([40, 41]) and the author [8].

1. Poincaré Pairs

A pair (X, Y) is called an *m-dimensional Poincaré pair over R*, (R a ring) if there is an element $[X] \in H_m(X, Y; R)$ such that

$[X]\cap: H^q(X;R)\to H_{m-q}(X,Y;R)$ is an isomorphism for all q. If $R=\mathbb{Z}$, the ring of integers, we call (X,Y) a Poincaré pair, and if $Y=\phi$ we call X a Poincaré complex over R. (Recall that $x\cap y$ is defined on the chain level by the formula $x\cap y=\Sigma_i(y(x_i'))x_i$, where $y\in C^*$, $x\in C_*$, and $\Delta x=\Sigma_i x_i\otimes x_i'$, $\Delta: C_*\to C_*\otimes C_*$ being a chain approximation to the diagonal.) The element $[X]\in H_m(X,Y;R)$ is called the *fundamental class* of X, or the *orientation class* if $R=\mathbb{Z}$.

We recall some simple properties of Poincaré pairs developed in [7]:

(1.1) The diagram below is commutative (up to a sign) and all the vertical arrows are isomorphisms.

$$\begin{array}{ccccccccc}
\cdots\longrightarrow & H^{q-1}(Y;R) & \xrightarrow{\delta} & H^q(X,Y;R) & \xrightarrow{j^*} & H^q(X;R) & \xrightarrow{i^*} & H^q(Y;R) & \longrightarrow\cdots\\
& \downarrow{\scriptstyle(\partial[X])\cap} & & \downarrow{\scriptstyle[X]\cap} & & \downarrow{\scriptstyle[X]\cap} & & \downarrow{\scriptstyle(\partial[X])\cap} & \\
\cdots\longrightarrow & H_{m-q}(Y;R) & \xrightarrow{i_*} & H_{m-q}(X;R) & \xrightarrow{j_*} & H_{m-q}(X,Y;R) & \xrightarrow{\partial} & H_{m-q-1}(Y;R) & \longrightarrow\cdots
\end{array}$$

(In particular Y is a Poincaré complex over R of dimension $m-1$ with fundamental class $[Y]=\partial[X]$.)

Now we shall describe how to "add" Poincaré pairs over a Poincaré pair contained in the boundary. One must have a compatibility relation among the fundamental classes.

Let $X=X_1\cup X_2$, $X_0=X_1\cap X_2$, $Y_i=Y\cap X_i$. Let $x\in H_m(X,Y;R)$, and let $x_i\in H_m(X_i,X_0\cup Y_i;R)$ be the image under the composite

$$H_m(X,Y;R)\longrightarrow H_m(X,X_{i+1}\cup Y;R)\xleftarrow{\cong} H_m(X_i,X_0\cup Y_i;R),\ i=1,2,$$

$$(i+1=1 \text{ for } i=2),\ x_0\in H_{m-1}(X_0,Y_0;R),\ x_0=\partial_1 x_1=-\partial_2 x_2,$$

$\partial_i: H_m(X_i,X_0\cup Y_i;R)\to H_{m-1}(X_0,Y_0;R)$ being the boundary operator in the triple $(X_i,X_0\cup Y_i,Y_i)$, composed with the inverse of the excision $H_{m-1}(X_0,Y_0;R)\xrightarrow{\cong} H_{m-1}(X_0\cup Y_i,Y_i;R)$.

(1.2) (Addition property) Two of the following three statements imply the third:

a) (X,Y) is a Poincaré pair over R with fundamental class x.

b) $(X_i,X_0\cup Y_i)$ is a Poincaré pair over R with fundamental class x_i, $i=1$ and 2.

c) (X_0,Y_0) is a Poincaré pair over R with fundamental class x_0.

The theorem (1.2) may be used in various directions for various purposes: for example, in the form (a), (c) imply (b) it is an important algebraic step in the proof of [9, Theorem 1.1]. In this account we shall use it to define sums of Poincaré complexes, i.e., we use (b), (c) imply (a).

If (X,Y) and (X',Y') are Poincaré pairs (over R), a map $f:(X,Y)\to(X',Y')$ is said to be *of degree* 1 if $\bar{f}_*[X]=[X']$, where $\bar{f}_*$ denotes the induced homology map $\bar{f}_*: H_m(X,Y;R)\to H_m(X',Y';R)$. If f is a map of degree 1, then we may define inverses for $\bar{f}_*: H_i(X, Y; R) \to H_i(X', Y'; R)$ and $f_*: H_i(X; R) \to H_i(X'; R)$ and the analogous cohomology maps as follows:

Consider the diagram

$$\text{(1.3)} \qquad \begin{array}{ccc} H^{m-i}(X;R) & \xleftarrow{f^*} & H^{m-i}(X';R) \\ {\scriptstyle [X]\cap}\downarrow\cong & & {\scriptstyle [X']\cap}\downarrow\cong \\ H_i(X,Y;R) & \xrightarrow{\bar{f}_*} & H_i(X',Y';R) \end{array}$$

It follows easily from the properties of $\cap$-products that $\bar{f}_*([X]\cap f^*(x')) = [X']\cap x'$, for $x' \in H^{m-i}(X'; R)$. If we define $\bar{\alpha}_*: H_i(X', Y'; R) \to H_i(X, Y; R)$ by $\bar{\alpha}_*([X']\cap x') = [X]\cap f^*(x')$, then $\bar{f}_* \bar{\alpha}_* = 1$. A similar definition defines $\alpha_*: H_i(X'; R) \to H_i(X; R)$ such that $f_* \alpha_* = 1$ and similar splitting maps in cohomology, $\bar{\alpha}^*$, α^*, as follows: $\alpha^*: H^{m-i}(X; R) \to H^{m-i}(X'; R)$ is defined by $[X']\cap \alpha^*(x) = \bar{f}_*([X]\cap x)$ for $x \in H^{m-i}(X; R)$, and $\bar{\alpha}^*: H^{m-i}(X, Y; R) \to H^{m-i}(X', Y'; R)$ is defined similarly, so that $\alpha^* f^* = 1$ on $H^*(X'; R)$, $\bar{\alpha}^* \bar{f}^* = 1$ on $H^*(X', Y'; R)$.

Clearly the same definitions work for Y to provide $\beta_*: H_*(Y'; R) \to H_*(Y; R)$ and $\beta^*: H^*(Y; R) \to H^*(Y'; R)$ and we define:

$$\text{(1.4)} \qquad \begin{aligned} K_q(X;R) &= (\ker f_*)_q \subset H_q(X;R), \\ K^q(X;R) &= (\ker \alpha^*)^q \subset H^q(X;R), \\ K_q(X,Y;R) &= (\ker \bar{f}_*)_q \subset H_q(X,Y;R), \\ K^q(X,Y;R) &= (\ker \bar{\alpha}^*)^q \subset H^q(X,Y;R), \\ K_q(Y;R) &= (\ker (f|Y)_*)_q \subset H_q(Y;R), \\ K^q(Y;R) &= (\ker \beta^*)^q \subset H^q(Y;R). \end{aligned}$$

One may then prove the following properties of the K^* and K_*:

(1.5) There is a commutative (up to sign) diagram with exact rows and split columns:

$$\begin{array}{ccccccccc} & & 0 & & 0 & & 0 & & 0 & \\ & & \uparrow & & \uparrow & & \uparrow & & \uparrow & \\ \to & & K^{q-1}(Y;R) & \to & K^q(X,Y;R) & \to & K^q(X;R) & \to & K^q(Y;R) & \to \\ & & \uparrow & & \uparrow & & \uparrow & & \uparrow & \\ \to & & H^{q-1}(Y;R) & \to & H^q(X,Y;R) & \to & H^q(X;R) & \to & H^q(Y;R) & \to \\ & & \uparrow{\scriptstyle (f|Y)^*} & & \uparrow{\scriptstyle \bar{f}^*} & & \uparrow{\scriptstyle f^*} & & \uparrow & \\ \to & & H^{q-1}(Y';R) & \to & H^q(X',Y';R) & \to & H^q(X';R) & \to & H^q(Y';R) & \to \\ & & \uparrow & & \uparrow & & \uparrow & & \uparrow & \\ & & 0 & & 0 & & 0 & & 0 & \end{array}$$

A similar diagram exists in homology with similar properties.

(1.6) The functors K^* and K_* satisfy Poincaré duality, that is $[X]\cap$ carries $K^*(X,Y;R)$ into $K_*(X;R)$ etc. and $[X]\cap$ induces

$$[X]\cap : K^q(X,Y;R) \longrightarrow K_{m-q}(X;R),$$
$$[X]\cap : K^q(X;R) \longrightarrow K_{m-q}(X,Y;R),$$
$$(\partial[X])\cap : K^{q-1}(Y;R) \longrightarrow K_{m-q}(Y;R)$$

which are isomorphisms for all q.

(1.7) K^* and K_* satisfy the formulas of Universal Coefficients. In particular, if $f:(X,Y)\to(X',Y')$ is a map of degree 1 of Poincaré pairs (over $\mathbb{Z}$) then

$$K^q(X,Y;R)=\mathrm{Hom}(K_q(X,Y;\mathbb{Z}),R)+\mathrm{Ext}(K_{q-1}(X,Y;\mathbb{Z}),R)$$

and

$$K_q(X,Y;R)=K_q(X,Y;Z)\otimes R+\mathrm{Tor}(K_{q-1}(X,Y;\mathbb{Z}),R),$$

and similar formulas for $K^*(X;R)$, $K^*(Y;R)$, etc. If R is a field and f is a map of degree 1 of Poincaré pairs (over R) then

$$K^q(X,Y;R)=\mathrm{Hom}(K_q(X,Y;R),R)$$

and similar formulas for $K^*(X;R)$, $K^*(Y;R)$.

Taking $R=\mathbb{Q}$, the rational numbers, suppose f is a map of degree 1 over $\mathbb{Q}$ and suppose that $(f|Y)^*:H^*(Y';\mathbb{Q})\to H^*(Y;\mathbb{Q})$ is an isomorphism. Then from (1.5) it follows that $K^*(X,Y;\mathbb{Q})\cong K^*(X;\mathbb{Q})$. Then from (1.8) and the identity relating cup and cap products: $z([X]\cap y)=(y\cup z)([X])$, $z\in H^*(X;R)$, $y\in H^*(X,Y;R)$, we may deduce that the pairing:

(1.8)
$$K^i(X,Y;R)\otimes K^{m-i}(X,Y;R)\to R,$$
$$(x,y)=(x\cup y)[X],$$

is for $R=\mathbb{Q}$, a non-singular pairing, and if $i=m-i$ is even, it is symmetric.

Thus suppose $f:(X,Y)\to(X',Y')$ is a map of degree 1 of Poincaré pairs over $\mathbb{Q}$, and that $(f|Y)^*:H^*(Y';\mathbb{Q})\to H^*(Y;\mathbb{Q})$ is an isomorphism, and that $m=4q$. Then we may define the index of f

(1.9) $$I(f)=\text{signature of } (\ ,\ ) \text{ on } K^{2q}(X,Y;\mathbb{Q}).$$

Now in the splitting

$$H^*(X,Y;\mathbb{Q})=\text{image}\,\bar{f}^*+K^*(X,Y;\mathbb{Q})$$

it is easy to prove that the two factors are orthogonal under the pairing $(\ ,\ )$ and the pairing on image $\bar{f}^*$ is isomorphic to that on $H^*(X',Y';\mathbb{Q})$. It follows that if we define $I(X)=$ signature of $(\ ,\ )$ on $H^{2q}(X,Y;\mathbb{Q})$,

(1.10) $$I(f)=I(X)-I(X').$$

Now suppose $m=4k+1$, so that dimension $Y=4k$ and suppose $f\colon (X,Y)\to(X',Y')$ is a map of degree 1 of Poincaré pairs over $\mathbb{Q}$. Then $I(f|Y)$ is defined. On the other hand in $K^{2k}(Y;\mathbb{Q})$ the image of $K^{2k}(X;\mathbb{Q})$ is a subspace of half the rank, by Poincaré duality (1.6) and (1.7), and annihilates itself under the pairing. It follows that

(1.11) $$I(f|Y)=0.$$

Now suppose the pairs (X,Y) and (X',Y') are the "sums" of pairs, $X=X_1\cup X_2$, $X_1\cap X_2=X_0$, $Y_i=Y\cap X_i$, (X,Y) and $(X_i,X_0\cup Y_i)$ are Poincaré pairs over R, $i=1,2$ with compatible orientations (see (1.2)), and similarly for (X',Y'). Suppose $f\colon (X,Y)\to(X',Y')$ is a map of degree 1 over R and that $f(X_i)\subset X_i'$. It follows easily that $f_i=f|X_i$: $(X_i,X_0\cup Y_i)\to(X_i',X_0'\cup Y_i')$, $i=1,2$, and $f_0=f|X_0,Y_0)\to(X_0',Y_0')$ are maps of degree 1 over R.

Suppose that $(f_i|Y_i\cup X_0)^*$ $i=1,2$ are isomorphisms and that f_0^* is an isomorphism (all with $\mathbb{Q}$ coefficients). Then it is not hard to see that $K^{2k}(X,Y;\mathbb{Q})$ splits as the direct sum of $K^{2k}(X_1,X_0\cup Y_1;\mathbb{Q})$ and $K^{2k}(X_2,X_0\cup Y_2;\mathbb{Q})$ and we get if $\dim(X,Y)=4k$:

(1.12) (Addition property of index) $I(f)=I(f_1)+I(f_2)$.

We will want to define an analogous invariant in $\mathbb{Z}_2$ in dimensions $4k+2$, using $\mathbb{Z}_2$ cohomology, but its definition requires much more structure that the definition of index, and we shall not give its explicit definition here, (see [7, 29 and 14]).

2. Normal Maps and Cobordisms

Let (X,Y) be a pair of spaces, ξ^k a linear k-plane bundle over X. Let $(M^m,\partial M^m)$ be an m-dimensional compact differential manifold with boundary, ν^k its normal bundle for a smooth embedding

$$(M,\partial M)\subset(D^{m+k},S^{m+k-1}),\quad k>m+1.$$

(In such dimensions it follows that embeddings exist and are unique up to isotopy, so that ν^k is also unique.)

A normal map of $((M,\partial M),\nu)$ into $((X,Y),\xi)$ will consist of two maps (f,b), where $f\colon (M,\partial M)\to(X,Y)$ is a continuous map, and $b\colon \nu\to\xi$ is a linear bundle map lying over f.

A cobordism of a map $f\colon (M,\partial M)\to(X,Y)$ is a cobordism W of $(M,\partial M)$, i.e., a manifold W^{m+1} with $\partial W=M\cup V\cup M'$, where $\partial V=\partial M\cup\partial M'$, together with a map $F\colon (W,V)\to(X,Y)$ such that $F|M=f$. If ω^k is the normal bundle of $(W,V)\subset(D^{m+k}\times I,\ S^{m+k-1}\times I)$ (where $(M,\partial M)\subset(D^{m+k}\times 0,\ S^{m+k-1}\times 0)$, $(M',\partial M')\subset(D^{m+k}\times 1,\ S^{m+k-1}\times 1)$), and if (f,b) is a normal map, $b\colon \nu\to\xi$ over f, then a normal cobordism of (f,b) will be a cobordism (W,F) of f together with a linear bundle map $B\colon \omega\to\xi$ lying over F such that $B|(\omega|M)=B|\nu=b$.

We shall say that the cobordism is rel Y if $V=(\partial M\times I)$, and $F|V=f|\partial M$, $B|(\omega|V)=b|(\nu|\partial M)$. (We note that if $Y=\emptyset$, this is automatic.)

Now there is a natural collapsing map
$C\colon (D^{m+k}, S^{m+k-1})\to(T(\nu), T(\nu|\partial M))$ where $T(\nu)$ is the Thom complex of ν, $T(\nu)=E(\nu)/E_0(\nu)$, where $E(\nu)$ is the unit disk bundle, $E_0(\nu)$ is the unit sphere bundle associated to ν. The linear map $b\colon \nu\to\xi$ induces a map of Thom complexes $T(b)$: $(T(\nu), T(\nu|\partial M))\to(T(\xi), T(\xi|Y))$, and $T(b)_*$: $\pi_{m+k}(T(\nu), T(\nu|\partial M))\to\pi_{m+k}(T(\xi), T(\xi|Y))$. Then the element $T(b)_*(\{C\})\in\pi_{m+k}(T(\xi), T(\xi|Y))$ will be called the *Thom invariant* of the normal map (f,b). Then as a standard application of the Thom transversality theorem (see [49, 1, 7]) we have

(2.1) Theorem. *The Thom invariant establishes a one to one correspondence between normal cobordism classes of maps of m-manifolds into (X,Y), ξ, and elements in $\pi_{m+k}(T(\xi), T(\xi|Y))$.*

This theorem allows one to apply the methods of homotopy theory to the study of normal cobordism of normal maps, to calculate upper bounds to the number of such classes etc.

Now suppose (X,Y) is a Poincaré pair of dimension m and that (f,b) is a normal map $f\colon(M^m,\partial M^m)\to(X,Y)$, $b\colon\nu\to\xi$ such that f is of degree 1. Suppose that ν and ξ are oriented and that b preserves the orientation. First we recall

(2.2) Thom Isomorphism Theorem. *If ξ is an oriented k-plane bundle over X, then there is a class $U\in H^k(T(\xi))$ such that U restricted to each fibre is the orientation class, and such that*

$$\begin{aligned}
\cup U\colon H^q(X) &\longrightarrow H^{q+k}(T(\xi))\\
\cup U\colon H^q(X,Y) &\longrightarrow H^{q+k}(T(\xi),T(\xi|Y))\\
\cap U\colon H_p(T(\xi)) &\longrightarrow H_{p-k}(X)\\
\cap U\colon H_p(T(\xi),T(\xi|Y)) &\to H_{p-k}(X,Y)
\end{aligned}$$

are isomorphisms.

(See [49, 33, 27].)

Here the cup and cap products are defined using the isomorphisms

$$H_*(T(\xi))\cong H_*(E(\xi),E_0(\xi)),$$
$$H_*(T(\xi),T(\xi|Y))\cong H_*(E(\xi),E_0(\xi)\cup E(\xi|Y)),$$

(and the corresponding isomorphisms in cohomology) and the appropriate products in the pair, $(E(\xi), E_0(\xi))$, etc.

It follows easily that if $b: \nu \to \xi$ preserves orientation, then the appropriate diagrams involving the Thom isomorphisms commute, for example:

$$\begin{array}{ccc} H_p(T(\nu), T(\nu|\partial M)) & \xrightarrow{T(b)_*} & H_p(T(\xi), T(\xi|Y)) \\ \downarrow \cap U_\nu & & \downarrow \cap U_\xi \\ H_{p-k}(M, \partial M) & \xrightarrow{f_*} & H_{p-k}(X, Y) \end{array}$$

Hence if f_* is of degree 1, then $T(b)_*$ is of "degree 1" also. In particular, since if g is a generator of $H_{m+k}(D^{n+k}, S^{m+k-1})$, $C_*(g) = x \in H_{m+k}(T(\nu), T(\nu|\partial M))$, such that $x \cap U_\nu = [M]$, it follows that the Thom invariant $\alpha \in \pi_{m+k}(T(\xi), T(\xi|Y))$ of the normal map (f,b) of degree 1 satisfies

$$h(\alpha) \cap U_\xi = [X], \tag{2.3}$$

where $h: \pi_* \to H_*$ is the Hurewicz homomorphism.

Using (2.1) and (2.3) we get:

(2.4) *Normal cobordism classes of normal maps of degree* 1 *of m-manifolds into* (X, Y), ξ^k *are in one-to-one correspondence with elements* $\alpha \in \pi_{m+k}(T(\xi), T(\xi|Y))$ *such that* $h(\alpha) \cap U_\xi = [X]$, *the correspondence being given by the Thom invariant.*

In particular two such elements differ by an element in kernel h so that we get:

Normal cobordism classes of normal maps of degree 1 *of m manifolds into* (X, Y), ξ *are in one-to-one correspondence with the elements of kernel h,*

$$h: \pi_{m+k}(T(\xi), T(\xi|Y)) \to H_{m+k}(T(\xi), T(\xi|Y)).$$

In the stable range, it is a well known result of homotopy theory that kernel h is finite. Thus there are only a finite number of normal cobordism classes of normal maps of degree 1.

Now if (f,b) is a normal map of degree 1, such that $(f|\partial M)_*: H_*(\partial M) \to H_*(Y)$ is an isomorphism, $m = 4k$, then we have:

(2.5) $\qquad$ *$I(f)$ is divisible by* 8.

This follows from the fact that $I(f)$ is the index of the intersection form on $K_{2k}(M; \mathbb{Z})$/torsion, and this form has determinant 1 and is even (i.e. $(x,x) \in 2\mathbb{Z}$ for all $x \in K_{2k}(M; \mathbb{Z})$/torsion). But a unimodular symmetric matrix over $\mathbb{Z}$ with even diagonal entries has determinant divisible by 8, (see [34]). That the form is even may be deduced from the fact that a normal map sends Wu classes to Wu classes, or in several other ways (c. f. [7]).

Then besides $I/8$ in dimension $4k$, for a normal map (f,b), $f:(M^m,\partial M)\to(X,Y)$, $m=4k+2$, such that $(f|\partial M)_*: H_*(\partial M;\mathbb{Z}_2)\to H_*(Y;\mathbb{Z}_2)$ is an isomorphism, we may define an invariant called the *Kervaire invariant*, or *Arf invariant* (it is given as the Arf invariant of a quadratic form on $(\ker f_*)_{2k+1}$ over $\mathbb{Z}_2$). Then we have the basic theorem:

(2.6) Invariant Theorem. *Let (f,b), $f:(M,\partial M)\to(X,Y)$, $b:\nu\to\xi$ be a normal map of degree* 1, *$f_*: H_*(\partial M)\to H_*(Y)$ an isomorphism, (X,Y) a Poincaré pair of dimension m, etc. Then there is an invariant σ defined with the following properties:*

(i) *$\sigma\in\mathbb{Z}$, if $m=4k$, $\sigma=I(f)/8$,*
$\sigma\in\mathbb{Z}_2$, if $m=4k+2$, $\sigma=$ Kervaire invariant,
$\sigma=0$, if m is odd.

(ii) *If (f,b) is normally cobordant* rel Y *to (f',b') such that $f'_*: H_*(M')\to H_*(X)$ is an isomorphism, then $\sigma=0$.*

(iii) *$\sigma(f|\partial M, b|(\nu|\partial M))=0$.*

(iv) *σ is additive, i.e. if $M=M_1\cup M_2$, $M_0^{m-1}=M_1\cap M_2$, $X=X_1\cup X_2$, $X_0=X_1\cap X_2$, etc., as in* (1.2), *$(M,\partial M)$ and (X,Y) are the sum of two manifolds and two Poincaré pairs, respectively, (f,b) is a normal map of degree* 1, *$f(M_i)\subset X_i$, so that $f_i=f|M_i$, $b_i=b|(\nu|M_i)$, etc., (f_i,b_i) $i=1,2$ are normal maps, and σ is defined for each. Then*

$$\sigma(f)=\sigma(f_1)+\sigma(f_2).$$

(v) *If $Y=\phi$ and $m=4k$, then $\sigma(f,b)=\frac{1}{8}$* (Index M – Index X).

In the case $m=4k$, most of these properties follow easily from the results of § 1. The definition and properties of the Kervaire invariant $(m=4k+2)$ are much more difficult to deduce, and we refer to [7] for the complete account in thc form given here.

It turns out that in the simply connected situation when dimension $m\geqslant 5$, the invariant σ is the only obstruction to finding a normal cobordism rel Y to a homotopy equivalence. Explicitly we have:

(2.7) Fundamental Theorem of Surgery. *Let (X,Y) be a Poincaré pair of dimension $m\geqslant 5$, X 1-connected, ξ^k a linear k-plane bundle over X, and let (f,b) be a normal map of degree* 1, *$f:(M,\partial M)\to(X,Y)$, $b:\nu\to\xi$, such that $f_*: H_*(\partial M)\to H_*(Y)$ is an isomorphism. Then (f,b) is normally cobordant* rel Y *to (f',b') with $f': M'\to X$ a homotopy equivalence if and only if $\sigma(f,b)=0$. In any case (f,b) is normally cobordant to (f',b') with f' $[m/2]$-connected, where $[a]=$ greatest integer $\leqslant a$.*

This theorem was first proved by KERVAIRE and MILNOR [29] in the case where X was a disk or a sphere, and subsequently generalized by S. P. NOVIKOV [40, 41] and the author [8]. A complete account in the present form is given in [7]. Extensions of these results for the non-simply connected case have been given by WALL [51, 52].

The following theorem of KERVAIRE-MILNOR was to appear in the second part of [29], which unfortunately has not been published to date. A proof is given in [7].

(2.8) Plumbing Theorem. *For* $(X,Y)=(D^m,S^{m-1})$, $\xi^k=$ *trivial bundle,* $m\geqslant 5$, *all values are realized for* $\sigma(f,b)$, *for* (f,b) *normal maps of degree* 1, $f:(M^m,\partial M)\to(D^m,S^{m-1})$ *with* $f:\partial M\to S^{m-1}$ *a homotopy equivalence.*

(2.9) Remark. For $m=6$, 14, 30 all values are realized for $\sigma(f,b)$, $f: M^m\to S^m$.

The proof is well known for $m=6$, 14, and for $m=30$ it is proved in [14].

One may now deduce some powerful corollaries such as the following due to S. P. NOVIKOV [41] and the author [8].

(2.10) Homotopy Type of Manifolds. *Let X be a* 1*-connected Poincaré complex of dimension* $m\geqslant 5$, ξ *a linear k-plane bundle over* X, $k>m+1$, $\alpha\in\pi_{m+k}(T(\xi))$ *such that* $h(\alpha)\cap U_\xi=[X]\in H_m(X)$. *If either*

(i) *m is odd,*

(ii) $m=4k$ *and condition H below holds*

or

(iii) $m=6$, 14 *or* 30,

then there is a homotopy equivalence f of a smooth m-manifold M^m with X, $f:M\to X$ *such that* $f^*(\xi)$ *is the normal bundle of* $M^m\subset S^{m+k}$. *In cases* (i) *and* (ii) *above, α represents the normal cobordism class of the homotopy equivalence f.*

The condition (H) is essentially that the Hirzebruch Index Theorem [24] hold. Namely, let ξ^{-1} be the bundle inverse of ξ, p_i be the Pontryajin class, and L_k be the Hirzebruch polynomial [24].

(H) $$\operatorname{Index} X = L_k(p_1(\xi^{-1}),\dots,p_k(\xi^{-1}))\,[X]\ .$$

Then (2.10) is an immediate consequence of (2.4), (2.7), in case $m=4k$ of (2.6) (v) and the Hirzebruch index theorem, and in case $m=6$, 14 or 30, of (2.9) and (2.6) (iv).

If (f_i,b_i), $f_i:(M_i^m,\partial M_i)\to(X_i,Y_i)$ are normal maps of degree 1, $i-1,2$, then we may define the sum of the two along a cell in the two boundaries, (see (1.2)). We choose a component V_i of Y_i, $i=1,2$ and let $V_i = V_i^0 \cup e_i^{m-1}$, with attaching map $\alpha_i: S^{m-2} \to V_i^0$, e_i^{m-1} being the top dimensional cell of V_i, so that $H_{m-1}(V_i^0) = 0$. Then the sum $X_1 \coprod X_2$ along V_1 and V_2 is defined $X_1 \coprod X_2 = X_1 \cup (e^{m-1} \times I) \cup X_2$, with

$$e^{m-1} \times 0 \text{ identified with } e_1^{m-1} \subset V_1,$$
$$e^{m-1} \times 1 \text{ identified with } e_2^{m-1} \subset V_2,$$

and the sum $Y_1 \# Y_2$ along V_1 and V_2 is defined by $Y_1^0 \cup (S^{m-2} \times I) \cup Y_2^0$ with $Y_i^0 = V_i^0 \cup$ other components of Y_i, $i = 1,2$ and

$$S^{m-2} \times 0 \text{ identified with } \alpha_1(S^{m-2}) \subset V_1^0,$$
$$S^{m-2} \times 1 \text{ identified with } \alpha_2(S^{m-2}) \subset V_2^0.$$

Then $(X_1 \coprod X_2, Y_1 \# Y_2)$ is the Poincaré pair which is the sum (see (1.2)).

The connected sum $M_1 \coprod M_2$ along components of ∂M_1 and ∂M_2 is defined similarly using differentiably embedded cells in ∂M_1 and ∂M_2, $\partial M_1 \# \partial M_2$ defined similarly. There is a canonical way to put a differential structure on the sum known as "straightening the angle" (see [19, Chapter I, § 3]).

If we choose the cells in ∂M_i so that f_i maps them into those chosen for Y_i, we may, if necessary changing f_i by a homotopy, and covering by a homotopy of b_i, get an induced map

$$f_1 \coprod f_2 : (M_1 \coprod M_2, \partial M_1 \# \partial M_2) \to (X_1 \coprod X_2, Y_1 \# Y_2),$$

of degree 1, and arrange a map $b_1 \coprod b_2$ covering $f_1 \coprod f_2$.

It follows from results of [18, 42] that the sum of manifolds depends only on the components and orientations of the cells chosen, and a similar fact is true for Poincaré pairs. On the other hand, the maps defined may depend on the choice of homotopies if Y_i is not simply connected.

(2.11) Theorem. *Let (X, Y) be an m-dimensional Poincaré pair with X 1-connected, $Y \neq \phi$, $m \geq 5$, and let (f,b), $f: (W^m, \partial W^m) \to (X, Y)$ be a normal map of degree* 1 *such that $(f|\partial W)_*: H_*(\partial W) \to H_*(Y)$ is an isomorphism. Then there is a normal map of degree* 1, *(g,c), $g: (U^m, \partial U^m) \to (D^m, S^{m-1})$ with $g|\partial U$ a homotopy equivalence, such that $(f,b) \coprod (g,c)$ is normally cobordant* rel Y *to a homotopy equivalence. In particular, (f,b) is normally cobordant to a homotopy equivalence.*

Proof. Let $\sigma = \sigma(f,b)$ be the obstruction to surgery rel Y in (2.6). Let $g: (U^{m+1}, \partial U^{m+1}) \to (D^{m+1}, S^m)$ (g,c) a normal map, with $\sigma(g,c) = -\sigma$, and $g|\partial U: \partial U \to S^m$ a homotopy equivalence, which exists by the Plumbing Theorem (2.8). Take the sum of the two normal maps (f,b) and (g,c) along a cell in ∂W and another in ∂U. Then by (2.6) (iv), $\sigma(f \coprod g, b \coprod c) = \sigma - \sigma = 0$, so by the Fundamental Theorem (2.7), $(f \coprod g, b \coprod c)$ is normally cobordant rel Y to a homotopy equivalence.

Recall that a cobordism $W, \partial W = M \cup M'$ is called an h-cobordism if the inclusions $M \subset W$ and $M' \subset W$ are homotopy equivalences. It is a result of SMALE [44] that if $\partial M = \phi$, $\dim W \geqslant 6$, and W is 1-connected then W is diffeomorphic to $M \times I$ and $M' \times I$, and in particular, M is diffeomorphic to M'.

As a consequence of (2.11) we have a classification theorem due to S. P. NOVIKOV [40, 41],

(2.12) Classification of Manifolds. *Let* (f_i, b_i) *be normal maps,* $f_i: M_i \to X$, $i=0,1$, *such that* f_i *are homotopy equivalences,* X *a 1-connected Poincaré complex of dimension* $m \geqslant 4$. *Suppose* (f_0, b_0) *and* (f_1, b_1) *are normally cobordant. Then* M_0 *is h-cobordant to* $M_1 \# \Sigma$, *where* Σ *is a homotopy m-sphere which bounds a parallelizable manifold. In particular if m is even* >4, *then* M_0 *is diffeomorphic to* M_1.

KERVAIRE and MILNOR [29] have shown that the group of h-cobordism classes of homotopy m-spheres which bound parallelizable manifolds $m \geqslant 4$ is a cyclic group of finite order, 0 if m is even, or $m = 5, 13$, order at most 2 if $m = 4k+1$, and for $m = 4k+3$ they calculated its order, up to a factor of 2 in some cases. In [17], it was shown the order is 2 for $m = 8k+1$, and in [14], it is shown that the order is 2 for $m = 4k+1$ except possibly for $m = 2^q - 3$, and for $m = 29$ the group is zero. This information together with (2.12) gives some upper bounds for the number of closed manifolds in a given normal cobordism class.

Proof of (2.12). If (F, B) $F: W \to X$ is the normal cobordism, consider F as a map

$$F: (W, M_0 \cup M_1) \to (X \times I, X \times 0 \cup X \times 1),$$

$F(M_i) \subset X \times i$, $i = 0, 1$. Then (F, B) satisfies the hypotheses of (2.11), so $(F, B) \coprod (g, c)$ (along M_1) is normally cobordant rel $X \times 0 \cup X \times 1$) to a homotopy equivalences, i. e. an h-cobordism between M_0 and $M_1 \# \partial U$, $(\partial U = \Sigma)$. Since $g: (U, \partial U) \to (D^{m+1}, S^m)$ and D^{m+1} is contractible, it follows that the stable normal bundle of U is trivial, and since U has non-empty boundary, it follows that U is parallelizable.

As another application of (2.11) we have the following theorem of WALL [53] which extends (2.10) and (2.12) to the case of bounded manifolds.

(2.13) Theorem. *Let* (X, Y) *be an m-dimensional Poincaré pair,* $Y \neq \phi, m \geqslant 6$, X *and* Y *1-connected,* ξ *a k-plane bundle over* X, $k > m+1$, *and* $\alpha \in \pi_{m+k}(T(\xi), T(\xi|Y))$ *such that* $h(\alpha) \cap U_\xi = [X] \in H_m(X, Y)$. *Then there is a normal map* (f, b) *in the class determined by* $\alpha, f: (M, \partial M) \to (X, Y)$, *such that* f *is a homotopy equivalence, and such an* (f, b) *(and hence* $(M, \partial M)$*) is unique up to diffeomorphism.*

Proof. Consider a representative (h,a) of the normal cobordism class of α, $h:(N,\partial N)\to(X,Y)$. Then by (2.6) (iii) and (2.7), $(h|\partial N, a|\partial N)$ is normally cobordant to a homotopy equivalence. This normal cobordism extends in an obvious way to a normal cobordism of (h,a) to (h',a') such that $h':(N',\partial N')\to(X,Y)$, $h'|\partial N':\partial N'\to Y$ is a homotopy equivalence. Then (2.11) implies (h',a') is normal cobordant rel Y to a homotopy equivalence.

To prove uniqueness we consider (f_i,b_i), $f_i:(M_i,\partial M_i)\to(X,Y)$ homotopy equivalences, $i=0,1$, $F:(W^{m+1},V^m)\to(X,Y)$, (F,B) a normal cobordism between (f_0,b_0) and (f_1,b_1). As in the proof of (2.12), we consider F as a map $F:(W,V,M_0,M_1)\to(X\times I, Y\times I, X\times 0, X\times 1)$ (we recall that $\partial W=M_0\cup V\cup M_1$, $\partial V=\partial M_0\cup\partial M_1$. Then $\sigma(F|\partial W)$ $=\sigma(F|V)+\sigma(F|M_0)+\sigma(F|M_1)$ by (2.6) (iv), $\sigma(F|\partial W)=0$ by (2.6) (iii), and $\sigma(F|M_i)=0$, $i=0,1$, since $F|M_i=f_i$ is a homotopy equivalence. Hence $\sigma(F|V)=0$ so there is by (2.7) a normal cobordism of $F|V$ rel $Y\times 0\cup Y\times 1$ to a homotopy equivalence. Hence we get a new normal cobordism (F',B') between (f_0,b_0) and (f_1,b_1) which is an h-cobordism between ∂M_0 and ∂M_1 in $Y\times I$. Then as in (2.11) we may add a normal map (g,c), $g:(U^{m+1},\partial U^{m+1})\to(D^{m+1},S^m)$ to (F',B') so that the result is normally cobordant to a homotopy equivalence, i.e. h-cobordism. But if (g,c) is added to (F',B') along the part of the boundary of W' between ∂M_0 and ∂M_1, i.e. away from M_0 and M_1, then $(F'\coprod g, B'\coprod c)$ is still a normal cobordism between (f_0,b_0) and (f_1,b_1). Hence we arrive at an h-cobordism between them, and applying the h-cobordism theorem of SMALE [44] twice, first to the boundary, and then to the interior, we arrive at the result.

Now we give an example of an embedding theorem that can be proved with these methods. The theorem is similar to a theorem of the author in [10] and the proof is identical.

Suppose the total space of a q-plane bundle α^q over a space A is contained as an open set in a space X, and let ξ^k be a k-plane bundle over X. If (f,b), $f:M^m\to X$, $b:\nu\to\xi$ is a normal map, then by making f transverse regular to $A\subset X$, we may suppose $f^{-1}(A)=N^n$, N^n has normal bundle η^q in M^m, $m-n=q$, and $f|E(\eta)$ is a linear bundle map, $c:\eta\to\alpha$, where $E(\eta)$ is the total space of the vector bundle η which is a tubular neighborhood of N in M. Then $(f|N, c+b|(\nu|N))$ is a normal map into A, with bundle $\alpha+(\xi|A)$, (since the normal bundle of N in D^{m+k} is the sum of its normal bundle in M and the restriction of M's normal bundle in D^{m+k}, i.e. $\eta+(\nu|N)$.)

Suppose X, A Poincaré complexes are, and the collapsing map $X\to T(\alpha)$ is of degree 1, and suppose M a closed manifold and f is of degree 1. Then $(f|N, c+b|(\nu|N))$ is a normal map of degree 1, and hence $\sigma(f|N, c+b|(\nu|N))$ is defined. We will set $\sigma(f|N, e+b|(\nu|N))=\sigma_A(f,b)$

to emphasize that its value does not depend on the particular way in which f was made t-regular on A (see (2.6) (iii)).

Now suppose $f: M^m \to M'^m$ is a homotopy equivalence of smooth manifolds. Then if we take over M' the bundle ξ^k such that $f^*(\xi) = \nu$, the normal bundle of M in D^{m+k}, $k > m+1$, then choosing a bundle map $b: \nu \to \xi$, (f, b) is a normal map. If N'^n is a smooth closed submanifold of M'^m, then as above $\sigma_{N'}(f, b)$ is defined, and it can be shown to be independent of the choice of b, above, so we may write in this case $\sigma_{N'}(f, b) = \sigma_{N'}(f)$.

(2.14) Theorem. *Let $f: M^m \to M'^m$ be a homotopy equivalence of closed smooth 1-connected manifolds, $N'^n \subset M'^m$ a 1-connected smooth submanifold, α' its normal bundle, such that $M' - N'$ is 1-connected, $n \geqslant 5$. If $\sigma_{N'}(f) = 0$, then (f, b), for any choice of b, is normally cobordant to (f', b'), $f': M \to M'$ a homotopy equivalence, such that, if $f'^{-1}(N') = N$, $f': (M, N, M-N) \to (M', N', M'-N')$ is a homotopy equivalence on each term. In particular, M has a submanifold homotopy equivalent to N', with complement homotopy equivalent to $M' - N'$ and normal bundle induced from α'.*

This theorem is a special case of a very general theorem about submanifolds and "supermanifolds" which has many applications, to existence and isotopy of embeddings, to study of manifolds with free or free abelian fundamental group (c.f. [11, 10, 9], and a general treatment will be given in a later paper.

Proof. If $\sigma_{N'}(f, b) = 0$, then $f^{-1}(N') = N''$ is normally cobordant to an N homotopy equivalent to N'. Let (G, B), $G: (U^{n+1}, N'' \cup N) \to (N' \times I, N' \times 0 \cup N' \times 1)$ be the normal cobordism, $B: \omega \to \alpha' + (\xi | N')$, ω the normal bundle of $U \subset S^{m+k} \times I$. Then $B^{-1}(\alpha') = \alpha_0$ is a q-plane bundle over U, $\alpha_0 | N'' = \alpha''$, the normal bundle of $N'' \subset M$. Then $M \times I \cup E(\alpha_0)$ with $E(\alpha'') \times 1$ identified with $E(\alpha_0 | N'')$, defines a cobordism W of M, and f on $M \times I$ together with $B | \alpha_0$ defines a map $F: W \to M'$ with $F^{-1}(N') = U$, (see figure). Then b over $M \times I$ and B

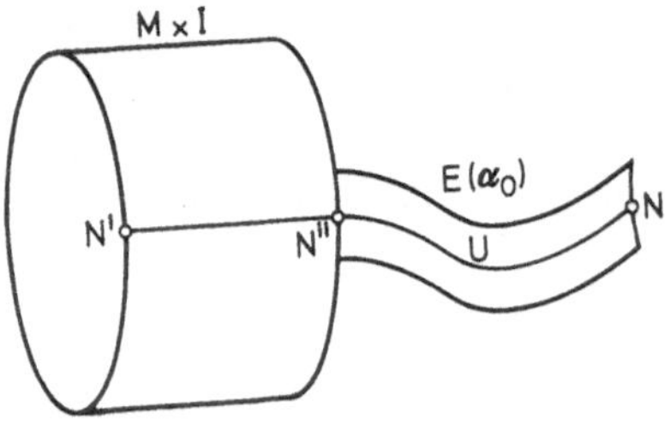

Fig.

restricted to the orthogonal complement of α_0 defines a map of the normal bundle of W into ξ, to get a normal cobordism of (f,b) with (f_1,b_1). Now $f_1:M_1\to M'$ is no longer a homotopy equivalence, but $f_1^{-1}(N')=N$, $f_1|N:N\to N'$ is a homotopy equivalence, $f_1^*(\alpha')=\alpha=$ normal bundle of $N\subset M_1$, so $f_1|\partial E(\alpha):\partial E(\alpha)\to\partial E(\alpha')$ is also a homotopy equivalence. Hence (f_1,b_1) is the sum of two maps

$$f_1|E(\alpha):(E(\alpha),\partial E(\alpha))\to(E(\alpha'),\partial E(\alpha'))$$

and

$$f_1|(M_1-E(\alpha)),(M_1-E(\alpha),\partial E(\alpha))\to(M'-E(\alpha'),\partial E(\alpha'))$$

along $\partial E(\alpha)$. Hence

$$\sigma(f_1)=\sigma(f_1|E(\alpha))+\sigma(f_1|M_1-E(\alpha))$$

by (2.6) (iv), and since $f_1|E(\alpha)$ is a homotopy equivalence, $\sigma(f_1|E(\alpha))=0$ and

$$\sigma(f_1)=\sigma(f_1|M_1-E(\alpha)).$$

Now f_1 is normally cobordant (by the construction) to f, and f is a homotopy equivalence, so $\sigma(f_1)=\sigma(f)=0$, by (2.6) (iii), and hence $\sigma(f_1|M_1-E(\alpha))=0$. Now $M'-E(\alpha')$ is homotopy equivalent to $M'-N'$ which is 1-connected by hypothesis, so from (2.7) it follows that $f_1|M_1-E(\alpha)$ is normally cobordant rel $\partial E(\alpha')$ to a homotopy equivalence. This gives a normal cobordism of f_1 with $f_2:M_2\to M'$, $f_2=f_1$ on $E(\alpha)$ and $f_2|M_2-E(\alpha)$ is a homotopy equivalence with $M'-E(\alpha')$. It follows from the Mayer-Vietoris theorem that $f_{2*}:H_*(M_2)\to H_*(M')$ is an isomorphism, and hence a homotopy equivalence, by the theorem of J. H. C. WHITEHEAD, and $f_2:(M_2,N,M_2-N)\to(M',N',M'-N')$ is a homotopy equivalence on each term. Now (f_2,b_2) is normally cobordant to (f,b), so by NOVIKOV's theorem (2.12), (f,b) is diffeomorphic to $(f_2,b_2)\#(g,c)$, where $g:\Sigma^m\to S^m$ is a homotopy equivalent. But if Σ is added to M_2-N, i.e. away from N, then $M=M_2\#\Sigma$ still contains N with normal bundle α, complement homotopy equivalent to $M'-N'$, etc. This completes the proof of (2.14).

Next we use the methods of surgery to construct exotic diffeomorphisms of simply connected manifolds. This method was developed in [13] and we review it in the special case of diffeomorphisms homotopic to the identity.

Let W^{m+1} be a manifold, $M^m\subset W^{m+1}$ a smooth submanifold, M 1-connected, and let $h:W^{m+1}\to M^m\times S^1$ be a homotopy equivalence with the property that $h^{-1}(M\times s)=M\subset W$, some $s\in S^1$; and $h'=h|M:M\to M\times s$ is given by $h'(x)=(x,s)$, $x\in M$. Let $m\geqslant 5$.

(2.15) *There is a diffeomorphism* $f: M \to M$ *such that* f *is homotopic to the identity and* W *is diffeomorphic to the* mapping torus M_f *of* f, $M_f = M \times I$ *modulo the identification* $(x,0) = (f(x),1)$, $x \in M$.

If we "cut W along M", i.e. remove a neighborhood $M \times I$ of M in W, we obtain a manifold $\bar{W}$, $\partial \bar{W} = M \times 0 \cup M \times 1$, and f restricts to a map

$$\bar{f}: (\bar{W}, M \times 0 \cup M \times 1) \to (M \times I, M \times 0 \cup M \times 1)$$

with $\bar{f}|M \times i = \text{identity}$, $i = 0,1$, (thinking of $M \times S^1 = (M \times I) \cup (M \times I)$ and assuming $f|M \times I = \text{identity}$, $M \times I \subset W$). Since M is 1-connected and f is a homotopy equivalence, it is easy to deduce that $\bar{W}$ is 1-connected and $\bar{f}$ is a homotopy equivalence, and hence $\bar{W}$ is an h-cobordism between $M \times 0$ and $M \times 1$. Since $m \geqslant 5$, SMALE's h-cobordism theorem implies that there is a diffeomorphism $F: \bar{W} \to M \times I$ such that $F|M \times 0 = \text{identity}$. Define $f = F|M \times 1: M \to M$, which is a diffeomorphism. Then $W = \bar{W} \cup (M \times I)$ with $M \times i \subset \bar{W}$ identified by the identity with $M \times i \subset M \times I$, $i = 0,1$. Then $G = F \cup \text{identity}$: $W = \bar{W} \cup (M \times I) \to M_f = (M \times I) \cup (M \times I)$ (with $M \times 0$ identified with $M \times 0$ by the identity, $M \times 1$ identified with $M \times 1$ by f) defines a diffeomorphism of W with M_f = the mapping torus of f.

Now $hG^{-1}: M_f \to M \times S^1$ is a homotopy equivalence and $hG^{-1}|M = \text{identity}: M \to M \times s$, $s \in S^1$. "Cutting along M" again we get a map $H: M \times I \to M \times I$, $H|M \times 0 = \text{identity}$, $H|M \times 1 = hG^{-1}|M \times 1 = (h|M \times 1) f^{-1} = f^{-1}$ since $h|M \times 1 = \text{identity}$, $M \times 1 \subset W$. Hence H defines a homotopy between f^{-1} and the identity, and hence f is homotopic to the identity, which proves (2.15).

We note that the analog of (2.15) is true with a similar proof if we take another bundle over S^1 instead of $M \times S^1$. Also similar results hold for bounded M (see [13]).

Now we investigate the problem of finding such manifolds W as in (2.15), using the methods of surgery.

Let ξ^k be a linear k-plane bundle over $M^m \times S^1$ where M^m is a smooth 1-connected manifold of dimension $m \geqslant 5$, $k > m+2$. Let ν^{k+1} be the normal bundle of M^m in S^{m+k+1} and let $b: (\xi^k|M \times s) + \varepsilon^1 \to \nu^{k+1}$ be a linear bundle equivalence, $s \in S^1$. Let $\eta: T(\xi) \to T(\xi|M \times s) + \varepsilon^1)$ be the natural map which collapses $T(\xi|M \times I)$ to a point, where $M \times I \subset M \times S^1$ is disjoint from the fibre $M \times s$. Let $c \in \pi_{m+k+1}(T(\nu))$ be the homotopy class of the natural collapsing map.

(2.16) Theorem. *Let* $\alpha \in \pi_{m+k+1}(T(\xi))$ *be such that* $T(b)_* \eta_*(\alpha) = c \in \pi_{m+k+1}(T(\nu))$, *and let* (f,b) *be the normal map* $f: W \to M \times S^1$ *corresponding to* α. *Then* (f,b) *is normally cobordant to* (f',b'),

$f': W' \to M \times S^1$, $W' = M_g$, $g: M \to M$ a diffeomorphism, $f'|M = identity$, if and only if $\sigma(f,b) = 0$.

This theorem is essentially a special case of a theorem about manifolds with $\pi_1 = \mathbb{Z}$ proved in [11]. The proof is similar to that of (2.14).

Proof. Since $T(b)_* \eta_*(\alpha) = c$, it follows by a transversality argument that $f|f^{-1}(M \times s)$ is normally cobordant to the identity $M \to M \times s$. By the same argument as in the proof of (2.14) we may find a normal cobordism of f to $f': W \to M \times S^1$, $M \times I \subset W$ and $f'|M \times I: M \times I \to M \times I \subset M \times S^1$ is the identity. Let $U = W - \operatorname{int} M \times I$ so that $f'' = f'|U: U \to M \times I$ (the other half of $M \times S^1 = (M \times I) \cup (M \times I)$). Then f'' is a normal map, $f''|\partial U$ is the identity, so $\sigma(f'',b'')$ is defined, and by (2.6) (iv), $\sigma(f'',b'') = \sigma(f',b')$, while $\sigma(f',b') = \sigma(f,b)$ since they are normally cobordant. Hence if $\sigma(f,b) = 0$, then $\sigma(f'',b'') = 0$, and since $M \times I$ is 1-connected, dimension $\geqslant 5$, it follows from (2.7) that (f'',b'') is normally cobordant rel $M \times 0 \cup M \times 1$ to a homotopy equivalence, $f''': (U', \partial U') \to (M \times I, M \times \{0,1\})$, $\partial U' = M \times 0 \cup M \times 1$ and $f'''|M \times 0 \cup M \times 1 = \text{identity}$. Then the union $U' \cup M \times I = W'$, and $f''' \cup \text{id}: W' \to M \times S^1$ is a homotopy equivalence satisfying the hypotheses of (2.15) and the result follows.

Let $\varphi: W^{m+1} \to M^m \times S^1$ be a homotopy equivalence such that $M \subset W$, $\varphi|M = \text{identity}: M \to M \times s$, and let ξ^k be the normal bundle of W in S^{m+d+1}, $k > m+2$. Then it follows that $\xi|M + \varepsilon^1 = \nu^{d+1}$ is the normal bundle of M in S^{m+k+1}, since $M \subset M \times I \subset W$, and it follows from the theorem of Hirsch [23] and Atiyah [2] that there is a fibre homotopy equivalence $\beta: \xi \to \nu \times \varepsilon^0$ covering φ such that $\beta|(\xi|M)$ is a linear equivalence.

Thus in looking for bundles ξ to use in applying (2.16), we may restrict our attention to bundles ξ over $M \times S^1$ such that there is a fibre homotopy equivalence $\beta: \xi \to \nu \times \varepsilon^0$ which restricted to $\xi|M$ is a linear equivalence. Now $M \times S^1/M \times s$ is homeomorphic to $\Sigma(M_+)$, the reduced suspension of M_+, where $M_+ = M \cup (\text{point})$. Then the virtual bundle $\xi - (\nu \times \varepsilon^0)|M \times s$ is trivial (as a linear bundle), so $\xi - (\nu \times \varepsilon^0) = h^*(\gamma)$ where $h: M \times S^1 \to \Sigma(M_+)$ is the collapsing map, $\gamma \in KO(\Sigma(M_+))$. Since $M \times s \subset M \times S^1$ is a retract it follows that

$$0 \to KO(\Sigma(M_+)) \to KO(M \times S^1) \to KO(M \times s) \to 0$$

so γ is unique, and it follows similarly that γ is fibre homotopically trivial. We denote by $L(X) \subset KO(X)$ the subgroup of fibre homotopically trivial bundles.

Conversely if $\gamma \in L(\Sigma(M_+))$, then one may assume the fibre homotopy trivialization of γ is linear at the base point. Then this induces a fibre homotopy equivalence $\beta: h^*(\gamma^q) \to \varepsilon^q$, such that $\beta|(h^*(\gamma)|M \times s)$ is linear. Hence $\alpha = \mathrm{id} + \beta: \nu \times \varepsilon^0 + h^*(\gamma) \to (\nu \times \varepsilon^0) + \varepsilon^q = (\nu^k + \varepsilon^q) \times \varepsilon^0$ is a fibre homotopy trivialization, linear over $M \times s$, and $\nu^k + \varepsilon^q$ is the normal bundle of M in S^{m+k+q}. Let $c \in \pi_{m+k+q+1}(T((\nu^k + \varepsilon^q) \times \varepsilon^0))$ be the collapsing map for $M \times S^1$, where $(\nu^k + \varepsilon^q) \times \varepsilon^0$ is the normal bundle of $M \times S^1$ in $S^{m+k+q+1}$. Then $\xi = \nu \times \varepsilon^0 + h^*(\gamma)$ and $T(\alpha^{-1})_*(c)$ satisfy the hypotheses of (2.16).

(2.17) Lemma. *The Pontryagin class defines a linear map* $P: KO(\Sigma X) \to H^{4*}(\Sigma X)$, *i.e.* $P(x+y) = P(x) + P(y) - 1$, *where* $x, y \in KO(\Sigma X)$, $P = 1 + p_1 + p_2 + \cdots$ *is the total Pontryagin class,* $p_i \in H^{4i}$.

Proof. In general $P(\alpha + \beta) = P(\alpha) P(\beta)$ (see [33, 27]), but ΣX being a suspension, products of positive dimensional classes are zero, and the result follows.

If $I = (i_1, i_2, \ldots, i_n)$, define the Pontryagin number of a bundle δ over a manifold N, by

$$P(I,\delta)[N] = \langle p_{i_1}(\delta) p_{i_2}(\delta) \ldots p_{i_n}(\delta), [N] \rangle .$$

(2.18) Lemma. *The map* $R: KO(\Sigma(M_+)) \to \mathbb{Q}$ *given by* $R(\gamma) = \Sigma_k \lambda_k P(I_k, (\nu \times \varepsilon^0) + h^*(\gamma))[M \times S^1]$, $\lambda_k \in \mathbb{Q}$ *is a linear map.*

Proof. Since all products of positive dimensional elements of $H^*(\Sigma(M_+))$ are zero, and $h^*(P(\gamma)) = P(h^*(\gamma))$, it follows that for any polynomial $G(x_1, \ldots, x_s)$, $G(a_1, a_2, \ldots, p_1(h^*(\gamma)), \ldots, p_t(h^*(\gamma)))$ is linear in the elements $p_i(h^*(\gamma))$, and the coefficients depend only on the a_i's and the coefficients of G. Since $P(\nu \times \varepsilon^0 + h^*(\gamma))$ is a polynomial, $P(I_k, (\nu \times \varepsilon^0) + h^*(\gamma))$ is a polynomial, and the result follows.

(2.19) Theorem. *Let* M^m *be a 1-connected manifold,* $m \geqslant 5$, *m even or* $m = 5$, 13 *or* 29. *Then for each element* $\gamma \in L(\Sigma(M_+))$ *there is a diffeomorphism* $f: M \to M$ *such that* f *is homotopic to the identity, and the bundle over* $M \times S^1$ *corresponding to the normal bundle of the mapping torus* M_f *is* $\nu \times \varepsilon^0 + h^*(\gamma)$, $h: M \times S^1 \to \Sigma(M_+)$, ν *the stable normal bundle of* M. *If* $m \equiv 3 \bmod 4$, *this is true for a submodule* L' *of* $L(\Sigma(M_+))$, *where* $L' =$ *kernel of the linear map* $L(\Sigma(M_+)) \to \mathbb{Q}$ *induced by the Hirzebruch L-genus.*

This follows immediately from (2.16), the discussion following and (2.18).

Now let us consider the question of whether the diffeomorphisms corresponding to different elements of $L(\Sigma(M_+))$ are actually different, up to isotopy or pseudo-isotopy. If $\mathscr{D}(M) =$ group of pseudo-isotopy classes of orientation preserving diffeomorphisms $f:M\to M$, and let $\mathscr{D}_0(M)$=subgroup of f which are homotopic to the identity. One would like to define a map $\mathscr{D}_0(M)\to L(\Sigma(M_+))$ inverse to the construction above but there is a difficulty created by the fact that there may be several different ways of defining a homotopy of $f\in\mathscr{D}_0(M)$ to the identity.

Let $f:M\to M$ be a map, $h_i:M\times I\to M\times I$, $i=1,2$, two homotopies of f with the identity, i.e. $h_i(x,0)=(x,0)$, $h_i(x,1)=(f(x),1)$, $i=1,2$. Define $\hat{h}_i:M\times S^1\to M_f$ by $\hat{h}_i(x,t)=(h_i(x,t),t)$, which respects the identifications of $M\times I$ to $M\times S^1$ and M_f. Thus $\hat{h}_i$ has the properties:

(i) $\hat{h}_i$ is a homotopy equivalence,
(ii) $\hat{h}_i|M\times s$=inclusion of the fibre $M\subset M_f$,
(iii) $\bar{p}\hat{h}_i=p$, where $\bar{p}:M_f\to S^1$, $p:M\times S^1\to S^1$

are the natural projections induced by $(x,t)\to(t)$.

Let us consider the induced homotopy equivalence $g=\hat{h}_2^{-1}\hat{h}_1$: $M\times S^1\to M\times S^1$. Let $\sim$ mean "homotopic to".

(2.20) Lemma. *Let $g:M\times S^1\to M\times S^1$ be the homotopy equivalence induced by two different homotopies of a map f to the identity (as above). Then*

a) *$g|M\times s\sim$ inclusion,*
b) *$pg\sim p$, where $p:M\times S^1\to S^1$* is *projection.*

The proof follows immediately from the properties (ii) and (iii) of the $\hat{h}_i$, $i=1,2$.

Let $\mathscr{T}(\omega)$=group of homotopy classes of bundle maps $b:\omega\to\omega$, ω some stable bundle over M, b covering $\rho(b):M\to M$, the induced map of base spaces, which is assumed a homotopy equivalence. Then $\rho:\mathscr{T}(\omega)\to\mathscr{H}(M)$=the group of homotopy classes of homotopy equivalences, is a homomorphism of groups. There is an easy argument that the groups $\mathscr{T}(\omega)$, $\mathscr{T}(\omega^{-1})$ and $\mathscr{T}(\varepsilon^q)$ are all isomorphic, ε^q the trivial bundle, provided they are all stable (i.e. fibre dimension $\gg\dim M$).

Let $\mathscr{T}_0(\omega)=\ker\rho$, and define $\mathscr{T}_{\#}(\omega)$=group of homotopy classes of bundle maps $c:\omega\to\omega$ which lie over the identity of M (as do the homotopies). Then there is a natural homomorphism $\zeta:\mathscr{T}_{\#}(\omega)\to\mathscr{T}_0(\omega)$, which is clearly onto by the bundle covering homotopy theorem, but may not in general be $1-1$.

If $b:\omega\to\omega$ is a bundle map covering the map $f:M\to M$ of base spaces, then we define a bundle $\bar{\omega}$ over M_f by identifying $\omega\times 0$ and $\omega\times 1$ in $\omega\times I$ over $M\times I$, using the bundle map b. Then $\bar{\omega}|M\times s=\omega$. Suppose f=identity, so that $\bar{\omega}$ is a bundle over $M\times S^1$ and $\bar{\omega}|M\times s=\omega\times\varepsilon^0$. Then $(\bar{\omega}-\omega\times\varepsilon^0)|M\times s=0$ in $KO(M\times s)$ so that $\bar{\omega}-\omega\times\varepsilon^0$

$=h^*(\gamma)$, $\gamma \in KO(\Sigma(M_+))$ where $\Sigma(M_+)=M\times S^1/M\times s$, and γ is unique, since $M\times s$ is a retract of $M\times S^1$.

(2.21) Proposition. *The correspondence $b\to\gamma$ above, depends only on the class of $b\in\mathcal{T}_{\#}(\omega)$, and defines a homomorphism $\beta:\mathcal{T}_{\#}(\omega)\to KO(\Sigma(M_+))$.*

Proof. To check that the definition depends only on the class in $\mathcal{T}_{\#}(\omega)$ is routine.

To show that β is a homomorphism, we use the fact that for a stable bundle ω, $\mathcal{T}_{\#}(\omega)\cong\mathcal{T}_{\#}(\omega+\varepsilon^q)$, any q. If $b_1,b_2\in\mathcal{T}_{\#}(\omega)$, $\beta b_i=\bar{\omega}_i-\omega\times\varepsilon^0$, then consider $1+1+b_2$ and $b_1+1+1\in\mathcal{T}_{\#}(\omega+\omega^{-1}+\omega)$. Clearly b_1 and b_1+1+1 have the same image in $KO(\Sigma(M_+))$, similarly for b_2. Also $b_1b_2\in\mathcal{T}_{\#}(\omega)$ corresponds to $(b_1+1+1)(1+1+b_2)=b_1+1+b_2\in\mathcal{T}_{\#}(\omega+\omega^{-1}+\omega)$. But the bundle corresponding to b_1+1+b_2 over $M\times S^1$ is clearly $\bar{\omega}+\omega^{-1}\times\varepsilon^0+\bar{\omega}_2$. Then since $\omega^{-1}=-\omega$ in KO, $\beta(b_1+1+b_2)=\bar{\omega}_1+\omega^{-1}\times\varepsilon^0+\bar{\omega}_2-\omega\times\varepsilon^0=\bar{\omega}_1-\omega\times\varepsilon^0+\bar{\omega}_2-\omega\times\varepsilon^0$ $=\beta(b_1)+\beta(b_2)$, which proves (2.21).

(2.22) Lemma. *Let $x_0,x_1\in\mathcal{T}_{\#}(\omega)$ such that $\zeta(x_0)=\zeta(x_1)\in\mathcal{T}_0(\omega)$, and let $\omega_0,\omega_1\in KO(M\times S^1)$ be the elements corresponding to x_0 and x_1. Then there is a homotopy equivalence $g:M\times S^1\to M\times S^1$ such that* (a) *$g|M\times s\sim$ inclusion,* (b) *$pg\sim p$, where $p:M\times S^1\to S^1$ is projection, and such that $g^*\omega_1=\omega_0$, and if $\bar{g}:\Sigma(M_+)\to\Sigma(M_+)$ is the homotopy equivalence induced by g, $\bar{g}^*(\beta(x_1))=\beta(x_0)$.*

The lemma follows from the bundle covering homotopy theorem and (2.20).

Proof. Let $b_i:\omega\to\omega$ be representatives of $x_i\in\mathcal{T}_{\#}(\omega)$, $i=0,1$, so that b_i lies over $1:M\to M$. Then $\zeta(x_0)=\zeta(x_1)$ implies there is a bundle map $B:\omega\times I\to\omega\times I$ covering $h:M\times I\to M\times I$ such that $B|\omega\times i=b_i$, so that $h|M\times i=$identity, $i=0,1$. Taking the bundle map $B'=B(b_0^{-1}\times 1)$: $\omega\times I\to\omega\times I$ over h, we find $B'|\omega\times 0=$identity, $B'|\omega\times 1=b_1b_0^{-1}$, and thus defines a map of the identified bundles $\omega_0\to\omega_1$, where $\omega_i=\omega\times I$ with $(v,0)$ identified with $(b_i(v),1)$, $v\in E(\omega)$. For $B'(v,0)=(v,0)$ and $B'(b_0(v),1)=(b_1b_0^{-1}b_0(v),1)=(b_1(v),1)$, so preserves identification. Hence the map g induced by h, $g:M\times S^1\to M\times S^1$ has properties (a) and (b) and $g^*(\omega_1)=\omega_0$.

With strong hypotheses on M we may define some strong invariants.

Condition K. *A manifold satisfies condition K if for any homotopy equivalence $g:M\times S^1\to M\times S^1$ such that* a) *$g|M\times s\sim$ inclusion and* b) *$pg\sim p$, we have $g^*:H^{4*}(M\times S^1;\mathbb{Q})\to H^{4*}(M\times S^1;\mathbb{Q})$ is the identity.*

(2.23) Lemma. *If M satisfies condition (K) above then the formula $A(y)=P(\beta x)$, where $\zeta x=y$, defines a linear map $\mathcal{T}_0(\omega)\to H^{4*}(\Sigma(M_+);\mathbb{Q})$.*

Proof. A is defined by the digram

$$\begin{array}{ccccc} \mathcal{T}_{\#}(\omega) & \xrightarrow{\beta} & KO(\Sigma(M_+)) & \xrightarrow{P} & H^{4*}(\Sigma(M_+);\mathbb{Q}) \\ \zeta\big\downarrow & & A & \nearrow & \\ \mathcal{T}_0(\omega) & & & & \end{array}$$

Since ζ is a homomorphism, β is a homomorphism by (2.21) and P is linear by (2.17), and ζ is onto, the result will follow if A is well defined. By (2.22), if $\zeta(x_1)=\zeta(x_2)$ then $\beta(x_2)=\bar{g}^*\beta(x_1)$, and hence $P(\beta(x_2)) =P(\bar{g}^*\beta(x_1))=\bar{g}^*P(\beta(x_1))=P(\beta(x_1))$, since $g^*=\text{identity}$ implies that $\bar{g}^*=\text{identity}$ since $h^*\colon H^*(\Sigma(M_+))\to H^*(M\times S^1)$ is a monomorphism. This proves (2.23).

We may now sum up with the following theorem which we will apply in § 6.

(2.24) Theorem. *Let M^m be a closed 1-connected manifold of dimension $m\geqslant 5$, which satisfies condition (K) above. Then there is a homomorphism $A_0\colon \mathcal{D}_0(M)\to H^{4*}(\Sigma(M_+);\mathbb{Q})$ such that*

a) *If m is even or $m=5,13,29$, then A_0 is onto the image $P(L(\Sigma(M_+)))$.*

b) *If $m=4n-1$, $A_0(\mathcal{D}_0(M))\subset P(L(\Sigma(M_+)))$ is a submodule of rank one less.*

Proof. The differential $f\to df$ defines a homomorphism $d\colon\mathcal{D}_0(M)\to\mathcal{T}_0(\tau)$, $\tau=$tangent bundle of M. Then A of (2.23) composed with d defines $A_0=Ad$. But $d(f)$ is easily seen to induce the tangent bundle of M_f in $S^{m+k+q+1}$. Then (2.19), (2.23) complete the proof.

We note that if $m=4n-1$ the Hirzebruch polynomial is not zero on $L(\Sigma(M_+))$, because it contains p_n with a non-zero coefficient, so that the rank is always reduced by 1 in applying (2.19), to satisfy the index condition, (H).

(2.25) Lemma. *$KO(X)$ modulo torsion is isomorphic to $\sum_{\ell} H^{4\ell}(X)$ modulo torsion, and the Pontryagin classes are a complete set of invariants, modulo torsion, (X finite complex).*

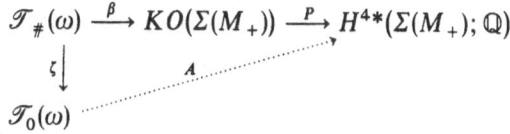

This is a standard fact, a special case of a very general theorem that maps into H-spaces are determined modulo torsion by their induced maps in rational cohomology. For the special case of K-theory we refer to [3] or [4].

Now we recall that for any space X

(2.26) Lemma. *$L(X) \subset KO(X)$ is a submodule of maximal rank.*

Proof. $L(X)$ is defined to be the kernel $J: KO(X) \to J(X)$ (see [2]). But $J(X)$ is finite [2], so $L(X)$ is of maximal rank.

Combining (2.24), (2.25) and (2.26) we get the result:

(2.27) Corollary. *With hypotheses of* (2.24) *if m is even or $m=5$, 13, or 29, $A_0(\mathscr{D}(M))$ has rank$=\operatorname{rank} H^{4*}(\Sigma(M_+))$, while $\operatorname{rank} A_0(\mathscr{D}_0(M)) = \operatorname{rank} H^{4*}(\Sigma(M_+))-1$ if $m=4n-1$.*

Without special assumptions on M we may define invariants using Pontryagin numbers.

Let $f_i: M_i \to M_i$ be diffeomorphisms, $i=1,2$. We shall say that f_1 is cobordant to f_2 if there is a W, cobordism between M_1 and M_2, and a diffeomorphism $F: W \to W$ such that $F|M_i = f_i$. In the usual way one may speak of oriented cobordism (of orientation preserving diffeomorphisms of oriented manifolds) etc., just as in the usual theory of cobordism. Clearly, if f_1 is cobordant to f_2, then M_{f_1} is cobordant to M_{f_2} (W_F being the cobordism). As usual the Pontryagin numbers depend only on the cobordism class and we get:

(2.28) Theorem. *The mapping torus construction defines a map $\theta: \mathscr{D}(M^m) \to \Omega_{m+1}$ such that*

(i) *$\theta(f)$ depends only on the cobordism class of f,*
(ii) *$\theta(f^n) = n\theta(f)$,*
(iii) *If $f_1, f_2 \in \mathscr{D}_0(M)$, $\theta(f_1 f_2) = \theta(f_1) + \theta(f_2)$.*

Proof. (i) is obvious. (ii) follows from the fact that M_{f^n} is the n-fold cover of M_f. (iii) follows from (2.18), (2.22) and the fact that such a g is of degree 1.

II. Free Actions of S^1 and S^3 on Homotopy Spheres

The first strong application of the modern methods of differential topology to transformation groups was the paper of W. C. and W. Y. HSIANG [26] on S^1 and S^3 actions on S^{11}, followed soon by the paper of MONTGOMERY and YANG [35], on S^1 actions on homotopy seven spheres. In the former, the special results of EELLS and KUIPER [20] on 8-manifolds were used, while in the latter, special properties of dimension 7 were used to translate the classification of free S^1 actions into a question about knotted S^3's in S^6, and then the results of HAEFLIGER [22] were applied.

In order to generalize the theorems of the HSIANGS and MONTGOMERY and YANG to higher dimensions it is necessary to use the techniques of surgery, using the results of Chapter I. This was first done by W. C. HSIANG [25], as well as in unpublished work of M. ROTHENBERG. More recently SULLIVAN [47, 48] obtained some interesting examples, and some very comprehensive piecewise linear results, but we will not discuss them here.

In § 3, we shall discuss the construction of free S^1 and S^3 actions, studying the orbit space with the results of Chapter I, and in § 4 we study the question of finding invariant embedded spheres.

3. Construction of Free S^1 and S^3 Actions

Let us denote by (M^m, φ, G) a differentiable action of the Lie group G on the smooth manifold M, i.e. $\varphi: G\times M \to M$ such that, if $m \in M$, $x, y \in G$,

(i) $\varphi(x, \varphi(y, m)) = \varphi(x\,y, m)$,

(ii) $\varphi(e, m) = m$, $e =$ identity of G,

(iii) φ is a C^∞ map.

The action φ is called *free* if for any $m \in M$

$$\varphi(g, m) = m \text{ implies } g = e.$$

We recall that if $M^m/\varphi = N$ is the orbit space, $N = \{G\,m, m \in M\}$, then for a free G-action, $M^m \to N$ is a principal G-bundle, N is a differentiable n-manifold, $m - n =$ dimension of G.

Now if M^m is a homotopy m-sphere, then M is $(m-1)$-connected and thus by [45, (19.4)], the principal G-bundle $M \to N$ is m-universal, and N is the homotopy type of the classifying space of G, up to dimension m. If $G = S^i$, $i = 0, 1, 3$, a classifying space for G is the projective space over the reals, complexes, or quaternions, respectively, and the fibrations of spheres, $S^n \to \mathrm{RP}^n$, $S^{2n+1} \to \mathrm{CP}^n$, $S^{4n+3} \to \mathrm{HP}^n$, are k-universal for $k = n$, $2n+1$, $4n+3$ respectively, for the different groups.

Let (Σ^m, φ, S^1) be a free action of the circle group S^1 on a homotopy m-sphere Σ^m. By the Lefschetz fixed point theorem, since S^1 is connected and the action is free (so without fixed points), it follows that m must be odd, $m = 2n+1$. Then the orbit space $N = \Sigma/\varphi$ is of dimension $2n$, and the principal bundle $\Sigma^m \to N^{2n}$ is classified by a map $f: N^{2n} \to \mathrm{CP}^n$, since $S^{2n+1} \to \mathrm{CP}^n$ is $(2n+1)$-universal. Since for the m-classifying space B of a group G, $\pi_i(B) \cong \pi_{i-1}(G)$, $i \leqq m-1$, [45, (19.9)], it follows that $f_*: \pi_i(N) \to \pi_i(\mathrm{CP}^n)$ is an isomorphism for $i \leqq 2n$, and it follows that since N and CP^n are $2n$-dimensional that N is homotopy equivalent to CP^n. Similar arguments in the case of S^0 and S^3 may be made, so that we get:

(3.1) Proposition. *Let (Σ^m, φ, S^i), $i=0,1$ or 3 be a free action of a sphere group on a homotopy m-sphere Σ^m. Then the orbit space $N = \Sigma/\varphi$ is the homotopy type of projective space, real, complex or quaternionic, respectively, for S^0, S^1 or S^3.*

On the other hand if two actions $(\Sigma_1^m, \varphi_1, S^i)$ and $(\Sigma_2^m, \varphi_2, S^i)$ are equivalent, i.e. there is an equivariant diffeomorphism $f: \Sigma_1^m \to \Sigma_2^m$, then f induces a diffeomorphism of the orbit spaces $\bar{f}: N_1 \to N_2$. On the other hand a diffeomorphism $\bar{f}: N_1 \to N_2$ induces a diffeomorphism of the principal bundles $f: \Sigma_1 \to \Sigma_2$ which is a G-bundle map, i.e. an equivalence of the actions. It follows that:

(3.2) Proposition. *Equivalence classes of actions (Σ^m, φ, S^i) are in $1-1$ correspondence with diffeomorphism classes of manifolds homotopy equivalent to projective space, over $\mathbb{R}$, $\mathbb{C}$ or $\mathbb{H}$ respectively, for $i=0$, 1 or 3.*

This problem is exactly of the type which may be studied using the surgery techniques of Chapter I, in the case of S^1 and S^3.

We recall the theorem of HIRSCH [23] and ATIYAH [2]:

(3.3) Theorem. *Let M_1 and M_2 be two closed m-manifolds embedded in S^{m+k} with normal bundles ν_1 and ν_2 respectively, $k > m+1$. If $f: M_1 \to M_2$ is a homotopy equivalence, then there is fibre homotopy equivalence $b: \nu_1 \to \nu_2$ lying over f.*

Thus the stable normal bundle of a manifold homotopy equivalent to projective space has its stable normal bundle fibre homotopy equivalent to that of projective space.

Let ξ^k be a linear k-plane bundle over $X = \mathrm{CP}^n$ (or QP^n) which is fibre homotopy equivalent to the stable normal bundle ν^k of X in S^{m+k}, $m = \dim X = 2n$ for CP^n ($m = 4n$ for QP^n). Then it follows that the fibre homotopy equivalence $f: \nu \to \xi$ induces a homotopy equivalence of the Thom complexes $T(f): T(\nu) \to T(\xi)$. Also $T(f)_*$ commutes with the Thom isomorphism since $T(f)^*(U_\xi) = U_\nu$, so it follows that if $c \in \pi_{2n+k}$

$(T(\nu))$ (or $\pi_{4k+k}(T(\nu))$) is the class of the natural collapsing map (see § 2), then $h(c) \cap U_\nu = [X]$ and it follows that $h(T(f)_*(c)) \cap U_\xi = [X]$.

Then if $X = \mathrm{CP}^n$ and $n = 3$, 7 or 15 we may immediately apply (2.10) to get a smooth manifold M^{2n}, homotopy equivalent to CP^n with normal bundle ξ^k in S^{2n+k}.

A simple way to construct such bundles ξ^k over CP^n, $n = 3$, 7 or 15, is the following: Let η^k be a bundle over S^{2n-2}, $2n-2=4q$, let $g: \mathrm{CP}^{n-1} \to S^{2n-2}$ be of degree 1 and suppose that $p_q(\eta) \neq 0$ and η is fibre homotopically trivial. Since $n-1$ is divisible by 2, if $x \in H^2(\mathrm{CP}^n; \mathbb{Z}_2)$, then $Sq^2(x^{n-1}) = 0$, by the Cartan formula, and therefore, since x^{n-1} generates $H^{2n-2}(\mathrm{CP}^n; \mathbb{Z}_2)$, the map $g: \mathrm{CP}^{n-1} \to S^{2n-2}$ extends to $f: \mathrm{CP}^n \to S^{2n-2}$ (see [46]) and $f^*(\eta)$ is a fibre homotopically trivial bundle with q-th Pontryagin class $p_q(f^*(\eta)) \neq 0$. We note that such bundles η over S^{4q} exist since $\pi_{4q}(B_{s_0}) = \mathbb{Z}$ and the number of fibre homotopy classes of spherical fibre spaces is finite, being the order of $\pi_{4q+k}(S^k)$, (k large), (see [2]).

Then if we set $\xi = \nu + f^*(\eta)$, since $f^*(\eta)$ is fibre homotopically trivial, ξ is fibre homotopy equivalent to $\nu + \varepsilon^\ell$, ε^ℓ = the trivial bundle, which is again the normal bundle of CP^n in $S^{2n+k+\ell}$. Hence we may apply (2.10) to get a manifold M^{2n} homotopy equivalent to CP^n but with $p_q(M)$ different from $p_q(\mathrm{CP}^n)$, ($n = 3$, 7 or 15, $q = 1$, 3 or 7). The different possible η's each give different M^{2n} with different p_q, so that we get a different action of S^1 on S^{2n+1} for each of these, and we have proved:

(3.3) Theorem. *One can construct infinitely many different free S^1 actions on homotopy $(2n+1)$-spheres for $n = 3$, 7 or 15, distinguished by the Pontryagin class p_q of the orbit space, $q = 1$, 3 or 7, with p_i for $i < q$, being the same as for* CP^n.

In dimension 6 we can get a slightly finer result to reconstruct all the Montgomery-Yang examples [35]. Namely if $\alpha \in \pi_4(B_{s_0})$ is a generator, then the smallest multiple of α which is fibre homotopy trivial is 24α. However, if $g: \mathrm{CP}^2 \to S^4$ is of degree 1, it can be shown that $g^*(12\alpha)$ is fibre homotopy trivial. Then the extension of $g^*(12\alpha)$ to CP^3 is still fibre homotopy trivial since $\pi_{5+k}(S^k) = 0$ (see [50]). Using all multiples of this bundle yields the Montgomery-Yang examples, which are all of the actions of S^1 on homotopy 7-spheres [35].

For S^1 actions on homotopy $(4n+1)$-spheres and S^3 actions on homotopy $(4n+3)$-spheres, one may proceed in a similar way, trying to find bundles ξ over CP^{2n} or QP^n which are fibre homotopy equivalent to the normal bundle ν, and with the additional restriction that condition (H) holds, i.e. that the Hirzebruch index formula is true for ξ in place of v. This was shown by W. C. HSIANG [25], and we refer to that paper for the explicit construction of these bundles, by which he proves:

(3.4) Theorem. *There are infinitely many free S^1 (or S^3) actions on homotopy $(2n+1)$-spheres ($(4n+3)$-spheres) $n>3$ ($n>1$) which are inequivalent and distinguished by the rational Pontryagin classes of the quotient space.*

We note that the exotic S^1 actions on $(4n+3)$-spheres are constructed as the restrictions of the exotic S^3 actions (see [25]).

4. Characteristic and Invariant Spheres of Free S^1 and S^3 Actions on Homotopy Spheres

Let (Σ^m, φ, S^i), $i=0, 1$ or 3 be a free S^i action on the homotopy sphere Σ^m. As we saw in § 3, the quotient space $\Sigma^m/\varphi = N^n$, $m-n=i$, is a differentiable manifold, homotopy equivalent to the corresponding projective space P^k (real, complex or quaternionic according to whether $i=0$, 1 or 3). Let $f: N \to P^k$ be the homotopy equivalence, and suppose f is transverse regular on $P^\ell \subset P^k$, so that $Q = f^{-1}(P^\ell)$ is a smooth submanifold of N with normal bundle induced from that of $P^\ell \subset P^k$. We call Q a characteristic submanifold of N and the induced S^i principal bundle $\tilde{Q} \subset \Sigma^m$, a characteristic submanifold of Σ^m. Note that the codimensions of Q and $\tilde{Q}$ are the same, and are a multiple of $i+1$. We consider the question: *When does Σ^m have a characteristic homotopy sphere of a given codimension, i.e., is f homotopic to f' such that $\tilde{Q}$ is a homotopy sphere (or equivalently Q is homotopy equivalent to P^ℓ)?* This question for $i=0$ and codimension 1 was studied by G. R. LIVESAY and the author [16], and an obstruction was defined. Though the results were similar, the methods go outside the scope of the present context, and we will restrict ourselves here to $i=1$ or 3, i.e., free actions of S^1 and S^3 (see [31 and 37]).

A smooth submanifold $X \subset \Sigma^m$ is called *invariant* if $\varphi(G \times X) \subset X \subset \Sigma^m$.

Now the smooth submanifolds V of the orbit space N, yield invariant submanifolds $\tilde{V}$ of Σ^m, where $\tilde{V}$ is the induced principal S^i bundle over V, and conversely invariant submanifolds of Σ^m under the action φ, yield smooth submanifolds of N. We may also ask if (Σ^m, φ, S^i) has invariant homotopy spheres of various dimensions, which is the same thing as asking if the mapping $P^\ell \longrightarrow P^k \xrightarrow{f^{-1}} N$ is homotopic to an embedding of a homotopy P^ℓ into N. This may in general be much easier than finding a characteristic sphere of this dimension; for example the freedom in choosing the normal bundle should make the problem considerably easier. Thus if $k>2\ell$, the map $P^\ell \to N$ is homotopic to an embedding, and thus every free action has all the standard actions of less than half the dimension embedded in it as *invariant* spheres. This will not be true for characteristic spheres, as we shall see later.

(4.1) Proposition. *Let (Σ^m, φ, S^i) be a free S^i action, $i=1$ or 3, and let $V \subset \Sigma^m$ be an invariant manifold of codimension $i+1$, corresponding to $\tilde{V} \subset N^n$, such that $i_*: H_{n-i-1}(V) \xrightarrow{\cong} H_{n-i-1}(N)$. Then with appropriate choice of orientation, the normal bundle of V in N has Euler class $\chi = i^*(\alpha) \in H^{i+1}(V)$, where $\alpha \in H^{i+1}(N)$ is a generator which is the Euler class of the canonical bundle over N, the linear bundle associated with $\Sigma^m \to N$, $(i: V \to N$, the inclusion$)$.*

Proof. Recall that if $j: V \to T(\nu)$ is the inclusion of the zero-cross-section in the Thom complex, then $\chi = j^*(U)$, where $U \in H^*(T(\nu))$ is the Thom class (see [33, 27]). Now j factors, $V \xrightarrow{i} N \xrightarrow{\eta} T(\nu)$, $j = \eta i$, where η is the natural collapsing map. Hence $\chi = i^*(\eta^* U)$.

Consider the inclusion of pairs, where E is a tubular neighborhood of V in N:

$$j_0: (E, \partial E) \to (N, N_0)$$

where $N_0 = N - \operatorname{int} E$, so j_0 is an excision, and consider the inclusion $k: N \to (N, N_0)$. Then $j_0^*: H(N, N_0) \to H(E, \partial E)$ is an isomorphism, and $\eta^* = k^* j_0^{*-1}$. Also $k^*: H^n(N, N_0) \to H^n(N)$ is an isomorphism. Now we have the commutative diagrams with cup products

a)
$$\begin{array}{ccc} H^q(N,N_0) \otimes H^{n-q}(N) & \longrightarrow & H^n(N,N_0) \\ \downarrow{\scriptstyle j_0^* \otimes i^*} & & \downarrow{\scriptstyle j_0^*}\,\cong \\ H^q(E,\partial E) \otimes H^{n-q}(E) & \longrightarrow & H^n(E,\partial E) \end{array}$$

b)
$$\begin{array}{ccc} H^q(N,N_0) \otimes H^{n-q}(N) & \longrightarrow & H^n(N,N_0) \\ \downarrow{\scriptstyle k^* \otimes 1} & & \downarrow{\scriptstyle k^*}\,\cong \\ H^q(N) \otimes H^{n-q}(N) & \longrightarrow & H^n(N) \end{array}$$

Now $i^*: H^s(N) \to H^s(E)$ is an isomorphism for $s = n-(i+1)$, since i_* is. Hence, if $j^*(U') = U \in H^{i+1}(E, \partial E)$, $U \cup : H^{n-q}(E) \to H^n(E, \partial E)$ is an isomorphism by the Thom isomorphism, so that $U' \cup : H^{n-q}(N) \to H^n(N, N_0)$ is also an isomorphism, using diagram (a). Using (b) we get $k^*(U') \cup : H^{n-q}(N) \to H^n(N)$ is an isomorphism. But if $\alpha \in H^{i+1}(N)$

is the generator, $\alpha \bigcup$ is an isomorphism, so $k^*(U') = \pm\alpha$ and it follows that $\chi = i^*\eta^*(U) = i^*k^*(j_0^{*-1}U) = i^*k^*(U') = \pm i^*\alpha$, and with appropriate choice of orientation, the proposition is proved.

(4.2) Proposition. *Let (Σ^m, φ, S^1) be a free S^1 action, $\tilde{V} \subset \Sigma^m$ a codimension 2 invariant manifold such that $i_*: H_{n-2}(V) \to H_{n-2}(N)$ is an isomorphism (with notation as in (4.1)). Then $\tilde{V}$ is characteristic.*

Proof. First we note that an oriented 2-plane bundle is determined by its Euler class. If we take the map of V into CP^k $(m = 2k+1, n = 2k)$ coming from the inclusion of V into N followed by the homotopy equivalence N to CP^k, this map factors through CP^{k-1} since V is $2k-2$ dimensional. Then from (4.1) and the remark above it follows that this map $s: V \to \mathrm{CP}^{k-1}$ induces the normal bundle of V in N from that of CP^{k-1} in CP^k. Now the complement of CP^{k-1} (or a tubular neighborhood of it) in CP^k is a $2k$-cell D^{2k}, so that the bundle map of normal bundles extends to the complement of V in N to D^{2k}. The union of these maps induces a map $t: N^{2k} \to \mathrm{CP}^k$, which is still a homotopy equivalence, since the map induces an isomorphism on H^2, as $t^* = s^*$ on H^2, and $t^{-1}(\mathrm{CP}^{k-1}) = V$. Hence V is characteristic.

The analogous theorem is true for S^0 actions, but seems less likely for S^3 actions. For 4-plane bundles are not characterized by the Euler class, and in fact submanifolds of codimension 4 correspond to maps $N \to \mathrm{MSO}(4)$, rather than into HP^∞, the quaternionic projective space. In fact it is false in the piecewise linear case as we shall see later.

Now we shall consider the question of finding characteristic spheres, using (2.14). We consider the homotopy equivalence of N with the projective space P^k over C or H, for S^1 or S^3 actions. Utilizing the notation of (2.14), we consider the homotopy equivalence $f: N \to P^k$, let ξ be a bundle over P^k such that $f^*(\xi) = \nu$ is the normal bundle of N in a high dimensional Euclidean space, and we consider bundle maps $b: \nu \to \xi$ covering f.

(4.3) Theorem. *If $\sigma_{p^{k-\ell}}(f) = 0$ then N has a characteristic homotopy $(i+1)(k-\ell)+i$ sphere, provided $(i+1)(k-\ell) > 4$.*

This is a direct application of (2.14). It was first proved by ROTHENBERG in unpublished work.

We recall that for dimension of $P^{k-\ell} = 4q$, the definition of $\sigma_{p^{k-\ell}}(f, \mathrm{b})$ is simply the difference of two indices, namely index $P^{k-\ell}$ − index $V = 1 -$ index V, where V is the characteristic submanifold. Now index V may be calculated using the Hirzebruch Index Theorem [24].

The normal bundle of V in Euclidean space $\mathbb{R}^Q$ is the sum of the normal bundle of N in $\mathbb{R}^Q$ restricted to V, plus $\ell\eta$ where η is the canonical bundle ρ over N restricted to V. Hence the stable tangent bundle of V is

$$\tau_V=(\nu|\zeta)^{-1}+(\ell\eta)^{-1}=\tau_N|V+\ell\eta^{-1}.$$

Suppose $(i+1)(k-\ell)=4q=\dim V$. Then

$$\begin{aligned}\sigma(V)&=\langle L_q(p_1(\tau_N|V+\ell\eta^{-1}),\ldots),[V]\rangle\\&=\langle i^*L_q(p_1(\tau_N+\ell\rho^{-1}),\ldots),[V]\rangle\\&=\langle L_q(p_1(\tau_N+\ell\rho^{-1}),\ldots),\chi_{k-\ell}\rangle\end{aligned}$$

where $i_*[V]=\chi_{k-\ell}$, $f_*(\chi_{k-\ell})=\text{image}[P^{k-\ell}]$ in $H_*(P^k)$. Hence we get

(4.4) Corollary. *Let $(\Sigma^{2k+1},\varphi,S^1)$ (or $(\Sigma^{4k+1},\varphi,S^3)$) be a free differentiable action. Then the action has a characteristic homotopy $(4q+1)$-sphere ($(4q+3)$-sphere) $q>1$ if and only if $<L_q(p_1(\tau_N+\ell\rho^{-1}),\ldots),\chi>=1$ where ρ is the canonical bundle over N, χ is the generator of $H_{4q}(N)$, $\ell=k-2q$ (or $\ell=k-q$).*

Applying (4.4) to the actions on $4q+3$ homotopy spheres constructed in (3.3), we find that since $p_i(N)=p_i(\mathrm{CP}^k)$ for $i<q$ and $p_q(N)\neq p_q(\mathrm{CP}^k)$, $(2k=4q+2)$, it follows easily that none of these actions have characteristic homotopy $(4q+1)$-spheres, while they all have them of dimension $(4i+1)$, $i<q$.

We now construct similar examples to those of (3.3), as follows. Recall that the ring of complex vector bundles, $K(P^k)\cong H^*(P^k)$, $P^k=\mathrm{CP}^k$ or HP^k, and the Chern character $(\mathrm{ch}:K(P^k)\to H^*(P^k;\mathbb{Q}))$ is a monomorphism (see [3]). It follows that we can find complex bundles θ over P^k with $c_1=0$ and $c_2\neq 0$. It follows from the formula relating Pontryagin and Chern classes that $p_1(\theta)\neq 0$, considering θ as a real bundle (see [33, 27]). Since $J(P^k)$ is finite, some multiple $n\theta=\theta'$ is fibre homotopically trivial (see [2]). Setting $\xi=\nu+n\theta'$, and suppose $k=7$ or 15, we may proceed as in the proof of (3.3) to prove:

(4.5) Theorem. *There are infinitely many different actions of S^1 on homotopy 15 or 31 spheres, distinguished by p_1 of the orbit space, so that none of them has a characteristic homotopy 5-sphere.*

This then demonstrates that in high codimensions invariant and characteristic manifolds are quite different.

We end with a few remarks about the piecewise linear (p.ℓ.) situation. Considering the analogous questions about p.ℓ. submanifolds of the orbit space N, we can show:

(4.6) Remark. *Any S^3 action on S^{4k+3} has an invariant p.ℓ. S^{4k-1}, i.e. the orbit space N^{4k} has HP^{k-1} embedded in it.*

Proof. It is easy to see that $N-$(point) is homotopy equivalent to HP^{k-1}. Now HAEFLIGER (see [22]) and CASSEN and SULLIVAN have shown that:

If $f: N^n \to M^m$ is a homotopy equivalence of p.ℓ. manifolds, $m-n \geqslant 3$, N 1-connected, closed, $m \geqslant 6$, then f is homotopic to a p.ℓ. embedding.

Thus HP^{k-1} embeds p.ℓ. in N^{4k}, so S^{4k+3} has the standard action on S^{4k-1} embedded in it piecewise linearly

On the other hand, a characteristic submanifold has a linear normal bundle, so a p.ℓ. characteristic submanifold of a smooth orbit space N can be smoothed, using the smoothing theory of HIRSH-MAZUR, LASHOF-ROTHENBERG (see [30]). Hence the problem for characteristic submanifolds is the same, p.ℓ. or smooth.

We remark that MONTGOMERY-YANG [36] have shown that characteristic spheres of codimension 2 of S^1 actions are unknotted.

III. Semi-free S^1 Actions

An action (M^m, φ, G) is called *semi-free* if it is free off of the fixed point set, i.e. there are two types of orbits, fixed points and G. We shall study the situation where (Σ^m, φ, S^1) is a semi-free differentiable action of S^1 on a homotopy sphere Σ^m, and the fixed point set F^q is a homotopy sphere. For a general discussion of semi-free S^1 actions we refer to [6].

We shall show how to use the results of surgery to construct many such exotic actions, generalizing the constructions of MONTGOMERY-YANG [35] in dimension 6, and in other dimensions [36].

In § 5 we describe the reduction of the problem to conventional problems in differential topology. In § 6 we use these results to construct actions with exotic spheres as fixed point sets, and then show how to get infinitely many exotic actions with the standard sphere as fixed point set. The most powerful theorem proved is (6.22).

5. Constructions of Semi-free Actions of S^1

Let (Σ^m, φ, S^1) be a semi-free action, with fixed point set $F^q \subset \Sigma^m$, F^q a homotopy q-sphere (i.e. S^1 acts freely outside F). Then S^1 acts freely and linearly on the normal space to F at each point of F (see [38] or [19, page 58]). Considering $S^1 = \{z \in C, |z| = 1\}$, this action defines a complex structure on the normal space, since it defines multiplication by $i \in S^1$ with the right properties. The action of S^1 is then linear with respect to the structure of a complex vector space, and it follows that the action of S^1 is exactly the action of the complex numbers of unit modulus in a complex vector space. Hence the normal bundle of F has a complex structure, and the action of S^1 on it, is just that induced by the complex structure*. In particular the codimension $m-q=2k$.

Let η be the complex bundle over F defined by the action.

(5.1) Theorem. *If* (Σ^m, φ, S^1) *and* $(\Sigma^m, \varphi', S^1)$ *are equivalent, then* F *is diffeomorphic to* F' *and* η *is equivalent to* η'.

Proof. That F is diffeomorphic to F' is evident, and clearly the equivalence $f:\Sigma \to \Sigma$ defines a complex map of the normal bundles of F and F' so that they are equivalent as complex bundles.

Numerous examples in which F is an exotic sphere have been constructed (see [36] and [6]), and we shall construct some below.

Now let E be an equivariant tubular neighborhood of F in Σ (see [19, page 57]), and let S^{2k-1} be the boundary of a fibre of E. Now it has been shown** by MONTGOMERY-YANG [36] that if $k=1$, then $\Sigma^m = S^m$, $F = S^{m-2}$ embedded as usual, and the action is linear. Therefore we restrict ourselves to $k>1$. It follows from a homology argument that if $k>1$, then $S^{2k-1} \subset \Sigma - F$ is a homotopy equivalence. Now let $N = \Sigma - E_0$, where E_0 is the interior of an equivariant tubular neighborhood of F, with $\bar{E}_0 \subset \operatorname{int} E$. Then S^1 acts freely on N, and on $S^{2k-1} \subset N$, and S^{2k-1} is homotopy equivalent to N. It follows from the exact homotopy sequence of the fibre maps, using the diagram

$$\begin{array}{ccc} S^{2k-1} & \longrightarrow & N \\ \downarrow & & \downarrow \\ S^{2k-1}/S^1 & \longrightarrow & N/S^1 \end{array}$$

that $S^{2k-1}/S^1 \to N/S^1$ is a homotopy equivalence. Set $\bar{N} = N/S^1$. Since the action of S^1 on S^{2k-1} is linear, $S^{2k-1}/S^1 = \mathrm{CP}^{k-1}$, and since S^{2k-1} is the fibre of E over F it follows that its normal bundle is equivariantly trivial, so that we get an embedding $\mathrm{CP}^{k-1} \times D^{q+1} \subset \bar{N}^{m-1}$, and it is a homotopy equivalence. It follows similarly that the region between $\partial \bar{N}$ and $\mathrm{CP}^{k-1} \times S^q$ is an h-cobordism, so if $m>6$, by the h-cobordism theorem of SMALE if $q>1$ (see [44]) or its generalization, the s-cobordism

* This observation was made to me by G. BREDON, who also conjectured (5.5) below.

** (Added in proof) This result first appears in WU-YI HSIANG. On the unknottedness of the fixed point set of differentiable circle group actions on spheres – PA Smith conjecture, Bull. AMS **70** (1964) 678—680.

theorem [28] if $q=1$, it is diffeomorphic to the product $\mathbf{C}P^{k-1}\times S^q\times I$, and hence $\bar{N}$ is diffeomorphic to $\mathbf{C}P^{k-1}\times D^{q+1}$, and $N\to\bar{N}$ is equivalent to $h\times 1: S^{2k-1}\times D^{q+1}\to \mathbf{C}P^{k-1}\times D^{q+1}$, where $h: S^{2k-1}\to\mathbf{C}P^{k-1}$ is the Hopf map, i.e. the principal bundle $N\to\bar{N}$ is induced by the map $\bar{N}\to\mathbf{C}P^{k-1}$.

Hence we have shown

(5.2) Theorem. *Let (Σ^m,φ,S^1) be a semi-free action on a homotopy sphere Σ^m, with fixed point set F a homotopy q-sphere, $m-q=2k$, $k\geqslant 1$, $m\geqslant 6$. If N is the complement of an open tubular neighborhood of F in Σ^m, then N is equivariantly diffeomorphic to $S^{2k-1}\times D^{q+1}$, with the standard action on S^{2k-1}, trivial action on D^{q+1}. In particular the projective space bundle P associated to η, which is $\partial N/S^1$, is diffeomorphic to* $\mathbf{C}P^{k-1}\times S^q$.

Now we study the bundle η.

(5.3) Theorem. *The bundle of projective spaces $\pi: P\to F$ with fibre $\mathbf{C}P^{k-1}$ associated with $E\to F$ is fibre homotopically trivial, i.e., there is a map $t: P\to\mathbf{C}P^{k-1}\times F$ such that t is a homotopy equivalence and commutes with the projections, $p_2 t=\pi$, ($p_2:\mathbf{C}P^{k-1}\times F\to F$ is projection on the second factor). Further, $t^*(\alpha\times\varepsilon^0)=\alpha'$ the canonical line bundle over P, where α is the canonical line bundle over* $\mathbf{C}P^{k-1}$.

Proof. Clearly $P=\partial\bar{N}$ and by (5.2) there is a diffeomorphism $h: P\to\mathbf{C}P^{k-1}\times S^q$ such that $h^*(\alpha\times\varepsilon^0)=\alpha'$, the canonical line bundle over P, where α is the canonical line bundle over $\mathbf{C}P^{k-1}$. Let $t: P\to\mathbf{C}P^{k-1}\times F$ be defined by $t(z)=(p_1 h(z),\pi(z))$. Clearly $p_2 t=\pi$, so t preserves fibres.

Now $p_1^*(\alpha)=\alpha\times\varepsilon^0$, and hence $t^*(\alpha\times\varepsilon^0)=t^*p_1^*(\alpha)=(p_1 t)^*\alpha=(p_1 h)^*(\alpha)=h^*p_1^*(\alpha)=h^*(\alpha\times\varepsilon^0)=\alpha'$, by (5.2), so the last condition of the theorem is satisfied for t. On the other hand, if $i:\mathbf{C}P^{k-1}\to P$ is the inclusion of a fibre, then $i^*(\alpha')=\alpha$. It follows easily that if t' is the restriction of t to the fibres $t':\mathbf{C}P^{k-1}\to\mathbf{C}P^{k-1}$, then $t'^*(\alpha)=\alpha$. Hence t' is a homotopy equivalence on the fibres and hence t is a homotopy equivalence, which completes the proof.

(5.4) Question. *Are there non-trivial complex bundles whose projective space bundles are fibre homotopy trivial by a trivialization t sending the canonical line bundles into each other, as in* (5.3)? *Such bundles are stably trivial* (*see* (5.5)).

As another application of (5.2) we get:

(5.5) Theorem. *The Chern classes of η are 0 so that η is stably trivial (as a complex bundle).*

Proof. We recall the definition of the Chern classes after GROTHENDIECK (see [4]):

Let $P \to F$ be the CP^{k-1} bundle associated to η, and let $x \in H^2(P)$ be the first Chern class of the canonical line bundle over P. Then the powers of x, i.e., $1, x, x^2, \ldots, x^{k-1}$ are a basis for $H^*(P)$ over $H^*(F)$ and hence we have a relation $x^n + \sum_{i=1}^{n} c_i x^{n-1} = 0$. Then $c_i = c_i(\eta)$ are the Chern classes.

Now by (5.2), P is diffeomorphic to $\mathrm{CP}^{k-1} \times S^q$, and the canonical bundle over P is induced from that over CP^{k-1}. Hence if $h: P \to \mathrm{CP}^{k-1} \times S^q$, $x = h^*(y)$, $y \in H^2(\mathrm{CP}^{k-1}) \subset H^2(\mathrm{CP}^{k-1} \times S^q)$, and $x^k = h^*(y^k) = 0$. Hence $c_i(\eta) = 0$ for $i > 0$. But a stable bundle over a homotopy sphere is determined by its Chern classes, so η is stably trivial.

(5.6) Corollary. *If $q \leqslant 2k$, then η is trivial.*

Proof. In these dimensions, η is already stable, hence trivial by (5.5).

It would be interesting to know if there are examples of semi-free S^1 actions (Σ^m, φ, S^1) with η non-trivial. In fact we may characterize the bundles η which occur by the following theorem, which describes how to construct semi-free S^1 actions.

Let F^q be a homotopy q-sphere, η a complex k-plane bundle over F. Let $\pi: P \to F$ be the associated CP^{k-1} bundle to η, and suppose $h: P \to \mathrm{CP}^{k-1} \times S^q$ is an orientation preserving diffeomorphism such that $h^*(y) = x$, where $y = p_1^*(c_1)$, $p_1: \mathrm{CP}^{k-1} \times S^q \to \mathrm{CP}^{k-1}$, c_1 is the Chern class of the canonical bundle over CP^{k-1} and x is the first Chern class of the canonical line bundle over P.

(5.7) Theorem. *There is a semi-free S^1 action (Σ^m, φ, S^1) with fixed point set F embedded in Σ^m with (complex) normal bundle η, and such that the orbit space is $C_\pi \bigcup_h \mathrm{CP}^{k-1} \times D^{q+1}$, where C_π is the mapping cylinder of π, and $\bigcup_h$ means we identify $P \subset C_\pi$ with $\mathrm{CP}^{k-1} \times S^q \subset \mathrm{CP}^{k-1} \times D^{q+1}$ via the diffeomorphism h. Every semi-free action of S^1 on a homotopy sphere of dimension >6 with fixed point set a homotopy sphere is given this way.**

---

* One can remove the condition that dimension >6 if one substitutes h-cobordism for diffeomorphism in the statement.

(5.8) Theorem. *Let $(\Sigma_i^m, \varphi_i, S^1)$ be semi-free S^1 actions constructed as in (5.7), $i=0,1$, with F_i, η_i, h_i, $i=0,1$. Then $(\Sigma_0^m, \varphi_0, S^1)$ is equivalent to $(\Sigma_1^m, \varphi_1, S^1)$ if and only if there are*

(i) *A diffeomorphism $f: F_0 \to F_1$;*
(ii) *A complex bundle equivalence $b: \eta_0 \to \eta_1$ covering the diffeomorphism f;*
(iii) *A diffeomorphism $g: \mathrm{CP}^{k-1} \times D^{q+1} \to \mathrm{CP}^{k-1} \times D^{q+1}$ such that the diagram below commutes:*

$$\begin{array}{ccc} P_0 & \xrightarrow{P(b)} & P_1 \\ {\scriptstyle h_0}\downarrow & & \downarrow{\scriptstyle h_1} \\ \mathrm{CP}^{k-1} \times S^q & \xrightarrow{\partial g} & \mathrm{CP}^{k-1} \times S^q \end{array}$$

where P_i is the associated CP^{k-1} bundle to η_i, $P(b)$ is the map induced by b, and ∂g is the restriction of g.

These two theorems give us the necessary tools to attempt to construct and distinguish semi-free S^1 actions, using the surgery results of Chapter I.

Proof of (5.7). Consider the semi-free S^1 action on the total space of η, $E(\eta)$, defined by the complex structure, and the free S^1 action on $S^{2k-1} \times D^{q+1}$ defined by the linear free action on S^{2k-1}. Then the diffeomorphism $h: P \to \mathrm{CP}^{k-1} \times S^q$ induces an equivariant diffeomorphism $h': E_0(\eta) \to S^{2k-1} \times S^q$, where $E_0(\eta) = \partial E(\eta)$ is the associated sphere bundle. Then $M = E(\eta) \bigcup_{h'} (S^{k-1} \times D^{q+1})$ is then a manifold with a semi-free S^1 action, fixed point set F with normal bundle η. It remains to show that M is a homotopy sphere.

Since we have assumed $k > 1$ throughout this section, it follows that $\pi_1(E_0(\eta)) \cong \pi_1(F)$, $S^{2k-1} \times D^{q+1}$ is 1-connected and $\pi_0(E_0(\eta)) \cong \pi_0(F)$. Therefore, by Van Kampen's theorem, M is 1-connected.

Consider the commutative diagram

$$\begin{array}{ccccc} E_0(\eta) & \xrightarrow{h'} & S^{2k-1} \times S^q & & \\ {\scriptstyle \pi}\downarrow & & \downarrow & \searrow^{p_2} & \\ P & \xrightarrow{h} & \mathrm{CP}^{2k-1} \times S^q & \xrightarrow{p_2} & S^q \end{array}$$

Let $a \in H_{2k-1}(E_0(\eta))$ be the image of the generator of the homology of the fibre $H_{2k-1}(S^{2k-1})$. Then $\pi_*(a) = 0$, so $(p_2 h \pi)_*(a) = 0$, and hence $(p_2 h')_*(a) = 0$. Hence $(h')_*(a) = \lambda(g \otimes 1) \in H_{2k-1}(S^{2k-1}) \otimes H_0(S^q)$

$\subseteq H_{2k-1}(S^{2k-1} \times S^q)$, $\lambda \in \mathbb{Z}$, $g \in H_{2k-1}(S^{2k-1})$, a generator. Since h_1 is a diffeomorphism, $\lambda = \pm 1$. If $j: S^{2k-1} \times S^q \to S^{2k-1} \times D^{q+1}$ is inclusion we deduce:

(5.9) $$j_* h'_*(a) = \pm g.$$

Now consider the Mayer-Viettoris sequence for M:

(5.10) $$\cdots \longrightarrow H_{s+1}(M) \longrightarrow H_s(E_0(\eta)) \xrightarrow{i_* + \rho_*} H_s(E(\eta)) + H_s(S^{2k-1} \times D^{q+1}) \longrightarrow H_s(M) \longrightarrow \cdots$$

where $\rho = jh'$. Then $(i_* + \rho_*)a = i_*(a) + \rho_*(a) = 0 \pm g$. Assume the sign is $+$. Let $b \in H_q(E_0(\eta))$ be another generator for the homology of $E_0(\eta)$, such that $\pi_*(b)$ is a generator of $H_q(F)$, so that a and b are a basis for $H_*(E_0(\eta))$ in dimensions between 0 and $2k+q-1$. Let $h'_*(b) = \mu g \otimes 1 + \gamma 1 \otimes c$, c a generator of $H_q(S^q) \subset H_q(S^{2k-1} \times S^q)$. Choose as a new generator $b' = b - \mu a$. Then $(h')_*(b') = (h')_*(b - \mu a) = \mu g \otimes 1 + \gamma 1 \otimes c - \mu g \otimes 1 = \gamma 1 \otimes c$. Since b' is a generator, $h_{1*}(b')$ is a generator and $\gamma = \pm 1$. Now $\pi_*(b') = \pi_*(b) - \pi_*(\mu a) = \pi_*(b)$. Hence $i_* + \rho_*$ is an isomorphism between $H_s(E_0(\eta))$ and $H_s(E(\eta)) + H_s(S^{2k-1} \times D^{q+1})$ for $0 < s < 2k+q-1$, and it follows from (5.10) that M is a homology sphere, hence a homotopy sphere. This completes the construction.

On the other hand (5.2) shows that all such semi-free actions arise from this construction. Thus (5.7) is proved.

Proof of (5.8). It follows from condition (iii) of (5.7) that the diagram commutes:

$$\begin{array}{ccc} E_0(\eta_0) & \xrightarrow{b_0} & E_0(\eta_1) \\ {\scriptstyle h'_0}\downarrow & & \downarrow{\scriptstyle h'_1} \\ S^{2k-1} \times S^q & \xrightarrow{\partial g'} & S^{2k-1} \times S^q \end{array}$$

where $'$ denotes the map induced on principal bundles by the maps of base spaces, $\partial g'$ is the restriction of g'. Thus the maps $b: E(\eta_0) \to E(\eta_1)$ and $g': S^{2k-1} \times D^{q-1} \to S^{2k-1} \times D^{q+1}$ respect the identifications by h'_0 and h'_1 and are equivariant, so define an equivariant diffeomorphism of Σ_0^m to Σ_1^m. Hence conditions (i), (ii) and (iii) yield an equivalence of actions.

Now suppose $\psi: \Sigma_0 \to \Sigma_1$ is an equivalence of actions, i.e. an equivariant diffeomorphism. As in (5.1), $f = \psi | F_0 : F_0 \to F_1$ is a diffeomorphism and ψ restricted to an equivariant tubular neighborhood N_0 of F_0 induced a complex bundle equivalence $b: \eta_0 \to \eta_1$ covering f. Let $N_1 = \psi(N_0)$, and $V_i = \Sigma_i - \operatorname{int} N_i$. Then since $\psi(N_0) = N_1$ and ψ is a diffeomorphism, $\psi(V_0) = V_1$, and $\psi | V_0 = g': V_0 \to V_1$ is an equivariant diffeomorphism, and $\partial V_i = E_0(\eta_i)$, $g' | \partial V_0 = b_0 = b | E_0(\eta_0)$, and let $\bar{g}: V_0/S^1 \to V_1/S^1$ be the induced diffeomorphism. By (5.2) there are diffeomorphisms $k_i: V_i/S^1 \to \mathrm{CP}^{k-1} \times D^q$, so we define $g: \mathrm{CP}^{k-1} \times D^q \to \mathrm{CP}^{k-1} \times D^q$ by $g = k_1 \bar{g} k_0^{-1}$. Since $g'|\partial V_0 = b_0$, it follows that $\bar{g}|(\partial V_0/S^1) = \bar{g}|P_0 = P(b)$, and it follows that, since $h_i = k_i|P_i$, $(P_i = \partial V_i/S^1)$, the diagram of (iii) commutes, which proves (5.8).

6. Applying Surgery to Construct Semi-free Actions

In this paragraph we apply results of surgery together with (5.7) and (5.8) to construct semi-free actions.

By (5.7), the problem is to find over a homotopy sphere F^q a complex k-plane bundle η and a diffeomorphism $h: P(\eta) \to \mathrm{CP}^{k-1} \times S^q$, where $P(\eta)$ is the associated bundle with fibre CP^{k-1}, satisfying the condition on Chern classes. If $q \neq 2$, then any diffeomorphism h can be made to have this property by composing h if necessary with $\alpha \times \varepsilon: \mathrm{CP}^{k-1} \times S^q \to \mathrm{CP}^{k-1} \times S^q$ where $\alpha^*(c_1) = -c_1 \in H^2(\mathrm{CP}^{k-1})$ and ε = identity if k is even, ε is orientation reversing if k is odd.

As I do not know of any non-trivial bundles η with the above property, I will always take η to be the trivial complex k-plane bundle. We will say in this case F is untwisted. We shall first consider the problem of constructing actions with exotic spheres as fixed point set, and we get some results extending those of BREDON [5] and MONTGOMERY-YANG [36].

(6.1) Theorem. *Let F^{2n-1} be a homotopy sphere which bounds a parallelizable manifold, $n \geq 2$ ($F^{2n-1} \in \mathrm{bP}^{2n}$ in the notation of [29]). Then for each even $k > 1$, there is a semi-free action of S^1 on a homotopy sphere $\Sigma^{2n-1+2k}$ with F as untwisted fixed point set.*

If M^m is a closed manifold, let $I(M) = \{\Sigma \in \theta^m$ such that $M \# \Sigma$ is diffeomorphic to $M\}$. $I(M)$ is a subgroup and is called the *inertia group* of M. Let $I_0(M) = I(M) \cap \mathrm{bP}^{m+1}$. We recall from [29] that bP^{4s} is a finite cyclic group, and we denote its order by m_s.

(6.2) Theorem. *Suppose $q = 4t-1$ and $k > 1$, k odd, and let ℓ = order of $I_0(\mathrm{CP}^{k-1} \times S^q)$. Then an element $F \in \mathrm{bP}^{4t}$, $t > 1$, occurs as an untwisted fixed point set of a semi-free action of S^1 on a homotopy sphere Σ^{4s+1}, where dimension $4s+1 = 4t-1+2k$ if and only if $F \in (m_s/\ell)\mathrm{bP}^{4t}$.*

(6.3) Theorem. *Suppose $q = 4n+1$, k odd, $k > 1$, and $F \neq 0$ in bP^{4n+2}. Then F occurs as an untwisted fixed point set of a semi-free S^1 action on a homotopy sphere Σ of dimension $4n+1+2k = 4s+3$ if and only if $I_0(\mathrm{CP}^{k-1} \times S^q) = \mathrm{bP}^{4s+2}$.*

In case $\mathrm{bP}^{4s+2} = 0$ for example when $s = 1$, 3 or 7 (see [14]) then the condition is of course satisfied for any F.

The proofs of these three theorems are very similar, utilizing surgery techniques to create a diffeomorphism of $\mathrm{CP}^{k-1} \times F$ with $\mathrm{CP}^{k-1} \times S^q$.

Let $F = \partial W^{2n}$, W parallelizable. We may consider $W_0 = (W - \operatorname{int} D^{2n})$ as a parallelizable cobordism between F and S^{2n-1}, and thus we may define a normal map $G: (W_0, F \cup S^{2n-1}) \to (S^{2n-1} \times I, S^{2n-1} \times 0 \cup S^{2n-1} \times 1)$ with $G|S^{2n-1} = \text{identity}$. By (2.7), we may assume W_0 to be $(n-1)$-connected. Multiplying by CP^{k-1} we get a normal map

$$1 \times G: \ \mathrm{CP}^{k-1} \times (W_0, F \cup S^{2n-1}) \to \mathrm{CP}^{k-1} \times (S^{2n-1} \times I, S^{2n-1} \times 0 \cup S^{2n-1} \times 1)$$

with $1 \times G|\mathrm{CP}^{k-1} \times S^{2n-1} = \text{identity}$. The remainder of the proofs of the three theorems is computing the obstruction σ for this normal cobordism, and using this to determine if $\mathrm{CP}^{k-1} \times F$ is diffeomorphic to $\mathrm{CP}^{k-1} \times S^{2n-1}$. The three theorems correspond to the three cases:

(6.1) k is even,

(6.2) k is odd, n is even,

(6.3) k is odd, n is odd.

(6.4) Lemma. $\ker(1 \times G)_* = H_*(\mathrm{CP}^{k-1}) \otimes \ker G_*$.

Proof. By the Künneth formula, since $H_*(\mathrm{CP}^{k-1})$ is torsion free $H_*(\mathrm{CP}^{k-1} \times (W_0, F \cup S^{2n-1})) \cong H_*(\mathrm{CP}^{k-1}) \otimes H_*(W_0, F \cup S^{2n-1})$, and $(1 \times G)_* = 1 \otimes G_*$. Thus the lemma follows.

Now $\dim W_0 = 2n$, $\dim \mathrm{CP}^{k-1} = 2k-2$ and W_0 is $(n-1)$-connected, so that $\ker(1 \times G)_* = H_*(\mathrm{CP}^{k-1}) \otimes H_n(W_0)$. If k is even, then $k-1$ is odd, and thus $H_{k-1}(\mathrm{CP}^{k-1}) = 0$. Hence $(\ker(1 \times G)_*)_{n+k-1} = 0$ and hence $\sigma(1 \times G) = 0$ and by (2.7), the Fundamental Theorem, $1 \times G$ is normally cobordant $\operatorname{rel} \mathrm{CP}^{k-1} \times S^{2n-1} \times \{0,1\}$ to a homotopy equivalence, i.e. an h-cobordism between $\mathrm{CP}^{k-1} \times F$ and $\mathrm{CP}^{k-1} \times S^{2n-1}$. Now $\dim(\mathrm{CP}^{k-1} \times S^{2n-1}) = 2k-2+2n-1 = 2n+2k-3$ and $n \geqslant 2$, $k > 1$ so that $n+k \geqslant 4$ so that $2n+2k-3 \geqslant 5$. Hence Smale's h-cobordism theorem applies, and $\mathrm{CP}^{k-1} \times F$ is diffeomorphic to $\mathrm{CP}^{k-1} \times S^{2n-1}$. Applying (5.7), it follows that there is a semi-free action of S^1 on some homotopy sphere Σ^m with F as untwisted fixed point set, $m = 2n-1+2k$. This proves (6.1).

In case $k = 2q+1$ then $(\ker(1 \times G)_*)_{n+2q} = H_{2q}(\mathrm{CP}^{2q}) \otimes H_n(W_0)$ and it follows easily, in the notation of § 1, that $K^{n+2q}(\mathrm{CP}^{k-1} \times W_0) = H^{2q}(\mathrm{CP}^{2q}) \otimes H^n(W_0)$. If $x \in H^2(\mathrm{CP}^{2q})$ is a generator, then x^q generates $H^{2q}(\mathrm{CP}^{2q})$ and x^{2q} generates $H^{4q}(\mathrm{CP}^{2q})$. Hence

$$\begin{aligned}(x^q \otimes a, x^q \otimes b) &= ((x^q \otimes a) \cup (x^q \otimes b))[\mathrm{CP}^{2q} \times W_0] \\ &= ((x^{2q})[\mathrm{CP}^{2q}])(ab)(W_0] = (ab)[W_0] = (a,b).\end{aligned}$$

Hence the pairing on $K^{n+2q}(\mathrm{CP}^{2q}\times W_0)$ is isomorphic to the pairing on $K^n(W_0)$, so it follows if n is even that $I(1\times G)=I(G)$. It follows as in the proof of (2.12) NOVIKOV's Classification Theorem and (2.11), that $\mathrm{CP}^{2q}\times F$ is diffeomorphic to $(\mathrm{CP}^{2q}\times S^{2n-1})\# L$ where $L\in \mathrm{bP}^{2n}$, and $L=\partial U$, $U\in P^{2n}$, index $U=$ index W.

Hence by (5.7) $\mathrm{CP}^{2q}\times S^{2n-1}$ is diffeomorphic to $(\mathrm{CP}^{2q}\times S^{2n-1})\# L=\mathrm{CP}^{2q}\times F$, i.e. $L\in I_0(\mathrm{CP}^{2q}\times S^{2n-1})$, if and only if F is the untwisted fixed point set of a semi-free action of S^1 on a homotopy $2q+2n+1$ sphere, where if $F=mg_n$, g_s a generator of bP^{2s}, s even, then $L=mg_{n+2q}$. It follows that F is the untwisted fixed point set if and only if m is divisible by r, where rg_{n+2q} generates

$$I_0(\mathrm{CP}^{2q}\times S^{2n-1})\subset \mathrm{bP}^{4q+2n},$$

and

$$r=\text{order } \mathrm{bP}^{4q+2n}/\text{order } I_0(\mathrm{CP}^{2q}\times S^{2n-1}),$$

which proves (6.2).

The proof of (6.3) is similar where one shows that $\sigma(1\times G)=\sigma(G)$ using properties of σ in dimension $\equiv 2 \pmod 4$ which we have not discussed here (such as the definition). We refer to [14] or [7] for the necessary techniques. In particular, SULLIVAN has proved the following formula (in unpublished work): Let $f:(M,\partial M)\to(X,Y)$, (f,b) a normal map of degree 1, $(f|\partial M)_*: H_*(\partial M)\to H_*(Y)$ an isomorphism as in § 2, $m=\dim M=4k+2$, N a smooth manifold $n=\dim N=4\ell$. Then $1\times f: N\times(M,\partial M)\to N\times(X,Y)$ and $1\times b$ give a normal map and

$$\sigma(1\times f)=\chi(N)\sigma(f).$$

One cannot say much about the homotopy sphere on which S^1 acts from this construction. However if one takes the "equivariant connected sum" of Σ with itself ℓ-times, i.e. connected sum along a cell which intersects the fixed point set F and is invariant, we get a semi-free S^1 action on $\Sigma\#\Sigma\#\cdots\#\Sigma=\ell\Sigma$ with $\ell F=F\#F\#\cdots\#F$ as fixed point set, (see [6; § 3]). Hence if $\ell\Sigma=S^m$ we arrive at an action on S^m with ℓF as fixed point set. For example, order $\theta_{11}=992$, order $\theta_7=28$ (see [29]), so starting from any $F\in\theta_7=\mathrm{bP}^8$, using (6.1), and the above we may obtain $992F$ as the fixed point set of a semi-free action on S^{11}, and $992\theta_7=4\theta_7$ so we get the result of MONTGOMERY-YANG [36] (see [6]):

(6.5) There are semi-free S^1 actions on S^{11} with every element of $4\theta_7$ as untwisted fixed point set.

In applying (6.2), unfortunately not much is known about calculating $I_0(\mathrm{CP}^{2q}\times S^{2n-1})$, (c.f. [12]). However m_s is certainly a multiple of m_s/ℓ so we may use the same numbers used above in deriving (6.5) to show that there are semi-free actions of S^1 on homotopy 13-spheres with every element of $4\theta_7$ as untwisted fixed point set. But since $\theta_{13}=\mathbb{Z}_3$, we may use the connected sum method above to show that

(6.6) There are semi-free S^1 actions on S^{13} with every element of $4\theta_7$ as untwisted fixed point set.

Now we shall apply the methods of § 2 to construct exotic semi-free S^1 actions on homotopy spheres with fixed point set the standard sphere. We will construct infinitely many inequivalent such actions using (2.19), (2.27), etc. To apply these results we must have a manifold M which satisfies condition (K) in § 2:

Condition (K). If $g: M\times S^1 \to M\times S^1$ is a homotopy equivalence such that

(i) $g|M\times s \sim$ inclusion,

(ii) $pg\sim p, p: M\times S^1\to S^1$, then $g^*: H^{4*}(M\times S^1;\mathbb{Q})\to H^{4*}(M\times S^1;\mathbb{Q})$ is the identity.

(6.7) Proposition. *$M=\mathrm{CP}^{k-1}\times S^q$, q odd satisfies condition (K).*

Proof. Let $g: \mathrm{CP}^{k-1}\times S^q\times S^1\to \mathrm{CP}^{k-1}\times S^q\times S^1$ be a homotopy equivalence satisfying (i) and (ii), above. Then as a map into the product

$$\begin{array}{ccc} M & \xrightarrow{\Delta_2} & M\times M\times M \\ & {}_{g}\searrow & \downarrow{\scriptstyle (p_1 g,\, p_2 g,\, p_3 g)} \\ & & \mathrm{CP}^{k-1}\times S^q\times S^1 \end{array}$$

commutes, where $\Delta_2(m)=(m,m,m)$, $m\in M$. Hence $g^*(x)=(g^*p_1^*x)\cup(g^*p_2^*x)\cup(g^*p_3^*x)$ for $x\in H^*(\mathrm{CP}^{k-1}\times S^q\times S^1)$. Now by (ii) $pg\sim p$, and $p=p_3:\mathrm{CP}^{k-1}\times S^q\times S^1\to S^1$, so $g^*p_3^*=p_3^*$. If $q>1$, then $j^*: H^2(\mathrm{CP}^{k-1}\times S^q\times S^1)\to H^2(\mathrm{CP}^{k-1})$ is an isomorphism, $j:\mathrm{CP}^{k-1}\to \mathrm{CP}^{k-1}\times S^q\times S^1$ inclusion. By (i) above, $j^*g^*=j^*$, so that $j^*g^*p_1^*=j^*p_1^*=$ identity on $H^2(\mathrm{CP}^{k-1})$. Hence $g^*p_1^*=p_1^*$ on $H^2(\mathrm{CP}^{k-1})$, and since $H^2(\mathrm{CP}^{k-1})$ generates $H^*(\mathrm{CP}^{k-1})$ under $\cup$-product, $g^*p_1^*=p_1^*$ on $H^*(\mathrm{CP}^{k-1})$.

Using condition (i), if $z\in H^q(S^q)$, $x\in H^2(\mathrm{CP}^{k-1})$, $y\in H^1(S^1)$ are the generators, then $p_2^*z=1\otimes z\otimes 1$ and $g^*p_2^*(z)=1\otimes z\otimes 1+\lambda x^k\otimes 1\otimes y$, for some $\lambda\in\mathbb{Z}$, $q=2k+1$, since $1\otimes z\otimes 1$ and $x^k\otimes 1\otimes y$ are a basis for $H^q(\mathrm{CP}^{k-1}\times S^q\times S^1)$. A basis of $H^{4i}(\mathrm{CP}^{k-1}\times S^q\times S^1)$ is given by $x^{2i}\otimes 1\otimes 1$, $x^{\ell}\otimes z\otimes y$, where $2\ell+q+1=4i$. Hence $g^*(x^{2i}\otimes 1\otimes 1)$ $=x^{2i}\otimes 1\otimes 1$, and $g^*(x^{\ell}\otimes z\otimes y)=(x^{\ell}\otimes 1\otimes 1)(1\otimes z\otimes 1+\lambda x^k\otimes 1\otimes y)$ $(1\otimes 1\otimes y)=x^{\ell}\otimes z\otimes y+(x^{\ell+k}\otimes 1\otimes y^2)=x^{\ell}\otimes z\otimes y$, since $y^2=0$. Hence $g^*=$ identity on H^{4i}, and condition (K) is satisfied, if $q>1$.

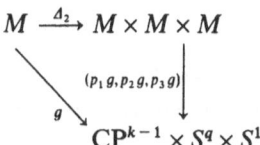

If $q=1$, then $g^* p_1^*(x)=x\otimes 1\otimes 1+\lambda(1\otimes g\otimes y)$. Since $x^k=0$, $g^* p_1^*(x^k)=0$, since $g^2=y^2=0$,

$$g^* p_1^*(x^k)=(g^* p_1^*(x))^k=(x\otimes 1\otimes 1+\lambda(1\otimes g\otimes y))^k$$
$$=x^k\otimes 1\otimes 1+k\lambda(x^{k-1}\otimes g\otimes y)=k\lambda(x^{k-1}\otimes g\otimes y)=0.$$

Hence $\lambda=0$, and $g^* p_1^*=p_1^*$. The rest of the proof proceeds as before, and therefore condition (K) is satisfied for any odd q.

(6.8) Lemma. rank $H^{4*}(\Sigma((\mathrm{CP}^{k-1}\times S^q)_+))=[(k-1)/2]+1$ *if q is odd and* $q+2k\equiv 1 \bmod 4$. *Also* rank $H^{4*}(\Sigma((\mathrm{CP}^{k-1}\times S^q)_+))=[k/2]$ *if q is odd and* $q+2k\equiv -1 \bmod 4$.

(6.9) Corollary. *If* $q>1$, *q odd*, $k>1$, *and* $m=q+2k-2=4n-1$ *then* $A_0(\mathscr{D}_0(\mathrm{CP}^{k-1}\times S^q))$ *has rank* $[(k-1)/2]$. *If* $m=5$, 13 *or* 29, *then* $A_0(\mathscr{D}_0(\mathrm{CP}^{k-1}\times S^q))$ *has rank* $[k/2]$.

This follows from (6.8), (6.7) and (2.27).

Now each element of $\mathscr{D}(\mathrm{CP}^{k-1}\times S^q)$, gives rise to a semi-free S^1 action on a homotopy $(q+2k)$-sphere with S^q as untwisted fixed point set, and we shall use the homomorphism $A_0\colon \mathscr{D}_0(\mathrm{CP}^{k-1}\times S^q)\to H^{4*}(\Sigma(\mathrm{CP}^{k-1}\times S^q)_+)$ and (5.8) to distinguish different actions.

By (5.8), if $f_i\colon \mathrm{CP}^{k-1}\times S^q\to \mathrm{CP}^{k-1}\times S^q$, $i=1,2$ are diffeomorphisms, then the induced S^1 actions are equivalent if and only if there are diffeomorphisms $d\colon S^q\to S^q$, $g\colon \mathrm{CP}^{k-1}\times D^{q+1}\to \mathrm{CP}^{k-1}\times D^{q+1}$ and a complex bundle map $b\colon C^k\times S^q\to C^k\times S^q$ covering d such that the diagram commutes:

$$\text{(U)}\qquad \begin{array}{ccc} \mathrm{CP}^{k-1}\times S^q & \xrightarrow{P(b)} & \mathrm{CP}^{k-1}\times S^q \\ \downarrow{\scriptstyle f_1} & & \downarrow{\scriptstyle f_2} \\ \mathrm{CP}^{k-1}\times S^q & \xrightarrow{\partial g} & \mathrm{CP}^{k-1}\times S^q \end{array}$$

Therefore we shall study such diffeomorphisms $P(b)$ and g.

(6.10) Lemma. *There is a map* $\alpha\colon S^q\to U(k)$ *such that* $P(b)=(1\times d)(f_\alpha)$ *where* $f_\alpha\colon \mathrm{CP}^{k-1}\times S^q\to \mathrm{CP}^{k-1}\times S^q$ *is the diffeomorphism* $f_\alpha(x,y)=(x\cdot\alpha(y),y)$, *where* $U(k)$ *acts on* CP^{k-1} *on the right.*

Proof. Any complex bundle map covering the identity is f_α for some $\alpha\colon S^q\to U(k)$. But $(1\times d)^{-1}P(b)$ is a complex bundle map covering the identity and the result follows.

(6.11) Lemma. *The correspondence which sends* $\alpha: S^q \to U(k)$ *into the diffeomorphism* $f_\alpha: \mathrm{CP}^{k-1} \times S^q \to \mathrm{CP}^{k-1} \times S^q$ *(defined above) defines a homomorphism of groups*

$$\varphi: \pi_q(U(k)) \to \mathscr{D}(\mathrm{CP}^{k-1} \times S^q).$$

Proof. Let $\alpha, \beta: S^q \to U(k)$ and let $h: S^q \times I \to U(k)$ be a homotopy of α to β. Then $H: \mathrm{CP}^{k-1} \times S^q \times I \to \mathrm{CP}^{k-1} \times S^q$, $H(x,y,t)=(x \cdot h(y,t), y)$ is a diffeomorphism for each t, and hence an isotopy. Hence the isotopy class of f_α depends only on the homotopy class of α.

Let $S^q = D_1^q \cup D_2^q$, $S^q - \operatorname{int} D_i^q \subset \operatorname{int} D_j$, $(i,j)=(1,2)$ and $(2,1)$. Since $\alpha|D_1$ is homotopic to a constant and $U(k)$ is connnected, using the homotopy extension theorem, α is homotopic to α' such that $\alpha'|D_1 = e$, the identity of $U(k)$. Similarly, β is homotopic to β', $\beta'|D_2 = e$. Hence $f_{\alpha'}|\mathrm{CP}^{k-1} \times D_1 = \text{identity}$, and $f_{\beta'}|\mathrm{CP}^{k-1} \times D_2 = \text{identity}$. If we take $\alpha'\beta': S^q \to U(k)$, $\alpha'\beta'(y) = \alpha'(y)\beta'(y)$, then $\alpha'\beta'|D_1 = \beta'$, $\alpha'\beta'|D_2 = \alpha'$, so that $f_{\alpha'\beta'} = f_{\beta'}$ on $\mathrm{CP}^{k-1} \times D_1^q$ and $f_{\alpha'\beta'} = f_{\alpha'}$ on $\mathrm{CP}^{k-1} \times D_2^q$. Also $f_{\alpha'} f_{\beta'} = f_{\beta'}$ on $\mathrm{CP}^{k-1} \times D_1$ and $= f_{\alpha'}$ on $\mathrm{CP}^{k-1} \times D_2$. Hence $f_{\alpha'\beta'} = f_{\alpha'} f_{\beta'}$. But $\alpha'\beta'$ is homotopic to $\alpha'+\beta'$ by a well known argument. Hence $f_{\alpha'\beta'}$ is isotopic to $f_{\alpha'+\beta'}$. Since $\alpha \sim \alpha'$, $\beta \sim \beta'$, then $\alpha'+\beta' \sim \alpha+\beta$, and $f_{\alpha'+\beta'}$ is isotopic to $f_{\alpha+\beta}$, $f_{\alpha'}$ is isotopic to f_α, $f_{\beta'}$ is isotopic to f_β. Hence $f_{\alpha+\beta}$ is isotopic to $f_\alpha f_\beta$. Hence $\varphi: \pi_q(U(k)) \to \mathscr{D}(\mathrm{CP}^{k-1} \times S^q)$ is a homomorphism.

(6.12) Proposition. *Let* $f: \mathrm{CP}^{k-1} \times S^q \to \mathrm{CP}^{k-1} \times S^q$ *be a map such that* $f^*(x \otimes 1) = x \otimes 1$, $x \in H^2(\mathrm{CP}^{k-1})$, *and suppose* f *is the restriction to the boundary of a map* $F: \mathrm{CP}^{k-1} \times D^{q+1} \to \mathrm{CP}^{k-1} \times D^{q+1}$. *Then* F *is homotopic to the identity.*

Proof. $\mathrm{CP}^{k-1} \times D^{q+1}$ is homotopy equivalent to CP^{k-1} and $j: \mathrm{CP}^{k-1} \to \mathrm{CP}^\infty$ is a homotopy equivalence up to dimension $2k-1$, the first non-zero homotopy group $\pi_i(\mathrm{CP}^{k-1})$ for $i>2$ being $\pi_{2k-1}(\mathrm{CP}^{k-1})$. The dimension of CP^{k-1} is $2k-2$, so the homology class of F is determined by $F^*|H^2(\mathrm{CP}^{k-1})$, which is determined in turn by f^*, so the result follows.

(6.13) Corollary. *Notation as in* (6.12), *the mapping torus* W_F *of* F *is homotopy equivalent to the product* $\mathrm{CP}^{k-1} \times S^1$ $(W = \mathrm{CP}^{k-1} \times D^{q+1})$.

(6.14) Proposition. *Let* $F: \mathrm{CP}^{k-1} \times D^{q+1} \to \mathrm{CP}^{k-1} \times D^{q+1}$ *be a diffeomorphism homotopic to the identity, so* $W_F \cong \mathrm{CP}^{k-1} \times S^1$. *Then* $P(W_F) = p_1^*(P(\mathrm{CP}^{k-1}))$, $p_1 =$ *projection of the first factor.*

Proof. Let $h\colon W_F \to \Sigma(W_+)$, $W = \mathrm{CP}^{k-1} \times D^q$, be the map collapsing $W \times s$ to a point, so that $\tau(W_F) - p_1^*(\tau(\mathrm{CP}^{k-1}) \times \varepsilon^0) \in \operatorname{im} h^*$. But $H^{4*}(\Sigma(\mathrm{CP}_+^{k-1})) = 0$, and the result follows.

(6.15) Corollary. *Suppose* $f \in \mathscr{D}_0(\mathrm{CP}^{k-1} \times S^q)$ *and* $f = F|\mathrm{CP}^{k-1} \times S^q$, F *a diffeomorphism of* $\mathrm{CP}^{k-1} \times D^{q+1}$. *Then* $A_0(f) = 0$.

Proof. The mapping torus $M_f = \partial W_F$, $M = \mathrm{CP}^{k-1} \times S^q$, $W = \mathrm{CP}^{k-1} \times D^{q+1}$. Hence $\tau(M_f) = \tau(W_F)|M_f$, and $P(\tau(M_f)) = i^* P(\tau(W_F))$ $i^* p_1^*(P(\mathrm{CP}^{k-1})) = P(\tau(\mathrm{CP}^{k-1} \times S^q) \times \varepsilon^0)$, as stable bundles, i.e. in $KO(\mathrm{CP}^{k-1} \times S^q \times S^1)$. Hence $P(\nu(M_f)) = P(\nu(\mathrm{CP}^{k-1} \times S^q) \times \varepsilon^0)$ and it follows from the definition of A_0 (see (2.24)), that $A_0(f) = 0$.

(6.16) Lemma. *Suppose in diagram (U),* $f_i \in \mathscr{D}_0(\mathrm{CP}^{k-1} \times S^q)$, $i = 1, 2$. *Then the maps* $P(b)$ *and* ∂g *are homotopic to the identity. Further* $1 \times d$ *and* f_α *are homotopic to the identity.*

Proof. Since $P(b)$ comes from a complex bundle map $P(b)^*(x) = x$ where $x \in H^2(\mathrm{CP}^{k-1} \times S^q)$ is the Chern class of the canonical complex line bundle over $\mathrm{CP}^{k-1} \times S^q$, and a generator. Since f_1 and $f_2 \in \mathscr{D}_0(\mathrm{CP}^{k-1} \times S^q)$, it follows that $\partial g = f_2(P(b)) f_1^{-1} \sim P(b)$, so that $(\partial g)^*(x) = x$. By (6.12), it follows that $g\colon \mathrm{CP}^{k-1} \times D^{q+1} \to \mathrm{CP}^{k-1} \times D^{q+1}$ is homotopic to the identity. It follows that if $i\colon \mathrm{CP}^{k-1} \times S^q \to \mathrm{CP}^{k-1} \times D^{q+1}$ is inclusion, then $i(\partial g) = g i \sim i$. If p_1 is projection on the first factor, then $p_1 i = p_1$, so $p_1(\partial g) = p_1 i(\partial g) = p_1 g i \sim p_1 i = p_1$.

Now since $P(b)$ is orientation preserving and $P(b)^*(x) = x$, it follows that $P(b)^*(y) = y$, y a generator of $H^q(\mathrm{CP}^{k-1} \times S^q) \cong H^q(S^q)$, in order to preserve orientation. Now since $P(b) = (1 \times d)(f_\alpha)$ by (6.10), $\alpha\colon S^q \to U(k)$, it follows that $d \sim 1$, and hence $p_2 P(b) \sim p_2$. Therefore we have $P(b) \sim \partial g$, $p_1(\partial g) \sim p_1$, $p_2(P(b)) \sim p_2$, so $p_2(\partial g) \sim p_2$. But $\partial g = (p_1 \partial g, p_2 \partial g) \sim (p_1, p_2) = \text{identity}$, so $\partial g \sim \text{identity}$ and hence $P(b) \sim \text{identity}$. We have shown $1 \times d \sim \text{identity}$, so $f_\alpha = (1 \times d)^{-1} P(b) \sim \text{identity}$, which completes the proof of (6.16).

(6.17) Lemma. $A_0(1 \times d) = 0$, $d \in \mathscr{D}_0(S^q)$, $1 \times d \in \mathscr{D}_0(\mathrm{CP}^{k-1} \times S^q)$.

Proof. If $t = 1 \times d$, then the mapping torus of t is $\mathrm{CP}^{k-1} \times (S^q_d)$. Now the mapping torus $(S^q_d) \cong S^q \times S^1$, so the normal bundle ν_d of $(S^q)_d$ is fibre homotopy equivalent to the trivial bundle, by the Atiyah-Hirsch Theorem [2], and is trivial along S^q, and along S^1 (since $(S^q)_d$ is orientable). Hence $\nu_d = h^*(\gamma)$, $\gamma \in KO(S^q \times S^1, S^q \vee S^1) \cong KO(S^q \wedge S^1) \cong KO(S^{q+1})$. Then since the index $(S^q)_d = 0$, it follows from the Index Theorem of HIRZEBRUCH [24], that $P(\gamma) = 0$ and (6.17) follows.

(6.18) Proposition. *Let* $\alpha \in \pi_q(U(k))$, $q > 1$. *If* $f_\alpha\colon \mathrm{CP}^{k-1} \times S^q \to \mathrm{CP}^{k-1} \times S^q$ *is homotopic to the identity, then* α *is of finite order.*

Proof. (c. f. (5.5)). The element α is the characteristic map (see [45]) of a complex vector bundle η^k over S^{q+1}, and if α is of infinite order, it follows that the Chern class $C(\eta)\neq 1$, (see [3]), so $c_n(\eta)\neq 0$, where $q+1=2n$. Recall the relation [4], that if $x\in H^2(P(\eta))$ is the Chern class of the canonical line bundle over $P(\eta)=$ the projective space bundle associated to η, then $\sum_{t=0}^{k} c_{k-t}(\eta)x^t=0$, so that $x^k+c_n(\eta)x^{k-n}=0$. Now $P(\eta)=\mathrm{CP}^{k-1}\times D^{q+1}\cup \mathrm{CP}^{k-1}\times D^{q+1}$ with boundaries identified by f_α. If f_α is homotopic to the identity, then it follows that there is a homotopy equivalence $\psi: P(\eta)\to\mathrm{CP}^{k-1}\times S^{q+1}$, and since $q>1$ $\psi^*(x')=\pm x$, $x'\in H^2(\mathrm{CP}^{k-1}\times S^{q+1})$ a generator. Now $x'^k=0$, so $\psi^*(x'^k)=\pm x^k=0$, so $c_n(\eta)=0$. Hence if f_α is homotopic to the identity, then $\alpha\in\pi_q(U(k))$ is of finite order.

(6.19) Corollary. $A_0(\pi_q(U(k))\cap\mathscr{D}_0(\mathrm{CP}^{k-1}\times S^q))=0$ *for* $q>1$.

(6.20) Corollary. *The natural map of* $\pi_q(U(k))$ *into homotopy equivalences of* $\mathrm{CP}^{k-1}\times S^q$ *has kernel a torsion group for* $q>1$.

(6.21) Theorem. *Let* $f_1, f_2\in\mathscr{D}_0(\mathrm{CP}^{k-1}\times S^q)$, $q>1$, *and suppose that* f_1 *and* f_2 *define equivalent* S^1 *actions, constructed using* (5.7). *Then* $A_0(f_1)=A_0(f_2)$.

Proof. By (5.8) we have that there are b and g such that diagram (U) above is commutative, where $P(b)=(1\times d)(f_\alpha)$, $\alpha: S^q\to U(k)$, $g:\mathrm{CP}^{k-1}\times D^{q+1}\to\mathrm{CP}^{k-1}\times D^{q+1}$, a diffeomorphism, etc., that is $f_2(1\times d)f_\alpha f_1^{-1}=\partial g$. By (6.16), all these maps are homotopic to the identity, so that we may apply the homomorphism $A_0:\mathscr{D}_0(\mathrm{CP}^{k-1}\times S^q)\to H^{4*}(\Sigma(\mathrm{CP}^{k-1}\times S^q)_+)$. Hence $A_0(f_2)+A_0(1\times d)+A_0(f_\alpha)-A_0(f_1)=A_0(\partial g)$. By (6.15), $A_0(\partial g)=0$, by (6.17), $A_0(1\times d)=0$, and by (6.19), $A_0(f_\alpha)=0$, so $A_0(f_2)=A_0(f_1)$.

(6.22) Corollary. *Let q be odd,* $q>1$. *If* (a) $q+2k\equiv 1 \bmod 4$ *and* $k>2$, *or* (b) *if* $q+2k=7$, 15 *or* 31 *and* $k>1$, *then there are an infinite number of distinct semi-free* S^1 *actions on homotopy* $(q+2k)$*-spheres with* S^q *as untwisted fixed point set.*

The case $q=3$, $k=2$ is a theorem of MONTGOMERY and YANG [35].

Proof. By (6.9), $A_0(\mathscr{D}_0(\mathrm{CP}^{k-1}\times S^q))\neq 0$ under the given conditions on q, k, and $q+2k$, and therefore there are an infinite number of elements of $\mathscr{D}_0(\mathrm{CP}^{k-1}\times S^q)$ with different images under A_0. Then the result follows from (6.21).

References

1. ABRAHAM, R., and J. ROBBIN: Transversal mappings and flows, W. A. Benjamin, New York, 1967.
2. ATIYAH, M.: Thom complexes, Proc. L. M. S. **11**, 291–310 (1961).
3. —, and F. HIRZEBRUCH: Vector bundles and homogeneous spaces, Proceedings of Symposia in Pure Mathematics, **3** Differential Geometry, A. M. S., 7–38 (1961).
4. BOTT, R.: Notes on K-theory, Harvard, 1962.
5. BREDON, G.: Examples of differentiable group actions, Topology **3**, 115–122 (1965).
6. — Exotic actions on spheres, these proceedings.
7. BROWDER, W.: Surgery on simply connected domains (in preparation).
8. — Homotopy type of differentiable manifolds, Proceedings of the Aarhus Symposium on Algebraic Topology, Aarhus, 42–46 (1962).
9. — Embedding 1-connected manifolds, Bull A. M. S. **72**, 225–231 (1966).
10. — Embedding smooth manifolds, Proceedings of the International Congress of Mathematicians, Moscow 1966 (to appear).
11. — Manifolds with $\pi_1 = \mathbb{Z}$, Bull. A. M. S. **72**, 238–244 (1966).
12. — On the action of $\theta^n(\partial\pi)$, from Differential and Combinatorial Topology, A symposium in honor of Marston Morse, Princeton, N. J., 1965.
13. — Diffeomorphisms of 1-connected manifolds, Trans. A. M. S. **128**, 155–163 (1967).
14. — The Kervaire invariant of framed manifolds and its generalization, Annals of Math. (to appear).
15. —, and M. HIRSCH: Surgery on PL-manifolds and applications, Bull. A. M. S. **72**, 959–964 (1966).
16. —, and G. R. LIVESAY: Fixed point free involutions on homotopy spheres, Bull. A. M. S. **73**, 242–245 (1967).
17. BROWN, E. H., and F. P. PETERSON: The Kervaire invariant of $(8k+2)$-manifolds, Bull. A. M. S. **71**, 190–193 (1965).
18. CERF, J.: Topologie de certains espaces de plongements, Bull. Soc. Math. France, **89**, 227–380 (1961).
19. CONNER, P. E., and E. E. FLOYD: Differentiable Periodic Maps, Berlin-Göttingen-Heidelberg-New York: Springer 1964.
20. EELLS, J., and N. KUIPER: Manifolds which are like projective planes, Publ. Math. I. H. E. S., No. 14.
21. HAEFLIGER, A.: Knotted $(4k-1)$-spheres in $6k$-space, Annals of Math. **75**, 452–466 (1962).
22. — Knotted spheres and related geometric problems, Proceedings of the International Congress of Mathematicians, Moscow, 1966 (to appear).
23. HIRSCH, M.: On the fibre homotopy type of normal bundles, Michigan Math. J. **12**, 225–230 (1965).
24. HIRZEBRUCH, F.: New Topological Methods in Algebraic Geometry, 3rd Edition. Berlin-Heidelberg-New York: Springer 1966.
25. HSIANG, W.-C.: A note on free differentiable actions of S^1 and S^3 on homotopy spheres, Annals of Math. **83**, 266–272 (1966).

26. —, and W.-Y. Hsiang: Some free differentiable actions on 11-spheres, Quat. J. Math. (Oxford) **15**, 371–374 (1964).
27. Husemoller, D.: Fibre Bundles, McGraw Hill, New York 1966.
28. Kervaire, M.: Proof of the theorem of Barden-Mazur-Stallings, Comment. Math. Helv. **40**, 31–42 (1966).
29. —, and J. Milnor: Groups of homotopy spheres I, Annals of Math. **77**, 504–537 (1963).
30. Lashof, R., and M. Rothenberg: Microbundles and smoothing, Topology **3**, 357–388 (1965).
31. Lopez de Medrano, S.: Some results on involutions of homotopy spheres, these proceedings.
32. Milnor, J.: Microbundles and differentiable structures, mimeographed notes, Princeton Univ. 1961.
33. — Characteristic classes, mimeographed notes, Princeton Univ. 1957.
34. — On simply connected 4-manifolds, Symposium Topologia Algebraica, Mexico, 122–128 (1958).
35. Montgomery, D., and C. T. Yang: Differentiable actions on homotopy seven spheres (I), Trans. A. M. S. **122**, 480–498 (1966), (II) these proceedings.
36. — — Differentiable transformation groups on homotopy spheres, Michigan Math. J. **14**, 33–46 (1967).
37. — — Free differentiable actions on homotopy spheres, these proceedings.
38. —, and L. Zippin: Topological Transformation Groups, Interscience, New York 1955.
39. Morlet, C.: Les voisinages tubulaires des variétés semi-linéaires, C. R. Acad. Sci. Paris, **262**, 740–743 (1966).
40. Novikov, S. P.: Diffeomorphisms of simply connected manifolds, Soviet Math. Dokl. **3**, 540–543 (1962).
41. — Homotopy equivalent smooth manifolds, I, A. M. S. Translations **48**, 271–396 (1965), = Izv. Akad. Nauk. S. S. S. R. Ser. Mat. **28**, 365–474 (1964).
42. Palais, R.: Extending diffeomorphisms, Proc. A. M. S. **11**, 274–277 (1960).
43. Rourke, C., and B. Sanderson: Block Bundles I, Annals of Math. **87**, 1–28 (1968), II, III to appear.
44. Smale, S.: On the structure of manifolds, Amer. J. Math. **84**, 387–399 (1962).
45. Steenrod, N.: The Topology of Fibre Bundles, Princeton Mathematical Series 14, Princeton Univ. Press, 1951.
46. —, and D. B. A. Epstein: Cohomology operations, Annals of Math. Studies No. 50, Princeton Univ. Press, 1962.
47. Sullivan, D.: Triangulating homotopy equivalences, Thesis Princeton University 1965.
48. — Triangulating and smoothing homotopy equivalences and homeomorphisms, mimeographed notes, Princeton Univ., 1967.
49. Thom, R.: Quelques propriétés globales des variétés differentiables, Comment. Math. Helv. **28**, 17–86 (1954).
50. Toda, H.: Composition Methods in Homotopy Groups of Spheres, Annals of Math. Studies No. 49, Princeton Univ. Press, 1962.

51. WALL, C. T. C.: Surgery of non-simply connected manifolds, Annals of Math. **84**, 217–276 (1966).
52. — Surgery on compact manifolds (to appear).
53. — An extension of results of Novikov and Browder, Amer. J. Math. **88**, 20–32 (1966).

PART I

Differentiable Transformation Groups

Exotic Actions on Spheres

GLEN E. BREDON*

In this review article I attempt to give a compendium of examples of exotic differentiable actions of compact Lie groups on homotopy spheres. It was written mainly for the benefit of the participants in the Tulane Conference on Compact Transformation Groups, May 8–June 2, 1967.

The basic arrangement of the examples is done through the linear representations that they resemble. Thus Section 1 treats examples resembling the standard representation of $O(n)$ plus a trivial representation. The examples in Section 2 resemble twice the standard representation of $O(n)$ and some of the many varied examples arising from these are given in Section 3. Section 4 treats examples resembling twice the standard representation plus a trivial 2-dimensional representation. In Section 5 we discuss analogues of the product of several standard representations plus a trivial representation. In Section 6, we consider semifree actions of $S^1=U(1)$ and $S^3=Sp(1)$. These are actions which are free outside the fixed point set and hence resemble some multiple of the standard representation plus a trivial one. The case of free actions of S^1 and S^3 is discussed in Section 7 and free involutions and cyclic actions are discussed in Sections 8 and 9.

Some actions cannot be said to resemble any linear representation, and we discuss some of these in Section 10.

We confine ourselves almost exclusively to the differentiable case, although some comments are made about other cases, and differentiability assumptions will be taken for granted throughout the article unless there is specific mention to the contrary.

Although we are mainly concerned here with various properties of examples, we also discuss a few theorems of a positive nature since these are sometimes of basic importance for adequate discussion of the examples.

There are many examples discussed here, due to myself as well as to others, which are quite recent and have not yet been published. There are also a few examples given which have more of a "folklore" nature.

---

* The author is supported by an ALFRED P. SLOAN fellowship and by National Science Foundation grant GP 3990.

I had several very helpful conversations, during the writing of this article, with KLAUS JÄNICH, DEANE MONTGOMERY, FRANK RAYMOND and C. T. YANG, and I am very grateful to them for clarifying some foggy areas of my knowledge. Lectures at the Institute for Advanced Study, during the preceding year, by W. BROWDER, W.-Y. HSIANG, JÄNICH, LOPEZ DE MEDRANO, YANG and others were also very helpful.

1. Actions Resembling $\rho_n + k\theta$

The linear model for these actions is given by the canonical inclusion $SO(n) \to SO(n+k)$ and the resulting action of $SO(n)$ on S^{n+k-1}. We put $m = n+k-1$. More generally let us restrict our attention to the induced action of a connected subgroup $G \subset SO(n)$ requiring, however, that G be transitive on the $(n-1)$-sphere.

In this example each non-fixed orbit is an $(n-1)$-sphere, the fixed point set $F(G, S^m)$ is a $(k-1)$-sphere, and the orbit space S^m/G is a k-disk with boundary corresponding to the fixed point set. Moreover the action is equivalent to the product action on the boundary (with corner straightened) of

$$D^n \times D^k$$

where $G \subset SO(n)$ acts as usual on D^n and trivially on D^k. Topologically, one might think of this action as the product action on

$$S^n \times D^k$$

with $S^n \times \{x\}$ identified to a point for each $x \in \partial D^k$.

The crucial point about this model is that

(1.1) dim. principal orbit + dim. fixed point set + 1 = dim. total space.

Actions satisfying this property (on spheres) have been extensively studied and, in some sense, are completely known even in the topological case. In fact, in the topological case, it is shown in [3], under the sole hypothesis (1.1), that the orbit space is a "generalized k-cell" X, that G is a subgroup of $SO(n)$ transitive on S^{n-1} for some n, and that the action is equivalent to the obvious action on

$$S^n \times X / \{S^n \times \{x\} : x \in \partial X\}.$$

(The converse construction from any given generalized k-cell gives such an action on some generalized sphere.)

In the differentiable case with total space a homotopy sphere, (1.1) implies that the orbit space is a contractible k-manifold X^k and that the action is equivalent to the obvious action on

$$\partial [D^n \times X^k]$$

with the corner straightened. (Note that the orbit space of this latter action is $X^k \cup$ a collar.) See [19] for this. If $n \geq 1$ and $n+k \geq 6$ then $D^n \times X^k \approx D^{n+k}$ by SMALE's work, so that the total space of such actions is necessarily the standard sphere. From these facts it is clear that such actions are classified by their orbit spaces X^k [19]. In particular, if $X^k \approx D^k$, then the action is equivalent to our linear model.

Now there are, for example, infinitely many contractible 5-manifolds X^5 distinguished by $\pi_1(\partial X^5)$ (see [30]). For each of these we have the differentiable action of $SO(n)$ on $\partial(D^n \times X^5) \approx S^{n+4}$ whose fixed point set is diffeomorphic to ∂X^5. Moreover, this action is embedded in the action of $SO(n)$ on $\partial(D^n \times X^5 \times D^1) \approx S^{n+5}$ whose orbit space is $X^5 \times D^1 \approx D^6$ (by SMALE) and, since this larger action still satisfies (1.1) it must be equivalent to the linear action $\rho_n + 6\theta \colon SO(n) \to SO(n+6)$.

If $H \subset SO(n)$ has no stationary points on S^{n-1}, then H also has ∂X^5 as its fixed point set on $\partial(D^n \times X^5) \approx S^{n+4}$. Since every non-trivial compact Lie group H has such a representation as a subgroup of some $SO(n)$ we have:

(1.2) Theorem [30, 32]. *For any non-trivial compact Lie group H there is an integer m such that H admits an infinite number of differentiable actions on the standard m-sphere which are distinguished from one another by the fundamental group of their fixed point sets. Moreover, these actions may be chosen to be the induced actions of $H \subset SO(n)$, $n = m-4$, on certain invariant m-spheres in the linear action $\rho_n + 6\theta \colon SO(n) \to SO(n+6) = SO(m+2)$ on S^{m+1}.*

We note that the boundary of a contractible manifold is always a homology sphere (over the integers). Conversely, the HSIANG's have shown that given a differentiable homology sphere we may change its differentiable structure in such a way that the new manifold bounds a contractible manifold. Thus, up to a change of differentiable structure, every homology sphere occurs as the fixed point set of an action of $SO(n)$ on some standard sphere. For all this see [20].

2. Actions Resembling $2\rho_n$

The linear model for these actions is given by the representation $\rho_n + \rho_n \colon SO(n) \to SO(2n)$ taking

$$A \mapsto \begin{bmatrix} A & 0 \\ 0 & A \end{bmatrix}.$$

This linear action is, in other words, the diagonal action of $SO(n)$ on $S^{2n-1} \subset E^n \times E^n$.

If $\langle x,y\rangle \in S^{2n-1}$ then the isotropy subgroup of $SO(n)$ at $\langle x,y\rangle$ is clearly just the subgroup leaving the linear span of x and y (in E^n) pointwise fixed. If x and y are dependent then this is a conjugate of $SO(n-1)$ in $SO(n)$ and the orbit of $\langle x,y\rangle$ is then an $(n-1)$-sphere. If x and y are independent then this is a conjugate of $SO(n-2)$ and the orbit of $\langle x,y\rangle$ is a Stiefel manifold $V_{n,2}$. The latter are the principal orbits and have codimension two in S^{2n-1}. From general principles the orbit space is a 2-disk with the principal orbits corresponding to points of the interior and the singular (S^{n-1}) orbits corresponding to points on the boundary.

Exotic actions resembling this linear model were first constructed in [2] and have attracted a considerable amount of attention in recent years. Such actions, under quite general hypotheses, were classified in [6] in both the topological and differentiable cases. We will remark on this later.

In order to construct the exotic analogues of this action we shall first describe the action in another way. Consider the invariant subsets

$$X_1 = \{\langle x,y\rangle \in S^{2n-1} \ \|x\| \leq \|y\|\},$$
$$X_2 = \{\langle x,y\rangle \in S^{2n-1} \ \|x\| \geq \|y\|\}.$$

Then X_1 is diffeomorphic to $D^n \times S^{n-1}$, and $X_2 \approx S^{n-1} \times D^n$ with the action in both cases being the diagonal action $g\langle x,y\rangle = \langle gx, gy\rangle$. The original action is obtained by attaching $D^n \times S^{n-1}$ to $S^{n-1} \times D^n$ by means of the identity $S^{n-1} \times S^{n-1} \to S^{n-1} \times S^{n-1}$. The idea involved in constructing exotic actions is just to attach these two spaces by some other equivariant diffeomorphism $S^{n-1} \times S^{n-1} \to S^{n-1} \times S^{n-1}$. (This is not how the examples were originally discovered. In fact the classification of [6] was proved first with the examples resulting out of necessity. It is, however, a reasonable facsimile of the procedures used in the classification.)

To describe the necessary equivariant diffeomorphisms we first consider the map $\theta_x \in O(n)$, for $x \in S^{n-1}$, which is the reflection through the line Rx, that is $\theta_x(y) = 2(x \cdot y)x - y$. Then $x \mapsto \theta_x$ defines a differentiable map

$$\theta : S^{n-1} \to O(n).$$

($-\theta$ is just the characteristic map of the tangent bundle of S^{n-1}.) Now we consider the maps

$$\psi_k : S^{n-1} \times S^{n-1} \to S^{n-1} \times S^{n-1}$$

defined by

$$\psi_k\langle x,y\rangle = \langle (\theta_x\theta_y)^k x, (\theta_x\theta_y)^k y\rangle.$$

It is easily checked that ψ_k is equivariant with respect to the diagonal action of $O(n)$. Also it is easily observed that

$$\psi_0 = 1, \quad \psi_{k+\ell} = \psi_k \circ \psi_\ell.$$

Since ψ_k and $\psi_{-k}=\psi_k^{-1}$ are differentiable, each ψ_k is a diffeomorphism. We denote the $O(n)$-manifold obtained by attaching $D^n\times S^{n-1}$ to $S^{n-1}\times D^n$ by means of ψ_k by

(2.1) $$M_k^{2n-1}=(D^n\times S^{n-1})\cup_{\psi_k}(S^{n-1}\times D^n).$$

Let us now observe an elementary fact. Suppose that $H\subset O(n)$ has as stationary point set $F(H,E^n)$ in E^n a subspace of dimension r. Then $F(H,D^n)=D^r$ and clearly

(2.2) $$F(H,M_k^{2n-1})=(D^r\times S^{r-1})\cup_{\psi_k}(S^{r-1}\times D^r)=M_k^{2r-1}.$$

We wish to identify the manifolds M_k^{2n-1}. First, it is easy to see that they are simply connected for $n\geqslant 3$. The Mayer-Vietoris theorem suffices to compute their homology. We shall omit the details, which are given in [2]. The result of the calculation is that M_k^{2n-1} is a homology sphere for n odd and that

(2.3) $$H_i(M_k^{2n-1};Z)\approx\begin{cases}Z, & i=0,\ 2n-1,\\ Z_{2k+1}, & i=n-1,\\ 0 & \text{otherwise.}\end{cases}\qquad (n\text{ even})$$

Thus the M_k^{4m+1} are homotopy spheres and the M_k^{4m-1} have Z_{2k+1} torsion in the middle dimension. Moreover it can be seen that

(2.4) $$M_k^3 \text{ is the lens space } L(2k+1,1).$$

These facts together with (2.2) yield, almost immediately, a number of interesting exotic actions. We shall, however, delay discussion of these until later.

We shall now discuss the classification theorem of [6] since this is important in order to identify these examples with examples obtained via other constructions. Before doing so we make the elementary observation that if $G\subset O(n)$ acts transitively on the unit tangent bundle $V_{n,2}$ of S^{n-1}, then the induced action of G on M_k^{2n-1} has the *same orbits* as does $O(n)$ itself. It can be seen that the connected subgroups of $SO(n)$ with this property are precisely the following:

(2.5) $$\begin{cases}SO(n) & n\geqslant 3,\\ \mathrm{Spin}(7)\subset SO(8),\\ G_2\subset SO(7).\end{cases}$$

We shall denote the linear action of G described at the beginning of this section by Φ and shall denote the action on M_k^{2n-1} by Ψ_k. Thus $\Phi=\Psi_0$.

We now state the classification theorem of [6] in the differentiable case. (However, the topological classification turns out to be precisely the same as the differentiable one!)

(2.6) Theorem [6]. *Let G be a connected compact Lie group acting effectively and differentiably on an integral homology m-sphere Σ^m in such a way that there are precisely two types of orbits, one type consisting of $(m-2)$-dimensional orbits and the other of orbits of smaller positive dimension. Then Σ^m is a homotopy m-sphere and $m=2n-1$ for some $n \geqslant 3$. Moreover, G is a subgroup of $SO(n)$ which is transitive on $V_{n,2}$ and the action is equivalent to exactly one of the following:*

$$\begin{matrix}\textit{(non-}\\ \textit{oriented)}\end{matrix}\begin{cases} n \text{ even: } \Phi \\ n \text{ odd: } \Psi_k \text{ for } k \geqslant 0, \text{ with } k=0 \text{ unless } G=SO(n) \text{ and } n \geqslant 5. \end{cases}$$

[Note the important fact that these actions are distinguished by the fixed point sets of $G \cap SO(n-2)$ which, by (2.2) and (2.4), is $L(2k+1,1)$ for the action Ψ_k. Concerning negative values of k, one can easily write down an equivariant diffeomorphism

$$M_k^{2n-1} \to M_{-k-1}^{2n-1}$$

which preserves orientation when n is odd and reverses it when n is even [2].

As indicated, the above classification is a non-oriented classification. If we orient our manifolds according to the natural orientation of $D^n \times S^{n-1}$, then it is easy to derive the following oriented classification:

$$\begin{matrix}\text{(oriented}\\ \text{total space)}\end{matrix}\begin{cases} n \text{ even: } \Phi, -\Phi, \\ n \text{ odd: } \psi_k \text{ for } k \geqslant 0, \text{ with } k=0 \text{ unless } G=SO(n) \text{ and } n \geqslant 5. \end{cases}$$

(In fact the reflection $\langle x,y\rangle \to \langle x,-y\rangle$ in $E^n \times E^n$ induces an equivariant autodiffeomorphism of M_k^{2n-1} which reverses orientation when n is odd. This is not possible when n is even since it would imply the existence of an orientation reversing autodiffeomorphism of CP_n which would contradict the structure of its cohomology ring.)

Actually it is also appropriate in these matters to orient the base space (or a principal orbit) as well. If this is done the classification becomes: (The first sign indicates orientation of the total space and the second, that of the base.)

$$\begin{matrix}\text{(oriented}\\ \text{base and}\\ \text{total space)}\end{matrix}\begin{cases} n \text{ even: } \Phi, (-,+)\Phi, \\ n \text{ odd: } \Psi_k, (-,+)\Psi_k \text{ for } k>0, \text{ with } k=0 \text{ unless} \\ \qquad G=SO(n),\ n \geqslant 5. \end{cases}$$

In fact $\langle x,y\rangle \to \langle x,-y\rangle$ gives an autodiffeomorphism of M_k^{2n-1} which preserves total and reverses base orientations when n is even and which reverses both orientations when n is odd. (It is not possible to reverse just one orientation when n is odd since $V_{n,2}$ does not admit an equivariant orientation reversing homeomorphism.)

The basic difficulty in proving (2.6) is to show that G is a subgroup of $SO(n)$ transitive on $V_{n,2}$ and that the isotropy groups are $G \cap SO(n-1)$ and $G \cap SO(n-2)$. If one restricts one's attention to the case $G = SO(n)$, the proof can be simplified considerably. Such a proof is given (with somewhat different hypotheses) by the HSIANG brothers in [20]. They also give analogous classification theorems for $SU(n)$ with base space a homotopy D^3 and for $Sp(n)$ with base space a homotopy D^5. If the base spaces are actually the disks D^3 or D^5, then these cases provide only the linear actions with no exotic examples arising! More general classification problems have been considered by JÄNICH [23, 24] and this work and that of the HSIANG's should be studied by anyone interested in these matters.

Jänich's classification theorem allows one to drop the assumption in (2.6) that the total space is a homology sphere provided one as assumes the orbit types are what they should be. This is also done in [6] provided one assumes that the fixed point set of H is a circle (rather than two circles) where H is the larger type of isotropy group. For convenience we shall state the latter result (which is proved, but is not stated, in [6]).

(2.7) Theorem [6]. *Let $G \subset O(n)$ be one of the following groups*

$$G = \begin{cases} O(n) & n \geqslant 2, \\ SO(n) & n \geqslant 3, \\ \mathrm{Spin}(7) & n = 8, \\ G_2 & n = 7. \end{cases}$$

Let G act differentiably on a $(2n-1)$-manifold N^{2n-1} and assume that the isotropy groups of G are precisely the conjugates of $H = G \cap O(n-1)$ and $K = G \cap O(n-2)$. Assume that $F(H,N)$ is a circle. Also assume that N/G is a 2-disk. (This is always true if $H^1(N,Z) = 0$.) Then the action is equivalent to exactly one of the following

$$\begin{cases} n \text{ even: } \Psi_k \quad k \geqslant 0, \\ n \text{ odd: } \Psi_k \quad k \geqslant 0, \text{ with } k = 0 \text{ when } G \text{ is } SO(3) \text{ or } G_2. \end{cases}$$

In fact these actions are distinguished by $F(K,N)$ which is $L(2k+1,1)$ for Ψ_k and $G \neq SO(3)$ or G_2, and hence they are also distinguished by $H_1(F(K,N)) \approx Z_{2k+1}$ for Ψ_k and $G \neq SO(3)$ or G_2.

As JÄNICH proved, if one drops the assumption that $F(H,N)$ is a circle one gets other actions which are still distinguished by $F(K,N)$ which is some $L(m,1)$ (with fundamental group Z_m). Of course no spheres arise as total spaces in such actions.

The HSIANG brothers [22] and HIRZEBRUCH observed independently that the action Ψ_1 (on M_1^{2n-1}) can be described as follows: Let $O(n)$ act, as a subgroup of $O(n+1)$, on the unit tangent disk bundle to S^n via the differential. Let $D_+^n \subset S^n$ be an invariant disk around one of the fixed points of this action. The part of the disk bundle over D_+^n is trivial and is naturally represented as $D_+^n \times D^n$ and, moreover, the action of $O(n)$ on this portion is just the diagonal action on this product of disks. $\langle x,y\rangle \mapsto \langle y,x\rangle$ is then equivariant so that there is a natural action of $O(n)$ on the plumbing N_1^{2n} of two copies of this disk bundle. It then can be seen quite explicitly that the $O(n)$-manifold M_1^{2n-1} is diffeomorphic to the $O(n)$-manifold ∂N_1^{2n}. More generally, HIRZEBRUCH observed the following. Let N_k^{2n} be the result of (straight line) plumbing of $2k$ copies of the unit disk bundle to S^n. Fig. 1 illustrates N_2^{2n} (for $n=1$) and the dots indicate the fixed points of the $O(n)$-action on N_2^{2n}. It is

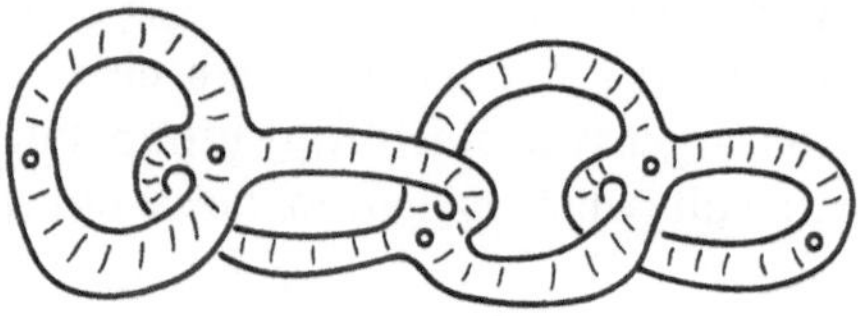

Fig. 1. The $O(n)$-manifold N_2^{2n} (for $n=1$)

clear that $F(O(n-i), N_k^{2n}) = N_k^{2i}$. It is also clear that ∂N_k^2 is a circle (see Fig. 1) and it can be seen (as is well-known) that $H_1(\partial N_k^4) \approx Z_{2k+1}$. It follows from the classification (2.7) that

$$M_k^{2n-1} \approx \partial N_k^{2n}$$

as $O(n)$-manifolds.

This description allows an easy determination of the differentiable structure on M_k^{2n-1} (n odd) since the Arf invariant of the parallelizable manifold N_k^{2n} is easily computed. In fact if $x_1, x_2, \dots, x_{2k}$ are homology classes represented by the bases of the original disk bundles *in order* and if n is odd and $\neq 1,3,7$, then in the notation of [25] we have $\psi_0(x_i) \equiv 1 \pmod 2$ since the representing sphere does not have trivial normal bundle. Letting $\lambda_1 = x_1$, $\lambda_2 = x_3 - x_1$, $\lambda_3 = x_5 - x_3 + x_1$, etc., and $\mu_i = x_{2i}$, then $\{\lambda_1, \dots, \lambda_k, \mu_1, \dots, \mu_k\}$ is a symplectic basis and one sees that $\psi_0(\lambda_i) \equiv i$ and $\psi_0(\mu_i) \equiv 1 \bmod 2$. Thus the Arf invariant is the mod 2 residue class

$$\Sigma_i \psi_0(\lambda_i)\psi_0(\mu_i) \equiv \Sigma_1^k i \equiv \begin{cases} 0 & \text{for} \quad k \equiv 3,4 \pmod 4, \\ 1 & \text{for} \quad k \equiv 1,2 \pmod 4. \end{cases}$$

Thus we see that (for n odd)

$$M_k^{2n-1} \text{ is the } \begin{Bmatrix}\text{standard sphere}\\ \text{Kervaire sphere}\end{Bmatrix} \text{ for } k \equiv \begin{Bmatrix}3, & 4\\ 1, & 2\end{Bmatrix} \bmod 4$$

(2.8)

$$\text{that is, for } 2k+1 \equiv \begin{Bmatrix}\pm 1\\ \pm 3\end{Bmatrix} \bmod 8.$$

(This was proved for $k \geqslant 1$ and is trivial for $k=0$. It follows also for negative k by use of the diffeomorphism $M_k^{2n-1} \approx M_{-k-1}^{2n-1}$ mentioned before.)

This determination of the differentiable structures was observed by HIRZEBRUCH [17]. Independently of HIRZEBRUCH, the HSIANG brothers proved the same result by somewhat different methods [22]. HIRZEBRUCH has also extended these results to apply to a more general class of manifolds containing, in particular, those due to BRIESKORN and MILNOR (see [7]).

We shall now give the Hirzebruch-Brieskorn-Milnor descriptions since they shed considerable light and also provide further exotic actions.

If $a_1, a_2, \ldots, a_{n+1}$ are integers we denote by

$$\Sigma^{2n-1}(a_1, a_2, \ldots, a_{n+1})$$

the $(2n-1)$-manifold which is the intersection of the complex hypersurface

$$z_1^{a_1} + \cdots + z_{n+1}^{a_{n+1}} = 0$$

in $\mathbb{C}^{n+1}$ with the unit sphere $\|z\|=1$. For the time being we shall restrict our attention to the case in which $a_1 = \cdots = a_n = 2$ and $a_{n+1} = d$ is *odd*. For this case we use the notation

$$\Sigma_d^{2n-1} = \Sigma^{2n-1}(\underbrace{2, 2, \ldots, 2}_{n}, d).$$

Consider the usual inclusion $O(n) \subset U(n) \subset U(n+1)$. Since $O(n)$ preserves $z_1^2 + \cdots + z_n^2$ as well as z_{n+1}^d we have an $O(n)$-action on Σ_d^{2n-1}.

Now it can be shown that $H_1(\Sigma_d^3) \approx Z_d$. It follows from the classification theorem (2.7) that there is an $O(n)$-equivariant diffeomorphism

$$\Sigma_{2k+1}^{2n-1} \approx M_k^{2n-1}.$$

Note that it follows that the actions on M_k^{2n-1} are all imbedded in this linear $O(n)$ action on $S^{2n+1} \subset \mathbb{C}^{n+1}$ which is the real representation $2\rho_n + 2\theta : O(n) \to O(2n+2)$.

A significant fact about this description is that one sees that there is an S^1 action on Σ_d^{2n-1} which commutes with the $O(n)$-action. In fact, if $|z|=1$ then

$$z \mapsto \begin{pmatrix} z^d & & & & 0 \\ & z^d & & & \\ & & \ddots & & \\ & & & z^d & \\ 0 & & & & z^2 \end{pmatrix} \in U(n+1) \tag{2.9}$$

is such a circle action. We have, therefore, the induced action of $O(n)\times S^1$ on Σ_d^{2n-1}. This action is not effective since $\begin{pmatrix} -I_n & 0 \\ 0 & 1 \end{pmatrix}$ is in both groups. The subgroup $K=\{\langle I,1\rangle,\langle -I,-1\rangle\}$ is the ineffective part of $O(n)\times S^1$. In the important case in which n is odd, K touches both components of $O(n)\times S^1$ and it follows that the effective quotient group $(O(n)\times S^1)/K$ is connected and, in fact, that the natural map

$$SO(n)\times S^1 \xrightarrow{\approx} (O(n)\times S^1)/K \quad (n \text{ odd}) \tag{2.10}$$

is an isomorphism. This fact will have some significance in the discussion to follow. CALABI was the first to notice the existence of this extra circle action, prior to the appearance of the Brieskorn-Hirzebruch descriptions. I always felt that this circle action should exist but was deterred from looking for it since its existence contradicts work of WANG. Subsequent to the discovery of it, the HSIANG'S discovered the error in WANG'S work; see [22].

3. Features of the Preceding Examples and Further Constructions

In this section we shall discuss some interesting aspects of the examples in Section 2 and shall indicate further examples which arise from these through some elementary constructions.

An immediate consequence of the classification theorem is the following:

(3.1) Theorem. *For any k the action of $SO(3)$ on $M_k^5 \approx S^5$ has the same orbits as does $O(3)$ (that is, $SO(3)$ is transitive on each orbit of $O(3)$) but unlike $O(3)$ each of these actions of $SO(3)$ is equivalent to the representation $2\rho_3$: $SO(3)\to SO(6)$. Similarly, the action of G_2 on $M_k^{13}\approx S^{13}$ has the same orbits as does $SO(7)$ (and $O(7)$) but the actions of G_2 are all equivalent to the representation $2\rho_7$: $G_2\subset SO(7)\to SO(14)$.*

Of course the actions of $O(3)$ on M_k^5, $k\geqslant 0$ and $SO(7)$ or $O(7)$ on M_k^{13} are all non-equivalent since the fixed point set of $O(1)$, respectively $SO(5)$ or $O(5)$, is the lens space $L(2k+1,1)$ with fundamental group Z_{2k+1}.

These are the only examples of such behavior that we know of.

Let $T \in O(2r+1)$ be the matrix

$$T = \begin{pmatrix} 1 & & & & & & 0 \\ & 1 & & & & & \\ & & -1 & & & & \\ & & & \cdot & & & \\ & & & & \cdot & & \\ & & & & & \cdot & \\ 0 & & & & & & -1 \end{pmatrix}$$

(for $r \geqslant 1$). Then T is an involution on the homotopy $(4r+1)$-sphere M_k^{4r+1} with fixed point set

$$F(T, M_k^{4r+1}) \approx L(2k+1,1).$$

(3.2) Theorem. *For every* $n \geqslant 6$ *and every* $k > 0$ *there is a differentiable involution* T *on* D^n *with* $F(T,D^n) \approx [L(2k+1,1) - D^3] \times D^1$. *Thus also* $F(T,S^{n-1}) \approx L(2k+1,1) \# -L(2k+1,1)$.

Proof. Let T act on M_k^5 as above and on D^m by the linear involution with $F(T,D^m) = D^1$. Then we have the product action on

$$(M_k^5 - D^5) \times D^m \approx D^{m+5}$$

with the corner straightened equivariantly. (Here D^5 is an invariant cell about a fixed point of T in M_k^5.) Clearly the fixed point set is as claimed.

Now, for $r \geqslant 1$ let $O(2r-1)$ act as a subgroup of $O(2r+1)$ on the homotopy sphere M_k^{4r+1} and also on D^m, $m \geqslant 1$, by the trivial action. Then as above we have an action of $O(2r-1)$ on

$$(M_k^{4r+1} - D^{4r+1}) \times D^m \approx D^{4r+m+1}.$$

Let F be the fixed point set of this action on D^{4r+m+1} and $\partial F = F \cap \partial D^{4r+m+1}$. One can readily compute the homology of F and of ∂F and note that in particular $H_*(F)$ and $H_*(\partial F)$ have $(2k+1)$-torsion and no other torsion. Now suppose we take k so that $2k+1$ runs through the odd primes p. Then for any prime $p \neq 2$ and any $n \geqslant 4r+1$ we have an action of $O(2r-1)$ on D^{n+1} such that F and ∂F have p-torsion and no other torsion. Now if $G \subset O(2r-1)$ has no fixed points on S^{2r-2} we see that the induced action of G on D^{n+1} has, by the construction, the same fixed point set as does $O(2r-1)$. Thus, in particular, we have

(3.3) Theorem. *Let* G *have an irreducible representation in* $O(2r-1)$ *for some* $r \geqslant 1$. *Then for every integer* $n \geqslant 4r+1$ *and every prime* $p \neq 2$

there is a differentiable action of G on D^{n+1} such that $F(G,D^{n+1})$ and $F(G,S^n)$ have p-torsion and no other torsion in their integral homology.

In connection with this we note that, as the HSIANG's have observed, every compact connected *non-abelian* Lie group has an irreducible odd dimensional representation so the the theorem applies to such groups. On the other hand it clearly does not apply to toral groups and, in fact, Smith theory shows that such examples are impossible for toral groups. The theorem also applies to any group G containing a subgroup of index two.

Now let us consider the actions on S^n constructed in the preceding proof. We note that the tangential representation of $O(2r-1)$ at a fixed point on M_k^{4r+1} is just $2\rho_{2r-1}+3\theta$ and hence is independent of k. It follows that the representations at fixed points on S^n are also all the same, in fact they are $2\rho_{2r-1}+(n-4r+2)\theta$. Thus we may form an equivariant connected sum of any of these actions with any other (with reversed orientation). Since there is more than one fixed point in each of these actions, we may iterate this procedure and get an action of $O(2r-1)$ on the (straight half line) infinite connected sum $(S^n)\#(-S^n)\#(S^n)\#(-S^n)\#\cdots$ (see Fig. 2) which is just the euclidean space E^n.

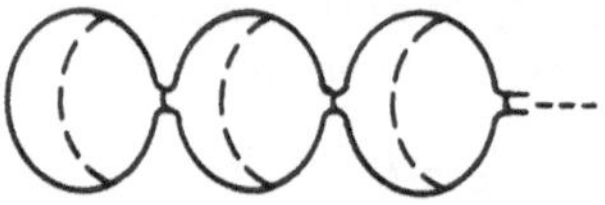

Fig. 2

Now on each of these factors we may take any of the actions we wish. Call the fixed point set on the i-th sphere F_i and that on the resulting connected sum F. Now we have $\tilde{H}_*(F)\approx\oplus_i\tilde{H}_*(F_i)$, since $F\approx(F_1)\#(-F_2)\#(F_3)\#\cdots$. It follows that we can arrange the actions so that $\tilde{H}_*(F)$ has p torsion for all p in some collection of odd primes and for no other p. Since there is a continuum number of such subsets of the primes, we have proved the following:

(3.4) Theorem. *Let G have an irreducible representation in $O(2r-1)$ for some $r\geqslant 1$. Then for any integer $n\geqslant 4r+1$ there is a continuum number of differentiable actions of G on E^n having homologically distinct fixed points sets.*

For example, there is a continuum number of involutions on E^5 with homologically distinct fixed point sets (in fact with distinct $H_1(F)$). We remark that there is at most a continuum number of distinct topological actions of G on a (second countable) manifold and that by a result of PALAIS, there is at most a countable number of differentiable actions of G on a compact manifold. Thus (3.4) is best possible.

To this point we have not made use either of the knowledge of the differentiable structures on the M_k^{4r+1} or of the descriptions of BRIESKORN-HIRZEBRUCH. We now do so. First note that $M_1^{13} \approx S^{13}$ and $M_1^9 \not\approx S^9$ since the Kervaire sphere is exotic in dimension 9 but not 13. Examining the action of $SO(2) \subset SO(7)$ on $M_1^{13} \approx S^{13}$ we see that it acts freely outside its fixed point set and that its fixed point set is M_1^9. Thus we have

(3.5) Theorem [2]. *There is a circle group action on S^{13} free outside the fixed point set F and with F the exotic Kervaire 9-sphere. For $x \in F$ the action $S^{13} - \{x\}$ is differentiably equivalent to a linear action on E^{13}. Thus the action on S^{13} is topologically equivalent to a linear action.*

The latter statements follow from a result of CONNELL, MONTGOMERY, and YANG [9]. There are similar actions on S^{11} as we shall remark later.

Now let us make use of the BRIESKORN-HIRZEBRUCH descriptions. Consider the involution on S^5 (or on any M_k^{4r+1}) with fixed point set $L(2k+1,1)$ (for example). Using (2.10) this involution is contained in a *connected* group action and, in particular, it extends to a circle action on S^5. This is a non-trivial fact. Similarly, the fix-point free involutions $-I \in O(2r+1)$ on M_k^{4r+1} extend to circle actions although the extensions given by this method are not free actions.

We shall take $k \geqslant 1$ for the remainder of this section.

Let V be the matrix

$$V = \begin{pmatrix} 1 & & & 0 \\ & 1 & & \\ & & 1 & \\ 0 & & & \varepsilon^2 \end{pmatrix}$$

where $\varepsilon^{2k+1} = 1$. This is in the group "S^1" operating on $\Sigma_{2k+1}^5 \approx S^5$ (see (2.9)). Clearly $F(V, S^5)$ is a principal orbit of $SO(3)$ and hence is $SO(3) \approx P^3$. V generates a subgroup of order $2k+1$. Thus there are periodic diffeomorphisms on S^5 of any odd period which have P^3 as their fixed point sets. This was observed by HIRZEBRUCH and answers a question of mine in [2]. YANG was the first to find such periodic maps using work of CALABI (prior to the Hirzebruch-Brieskorn examples).

Now, still on Σ_{2k+1}^5, we consider again the involution T given by the matrix

$$T = \begin{pmatrix} -1 & & & 0 \\ & 1 & & \\ & & 1 & \\ 0 & & & 1 \end{pmatrix}$$

which has $F(T, S^5) \approx L(2k+1,1)$.

Now T commutes with V so that TV generates a cyclic group of order $4k+2$. $F(TV,S^5)$ can be seen to be two circles by the equations and this can also be seen by general principles as follows: $F(TV,S^5) = F(T,F(V)) = F(T,P^3)$ which must be $pt+P^2$ or $P^1+P^1 = S^1+S^1$ (when non-empty) but also $F(TV,S^5) = F(V,F(T))$ which has odd dimension by dimensional parity results. Thus S^1+S^1 is the only possibility. ($+$ denotes disjoint union.) This was originally noted by YANG.

By equivariant connected sum operations we obtain actions of Z_{4k+2} $(k \geqslant 1)$ on S^5 with fixed point set being a disjoint union of any desired number of circles. There are two ways of forming infinite connected sums and these are illustrated in Fig. 3.

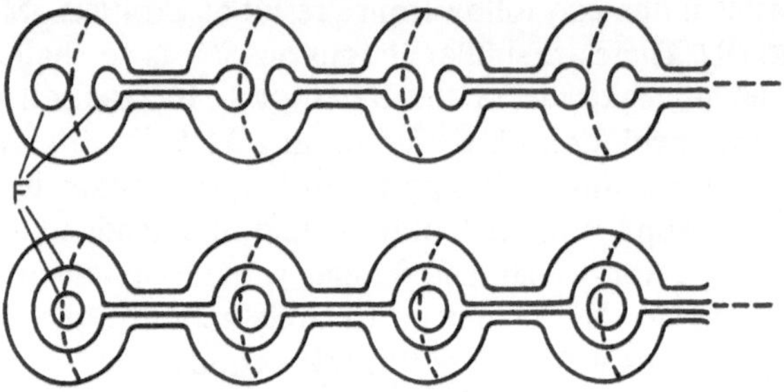

Fig. 3

The first of these methods constructs a Z_{4k+2}-action on E^5 with $F \approx S^1+S^1+\cdots$ and the second, with $F \approx E^1+S^1+S^1+\cdots$.

One will realize the great complexity of *topological* transformation groups if he ponders the actions of Z_{4k+2} of S^5 given by adding the point at infinity to these examples.

Again starting with this Z_{4k+2}-action on $\Sigma^5_{2k+1} \approx S^5$ with $F = S^1+S^1$ $(k \geqslant 1)$ let us delete a disk about a fixed point, thus obtaining an action on D^5 with $F = D^1+S^1$. Let Z_{4k+2} act trivially on D^m $(m \geqslant 1)$. Taking the product action gives an action

$$Z_{4k+2}: D^{m+5} \quad \text{with} \quad F = D^m \times (D^1+S^1) = D^{m+1} + D^m \times S^1.$$

Restriction to the boundary gives an action

$$Z_{4k+2}: S^{m+4} \quad \text{with} \quad F \approx S^m + (S^{m-1} \times S^1).$$

Similar actions of $SO(2r+1) \times Z_{2k+1}$; $r \geqslant 1$ and $O(2r+1) \times Z_{2k+1}$; $r \geqslant 0$ can clearly be constructed in the same manner.

One can certainly continue this game ad infinitum but we shall content ourselves with this. These cyclic actions are quite reminiscent of the examples of FLOYD [12] which were the first examples to show that Smith-type results do not hold for maps of composite period and also that for prime period p the use of Z_p coefficients is essential for the Smith theorems. FLOYD's method was to first construct the action on a contractible complex (of dimension three in fact) then to embed this equivariantly in a nice action and take an invariant regular neighborhood. Since the resulting examples are so similar to those above and since the method will be outlined in another case in Section 10, we shall not comment further on it.

Before going on to the next subject we shall briefly comment on some of the other Brieskorn-Hirzebruch examples. We let

$$\Sigma_{p,q}^{2n-1} = \Sigma^{2n-1}(\underbrace{2,2,\ldots,2}_{n-1},p,q)$$

where we always assume that p and q are odd and relatively prime. BRIESKORN proved that this is always a homotopy sphere ($n \geqslant 3$) and HIRZEBRUCH determined the differentiable structures [7, 17]. In particular he proved:

(3.6) $$\Sigma_{p,q}^{4m+1} \approx S^{4m+1} \quad (\text{for } n = 2m+1),$$

(3.7) $$\Sigma_{3,6k-1}^{4m-1} \approx k \text{ times the Milnor sphere} \quad (\text{for } n = 2m).$$

Thus these manifolds exhaust bP_{4m} and $\Sigma_{3,5}^{4m-1}$ is the Milnor sphere. (Also recall that the Σ_d^{4m+1} exhaust $bP_{4m+2} \approx 0$ or Z_2.)

(3.8) Theorem *There is a differentiable involution on S^{4m+1} which has any given element $\Sigma^{4r-1} \in bP_{4r}$ as a fixed point set (for $2 \leqslant r \leqslant m$).*

This results merely by taking the action of

$$\begin{pmatrix} -I_{2(m-r)+1} & 0 \\ 0 & I \end{pmatrix}$$

on $\Sigma_{3,6k-1}^{4m+1}$, $k = 1,2,\ldots$.

Similarly one has the following result, for example,

(3.9) Theorem. *There exists a differentiably S^1 action on S^{11}, free outside the fixed point set and such that the fixed point set is any given element of $4\Theta_7 \approx Z_7$.*

In fact one lets $SO(2)$ act on the first two variables of $\Sigma_{3,6k-1}^{11}$ which will give $\Sigma_{3,6k-1}^{7}$ as a fixed point set. Let k run through multiples of $992 = \text{order } \Theta_{11}$ and note that, modulo $28 = \text{order } \Theta_7$, k runs through multiples of 4. These observations clearly extend to other dimensions $\equiv 3 \pmod 4$.

The examples of (3.9) were first noted by MONTGOMERY and YANG [32] who constructed them by a more direct method. Their method is more general and yields some other information.

4. Actions Resembling $2(\rho_n+\theta)$

The linear model for these actions is obtained by adding a trivial 2-dimensional representation to the model of Section 2. More suggestively this is the restriction to $O(n)$ of the diagonal representation $2\rho_{n+1}$ of $O(n+1)$ on $E^{n+1}\times E^{n+1}\supset S^{2n+1}$.

Clearly one obtains exotic analogues of this action by restricting the $O(n+1)$ action on M_k^{2n+1} to $O(n)$ for *n even;* or in other words by restricting the $O(n+1)$ action on Σ_d^{2n+1}. Again the natural $O(n)$-action on $\Sigma_{p,q}^{2n+1}$ is an analogue of the linear model.

If one examines the linear model, or these analogues, one can see that the orbit types are $(O(n))$, $(O(n-1))$ and $(O(n-2))$ that is, fixed points, S^{n-1} and $V_{n,2}$. Moreover, the orbit space is a 4-disk D^4 with boundary ∂D^4 consisting of the singular orbits and such that the fixed point set F is a circle in ∂D^4.

One may ask whether or not F in knotted in ∂D^4. This is easily answered for the actions on $M_k^{2n+1}=\Sigma_{2k+1}^{2n+1}$ as follows: Consider the fixed point set $F(O(n-1),\, M_k^{2n+1})\approx M_k^3\approx L(2k+1,1)$. It is easy to see that the orbit map of $O(n)$ restricted to this set is a two sheeted covering of ∂D^4 outside F and, of course, is $1-1$ on F. Thus $L(2k+1,1)$ is the cyclic double covering of ∂D^4 branched along F which implies that F in knotted when $k\geqslant 1$. For $k=0$ we have the linear model so that $F(O(n-1),\, M_0^{2n+1})$ is a 3-sphere with $F\approx S^1$ unknotted in it and it follows that F is also unknotted in ∂D^4. A similar discussion of the actions on $\Sigma_{p,q}^{2n+1}$ is possible.

More precisely, HIRZEBRUCH has, in particular, shown that F is the torus knot $(2,d)$ (*resp.* (p,q)) in ∂D^4 for the action of $O(n)$ on Σ_d^{2n+1} (resp. $\Sigma_{p,q}^{2n+1}$).

One may ask if there are other examples of this type with F knotted in other ways in ∂D^4. JÄNICH answered this quuestion in [23] (also see [24]). He considered what he called knot-G-manifolds. To define this we let $G\subset O(n)$ be a group which is transitive on $V_{n,2}$. (That is $G=O(2)$ for $n=2$ or the identity component G_0 of G is $SO(n)$, $n\geqslant 3$; $G_2\subset SO(7)$; or $\mathrm{Spin}(7)\subset SO(8)$.) A knot-$G$-manifold is then a $(2n+1)$-manifold M^{2n+1} with an action of G such that the orbit types are (G), $(G\cap O(n-1))$ and $(G\cap O(n-2))$, such that the orbit space (which has dimension 4) is D^4, and such that the fixed point set $F=F(G,M)$ is a circle in ∂D^4. (Further restrictions, on the slice representations, are unnecessary.)

Now let K_G be the set of equivariant diffeomorphism classes of knot-G-manifolds and let K be the equivalence classes of knots in the three sphere. Then $M \mapsto (\partial(M/G), F)$ defines a map

$$\Delta_G: K_G \to K$$

which JÄNICH shows to be surjective. Assume now that $G = O(n)$, $n \geqslant 2$ or $G = SO(n)$, $n \geqslant 4$. Then JÄNICH proves that Δ_G is a one-one correspondence. Earlier the HSIANG's (who were the first to notice this type of example), restricting their attention to manifolds which are homotopy spheres, showed that Δ_G is *injective*. They also proved this for a somewhat larger class of transformation groups [20]. JÄNICH has also extended his result to more general situations [24].

Now for $k \in K$ and $G \subset O(n)$ given (as above) let $\gamma_n(k) = \Delta_G^{-1}(k)$ (it depends only on n and not on G). The question arises as to what the manifold $\gamma_n(k)$ is. HIRZEBRUCH has obtained detailed information on this question from which we mention only the following:

a) For n even $\gamma_n(k)$ is a homotopy sphere.

b) For n odd $\gamma_n(k)$ is $(n-1)$-connected and the order of $H_n(\gamma_n(k), Z)$ is the absolute value of the determinant of the knot, which is odd.

c) $\gamma_n(k)$ bounds a parallelizable manifold.

It should be remarked that the type of situation dealt with in these examples is not as special as it may appear at first glance. For example one can easily deduce the following result from (2.6):

(4.1) Theorem. *Let G be a compact connected Lie group acting differentiably on an m-manifold M^m with precisely three orbit types (G), (H), (K) such that $0 < \dim K < \dim H$. Assume that the fixed point set F has codimension three in M/G. Then we have the following:*

a) *$G \subset SO(n)$ for some $n \geqslant 4$ and is transitive on $V_{n,2}$.*

b) *$(H) = (G \cap SO(n-1))$, $(K) = (G \cap SO(n-2))$.*

c) *$m = \dim F + 2n$.*

d) *M/G is an $(m-2n+3)$-manifold with boundary consisting of the orbits of type (G) and (H) and F is an $(m-2n)$-submanifold of $\partial(M/G)$.*

Moreover, if M is $(n-2)$-connected then M/G is contractible. If M is an integral homology sphere then F is an integral homology sphere for n even and a mod 2 *homology sphere for n odd.*

The last two statements follow easily from the Vietoris mapping theorem and from Smith theory respectively and were noted by the HSIANG's under somewhat more general (and at the same time more special) hypotheses. These conclusions cannot be strengthened because of my examples of Section 2. The hypothesis that the codimension of F in M/G is three can certainly be dropped, with only a few more cases arising, but the details of this have not been given.

5. Actions Resembling $\bigtimes_{i=1}^{m} \rho_{k_i} + q\theta$

The linear model for these actions is the representation

$$\bigtimes_{i=1}^{m} \rho_{k_i} + q\theta : O(k_1) \times \cdots \times O(k_m) \to O(\Sigma k_i) \subset O(\Sigma k_i + q).$$

More generally we consider the restriction of this to $G_1 \times \cdots \times G_m$ where $G_i \subset O(k_i)$ is transitive on S^{k_i-1}.

Actions resembling this are treated rather completely in [4] even in the topological case! It is not hard to transfer the results to the differentiable case and we shall describe the classification for this case.

First we need some notation: If G acts on M and $K \subset G$ is a subgroup we put

$$M_K = \{y \in M | G_y \text{ is conjugate to } K\}$$

(i.e. the points on orbits of type (K)). Let

$$H = \text{a given principal isotropy group of } G \text{ on } M.$$

Now, assuming G has the form

$$G = G_1 \times \cdots \times G_m$$

then for any subset

$$\sigma \subset \{1, \ldots, m\}$$

we put

$$H_\sigma = \bigtimes_{i \in \sigma} (G_i \cap H) \times \bigtimes_{i \notin \sigma} G_i$$

and

$$M_\sigma = \bigcup_{\tau \subset \sigma} M_{H_\tau}.$$

Also let $\#(\sigma)$ denote the number of elements of $\sigma \subset \{1, 2, \ldots, m\}$.

We also use an asterisk to denote the image Y^* of any $Y \subset M$ in the orbit space $M/G = M^*$.

(5.1) Theorem [4]. *Let G_i be compact Lie groups and let $G = G_1 \times \cdots \times G_m$ act differentiably on a manifold M^n which is an integral homology sphere. Let $k_i - 1 = \max\{\dim G_i(y) | y \in M^n\}$ and put $k = \Sigma k_i$. Assume the following four things:*

1. *Each G_i is effective on M^n.*
2. *There are no indices (i,j) such that $G_i \approx Z_2 \approx G_j$ and such that the diagonal subgroup acts trivially on M^n.*
3. *There are no indices (i,j) such that the identity components G_i^0 and G_j^0 are both $SO(2)$ or both $Sp(1)$ and such that the action of $G_i^0 \times G_j^0$ has the diagonal subgroup as a principal isotropy group.*
4. $\dim F(G_i, M^n) = n - k_i$ *for all $i = 1, \ldots, m$.*

Then we have the following conclusions:

1. *The orbit types are exactly the* (H_σ), $\sigma \subset \{1,\ldots,m\}$.

2. $G_i/(G_i \cap H) \approx S^{k_i-1}$ *so that* $G/H_\sigma \approx \bigtimes_{i\in\sigma} S^{k_i-1}$.

3. M_σ^* *is an* $(n-k+\#(\sigma))$*-manifold with boundary* $\bigcup_{\tau<\sigma} M_\tau^*$ *(where* $<$ *denotes proper inclusion). Also it is acyclic for* $\sigma \neq \phi$ *and* $M_\phi^* = F(G,M)$ *is an integral homology* $(n-k)$*-sphere.*

4. *There exists a smooth cross-section* $f: M^* \to M$ *such that* $f(M_{H_\sigma}^*) \subset F(H_\sigma, M)$.

5. *If* M *is a homotopy sphere then* M^* *is contractible.*

6. *If* M^* *is contractible then* M *is a sphere (with the standard differentiable structure).*

Technically, in the differentiable case, we must consider $M^* = M/G$ to be a "manifold with corners" as treated by JÄNICH in [24] and others. Note that the orbit space structure (the information in 3.) resembles that of the join of an $(n-k)$-sphere with an $(m-1)$-simplex which is exactly the orbit space structure of the linear model (with $q = n-k+1$). (If $F(G,M) = \phi$ then M^* resembles an $(m-1)$-simplex.)

It is clear from 4. that the orbit space structure determines the action. It is also easy to see that all such structures arise from such actions (on spheres if M^* is contractible).

Thus such actions are completely classified by their orbit space structures. We remark that conditions (1), (2), and (3) of the theorem are merely technical and do not represent any real restrictions; see [4].

It is clear that one can vary the orbit space structure just as in Section 1. For example the "bad" part may occur in one face of the orbit space. In this way one can obtain exotic actions.

We shall outline one example without attempting precision. Let W^5 be a contractible 5-manifold which has $\pi_1(\partial W) \neq 1$. Let $V^4 \subset W^5$ be an imbedding of the union of four 4-faces of the standard 5-simplex in ∂W^5. This makes W^5 into a "pseudo 5-simplex" with "bad face" $\overline{\partial W^5 - V^4}$. Now double W^5 along its bad face (see Fig. 4a). This new 5-manifold has a boundary with the structure of the double of an ordinary 5-simplex along a face. Now add to this five more 5-simplices in the obvious way

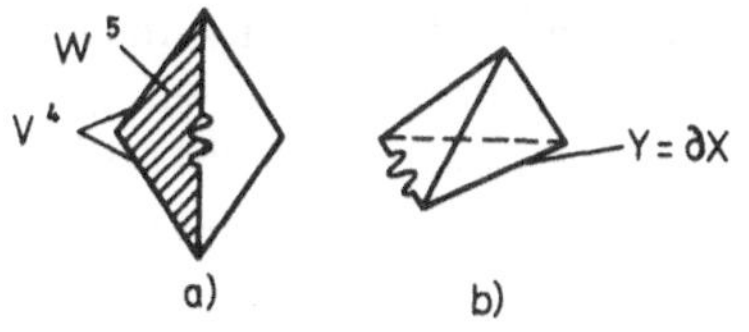

Fig. 4

to get a structure Y resembling the boundary of a 6-simplex (see Fig. 4b). Now Y, forgetting the face structure, is just a 5-sphere and hence bounds a 6-disk. Add this disk, producing a structure X resembling a 6-simplex. Then X occurs as the orbit structure of an action of

$$G=O(k_1)\times\cdots\times O(k_7)$$

for any $k_i\geqslant 1$, on S^{k-1} (where $k=\Sigma k_i$) with the following properties [where $G_i=O(k_i)$]:

a) $F(G,S^{k-1})=\phi$,

b) $F(G_i,S^{k-1})\approx S^{k-k_i-1}$,

c) $F(G_1\times G_2,S^{k-1})$ is a non-simply connected homology $(k-k_1-k_2-1)$-sphere,

d) $F(G_i\times G_j,S^{k-1})$ is a sphere for $(i,j)\neq(1,2)$,

e) $F(G_{i_1}\times G_{i_2}\times\cdots\times G_{i_j},S^{k-1})$ is a sphere for $j\geqslant 3$,

f) S^{k-1}/G is homeomorphic to the 6-disk.

g) The action of $G_1\times\cdots\times\hat{G}_i\times\cdots\times G_m$ on $F(G_i,S^{k-1})$ is linear for $i\geqslant 3$.

Clearly many other such exotic actions may be constructed but they all resemble the linear model so closely that there seems to be no point in pursuing them further.

A particular case of some interest is that of toral actions. Here one can find a simple criterion for the hypotheses of (5.1) to hold:

(5.2) Theorem [4]. *Let T^m be an m-torus acting effectively on an n-manifold M^n which is an integral homology sphere. Assume that the stationary point set has dimension $n-2m$. Then there exist circle subgroups $T_1,\ldots,T_m$ with $T^m=\times T_i$ and with* $\dim F(T_i,M)=n-2$ *for all $i=1,\ldots,m$. Thus the hypotheses of* (5.1) *are satisfied for this action.*

A similar result holds for the group $(Z_2)^m$. We also remark that the classification theorem in section one is just the special case $m=1$ of (5.1).

6. Semifree Actions

In this section we let G be either $U(1)=S^1$ or $Sp(1)=S^3$. We put $d=2$ when $G=U(1)$ and $d=4$ when $G=Sp(1)$. By a *semifree* action of G we mean an action which is free outside the fixed point set. We have already had examples of semifree actions on spheres. In Section 1 we mentioned examples in which the fixed point set is not simply connected and remarked that by a result of the HSIANG's the fixed point set can be *any* integral homology sphere up to the differentiable structure. In (3.5) and (3.9) we gave examples for $U(1)$ in which the fixed point sets have exotic differentiable structures. Similar examples for $Sp(1)$ are constructible by the same procedure using the standard embedding $Sp(1)\subset SO(4)$.

In this section we shall consider semifree actions for which the fixed point set F is diffeomorphic to the standard sphere.

The linear model for semifree actions on S^n with $F=S^{n-kd}$ is the representation $U(1)\to SO(n+1)$ (respectively $\mathrm{Sp}(1)\to SO(n+1)$) taking $A\in U(1)=SO(2)$ (respectively $A\in\mathrm{Sp}(1)\subset SO(4)$) into $\begin{pmatrix} B & 0 \\ 0 & I \end{pmatrix}$ where B is the block matrix

$$B=\begin{pmatrix} A & 0\cdots 0 \\ 0 & A & 0 \\ \vdots & \vdots & \vdots \\ 0 & 0\cdots & A \end{pmatrix}$$

with k copies of A down the diagonal. The case of codimension $2d$ is the easiest to deal with (codimension d is trivial and is dealt with in Section 1) since the orbit space S^n/G has the structure of a differentiable manifold and is, in fact, a homotopy $(n-d+1)$-sphere.

The case $G=S^1$ and $n=7$ with $F\approx S^3$ presents itself immediately since the orbit space is S^6 with F an embedded S^3 and according to work of HAEFLIGER the group of isotopy classes of S^3-knots in S^6 is infinite cyclic. The question arises immediately as to whether or not these knots are all represented as orbit spaces of semifree S^1-actions on homotopy seven spheres.

It was noted independently by myself and MONTGOMERY and YANG [31] that all such S^3-knots in S^6 are so represented. MONTGOMERY and YANG investigated this case rather fully and proved that the assignment of the knot $(\Sigma^7/S^1,F)$ to the semifree S^1-action yields a one-one correspondence between equivalence classes of such actions and isotopy (or h-cobordism) classes of such knots. In fact there are natural group structures arising from the connected sum and this correspondence is an isomorphism (of infinite cyclic groups). Regarding the differentiable structures on the total spaces Σ^7, MONTGOMERY and YANG constructed such an action on the Milnor 7-sphere and it follows, using the group structure, that every homotopy 7-sphere admits infinitely many such actions.

In order to construct some further exotic semifree actions we shall now consider a general method of changing a given (possibly linear) action. This method is mainly useful only for actions of torus groups, S^1 and S^3 groups or finite groups. For more details see [1].

Suppose we are given any action of a compact Lie group G on a manifold M^{n+k}. Let k be the maximum orbit dimension. Pick a point $x\in M$ which is on a *principal* orbit and pick a frame $\xi=\langle\xi_1,\ldots,\xi\rangle$ for the normal space to $G(x)$ at x. At $g(x)$ we pick the frame $(dg)(\pi)$. If $g(x)=g'(x)$ then $h=g^{-1}g'\in G_x$ which acts trivially on the normal space at x and hence $(dg')(\pi)=(dg)(dh)(\pi)=(dg)(\pi)$ so that this definition makes sense and gives a framing of $G(x)$ in M^{n+k}. By the THOM-PONTRYAGIN construction, this framed manifold represents a homotopy class $\alpha \in [M^{n+k}, S^n]$. If we choose an invariant riemannian metric on M then we may associate to this equivariant framing of $G(x)$ an equivariant tubular neighborhood, that is an equivariant embedding

$$G(x) \times D^n \subset M^{n+k}$$

where G acts as usual on $G(x)$ and trivially on D^n. Now let Σ^n be a homotopy sphere and let $B^n \subset \Sigma^n$ be an embedded disk. Let $f: B^n \to D^n$ be an orientation reversing diffeomorphism. Then we define a new manifold $\Sigma^n \circ M^{n+k}$ by attaching $G(x) \times (\Sigma^n - \operatorname{Int} B^n)$ to $M^{n+k} - (G(x) \times \operatorname{Int} D^n)$ by means of the map

$$1 \times f: G(x) \times \partial B^n \to G(x) \times \partial D^n.$$

Since this attaching is equivariant we obtain a natural G-action on $\Sigma^n \circ M^{n+k}$.

The question arises as to what is the differentiable manifold $\Sigma^n \circ M^{n+k}$. To discuss this we forget the G-action and associate, in the same way, to any framed k-manifold in M^{n+k} and any homotopy sphere Σ^n, the new manifold constructed above. Let us restrict our attention to the special case in which $M^{n+k} = S^{n+k}$. Then the framed k-manifold represents an element of $[S^{n+k}, S^n] = \pi_{n+k}(S^n)$ and Σ^n represents an element of Θ_n. The new $(n+k)$-manifold constructed above is then a homotopy sphere representing an element of Θ_{n+k}. One can show that this construction induces a map

$$\rho_{n,k}: \Theta_n \times \pi_{n+k}(S^n) \to \Theta_{n+k}$$

and, when $k < n-1$, this gives a pairing

$$\Theta_n \otimes \pi_k \to \Theta_{n+k}.$$

Now, one may compute the effect of this pairing in some cases and also may compute the element of π_* represented by the equivariantly framed principal orbit in certain linear actions. For all this and the following theorems see [1]. One obtains, in this way, the following specific information:

(6.1) Theorem. *If $m = 8$ or 12 then any element of Θ_{m-1} is represented by the orbit space of some semifree S^1 action on the standard S^m with $F \approx S^{m-4}$. Similarly if $m = 10$, 14, 16 or 18 then any element of Θ_{m-3} is represented by the orbit space of some semifree S^3 action on S^m with $F \approx S^{m-8}$.*

(6.2) Theorem. *For $m = 10, 15$ or 18 (resp. $m \equiv 1 \bmod 8$) there is a homotopy m-sphere Σ^m which does not bound a π-manifold (resp. a spin-*

manifold) but which admits a semifree S^1 *action with F of any codimension divisible by* 4.

(Note that one of the actions on Σ^{15} is free.)

(6.3) Theorem. *There is a* Σ^{13} *not bounding a* π*-manifold but which admits semifree* S^3 *actions with* $F\approx S^5$ *and* S^1. *There is a* Σ^{17} *not bounding a* π*-manifold but which admits semifree* S^3 *actions with* $F\approx S^9$ *and* S^1.

(Note that the S^1 actions on Σ^{13} obtained by restriction are not covered by (6.2).)

From certain other (non-semifree) representations we obtain the following two results:

(6.4) Theorem. *For* $m=10, 15$ *or* 18 *(resp.* $m\equiv 1 \bmod 8$*) there exists a* Σ^m *not bounding a* π*-manifold (resp. a spin-manifold) such that for any integer* j *with* $2\leqslant j\leqslant \frac{1}{2}(m+1)$ *there exist an infinite number of distinct actions (not semifree) of* S^1 *on* Σ^m *with stationary point set being a standard sphere of codimension* $2j$.

(6.5) Theorem. *There exists a* Σ^{18} *not bounding a* π*-manifold which has an infinite number of distinct effective* 2*-torus actions on it.*

Before we leave the topic of semifree actions we should mention the generalized SMITH conjecture which asks whether a periodic diffeomorphism on a homotopy sphere Σ^n can have a knotted Σ^{n-2} as its fixed point set. GIFFEN [14] showed that this is possible for $n>3$ and gives an extensive class of examples. Some rather trivial examples for period 2 are given by (3.8) since an exotic sphere Σ^{n-2} in S^n must be knotted. J. C. Su has recently given another class of such examples.

7. Free Actions of S^1 and S^3

Free actions are just semifree actions with F empty. The only exotic free action we have mentioned is an S^1-action on a homotopy 15-sphere not bounding a π-manifold (6.2). Free actions of S^1 and S^3 on homotopy spheres are classified by their orbit spaces which are homotopy complex and quaternionic projective spaces respectively.

In [18, 21] the HSIANGS constructed infinitely many free S^1-actions on certain homotopy $(2k+1)$-spheres, $k\geqslant 4$ and infinitely many free S^3-actions on homotopy $(4k+3)$-spheres, $k\geqslant 2$. These actions are distinguished by the rational PONTRYAGIN classes of their orbit spaces so that they are actually topologically inequivalent. Except for S^{11}, however, the HSIANGS were not able to determine the diffeomorphism type of the total spaces, and in particular could not show that any of the examples other than on S^{11} is on the ordinary sphere (although it is probable that infinitely many of them are).

A free action of S^1 on S^9 (standard structure) can be constructed, using the method of Section 6, such that $S^9/S^1 \approx CP_4 \# \Sigma^8$ where Σ^8 is the exotic 8-sphere. Moreover $CP_4 \# \Sigma^8$ is not diffeomorphic to CP_4, according to a communication from D. SULLIVAN.

Another free S^1-action on a homotopy 9-sphere comes from an example of SULLIVAN. He constructed an 8-manifold M^8 tangentially equivalent to CP_4 but not homeomorphic, or even topologically h-corbordant, to CP_4 (see [27]). By results of MAZUR and HIRSCH (see [15] $CP_4 \times D^k$ is diffeomorphic to $M^8 \times D^k$ for $k \geqslant 8$. This yields:

(7.1) Theorem. *There is a free S^1-action on some homotopy 9-sphere Σ^9 which is not topologically equivalent to the linear action but such that the induced action on $\Sigma^9 \times D^8$ (trivial on D^8) is differentiably equivalent to the standard action on $S^9 \times D^8$.*

It is of interest to ask whether or not the involution contained in this action desuspends to S^8. We suspect that it does not. SULLIVAN has informed me that this Σ^9 is the Kervaire sphere.

Certainly the best results so far on free actions are the very complete results of MONTGOMERY and YANG concerning actions on homotopy 7-spheres.

The following is one of their results (as yet unpublished):

(7.2) Theorem (MONTGOMERY, YANG). *Let $\Sigma_M^7 = \Sigma_{3,5}^7$ denote the Milnor sphere. Then a homotopy 7-sphere Σ^7 admits a free S^1-action iff $\Sigma^7 \approx k\Sigma_M^7$ for some* $k \equiv 0, \pm 4, \pm 6, \pm 8, \pm 10$ *or* $14 \pmod{28}$. *These all admit infinitely many such actions and all such actions are distinguished by the first Pontryagin class of the orbit spaces.*

Concerning the existence or non-existence of free S^1-actions on exotic homotopy spheres, the only result we know besides the above complete information on seven-spheres, the example (6.2) on a Σ^{15}, and an example on an exotic Σ^{11} due to the HSIANG's [21], is the following information obtained recently by R. LEE (a student of F. RAYMOND). He has observed that if Σ^{4k+1} admits a free S^1-action, then Σ^{4k+1} bounds a spin manifold. MILNOR has shown the existence of homotopy spheres in dimensions 9, 10, 17 and 18 not bounding spin-manifolds [28] and more recent work of ANDERSON, BROWN and PETERSON has extended this to dimensions $\equiv 1$ or $2 \bmod 8$. Thus LEE proved:

Theorem (R. LEE). *There exists a homotopy sphere in dimension* $8k+1$ $(k \geqslant 1)$ *which does not admit any free S^1-actions.*

It would be interesting to know whether or not there exist homotopy spheres which admit no S^1-action, free or not. Of course, by the Brieskorn-Hirzebruch examples, such a homotopy sphere could not bound a π-manifold.

8. Free Involutions

Concerning the existence of free involutions on homotopy spheres Σ^n we first make two elementary remarks:

(8.1) Proposition. *If Σ^{2n} admits a free involution T then $2\Sigma^{2n}=0$ in Θ_{2n}.*

Proof. T must reverse orientation since $2n$ is even and thus $\Sigma^{2n} \approx -\Sigma^{2n}$ and the result follows.

In particular, there are homotopy 10 and 18-spheres not admitting free involutions.

(8.2) Proposition. *If Σ^{2n+1} admits a free involution T and if Σ_0^{2n+1} is any other homotopy sphere then $\Sigma^{2n+1} \# 2\Sigma_0^{2n+1}$ admits a free involution.*

This is clear from Fig. 5 since T preserves orientation.

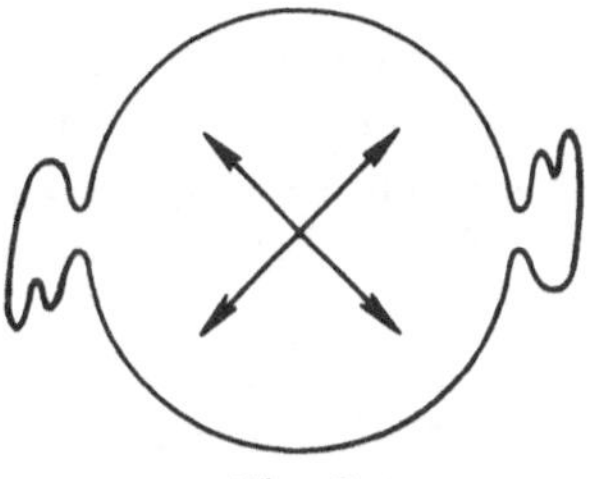

Fig. 5

The involution on S^7 resulting from this construction applied to $\Sigma = S^7$ and Σ_0 of order two in Θ_7 was shown by MILNOR to be exotic since, as he proved, $P^7 \# \Sigma_0 \not\approx P^7$; see [28]. Note that the involution desuspends to the standard involution on S^6.

Note that (8.2) is actually a special case of the considerations in Section 6 and that there are obvious analogues of both (8.1) and (8.2) in the non-free cases.

Similarly to the remarks surrounding (7.3), R. LEE has shown that if Σ^n admits a free involution and if $n \equiv 1, 2 \pmod 4$ then Σ^n bounds a spin manifold. Thus he proved:

(8.3) Theorem (R. LEE). *There exist homotopy spheres in dimensions* $\equiv 1$ *or* 2 mod 8 *which do not admit free involutions.*

Note that the element of order two in $\Theta_{10} \approx Z_6$ does not bound a spin-manifold so that, as LEE has noted, *no* exotic 10-sphere admits a free involution.

Now, as noted by MILNOR and HIRSCH [16], there is a free involution on the Milnor 7-sphere (which generates Θ_7). In fact this 7-sphere can be described as a certain S^3-bundle over S^4 and the antipodal involution in each fibre yields the desired involution.

By (8.2), since the zero and the generator of Θ_7 admit free involutions, so does every element of Θ_7.

Consider now the free S^1-actions on certain homotopy 7-spheres defined by MONTGOMERY and YANG (see (7.2)) and consider the involution T contained in this action. MONTGOMERY and YANG have recently shown that the Browder-Livesay obstruction [8] to desuspending T equals the Browder-Novikov obstruction to desuspending the original S^1-action and that it takes on infinitely many values for each of the spheres involved (see (7.2)). They also noted the obvious fact that the operation proving (8.2) does not change this obstruction. Hence:

(8.4) Theorem (MONTGOMERY-YANG). *There exist infinitely many free involutions on each element of* $2\Theta_7 \approx 2Z_{28} \approx Z_{14}$ *which are distinguished from one another by their Browder-Livesay desuspension obstructions, which take every multiple of* 8 *as value on each element of* $2\Theta_7$.

The first to find examples with non-zero Browder-Livesay obstruction was S. LOPEZ DE MEDRANO who recently constructed involutions on homotopy (homology if $n=1$) $(4n-1)$-spheres (the differentiable structures are unknown) for which the obstruction is any multiple of 8 (which is best possible).

Each Kervaire sphere admits a free involution, the action of $-I \in O(2r+1)$ on $M_1^{4r+1} = \Sigma_3^{4r+1}$. There is, of course, the generalization of this to $M_k^{4r+1} = \Sigma_{2k+1}^{4r+1}$ and specializing to $r=1$ gives a class of involutions on the 5-sphere. In [2] I asked whether any of these involutions on S^5 is exotic. This was answered by YANG [35] who identified these involutions with a class of involutions of MILNOR-HIRSCH [16]. These Milnor-Hirsch involutions are double desuspensions of certain involutions on homotopy 7-spheres, one of which is the involution on the Milnor sphere mentioned above. For this involution, which turns out [34] to be the involution on $M_2^5 = \Sigma_5^5$, MILNOR and HIRSCH showed that the orbit space is not diffeomorphic to P_5 and hence that the involution is exotic.

Recently C. T. C. WALL has achieved a complete classification of free involutions on $S^n (n \geqslant 5)$ in the *piecewise linear* case [34]. This coincides with the smooth case in dimensions 5 and 6 and it is also possible to handle dimension 7 in the smooth case. WALL defines certain invariants which provide a bijection of the *set* of PL-homeomorphism classes of manifolds homotopy equivalent to P^n (oriented if n is odd) with the set

$$Z_4 \oplus (Z_2)^{2i-2} \quad \text{if} \quad n=4i+1 \quad \text{or} \quad n=4i+2,$$
$$Z_4 \oplus (Z_2)^{2i-2} \oplus Z \quad \text{if} \quad n=4i+3,$$
$$Z_4 \oplus (Z_2)^{2i-2} \oplus Z_2 \quad \text{if} \quad n=4i+4.$$

Moreover, "suspension" from the case $n=4i+3$ to $4i+4$ is given by the epimorphism $Z \to Z_2$. Also, for $n=4i+3$ the Browder-Livesay desuspension invariant may be taken to be the projection onto Z.

In particular, there are 4 involutions on S^5 (each admits an orientation reversing equivalence) and it can be seen that these are the involutions on $M_k^5 = \Sigma_{2k+1}^5$ for $k=0,1,2$, and 3.

9. Free Z_p-actions for p odd

Using results of Stallings, Milnor observed in [29] that for any prime $p \geqslant 5$ and for L^{2n+1}, $n \geqslant 2$, a lens space with $\pi_1(L) \approx Z_p$, there exist an infinite number of $(2n+1)$-manifolds each h-cobordant to L, but having distinct simple homotopy types. The universal covering spaces of these are h-cobordant to S^{2n+1} and hence diffeomorphic to S^{2n+1}. Since there are only a finite number of distinct linear Z_p-actions on S^{2n+1} he deduced:

(9.1) Theorem (Milnor [29]). *For $p \geqslant 5$ prime and for $2n+1 \geqslant 5$ there exist infinitely many distinct free actions Z_p on S^{2n+1} all inequivalent to any linear action.*

With regard to this we remark that from their intimate knowledge of the 7-sphere, Montgomery and Yang have recently shown that there is a free Z_5-action on S^7 such that S^7/Z_5 is not h-cobordant to a lens space. It is obtained from one of their free S^1-actions by restricting to $Z_5 \subset S^1$ and then adding, equivariantly, 5 copies of some homotopy 7-sphere.

Montgomery and Yang have also shown that there exist exotic free Z_3-actions on S^7 which are extendible to free S^1-actions.

10. Very Exotic Actions

In this section we shall consider some examples which are so exotic that one cannot say that they resemble any linear action whatsoever. We have already seen some such actions, as byproducts of more reasonable examples, in Section 3.

We begin by describing the remarkable FLOYD and RICHARDSON example [13] of an action of the icosahedral group on a disk without stationary points.

Let $I \subset SO(3)$ denote the icosahedral group and let I' (the binary icosahedral group) be its inverse image in the covering group $\text{Spin}(3) \approx S^3$. It can be seen that $[I', I'] = I'$ so that the fundamental group of $SO(3)/I \approx \text{Spin}(3)/I'$ is perfect. Thus $H_1(SO(3)/I) = 0$ and POINCARÉ duality shows that $SO(3)/I$ is an integral homology sphere (it is the original "Poincaré sphere"). In fact $SO(3)/I$ is the only homogeneous (by a compact group) homology sphere which is not actually a sphere [5].

Now one examines the fixed point set $F(I, SO(3)/I)$. Clearly gI is fixed iff $IgI = gI$, that is iff $g \in N(I)$. But it is well-known that $N(I) = I$. Thus the coset I is the *only* fixed point of I on $SO(3)/I$.

We may assume that I acts simplicially in some simplicial structure of $SO(3)/I$. Let us remove the open star of the fixed point, thereby obtaining a complex K on which I acts simplicially without stationary points. Note that K is acyclic, but not simply connected.

The remainder of the construction does not use any properties of I or K other than that I acts simplicially and that K is a finite acyclic simplicial complex with no stationary points.

Let A be some finite complex on which I acts and let B be the set $F(I, A)$ of stationary points (e.g. $A = I$ with left translation and B empty or $A = B$ with trivial action).

Consider the join

$$K * A$$

with the diagonal action. It is well-known that $K * A$ is contractible and, clearly, $F(I, K * A)$ is a copy of B.

We now embed $K * A$ naturally in the simplex Δ^{m+1} having as vertices, the vertices of $K * A$. Clearly I acts on Δ^{m+1} with $K * A$ invariant. Let D^m be the second regular neighborhood in $\partial \Delta^{m+1}$ of $K * A$. Since $K * A$ is contractible, D^m is an m-disk (by J. H. C. WHITEHEAD) and clearly D^m is invariant under I. Now $F(I, D^m)$ is just the second regular neighborhood of B in $F(I, \partial \Delta^{m+1})$. (The latter is a sphere.) Thus B is a deformation retract of $F(I, D^m)$.

It is also reasonably clear that this action may be smoothed. Thus, for any finite complex B, we have produced a smooth action of I on some m-disk D^m such that $F(I, D^m)$ has B as a deformation retract.

When A merely consists of some number of points, k of which are fixed by I then, in fact, $F(I, D^m)$ consists of k disjoint r-disks (where $r = \dim F(I, \partial \Delta^{m+1})$) and $F(I, S^{m-1})$ consists of k disjoint $(r-1)$-spheres. The case $k = 0$ (i.e. B empty) is that of FLOYD and RICHARDSON [13].

We shall now remark briefly on some other examples, all on euclidean spaces.

In [11] CONNER and MONTGOMERY constructed an action of $SO(3)$ on E^n, $n \geqslant 12$, without stationary points. The action was not differentiable but can be smoothed, according to MONTGOMERY.

The ideas in this construction are due to earlier methods of CONNER and FLOYD [10] and these methods make basic use of the construction, by FLOYD of an equivariant map $f: S^4 \to S^4$ of degree zero, where $SO(3)$ acts on S^4 without stationary points (the irreducible 5-dimensional representation of $SO(3)$). FLOYD constructed similar maps $f: S^n \to S^n$ for $SO(m)$, m odd, but did not publish this work. This was rediscovered by the HSIANG'S [20] and they also noted that the same property follows for every compact *connected* non-abelian group. Thus, they remarked [20] that exactly the same constructions apply to give actions of any such group on sufficiently high dimensional euclidean spaces without stationary points.

KISTER [26] used these same ideas of CONNER and FLOYD [10] to construct a diffeomorphism $E^n \to E^n$ for any $n \geqslant 8$ which has any given period not a prime power and which has no stationary points.

References

1. BREDON, G.: A π_*-module structure for Θ_* and applications to transformation groups, Ann. of Math. **86,** 434—448 (1967).
2. — Examples of differentiable group actions, Topology **3,** 115—122 (1965).
3. — Fixed point sets and orbits of complementary dimension, Chap. XV of Seminar on Transformation Groups by A. Borel et al., Ann. of Math. Studies No. **46** (1960).
4. — On a certain class of transformation groups, Mich. Math. J. **9,** 385—393 (1962).
5. — On homogeneous cohomology spheres, Ann. of Math. **73,** 556—565 (1961).
6. — Transformation groups on spheres with two types of orbits, Topology **3,** 103—113 (1965).
7. BRIESKORN, E.: Beispiele zur Differentialtopologie von Singularitäten, Inventiones Math. **2,** 1—14 (1966).
8. BROWDER, W., and G. R. LIVESAY: Fixed point free involutions on homotopy spheres, Bull. A. M. S. **73,** 242—245 (1967).
9. CONNELL, E. H., D. MONTGOMERY, and C. T. YANG: Compact groups in E^n, Ann. of Math. **80,** 94—103 (1964), and **81** (1965).
10. CONNER, P., and E. FLOYD: On the construction of periodic maps without fixed points, Proc. A. M. S. **10,** 354—360 (1959).
11. —, and D. MONTGOMERY: An example for $SO(3)$, Proc. N. A. S. **48,** 1918—1922 (1962).
12. FLOYD, E.: Fixed point sets of compact abelian groups of transformations, Ann. of Math. **66,** 30—35 (1957).

13. —, and R. W. RICHARDSON, Jr: An action of a finite group on an n-cell without stationary points, Bull. A. M. S. **65,** 73—76 (1959).
14. GIFFEN, C. H.: The generalized Smith conjecture, Amer. J. Math. **88,** 187—198 (1966).
15. HIRSCH, M. W.: On tangential equivalence of manifolds, Ann. of Math. **83,** 211—217 (1966).
16. —, and J. MILNOR: Some curious involutions of spheres, Bull. A. M. S. **70,** 372—377 (1964).
17. HIRZEBRUCH, F.: $O(n)$-Mannigfaltigkeiten, exotische Sphären, kuriose Involutionen (to appear).
18. HSIANG, W. C.: A note on free differentiable actions of S^1 and S^3 on homotopy spheres, Ann. of Math. **83,** 266—272 (1966).
19. —, and W. Y. HSIANG: Classification of differentiable actions on S^n, R^n and D^n with S^k as the principal orbit type, Ann. of Math. **82,** 421—433 (1965).
20. — — Differentiable actions of compact connected classical groups, I, Amer. J. Math. (to appear).
21. — — Some differentiable actions of S^1 and S^3 on 11-spheres, Quart. J. Math. **15,** 371—374 (1964).
22. — — On compact subgroups of the diffeomorphism groups of Kervaire spheres, Ann. of Math. **85,** 359—369 (1967).
23. JÄNICH, K.: Differenzierbare Mannigfaltigkeiten mit Rand als Orbiträume differenzierbarer G-Mannigfaltigkeiten ohne Rand, Topology **5,** 301—320 (1966).
24. — On the classification of $O(n)$-manifolds (to appear).
25. KERVAIRE, M., and J. MILNOR: Groups of homotopy spheres: I, Ann. of Math. **77,** 504—537 (1963).
26. KISTER, J. M.: Differentiable periodic actions on E^8 without fixed points, Amer. J. Math. **85,** 316—319 (1963).
27. LASHOF, R., and M. ROTHENBERG: On the Hauptvermutung, triangulation of manifolds, and h-cobordism, Bull. A. M. S. **72,** 1040—1043 (1966).
28. MILNOR, J.: Remarks concerning spin manifolds, Differential and Combinatorial Topology, Princeton Univ. Press (1965).
29. — Some free actions of cyclic groups on spheres, Proc. Bombay Colloquium, 37—42 (1964).
30. MONTGOMERY, D., and H. SAMELSON: Examples for differentiable group actions on spheres, Proc. N. A. S. **47,** 1202—1205 (1961).
31. —, C. T. YANG: Differentiable actions on homotopy seven spheres (I), Trans. A. M. S. **122,** 480—498 (1966).
32. —, — Differentiable transformation groups on homotopy spheres, Mich. Math. J. **14,** 33—46 (1967).
33. ROSEN, R.: Examples of non-orthogonal involutions of euclidean spaces, Ann. of Math. **78,** 560—566 (1963).
34. WALL, C. T. C.: Free piecewise linear involutions on spheres (to appear).
35. YANG, C. T.: On involutions of the five sphere, Topology **5,** 17—19 (1966).

A Survey on Regularity Theorems in Differentiable Compact Transformation Groups*

WU-YI HSIANG

I. Introduction

In differentiable transformation groups, one mainly studies the geometric behavior of differentiable actions of compact Lie groups on certain given types of manifolds. Since the existence of some interesting differentiable action on a smooth manifold M is itself a *non-trivial fact* which is *not universally* enjoyed by general manifolds (i. e., not every manifold has interesting symmetries), it seems rather logical to consider firstly smooth actions on those manifolds with sufficiently rich natural symmetries, such as euclidean spaces, spheres, homogeneous spaces, etc. For manifolds with "natural actions", a simple-minded but rather fruitful approach is to *compare the behavoir of general differentiable actions with the behavior of "natural actions"*. From the above viewpoint, it is quite fair to say that euclidean spaces, spheres and discs are among the best testing spaces for the study of differentiable transformation groups. For these testing spaces, *linear actions* are clearly the "*natural actions*" and the comparisons between general differentiable actions and linear actions consist of the following two complementary efforts. Namely, one tries to prove more and more resemblances between differentiable actions and linear actions on the one hand, and on the other hand one tries to construct more and more varieties of differentiable examples to see how differentiable actions may differ from linear actions.

So far, most of the known "positive" results are of the form that, *to some degree, the geometric behavoir of differentiable actions on a given manifold M satisfying "suitable conditions" necessarily resembles the geometric properties of the "natural actions" on M*. We shall call results of the above type *regularity theorems*, and the "suitable conditions" *regularity conditions*. The purpose of this survey article is to try to summarize systematically the existing regularity theorems in differen-

* Dedicated to Professor DEANE MONTGOMERY.

tiable transformation groups. For the sake of motiviation for regularity theorems, we collect some simple but important facts about the geometric behavoir of linear actions in II. Since the linear actions are the natural models for the study of smooth actions on euclidean spaces and spheres, reasonable understanding of the geometric properties, such as orbit structures, etc., for linear actions of compact Lie groups is undoubtedly the logical first step towards the study of differentiable actions, Hence, it is of immediate interest to investigate more systematically the geometric invariants of linear actions (or representations), about which very little has been understood so far.

In III, we discuss some methods of constructing various differentiable actions. Since there is already an excellent review article about non-orthogonal actions on spheres by G. BREDON in these Proceedings, we try to organize this section from a viewpoint complementary to his. Hence, it seems to be a good idea to read both BREDON's article and III together.

IV is the central part of this article. We try to give a systematic summary of some classes of the existing regularity theorems. However, there are two major omissions based on the following reasons: a) Since the cohomology theory and the cobordism theory of actions of elementary abelian groups have already been systematically treated in [12] and [30] respectively we shall refer to the above two books for results in those two areas. b) From the point of view of differentiable transformation groups, free actions constitute only an extremely special class of actions. Technically, the study of free actions amounts to the study of differentiable principal bundles with *a given total space M*, and hence are essentially problems in differential topology. Due to the rapid development of differential topology in the past decade, the understanding of free actions of finite groups and circle groups has been greatly improved in recent years. However, it seems to be more suitable to treat them separately in another article rather than to include them in this article. Besides the above major omissions, it is very possible that the author may overlook some interesting and important results which should be included here. The author apologizes for any such omissions due to his inadequate knowledge.

In V, we indicate some applications of the regularity theorems that we summarized in IV. We also discuss some possibilities of further applications. Generally, in differentiable transformation groups, one studies the geometric properties of differentiable symmetries. Hence, one type of possible application of the results in differentiable transformation groups will be the investigation of the relationships between symmetries and other well-known geometric properties, such as curvatures, closed minimal submanifolds, etc.

We have inserted several problems in various places where they naturally occur. For further discussions of problems in differentiable transformation groups, we refer the reader to a joint paper by WU-CHUNG HSIANG and the author in these Proceedings.

II. Linear Models

Surely, *linear* transformation groups are among the simplest and most natural examples of differentiable transformation groups. In case that the group G is compact or semi-simple, the profound classical theory of group representations offers a complete *classification of their representations in terms of the associated character functions.* This type of classification seems good enough for most algebraic purposes. However, for the study of geometric problems such as differentiable transformation groups on euclidean spaces and spheres, we need some kind of classification theory *in terms* of various *geometric invariants* rather than in terms of the characters. For example, the so-called "orbit structure" constitutes an important set of natural geometric invariants. How to express the various data about the orbit structure in terms of the characters naturally presents an interesting and difficult problem. In this section, we make a rather humble attempt to work out some simple aspects of that general problem. The results and examples that we include here are simple and far from being systematic. However, as one will see later, they are the basic models, both for motivating theorems and for helping to overcome certain technical difficulties:

A. Low Dimensional Representations

Among all geometric invariants of a given representation φ, the total dimension, $\dim\varphi$, is surely the simplest one. It can be read off from the character function χ_φ and, for the irreducible case, the Weyl formula gives an effective way to compute $\dim\varphi$ in terms of its highest weight, namely,

$$\dim\varphi = \frac{\prod_{\alpha\in\Delta_+} (\alpha, \Lambda_\varphi + \delta)}{\prod_{\alpha\in\Delta_+} (\alpha, \delta)}$$

where Λ_φ is the highest weight of the irreducible representation φ, Δ_+ is the system of positive roots of the group G, and $\delta = \frac{1}{2}\Sigma\alpha$. One important geometric significance of the above formula is that $\dim\varphi$ *grows rapidly* when the *highest weight vector* Λ_φ *gets longer*. In other words, for a given bound N, the number of different irreducible representations of G with $\dim\varphi \leqslant N$ is *in general* quite small compared with N. Hence, for a given group G, if we set the bound N *reasonably low*, then the *linear models* of G-actions on $\mathbb{R}^m$, S^{m-1} or D^m with $m \leqslant N$ become quite simple. The above observation immediately suggests the following two types of problems in differentiable transformation groups:

1. For a given compact Lie group G, we have the following two well-defined positive integers:

$$r_G = \text{g.l.b. of } \{\dim \varphi \mid \varphi \text{ non-trivial representations of } G\},$$
$$f_G = \text{g.l.b. of } \{\dim \varphi \mid \varphi \text{ faithful representation of } G\}.$$

Problem 1. *Is* is true that there is no non-trivial differentiable action of G on $\mathbb{R}^m$ with $m < r_G$?

Problem 2. *Is* is true that there is no effective differentiable action of G on $\mathbb{R}^m$ with $m < f_G$?

It would appear likely that the answer to Problem 1 should be positive. It is clear that we need only to check it for simple Lie groups. However, Problem 2 may become very difficult for certain cases. For example, if $G = \mathrm{Spin}(n)$, n large, then f_G is huge and an affirmative answer of Problem 2 will be an extremely deep result. On the other hand, a counter example to Problem 2 in the case $G = \mathrm{Spin}(n)$ would be necessarily without fixed point, with many orbit types, and hence almost beyond imagination.

2. Let G be a large classical group, say $SO(n)$, $n > 11$ or $SU(n)$, $n \geqslant 8$, or $Sp(n)$, $n \geqslant 8$. We shall use ρ_n, μ_n, or ν_n to denote the natural real representation of $SO(n)$, $SU(n)$, or $Sp(n)$ respectively. It is well known that they are, respectively, the lowest dimensional non-trivial representation and $\dim \rho_n = n$, $\dim \mu_n = 2n$, $\dim \nu_n = 4n$. Since usually there are big "gaps" between the dimensions of irreducible representations, we see that the only possible representations φ with $\dim \varphi$ *below a reasonable bound* are of the form $k\rho_n \oplus r\theta$, or $k\mu_n \oplus r\theta$, or $k\nu_n \oplus r\theta$ respectively. Let us investigate the $SO(n)$ case in detail and omit the other two completely parallel cases.

For the case $G = SO(n)$, the next irreducible representation in dimension is $\Lambda^2 \rho_n = \mathrm{Ad}$ and $\dim \Lambda^2 \rho_n = \frac{1}{2}n(n-1)$. Hence, if φ is a representative of $SO(n)$ with $\dim \varphi = m < \frac{1}{2}n(n-1)$, then φ must be of the form $k\rho_n \oplus r\theta$, with $k \cdot n + r = m$. We may denote a general vector in $\mathbb{R}^m$ by

$$g(\mathfrak{x}_1, \mathfrak{x}_2, \ldots, \mathfrak{x}_k, \mathfrak{y}) \in \mathbb{R}^n \oplus \cdots \oplus \mathbb{R}^n \oplus \mathbb{R}^r = \mathbb{R}^m.$$

Then the action of $SO(n)$ is given by

$$g(\mathfrak{x}_1, \mathfrak{x}_2, \ldots, \mathfrak{x}_k, \mathfrak{y}) = (g\mathfrak{x}_1, g\mathfrak{x}_2, \ldots, g\mathfrak{x}_k, \mathfrak{y}).$$

Hence the isotropy subgroup G_X is exactly the subgroup of $SO(n)$ that leaves $\mathfrak{x}_1, \ldots, \mathfrak{x}_k$ fixed, namely, $G_X = SO\,(\langle \mathfrak{x}_1, \ldots, \mathfrak{x}_k \rangle^\perp)$. It is conjugate to the standard $SO(n-\ell)$ if ℓ is the rank of the subspace spanned by $x_1, \ldots, x_k$ in $\mathbb{R}^n$. Therefore, we have exactly $(k+1)$ orbit types, which are: point, S^{n-1}, the 2^n Stiefel manifold, ..., and the k-th Stiefel manifold. The fixed point set is $\mathbb{R}^r$. We note that its codimension is $(m-r)=n\cdot k$. A natural question that follows from the above observation is:

Problem 3. Let Φ be a differentiable action of $SO(n)$ on $\mathbb{R}^m$ with $n \geqslant 11$ and $m < \frac{1}{2}n(n-1)$. Is it true that

1. the orbit types are exactly point, $S^{n-1}, \ldots,$ and k-th Stiefel manifolds for certain $k<n$,
2. the fixed point set F is acyclic (with suitable coefficients) and is of codimension $n\cdot k$? The reader may formulate parallel questions for the $SU(n)$ and $Sp(n)$ cases.

Under slightly stronger dimensional restrictions, the above question has been proved affirmatively in [46]. We shall discuss this problem in § IV.

B. Representations with Non-generic Principal Orbit Types

For a given differentiable action of a compact Lie group, the biggest orbit type is called the *principal orbit type.* The principal orbit type theorem of MONTGOMERY-SAMELSON-YANG [70, 73] shows that the union of all principal orbits forms a *connected everywhere dense open submanifold* and *hence has a dominant influence on the behavior of the given action.* Now, suppose we look at the linear models and ask what are the possibilities of principal orbit types for a fixed group G with respect to all the possible representations of G. For such a problem, the following is an equation of basic importance.

Proposition 2.1. [51]. *Let φ be a real representation of G and H_φ be a principal isotropy subgroup of the linear action given by φ. Then, we have*

(2.1) $$\varphi|H_\varphi \oplus \mathrm{Ad}_{H_\varphi} = \mathrm{Ad}_G|H_\varphi \oplus \textit{trivial copies.}$$

Equation (2.1) is quite powerful and useful. For example, it follows immediately from (2.1) that the principal isotropy subgroups are the maximal tori of G if and only if $\varphi = \mathrm{Ad}_G + k\cdot\theta$.

It also follows from (2.1) readily that H_φ will be a discrete finite subgroup for the "generic" representation of G. For example, if $G=SO(n)$ or $SU(n)$, then we have the following precise answer proved in [p. 127–129, 51]:

Theorem 1. *Let φ be a real representation of $SU(n)$, $n \geqslant 6$ and (H_φ) be the conjugate class of the principal isotropy subgroups of the linear action given by φ. Then $\dim H_\varphi \neq 0$ only in the following cases:*

1. $\varphi = k(\varphi_1 \oplus \varphi_1^*) + \ell\theta$, $k < n$; $H_\varphi = SU(n-k+1)$;
2. $\varphi = \mathrm{Ad} \oplus \ell\theta$; $H_\varphi =$ *maximal torus*;
3. $\varphi = (\varphi_2 \oplus \varphi_2^*) \oplus \ell\theta$, $H_\varphi = SU(2) \times \cdots \times SU(2)$ *where* $\left[\frac{n+1}{2}\right]$

copies sit as diagonal blocks in $SU(n+1)$ and where φ_1, φ_2 are the first two basic representations of $SU(n+1)$.

Theorem 2. *Let φ be a representation of $SO(n)$, $n \geqslant 8$ and (H_φ) be the conjugate class of the principal isotropy subgroups of the linear action given by φ. Then $\dim H_\varphi \neq 0$ only in the following cases*

1. $\varphi = k \cdot \rho_n \oplus \ell\theta$, $k < (n-1)$ *and* $H_\varphi = SO(n-k)$,
2. $\varphi = \mathrm{Ad} \oplus \ell\theta$ *and* $H_\varphi =$ *maximal tori.*

Naturally, we shall say that the principal isotropy subgroups of a given representation ψ is *trivial* if $H_\psi = \ker(\psi)$. Then it is quite easy to give a general proof that there are only *finitely* many essentially different representations of a given *simple* compact connected Lie group with non-trivial principal isotropy subgroups. However, in order to be useful in the study of differentiable transformation groups, it is not *enough* simply to have such a *finiteness theorem;* but rather, one needs to know *exactly and explicitly what are those finite representations of a given group G that have non-trivial principal isotropy subgroups.* The task of working out such an explicit and exact list of such possibilities involves many tedious computations and case by case detailed discussions in representation theory. Since a result of this type is usually quite useful in the application of the differentiable slice theorem as well as in motivating possible problems, we shall include here the table of real irreducible representations of simple Lie groups with non-trivial principal isotropy subgroups that appeared in [47] and [63]. (See pp. 83–84.)

For further discussion of principal isotropy subgroups of linear actions of semi-simple compact Lie groups, see Chapter I of [47]. We mention only the following three theorems that reflect the "general behavior" of the principal orbit type of *linear actions* of compact connected Lie groups.

Theorem 3. *Let G be a semi-simple compact connected Lie group with at least three normal factors and ψ be an almost faithful real irreducible representation of G. Then $(H_\varphi^0) \neq \{e\}$ only when*

a) $G \sim SU(n) \times G'$ *and* $\psi | SU(n) = m \cdot [\mu_n]_{\mathbb{R}}$, $(n-m) \geqslant 2$;
b) $G \sim SO(n) \times G'$ *and* $\psi | SO(n) = m \cdot \rho_n$, $(n-m) \geqslant 2$;
c) $G \sim Sp(n) \times G'$ *and* $\psi | Sp(n) = m \cdot [\nu_n]_{\mathbb{R}}$, $(n-m) \geqslant 1$;

and $(H_\psi) = (SU(n-m))$ *or* $(SO(n-m))$ *or* $(Sp(n-m))$ *respectively.*

Table A of [47]

I. $\mathfrak{G}=A_r$: 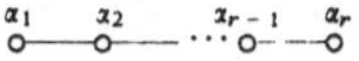

rank r	ψ	G	(H_ψ)
$r=1$	$S^2\varphi_1$ $S^4\varphi_1$	$SO(3)$ $SO(3)$	maximal tori maximal $\mathbb{Z}_2$-tori $\subseteq \mathbb{Z}_2\oplus\mathbb{Z}_2$
$r\geqslant 2$	$\varphi_1+\varphi_r$ $S^2\varphi_1+S^2\varphi_r$ $\varphi_1\times\varphi_5$	$SU(r+1)$ $SU(r+1)$ $SU(r+1)$	$(SU(r))$ maximal $\mathbb{Z}_2$-tori $\subseteq \mathbb{Z}_2^r$ maximal tori
$r=3$	φ_2 $S^2\varphi_2-\theta$	$SO(6)$ $SO(6)$	$(SO(5))$ maximal $\mathbb{Z}_2$-tori $\subseteq \mathbb{Z}_2^5$
$r\geqslant 4$	$\varphi_2+\varphi_{r-1}$	$SU(r+1)$	$(SU(2)\times\cdots\times SU(2);[(n+1)/2]$ *copies*
$r=4$	$2\varphi_3$	$SU(6)$	$T^2\subseteq SU(3)\times SU(3)\subseteq SU(6)$
$r=7$	φ_4	$SU(8)$	a finite group of order 8 $\subseteq SU(4)\times SU(4)$

II. $\mathfrak{G}=B_r$, $r\geqslant 2$: 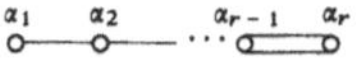

rank r	ψ	G	(H_ψ)
$r\geqslant 2$	φ_1 $S^2\varphi_1-\theta$ $\Lambda^2\varphi_1$	$SO(2r+1)$ $SO(2r+1)$ $SO(2r+1)$	$(SO(2r))$ maximal $\mathbb{Z}_2$-tori $\approx \mathbb{Z}_2^{2r}$ maximal tori
$r=2$	$2\varphi_2$	$Sp(2)$	$(Sp(1))$
$r=3$	φ_3	Spin(7)	(G_2)
$r=4$	φ_4	Spin(9)	Spin(7) with Spin(9)/Spin(7)$=S^{15}$
$r=5,6$	$2\varphi_5, 2\varphi_6$	Spin(11), Spin(13)	at most discrete
$r=7,8$	φ_7, φ_8	Spin(15), Spin(13)	at most discrete

III. $\mathfrak{G}=C_r$, $r\geqslant 3$ 

r	ψ	G	(H_ψ)
$r\geqslant 3$	$2\varphi_1$ $S^2\varphi_1$ $\Lambda^2\varphi_1-\theta$	$Sp(r)$ $Sp(r)$ $Sp(r)$	$(Sp(r-1))$ maximal tori $(Sp(1)^r)$
$r=3$	$2\varphi_3$	$Sp(3)$	discrete
$r=4$	φ_4	$Sp(4)$	discrete

IV. $\mathfrak{G} = D_r$, $r \geqslant 4$: 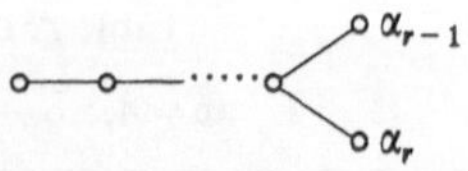

r	ψ	G	(H_ψ)
	φ_1	$SO(2r)$	$SO(2r-1)$
$r \geqslant 4$	φ_2	$SO(2r)$	maximal tori
	$S^2\varphi_2 - \theta$	$SO(2r)$	maximal $\mathbb{Z}_2$-tori
$r=5$	$\varphi_4 + \varphi_5$	Spin(10)	$SU(4)$
$r=6$	$2\varphi_r$ or $2\varphi_6$	Spin(12)	$SU(2) \times SU(2) \times SU(2)$
$r=7$	$\varphi_6 + \varphi_7$	Spin(14)	finite group
$r=8$	φ_7 or φ_8	Spin(16)	finite group

V. Exceptional Lie groups:

$\mathfrak{G}$	ψ	(H_ψ)
G_2: 	φ_1	maximal tori
	φ_2	$(SU(3))$
F_4: 	φ_1	maximal tori
	φ_4	$\text{Spin}(8) \subseteq \text{Spin}(9) \subseteq F_4$
E_6 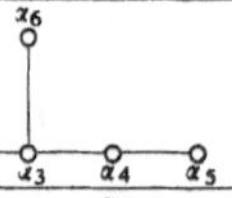	φ_6	maximal tori
	$\varphi_1 + \varphi_5$	
E_7 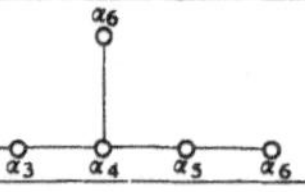	φ_6	maximal tori
	$2\varphi_1$	
E_8	Ad_{E_8}	maximal tori

Theorem 4. *For a given compact connected semi-simple Lie group G, there are only finitely many irreducible real representations φ of G with* $(H_\psi^0) \not\supseteq \ker(\psi)^0$.

Theorem 5. *For a given compact connected semi-simple Lie group G, if we consider all possible linear actions of G, there are only finitely many homogeneous spaces G/H with*

$$\dim(G/H) < \dim G$$

that can be realized as the principal orbit type of a given linear action of G.

In fact, it is clear in the proofs of the above theorems in [47] that if the group G consists of only a few simple normal factors, then the finite possibilities of principal orbit type for linear G-actions can be read off from Tables A and B of [47].

All the above results naturally suggest the following problems:

Problem 4. For a given group G, determine all the possibilities of principal orbit types for differentiable G-actions on euclidean spaces, spheres, or discs. Combining equation (2.1) and representation theory, it will not be difficult to work out a list of possibilities for the linear G-actions. Usually, *such a list for linear actions* will consist of *very few possibilities of principal orbit types.* Then a reasonable "conjecture" is that the possibilities of principal orbit types for the class of *differentiable* G-actions on euclidean spaces, spheres or discs are *almost the same as those provided by the much smaller subclass of linear G-actions.*

Problem 5. Suppose G/H is a possible principal orbit type of differentiable G-action on euclidean spaces with $\dim G/H < \dim G$. What are the possiblities of the singular orbit types for an action with the given G/H as its principal orbit type? What does the fixed point set look like for such an action?

A successful solution of the above problems will provide a good deal of insight into "regular" differentiable G-actions. The cases where G is any of the various classical groups have been treated in [46, 47, 51] and we shall discuss them in IV.

C. Representations with Few Orbit Types

In the study of differentiable transformation groups, the difficulties and complexities increase tremendously with the number of orbit types. Hence, if we want to start with the study of "easier" and "simpler" cases, we shall begin with those differentiable actions with few orbit types. We include here some observations:

Observation 2.1. *For a given differentiable G-action on euclidean space with (H) as the principal isotropy subgroups, there are at least $(\operatorname{rk} G - \operatorname{rk} H) + 1$ different orbit types.*

Proof. It is not difficult to see from P. A. Smith theory and slice representations that we must have at least one type of isotropy subgroup with rank j for every $\operatorname{rk} H \leqslant j \leqslant \operatorname{rk} G$.

As an easy consequence of Observation 2.1, we see that

Observation 2.2 *If we restrict ourselves to the investigation of differentiable G-actions on euclidean spaces with the number of orbit types less than $\operatorname{rk} G$, then the principal orbit type must be non-generic. In fact, among representations that we mentioned above in (A), (B), (C), we have the following relationship within a reasonable range. That is,*

$$(A) \supseteq (B) = (C).$$

Observation 2.3. *Let Φ be a differentiable G-action on M, G be compact connected and G_1 be a connected normal subgroup of G. If the number of orbit types of Φ is k, then the number of orbit types of the restricted G_1-action $\Phi|G_1$ has at most k orbit types.*

Observation 2.4. *For a faithful action of the r-dimensional torus T^r on a euclidean space, there are at least 2^r different orbit types.*

As an easy consequence of the above observation, we shall prove the following proposition:

Proposition 2.2. *Let G be a compact connected Lie group of rank $r \geqslant 6$ and φ be a faithful representation of G. If the linear action given by φ has exactly three orbit types, then $\varphi = 2\varphi_1 \oplus k\theta$ and φ_1 is a faithful representation of G such that G is transitive on the unit sphere. A complete list of such groups and representations was given by* MONTGOMERY-SAMELSON *and* BOREL [68, 10].

Proof. 1. If $\operatorname{rk} G \geqslant 3$ and $\varphi = \varphi_1 \oplus \varphi_2$ where φ_1, φ_2 are representations without trivial components, then at least one of them, say φ_1, is faithful, for otherwise, we shall have at least 4 orbit types. Further argument will show that φ_1 has only two types of orbits and $\varphi_2 = \varphi_1$. Hence in order to prove the proposition, we need only to show that it is impossibile for φ to be irreducible. If we combine Observations (2.3), (2.4), and (2.1) it is not difficult to show the above fact by contradiction.

Such a proposition naturally suggest the following problems:

Problem 6. In Chapter XIV of [12], A. BOREL proved that if G is a compact connected topological transformation group on $\mathbb{R}^n$ with exactly two orbit types, then G has a fixed point. If we try to generalize Borel's theorem, the following is probably a reasonable and workable formulation:

Conjecture. If G is a compact connected Lie group with $\operatorname{rk} G \geqslant 6$, then the fixed point set F of any effective differentiable action of G on a euclidean space with at most three orbit types is necessarily non-empty, and the orbit structure must model after those representations determined in Proposition (2.2).

Problem 7. Classify all possible differentiable actions of compact connected Lie groups G, with $\operatorname{rk} G \geqslant 6$, on homotopy spheres Σ^m with exactly two types of orbits. We list here the known information along this direction:

a) If one of the orbit types is *fixed point*, then the other orbit type must be *sphere* S^k, $k \neq 1,3$. A complete classification of such action is given in [43].

b) If G is a classical group and acts on Σ^m with two types of orbits neither of which is *fixed point*, then a complete classification can be deduced from the results of [46, 47].

Hence, the main point of the above problem is to show that there are no other possiblities besides a) and b). Of course, here, the main difficulty is the determination of the possibilities of G.

D. Linear Actions with Low Codimensional Principal Orbit Types

In a series of papers of MONTGOMERY-SAMELSON-YANG and H. C. WANG [112, 113, 108], a study is made of differentiable transformation groups on euclidean spaces or spheres with the codimension of the principal orbits equal to one or two. In order to push further along that direction, it seems to be a logical first step to investigate more carefully those linear actions with *low codimensional* principal orbit types. Here, again, the only possible complication comes from the fact that the group G is not fixed. Let us start with some simple observations:

Observation 2.5. *Let φ_1, φ_2 be two representations of G with (H_1), (H_2) as their principal isotropy subgroups, respectively. Let*

$$r_1 = \dim\varphi_1 - \dim(G/H_1), \qquad r_2 = \dim\varphi_2 - \dim(G/H_2)$$

and $r = \dim(\varphi_1 \oplus \varphi_2) - \dim(G/H)$, where G/H is the principal orbit type of the representation $\varphi_1 \oplus \varphi_2$. Then we have

$$r \geqslant r_1 + r_2 .$$

Proof. We may choose $H_1 \in (H_1)$ and $H_2 \in (H_2)$ such that $H = H_1 \cap H_2$. It follows from Proposition 2.1 that

$$\varphi_i|H_i = (\mathrm{Ad}_G|H_1 - \mathrm{Ad}_{H_i}) + r_i\theta \qquad (i=1,2).$$

$$\text{Hence } (\varphi_1+\varphi_2)|H = (2\cdot\mathrm{Ad}_G|H - \mathrm{Ad}_{H_1}|H - \mathrm{Ad}_{H_2}|H) + (r_1+r_2)\theta$$
$$= (\mathrm{Ad}_G|H - \mathrm{Ad}_H) + r\cdot\theta .$$

It follows, then, that

$$r = r_1 + r_2 + (\dim\mathfrak{G} + \dim(\mathfrak{H}_1 \cap \mathfrak{H}_2) - \dim\mathfrak{H}_1 - \dim\mathfrak{H}_2)$$
$$\geqslant r_1 + r_2 .$$

Observation 2.6. *Let φ be a faithful representation of G and H_1 be a principal isotropy subgroup of φ. We define a series of subgroups and representations as follows*

$$\varphi|H_1 = (\mathrm{Ad}_G|H_1 - \mathrm{Ad}_{H_1}) + r_1\theta = \varphi_1 + r_1\theta .$$

H_{i+1} is a principal isotropy subgroup of φ_i, which is a representation of H_i and $\varphi_i|H_{i+1}=(\mathrm{Ad}_{H_i}|H_{i+1}-\mathrm{Ad}_{H_{i+1}})+r_{i+1}\cdot\theta=\varphi_{i+1}+r_{i+1}\cdot\theta$. We observe the following properties:

a) *It is not difficult to see that H_i is a principal isotropy subgroup of the representation $i\cdot\varphi$ (by the slice theorem and the principal orbit type theorem). In fact, this is the main reason to introduce such a series.*

b) *Since φ is faithful, φ_i are all faithful and hence $\dim\varphi_i\neq 0$ unless $H_i=\{e\}$.*

c) $\dim\varphi_i=\dim\varphi_{i+1}+r_{i+1}$ *and if* $\dim\varphi_i\neq 0$, *then* $r_{i+1}>0$. *Hence* $\Sigma r_i\leqslant\dim\varphi$ *and* $H_i=\{e\}$, $\dim\varphi_i=r_i=0$ *for* $i>\dim\varphi$.

d) *The codimension of the principal orbits of $k\cdot\varphi$ is given by*

$$k\cdot r_1+(k-1)r_2+\cdots+r_k.$$

Observation 2.7. *Let $\varphi:G\to SO(n)$ be a given faithful representation with $(n-r)$-dimensional principal orbit type. If G' is a subgroup of $SO(n)$, such that $\varphi(G)\subseteq G'\subseteq SO(n)$, then the principal orbit type of the G'-action is of dimension bigger than or equal to $(n-r)$.*

Observation 2.8. *For a given homogeneous space G/H, G is said to be* irreducible transitive *if there is no proper normal subgroups of G which are also transitive on G/H.*

Observation 2.9. *Suppose φ is an irreducible real representation of a compact connected Lie group $G=G_1\times G_2$. Then*

$$\varphi|G_1=k_1\cdot\varphi_1,\qquad \varphi|G_2=k_2\cdot\varphi_2,$$

where φ_1, φ_2 are irreducible real representations of G_1, G_2 respectively and $\dim\varphi\leqslant\dim\varphi_1\cdot\dim\varphi_2\leqslant 4\cdot\dim\varphi$.

With the above observations, it is not difficult to determine all possible linear actions on $\mathbb{R}^n$ with *codimensional r principal orbit type for $r\leqslant\frac{1}{4}\sqrt{n}$.* We shall roughly indicate the procedure.

Determination of linear models with low codimensional principal orbit type:

a) By Observation (2.5), we may reduce the problem to the consideration of irreducible linear actions first, since at least one irreducible component must satisfy our condition.

b) Suppose φ is an irreducible real representation of a compact connected simple Lie group G with codimension r principal orbit type. If $r\leqslant\frac{1}{4}\sqrt{\dim\varphi}$, then the following are the only possibilities:

1. $G=SU(k)$, $k\geqslant 8$; $\varphi=\mu_k$, $\dim\mu_k=2k$; $r=1$.
2. $G=SO(k)$, $k\geqslant 16$; $\varphi=\rho_k$, $\dim\rho_k=k$; $r=1$.
3. $G=Sp(k)$, $k\geqslant 4$; $\varphi=\nu_k$, $\dim\nu_k=4k$; $r=1$.
4. $G=\mathrm{Spin}(9)$, $\varphi=\Delta_9$, $r=1$, $\dim\Delta_9=16$.

c) Suppose φ is an irreducible real representation of G. We shall use $r(\varphi)$ to denote the *codimension* of the principal orbit type of the linear action φ. If $G = G_1 \times G_2$, then, by Observation (2.9), we have two induced irreducible representations φ_1, φ_2 of G_1, G_2 respectively, such that

$$\varphi | G_i = k_i \cdot \varphi_i, \qquad \dim\varphi \leqslant \dim\varphi_1 \cdot \dim\varphi_2 \leqslant 4\dim\varphi.$$

Suppose that $\dim\varphi_1 \geqslant \dim\varphi_2$. Then $\dim\varphi_1 \geqslant \sqrt{\dim\varphi}$, $k_1 \leqslant \dim\varphi_1$. Let G' be the normalizer of $\varphi(G_1)$ in $SO(\dim\varphi)$. Then clearly, $G' \supseteq \varphi(G_1 \times G_2)$. Then, by Observation (2.7), one has $r(\varphi) \geqslant r(\varphi')$ where $\varphi' : G' \to SO(\dim\varphi)$. On the other hand, it is not difficult to compute G' by SCHUR's lemma. Here, we shall treat three different cases separately.

1. If the complexification of φ_1 is still irreducible, then

$$G'_0 \sim \varphi(G_1) \times SO(k)$$

and if we apply Observation (2.6), we get

$$\begin{aligned} r(\varphi') &\geqslant r(k_1 \cdot \varphi_1) - \dim SO(k_1) \\ &= [k_1 r_1 + (k_1 - 1) r_2 + \cdots + r_{k_1}] - \frac{k_1}{2}(k_1 - 1) \\ &= S_{k_1} + (S_{(k_1-1)} - (k-1)) + \cdots + (S_j - j) + \cdots + (S_1 - 1) \end{aligned}$$

where $S_j = \sum_{i=1}^{j} r_i \geqslant j$ and $S_j = j$ only when $r_i = 1$ for $1 \leqslant i \leqslant j$. Hence $\frac{1}{4}\sqrt{k_1 \cdot n_1} \geqslant r(\varphi) \geqslant S_{k_1} \geqslant k_1$ implies $n_1 = \dim\varphi_1 > k_1$. Moreover, $r_j = 0$ for certain j will imply, by Observation (2.6), $S_{k_1} \geqslant S_j = n_1$ which is impossible. Hence, we have

$$\tfrac{1}{4}\sqrt{k_1 \cdot n_1} \geqslant r(\varphi) \geqslant k_1 \cdot r_1, \qquad n_1 \geqslant 16 k_1$$

or, $r_1 < \frac{1}{4}\sqrt{n_1 / k_1}$, that is, either k_1 is comparatively much smaller than n_1 or r_1 is very small.

2. The complexification of φ, $\varepsilon^* \varphi = \psi \oplus \psi^*$ and $\psi \neq \psi^*$.
3. $\varepsilon^* \varphi = 2\psi$.

For both cases, a similar estimate may be obtained by SCHUR's lemma and Observation (2.6). Instead of including them here, we shall list all possible irreducible linear actions on $\mathbb{R}^n$ with codimension r principal orbit type for the range

$$r \leqslant 3 \quad \text{and} \quad n \geqslant 16 r^2.$$

The case $r = 1$ is already listed in b).

The case $r = 2$; $n \geqslant 64$.

1. $G = SO(k) \times SO(2)$; $\varphi = \rho_k \oplus \rho_2$; $\dim\varphi = 2k$,
2. $G = SU(k) \times SU(2)$; $\varphi = \varepsilon_*(\mu_k \oplus_{\mathbb{C}} \mu_2)$; $\dim\varphi = 4k$,
3. $G = Sp(k) \times Sp(2)$; $\varphi = \nu_k \oplus_{\mathbb{Q}} \bar{\nu}_2$; $\dim\varphi = 8k$.

The case $r=3$; $n \geqslant 144$.

1. $G=SO(k)\times SO(3)$; $\varphi=\rho_k \oplus \rho_3$; $\dim\varphi = 3k$,
2. $G=SU(k)\times SU(3)$; $\varphi=\varepsilon_*(\mu_k \oplus_{\mathbb{C}} \mu_3)$; $\dim\varphi = 6k$,
3. $G=Sp(k)\times Sp(3)$; $\varphi=\nu_k \oplus_{\mathbb{Q}} \bar{\nu}_3$; $\dim\varphi = 12k$.

It is not difficult to work out the next two cases for $r=4,5$. Based on the above knowledge of linear models with low codimensional principal orbit type, it will be quite natural to look at the following problems:

Problem 8. Classify all differentiable actions of compact connected Lie groups on $\mathbb{R}^n$ when the codimension of the principal orbit type is 3 or 4 for $n \geqslant 150$. For a problem of this kind, the proof of the existence of a fixed point usually constitutes a difficult central step.

Problem 9. Classify all differentiable actions of compact connected Lie groups on S^n with the codimension of the principal orbit type 2 or 3 for $n \geqslant 150$. If the group G is given beforehand to be classical, the answer of Problem 9 is included in the results of [46, 47].

E. Miscellaneous

We include here some linear actions which are useful for some problems in transformation groups.

1. $G=SO(n)$, $\varphi=(S^2\rho_n-\theta)$: The natural action of $SO(n)$ on quadratic forms of n real variables with trace zero. This is the only irreducible representation of $SO(n)$ with the $\mathbb{Z}_2$-maximal tori as its principal isotropy subgroups. Hence, in some sense, it is the $\mathbb{Z}_2$-adjoint representation of $SO(n)$. One of the important properties of such linear actions is that the induced action on the unit sphere, S^m, $m=(\dim\varphi-1)$, admits an equivariant map $F: S^m \to S^m$ with $\deg F = 1+(-1)^n$.
2. $G=SU(n)$, $\varphi=\mathrm{Ad}_{SU(n)}$: This also has the above property. For an explicit construction of F and its applications, see Chapter I, 46.

III. Construction of Differentiable Examples

A general picture that one may get from the discussion of the geometric behavior of linear actions in II is the following: There are many varieties of linear actions of a given compact Lie group G and the geometric structures of linear actions are generally quite complicated. However, linear actions satisfying "suitable" geometric conditions are rather few in number and their geometric structures are relatively simple and well unterstood.

Naturally, one expects that there are many, many more varieties of differentiable G-actions on euclidean spaces and spheres, and the geometric structure of differentiable actions is generally far more complicated than that of the linear actions. Hence, the construction of more and more varieties of differentiable examples satisfying various geometric properties constitutes an important, indispensable part of the understanding of transformation groups. In fact, this is exactly one of the two mutually complementary directions of pursuing the so-called comparison principle that we mentioned in the introduction. Although there are already many interesting non-orthogonal examples, they are still rather scattered. Hence, one needs a more systematic approach to the construction of new examples with considerable generality and deeper insights.

In this section, we shall discuss briefly some basic methods of constructing differentiable examples. Since there is already an excellent review article on various non-orthogonal differentiable actions on spheres by G. BREDON in these Proceedings, we refer to that article for the actual descriptions of those examples. In fact, BREDON'S article and this section are essentially written from two complementary viewpoints and hence it seems to be a good idea to read both of them together.

A. Equivariant Thickening

The idea of equivariant thickening was first used by FLOYD to construct actions of various finite groups on euclidean spaces without fixed points. Historically, FLOYD'S example [35] was the first example to show that the beautiful results of P. A. SMITH [10, 103] about actions of p-groups could not be generalized to other finite groups such as $\mathbb{Z}_6$. Later, by using the same idea of equivariant thickening but with more elaborate *group theoretical* arguments, examples of actions on euclidean spaces without fixed points were constructed for *all non-primary finite groups* and *all non-commutative connected compact Lie groups* [29, 114, 46].

The usefulness of equivariant thickening is based on the following observation: Suppose X is a *closed invariant subspace* of a *differentiable* G-space M. If it is possible to choose a suitable *equivariant* neighborhood $N(X)$ such that $X \subseteq N(X)$ is a homotopy equivalence, then $N(X)$ is called an *equivariant thickening* of X. Furthermore, if one chooses $N(X)$ small enough, then, by the semi-continuity of the orbit structure of M, every orbit in $N(X)$ is *less singular* or of the *same singularity as* some orbit in X. In particular, if X has no fixed point then one may choose $N(X)$ to be also without fixed points.

In the case of constructing differentiable examples on euclidean spaces without fixed points, one first constructs a contractible topological G-space without fixed points X and then one tries to embed X into a differentiable G-space M with an equivariant thickening $N(X)$. In this way, one gets an open contractible differentiable G-space, $N(X)$, without fixed points, and then it is quite easy to see that $N(X) \times R^k$ is a euclidean space with a differentiable G-action without fixed points.

B. Equivariant Plumbing and Equivariant Connected Sum

For certain interesting classes of manifolds, such as exotic spheres that bound parallelizable manifolds, one knows how to construct each element of the class from well understood manifolds by means of simple procedures such as plumbing, connected sum, etc. Hence, a natural way of constructing some actions on those manifolds is to perform the procedures equivariantly, if possible, with respect to certain group actions. For example, suppose M_1 and M_2 are two differentiable G-spaces of the same dimension, and p_1, p_2 are fixed points on M_1, M_2, respectively. Then the connected sum of M_1 and M_2 at p_1, p_2 can be performed equivariantly if and only if the local representations around p_1 and p_2 are equivalent.

Homotopy spheres in bP_{4k} or bP_{4k+2} can be constructed as the boundaries of suitable plumbing-manifolds. It is not difficult to see that the plumbing of finitely many tangent disc bundles of S^m, *say* $E_1^{2m}, \ldots, E_s^{2m}$, can be performed equivariantly with respect to the natural $O(m)$ action on E_i^{2m} if each E_i^{2m} only plumbs with at most two others; otherwise, the plumbing can be performed equivariantly with respect to the natural $O(m-1)$ action on E_i^{2m}. In fact, differentiable actions obtained by equivariant plumbing are very important for the study of transformation groups on homotopy spheres and they also possess many interesting properties. We refer to [16, 17, 40, 45, 57] for detailed discussion of their properties and significance.

Let us mention another simple but important example of such constructions. As a consequence of the h-cobordism theorem of SMALE, one knows that the product of a compact contractible manifold X^n with a disk D^k is diffeomorphic to D^{n+k} if $n+k \geqslant 6$. Hence, one may construct differentiable G-actions on $D^{n+k} = X \times D^k$ such that $g(x,y)=(x,gy)$. In particular, if G acts on D^k linearly via a representation φ and with only one fixed point, then the fixed point set of the above action restricted to $\partial(X \times D^k) = S^{n+k-1}$ is clearly ∂X. In this way, one gets differentiable actions on the standard sphere with various integral homology spheres as fixed point sets. (Notice that if $n \geqslant 4$, we have inifinitely many X^n with different $\pi_1(\partial X)$.) If one combines the above construction with equivariant connected sum, it is not difficult to see the following result: *If there is a differentiable action of a compact Lie group G on M^m, $m \geqslant 6$,* *with the dimension of the fixed point set* $\dim F \geqslant 3$, *then there are infinitely many differentiable G-actions on* M^m. See [115, 50] for details.

C. Equivariant Attachings

Consider the following general situation:

Let M be a differentiable G-space with k types of orbits, say G/H_1, $G/H_2, \ldots, G/H_k$. Suppose G/H_k is *one* of the most singular orbit types and $M_k = \{x \in M; G_x \sim H_k\}$ is the union of all orbits of G/H_k-type. Without loss of generality, one may assume that $M_k/G = X_k$ is connected. By choosing an invariant Riemannian metric on M (with respect to the given G-action), it is easy to see that there is an equivariant tubular neighborhood $N(M_k)$ of M_k such that $N(M_k) \to M_k$ is equivariantly diffeomorphic to the equivariant normal disc bundle $E_\varepsilon(v)$ of M_k in M. One observes that $M - N(M_k)$ is differentiable G-space with $(k-1)$ orbit types. Hence a *differentiable G-space M with k types of orbits can be considered as the equivariant attaching of the differentiable G-space,* $M - N(M_k)$, with $(k-1)$ *types of orbits and an equivariant disk bundle,* $E_\varepsilon(v)$, *over a differentiable G-space* M_k *with only one type of orbit.*

The above observation points out an inductive procedure for building differentiable G-spaces with more and more orbit types. In order to make the above statement clearer and more precise, let us introduce a few more observations.

1. Suppose M is a differentiable G-space with only one type of orbit, say G/H. Then $G/H \to M \to M/G$ is a smooth bundle with

$$N(H)/H \to F(H,M) \to M/G$$

as its principal bundle, where $N(H)$ is the normalizer of H in G and $F(H,M)$ is the fixed point set of H in M.

Conversely, if $G/H \to M \to X$ is a smooth bundle with $N(H)/H$ as its structural group, then M can be made into a G-space with X as its orbit space in a unique way. [See Chapter XII of 12.]

2. Let Y be a connected differentiable G-space with all orbits of the same type, say G/H. Suppose v: $\mathbb{R}^n \to E(v) \xrightarrow{\pi} Y$ is a G-equivariant vector bundle over Y in the sense that $E(v)$ is a differentiable G-space with G acting as bundle map and suppose π is equivariant. Let $F(H,Y)$ be the fixed point set of H in Y, and let $v' = v|F(H,Y)$ be the restriction of v over the part $F(H,Y)$. It is clear that v' is naturally an H-equivariant vector bundle over $F(H,Y)$ with respect to the trivial action of H on $F(H,Y)$. Since Y is connected, the linear actions of H on the fibres $\mathbb{R}^n_x$, over $x \in F(H,Y)$ are clearly independent of x, and hence give a unique representation of H:

$$\psi = \psi_x : H \to O(n).$$

Since ν' is H-equivariant with respect to the trivial action on the base space $F(H,Y)$, it is not difficult to see that the structural group of ν' is reducible to the centralizer, Z_ψ, of $\psi(H)$ in $O(n)$, or equivalently ν' is associated with a principal Z_ψ-bundle $\hat{\nu}'$: $Z_\psi \to E(\hat{\nu}') \to F(H,Y)$.

Clearly, the representation ψ and the principal Z_ψ-bundle $\hat{\nu}'$ constitute the complete set of invariants of the H-equivariant vector bundle ν', and ν' is in turn an invariant of the G-equivariant vector bundle ν over Y. In fact, ν' is already a complete invariant since one may recover the G-equivariant vector bundle ν from ν' uniquely as follows: Namely $E(\nu)=E(\nu')\times_{N(H)/H} G/H$.

3. Given a differentiable G-space M with k types of orbits, say $G/H_1,\ldots,G/H_k$, let M_j be the union of all orbits in M of G/H_j-type and $X_j=M_j/G$. Then by 1., one has the following fibre bundle ξ_j and its associated principal bundle $\hat{\xi}_j$:

$$\xi_j\colon G/H_j \to M_j \to X_j,$$
$$\hat{\xi}_j\colon N(H_j)/H_j \to F(H_j; M_j) \to X_j.$$

Obviously, the bundles $\{\xi_j\}$ (or equivalently, their associated principal bundles $\{\hat{\xi}_j\}$) are important invariants of the G-space M. However, there are far more complicated invariants that tell us how those "parts" $\{M_j\}$ are pieced together *equivariantly*. A *workable formulation* of *the invariants of "equivariant attachings"* in terms of bundle reductions, etc, is undoubtedly one of the important steps towards a proper understanding of the structure of differentiable G-spaces. In some special interesting cases with two types or three types of orbits, such a formulation was given in [46] and [57] together with some interesting applications.

Remarks. a) In the formulation of invariants for equivariant attachings, one may assume without loss of generality that the orbit types have an absolute minimal type, G/H_k, i.e., $G/H_i \geqslant G/H_k$.

b) As we observed in the beginning of this sub-section, M can be considered as the equivariant attaching of $\overline{M-N(M_k)}$ with an equivariant normal disk bundle $E_\varepsilon(\nu)$ along the boundary $\partial E_\varepsilon(\nu)$. On the other hand, $\overline{M-N(M_k)}$ has only $(k-1)$ orbit types and according to 2., the equivariant vector bundle ν is uniquely determined by the reduced slice representation $\psi\colon H\to O(n)$ and the associated Z_ψ-bundle ν'. Hence it seems rather natural to proceed inductively on the number of orbit types.

c) As a preliminary to the study of the k-th stage attaching, one shall study fixed point free orthogonal actions on spheres with $(k-1)$ orbit types. In general, such representations are very rare if k is small and the representation ψ is usually uniquely determined by its $(k-1)$ orbit types.

d) In the study of *differentiable G-spaces*, the point of view of equivariant attaching seems to be the most natural one and a workable formulation of attaching invariants will certainly offer a classification of G-spaces up to equivariant diffeomorphisms in terms of those invariants. However, in the study of *transformation groups*, as we have pointed out in the introduction, one is primarily interested in differentiable actions on a *given* total manifold M, or equivalently, the geometric behavior of compact subgroups of Diff(M) of a given M. Hence, one is mainly interested in constructing a G-space with *a certain given type of total space*. In general, if one just goes ahead and builds G-spaces with some orbit types over some orbit spaces by attaching them equivariantly, it is rather unlikely that one might accidentally end up with the given total space. In fact, the probability of such an accident should be zero theoretically. Hence, the study of possible orbit structures, such as the determination of orbit types and orbit spaces, etc., is still a major part of the study of differentiable transformation groups, and the understanding of equivariant attachings is useful at the final stage of verifying the possibilities. For concrete examples to illustrate the above point of view, see [43, 46, 50].

IV. Regularity Conditions and Regularity Theorems

In this section, we shall try to give a survey of an important part of the "positive" results in differentiable transformation groups that we propose to call *regularity theorems*. However, mainly due to the inadequate knowledge of the author and partially due to the lack of time, there will be inevitably many undesirable omissions. As we have explained in the introduction, the major omissions are a) results in the cohomology theory of actions of elementary abelian groups, b) results in the cobordism theory of periodic differentiable maps, and c) results on free action of finite groups and the circle group. We refer to [12] and [30] for systematic accounts of results in the cohomology theory and cobordism theory of group actions respectively.

Considerable emphasis will be given to results about actions on important natural testing spaces such as spheres, euclidean spaces and homogeneous spaces. Here, the central theme seems to be the following: *differentiable actions satisfying certain suitable geometric conditions must resemble some natural models up to a certain extent.* We shall call those geometric conditions *regularity conditions.* In fact, if one reflects on the complexities of *general* linear actions and the vast varieties of differentiable actions, it seems clear that the study of "regular" differentiable acrions on manifolds with natural symmetries is probably one of the most natural and hopeful approaches towards the unterstanding of differentiable transformation groups.

Suggested by the results of II, the following are some examples of regularity conditions:

a) The dimension of the total space is low as compared with the dimension of the group.

b) The number of orbit types is not too large.

c) The principal orbit type is of lower dimension than that of the group.

d) The codimension of the principal orbit type is low as compared with the dimension of the total space.

Surely, one may formulate many other regularity conditions, but there are two obvious criteria for the usefulness of a given regularity condition: Namely, 1. the validity of the given regularity conditions can be verified in quite a few interesting cases, 2. the possibility of deducing strong theorems from the given conditions. However, the second criterion is usually difficulat to foresee. A reasonable and workable substitute of 2. is; 2'. the possibility of deducing strong consequences for those *natural* actions satisfying the given conditions.

We shall organize the known regularity theorems according to their corresponding regularity conditions. This section will begin with a short summary of basic generalities on orbit structures and end with a brief discussion on some possible generalizations and further developments.

A. Basic Theorems on the Orbit Structure of Differentiable Actions of Compact Lie Groups

The basic reason that differentiable actions of *compact Lie groups* are far more *regular* than those of *non-compact Lie groups* is the *existence* of *invariant Riemannian metrics* for actions of compact groups. An important fundamental theorem that follows directly from the existence of invariant metrics is the following theorem of MONTGOMERY and YANG [72].

The Differentiable Slice Theorem: *Let* $G(x) \supseteq M^m$ *be the orbit of* $x \in M^m$, G_x *be the isotropy subgroup of* x, *and* ξ *the canonical* G_x*-bundle* $G_x \to G \to G/G_x = G(x)$. *We may assume that* M^m *is already equipped with an invariant Riemannian metric. Then the action of* G_x *induces a natural representation* ψ_x *on the normal vectors of* $G(x)$ *at* x. *(This representation is usually called the slice representation of* G_x *at* x.*) Then the normal bundle of* $G(x)$ *is given by*

$$\nu(G(x)) = \alpha_\xi(\psi_x)$$

and the usual exponential map will give an equivariant diffeomorphism of a sufficiently small disk bundle of $\alpha_{\xi}(\psi_x)$ *onto an equivariant tubular neighborhood of* $G(x)$ *in* M^m.

The geometric significance of the above theorem is that the local *behavior around an orbit* $G(x)$ is completely given by the *slice representation*. Hence, the slice theorem is, in a sense, an ultimate generalization of the BOCHNER theorem [5] and it completely solves the *local classification* of differentiable G-spaces. This is also one of the places that the theory of vector bundles and group representation theory naturally come in as tools. [See 43, 46, 50, 51.]

The following are some of the direct consequences of the slice theorem that are basic in the study of orbit structures.

1. *Finiteness of orbit types*. The number of orbit types of a compact differentiable G-space is always finite [85].

2. *Triangulation of orbit space*. Suppose M^m is a differentiable compact G-space and $(H_1),(H_2),\ldots,(H_k)$ are the finite orbit types of M^m. Then, it is clearly possible to arrange (H_i) in such order that $(H_j)\geqslant(H_i)$ implies $j\geqslant i$; and hence, the orbit space $X=M^m/G$ is filtered by closed subspaces $X=X_1 \not\supseteq X_2 \not\supseteq \cdots \not\supseteq X_k \not\supseteq \phi$ such that (X_i-X_{i+1}) is exactly the images of those orbits of type (H_i). It was proved in [111] that $X=M^m/G$ possesses a triangulation such that X_i are subcomplexes.

3. *Fary spectral sequence*. It has been shown in [51, p. 132] that if we mix the above "orbit type filtration" $\{X_i\}$ with the skeleton filtration $X^{(j)}$ of X with respect to suitable triangulation, we get a Fary spectral sequence with

$$E^1=\Sigma\, C_j(X_i, X_{i+1})\otimes H_*(G/H_i)$$

and

$$E^2=\Sigma\, H_j(X_i, X_{i+1}\,;\, \underline{H_*(G/H_i)}).$$

Such a spectral sequence should be useful in the study of global orbit structures. However, reasonable understanding of this spectral sequence is lacking even in the case of *linear actions on spheres*.

4. *Principal orbit type theorem* [70, 73]: Among all orbit types of a given connected differentiable G-space M, there is an absolute minimal orbit type $(H)\leqslant(H_x)$ for all $x\in M$. This orbit type is called the *principal orbit type* of the G-space M. Moreover, the *union of all principal orbits* forms an *open everywhere dense submanifold*.

It seems proper to mention here one classical origin of the above theorem. Namely, if one considers the adjoint action of a compact connected Lie group G on its Lie algebra $\mathfrak{G}$, then it is not difficult to see that a point $x\in\mathfrak{G}$ is regular if and only if the orbit of x, $G(x)$, is principal. Hence, in fact, the principal orbit type theorem is essentially a generalization of the maximal tori theorem of CARTAN.

Finally, let us mention the beautiful result of equivariant embeddings by MOSTOW and PALAIS which may be regarded as a profound application of the slice theorem.

Equivariant Embedding Theorem (MOSTOW [84], PALAIS [90]). *Let G be a compact Lie group and X be a separable, metrizable G-space of finite orbit type and finite dimension. Then there exists a high dimensional euclidean space $\mathbb{R}^N$ with a suitable linear G-action such that X admits an equivariant embedding into the linear G-space $\mathbb{R}^N$.*

Remarks. 1. The above result fully reveals the complexities of the set of all linear G-spaces. It shows that, generally, linear G-spaces are complicated enough to include all reasonable G-spaces as sub-G-spaces.

2. It is a beautiful general theory. However, there are almost no controls over the orbit structures and the dimension of the ambient linear space.

B. Differentiable Actions on Euclidean Spaces or Spheres with Low Co-dimensional Principal Orbit Types

1. *Differentiable G-spaces with codimension one principal orbit type:* Since the only *linear actions* with codimension one principal orbit type are those linear representations that are transitive on the unit sphere, it is then quite clear from the slice theorem that the *orbit space*, M/G, of a differentiable G-space with codimension one principal orbit type is naturally a *one-dimensional manifold* with the image of the singular orbits as boundary points. Hence, if M is compact, there are only two possibilities, M/G is a closed interval or a circle. In the latter case, M is a G/H-bundle over S^1 with $N(H)/H$ as its structural group, and in the former case M is completely determined by its principal orbit type G/H and the two singular orbits, say G/K_1 and G/K_2, corresponding to the two end points of M/G. If M is noncompact, then M/G is either a half open interval or an open interval. Again, it is quite clear that $M = G/H \times \mathbb{R}^1$ in the latter case, and in the former case M is equivalently diffeomorphic to the normal bundle of the *unique singular orbit*. (See [82] for detailed discussion).

Hence, in a sense, the above observation shows that smooth G-spaces are "classified" by their *orbit structures* up to equivariant diffeomorphism. However, as we have pointed out in the introduction, one is *mainly interested in group actions on a fixed total space.* Hence, in fact, we shall start with manifolds M of certain special types such as spheres, euclidean spaces, and then proceed to determine all possible groups G and their respective orbit types, say G/H, G/K_i, such that the total space we build in accordance with the above observation is the given space M. The cases that M are euclidean spaces and spheres have been worked out in [77] and [108] respectively. Their results may be briefly stated as follows:

a) *The case* $M \approx \mathbb{R}^m$ [76, 77]: Since $\mathbb{R}^m$ cannot be written as $G/H \times \mathbb{R}^1$, there must be a unique singular orbit, say G/K, and $\mathbb{R}^m$ is the *total space of the normal bundle of* G/K. It is clear that G/K must be homotopic to $\mathbb{R}^m$ and hence is *a point*. Then it is obvious that $G/H \approx S^{m-1}$ and the action is linear.

b) *The case* $M \approx S^m$ [108]: Since it is impossible to fibre S^m over S^1, S^m/G must be a closed interval. Hence, the job is to determine the *two singular orbits* and the *principal orbit type*. Naturally, it will involve many computations of Lie groups and the topology of homogeneous spaces in order to carry out such a determination. According to [108], if m is even and $\neq 4$ or is odd and $\geqslant 31$, then any differentiable compact connected transformation group with codimension one orbits is in fact linear. However, as was pointed out in [45], there exist infinitely many non-equivalent smooth non-linear actions of $SO(n) \times SO(2)$ on S^{2n-1} $(n \geqslant 3)$ with codimension one orbits. It is rather surprising that there are also infinitely many non-equivalent actions on Kervaire spheres with codimension one orbits [50]. The existence of such actions, on the one hand, shows that a simple construction in transformations groups may lead to some highly interesting manifold such as the Kervaire sphere, and on the other hand, shows that some "exotic" spheres are in fact quite symmetric [see 50, 40, for detailed discussion].

2. *Differentiable actions on euclidean spaces and spheres with codimension two principal orbit type:* a) It was proved in [112] that if a compact connected Lie group acts *differentiably* on $\mathbb{R}^m$ with an $(m-2)$-dimensional orbit then the G-action is differentiably equivalent to a suitable *linear action*. As usual, one of the central steps is to show the existence of a fixed point. Note that although the differentiability *was needed only in the last step* of its proof, *the above theorem is not true without differentiability*. For example, we may take any non-orthogonal differentiable action of $SO(n) \times SO(2)$ on the unit sphere S^{2n-1} of $\mathbb{R}^{2n}$ and extend it to a *topological action* on $\mathbb{R}^{2n}$ by the obvious cone construction. Then it is not difficult to see such an action is not topologically equivalent to any linear action.

b) If we combine the method of Chapters IV, V of [46] with the above result of [112], it is not difficult to show that a differentiable action of a compact connected Lie group on a sphere is equivalent to a suitable linear action if there *exist at least one fixed point* and an orbit of codimension two. However, it seems to me the above fact is by no means a trivial consequence of the result of [112].

In the general case, suppose G is a compact connected Lie group that acts on S^m, $m \geqslant 4$, with codimension two principal orbits. Then it is not difficult to see from the slice theorem and some cohomology arguments that the orbit space S^m/G is naturally a two-dimensional disk with the boundary points corresponding to the singular orbits. In general, the singular orbits may have several different types. The special case that all singular orbits are of the same type was classified by BREDON [16, 17]. In the case $m=4k+1$, there are infinitely many differentiably non-equivalent actions of $SO(2k+1)$ $(k>1)$ on S^{4k+1} as well as on $(4k+1)$-Kervaire spheres (see [45] for the determination of the diffeomorphy classes of BREDON's examples).

In view of the results of D., II., it seems to me a *classification* of all differentiable actions of compact connected Lie groups on S^m with codimension two principal orbit type should be possible *at least for* $m \geqslant 63$. We shall discuss the possible answer of such a classification at the end of this section.

3. *Other results.* There are also some scattered results about differentiable actions of compact connected Lie groups on euclidean spaces and spheres with codimension three principal orbit types [113]. The general picture seems to be the following:

If the codimension, r, of the principal orbit type of a differentiable action of a compact connected Lie group G on a euclidean space $\mathbb{R}^m$ is far smaller than m, such as $r<\frac{1}{4}\sqrt{m}$, then it seems to be possible to show the existence of at least one fixed point and to determine the possibilities of all such groups G (cf. D. of II).

C. Differentiable Actions on Spaces of Comparatively Low Dimensions

This type of *regularity condition* was first suggested to the author by the following theorem of classical Riemannian geometry:

Theorem (p. 33, 239). *Let M^m be an m-dimensional Riemannian manifold. Then the isometry group of M^m is a Lie group and*

$$\dim ISO(M^m) \leqslant \tfrac{1}{2} m(m+1).$$

We may reformulate the above theorem as follows:

Theorem 6. *Let M^m be an m-dimensional smooth manifold and G be any compact subgroup of* $\mathrm{Diff}(M^m)$. *Then*

$$\dim G \leqslant \tfrac{1}{2} m(m+1)$$

and $\dim G = \frac{1}{2} m(m+1)$ *only when* $M^m = S^m$ *or* P^m *and the action of G is equivalent to the natural action of the orthogonal group.*

The above theorem suggests that differentiable actions of large compact Lie groups are more *regular* than those of small compact Lie groups. Of course, we have to introduce more sophisticated tools such as group representation theory and vector bundle theory in order to exploit effectively the "advantages" of large group actions.

A trial along this direction was first worked out in [50] for the technically simplest case that M^m is a simply connected stably parallelizable manifold and $G=SO(n)$, $n>\frac{1}{2}(m-1)$. A satisfactory classification was obtained for such actions and quite a few applications may be deduced directly from the classification theorems. We refer the reader to [50] for details.

Later, in Chapters II and III of [46] more systematic applications of characteristic classes, representation theory and P. A. Smith theory to the study of large classical group actions were developed. We quote some of the main results as follows:

Theorem 7. *Let Φ be a differentiable action of $SO(n)$ on an m-dimensional manifold M^m, where $n\geqslant 11$ and $m\leqslant(n-1)^2/4$. If the first rational Pontryagin class of M^m, $P_1(M^m)$, vanishes, then the identity component of any isotropy subgroup, $G_x^0, x\in M^m$, is always conjugate to a standardly imbedded $SO(k)$ for $k\geqslant 2n/3$.*

Similarly, in the cases $G=SU(n)$ or $Sp(n)$ and $n\geqslant 8$, $m\leqslant(n-1)^2/2$ or $m\leqslant(n-1)^2$, respectively, if $P_1(M^m)=0$, then the identity component of any isotropy subgroup, $G_x^0, x\in M^m$ is always conjugate to a standardly imbedded $SU(k)$ *or* $Sp(k)$, $k\geqslant 2n/3$, respectively.

Theorem 8. *Let G be $SU(n)$, $(n\geqslant 8)$; or $Sp(n)$, $(n\geqslant 8)$; or $SO(n)$, $(n\geqslant 12$ and even$)$. If M^m is a compact differentiable G-space with* $\dim M^m = M<\frac{2}{5}(\dim G)$ *and $P_1(M^m)=0$, then*

$$F(G,M^m)=F(T,M^m) \quad \textit{and} \quad \chi(F(G,M^m))=\chi(M^m),$$

where T is a maximal torus of G. Furthermore, if M^m is totally non-homologous to zero in the fibration $M^m\to M^m\times_G E_G\to B_G$, then $\dim H^*(F(G,M^m);\mathbb{Q})=\dim H^*(M^m;\mathbb{Q})$. *In particular, $F(G,M^m)$ is necessarily non-empty.*

Theorem 9. *For a given differentiable action Φ of $SO(n)$ on a homotopy sphere Σ^m (respectively a euclidean space $\mathbb{R}^m$ or a disk D^m), where $n\geqslant 11$ and $m<\frac{1}{4}(n-1)^2$, we have that*

1. *all orbits are real Stiefel manifolds,*
2. *if $SO(n)/SO(k)$ is the principal orbit type and F is the fixed point set, then*

$$H^*(F;A)\cong H^*(S^r;A) \quad \textit{(respectively } H^*(F;A) \textit{ is acyclic)}$$

where $r = \dim F = m - n(n-k)$ *and* $A = \mathbb{Z}_2$ *if* n *is odd,* $A = \mathbb{Z}$ *if* n *is even,*

3. *the orbit space* M^m/Φ *is naturally a* $(k-1)$*-connected manifold* B *with boundary* ∂B*, where the image of singular orbits are boundary points.*

Theorem 10. *Let* Φ *be a differentiable action of* $SU(n)$ *on a homotopy sphere* Σ^m *(respectively a euclidean space* $\mathbb{R}^m$ *or a disk* D^m*) where* $n \geqslant 8$ *and* $m < \frac{1}{2}(n-1)^2$*. Then, we have*

1. *all orbits are complex Stiefel manifolds,*
2. *if* $SU(n)/SU(k)$ *is the principal orbit type and* F *is the fixed point set, then*

$$H^*(F;\mathbb{Z}) \cong H^*(S^r;\mathbb{Z}) \quad (\text{respectively } F \text{ is acyclic})$$

where $r = \dim F = m - 2n(n-k)$,

3. *the orbit space* Σ^m/Φ *is an at least* $(2k-1)$*-connected manifold* B *with boundary* ∂B*, where* ∂B *is the image of singulare orbits.*

Theorem 11. *Let* Φ *be a differentiable action of* $Sp(n)$ *on a homotopy sphere* Σ^m *(respectively a euclidean space* $\mathbb{R}^m$ *or a disk* D^m*). If* $n \geqslant 8$ *and* $m < (n-1)^2$*, then we have*

1. *all orbits are quaternionic Stiefel manifolds,*
2. *if* $Sp(n)/Sp(k)$ *is the principal orbit type and* F *is the fixed point set, then*

$$H^*(F;\mathbb{Z}) \cong H^*(S^r;\mathbb{Z}) \quad (\text{respectively } F \text{ is acyclic})$$

where $r = \dim F = m - 4n(n-k)$,

3. *the orbit space* Σ^m/Φ *is an at least* $(4k-1)$*-connected manifold* B *with its boundary* ∂B *being the image of singular orbits.*

The proofs of the above theorems are rather long. We refer to [46]. If one compares the discussions of A., II about low dimensional representations with the above results of [46], one sees that those differentiable actions on spaces of comparatively low dimension are quite regular. And under suitable topological conditions on the total spaces, such as vanishing first PONTRJAGIN class or acyclic cohomology, we can in fact show that they must resemble the low dimensional representations in many aspects. As one will see next in Section D., the results of [46] are not the best possible with respect to dimensional assumptions. It is not difficult to see from linear examples that the dimension restrictions on m may be improved by *at most* a factor of 2. In a succeeding paper [47], such an improvement was proved via a thorough discussion of the possibilities of principal orbit types of differentiable actions on euclidean spaces. (Cf. D. of IV.)

D. Differentiable Actions With Non-trivial Principal Isotropy Subgroups

In view of the principal orbit type theorem (cf. D. of 1. of IV), the principal orbits are, population-wise, the dominant type of orbits, and hence it is only natural that the type of principal orbits usually has great influence on all the other aspects of the differentiable action. Hence, in the study of differentiable actions of a *given group* G on a given smooth manifold M, a logical first step is to *determine the possibilities for the principal orbit type*. On the other hand, the strong results that we mentioned in B. of II show that there are only a few homogeneous spaces of a given group G which can be realized as the principal orbit type of suitable linear G-actions. Hence, if the total space M is in particular a euclidean space $\mathbb{R}^m$, one is tempted to hope that there are also only a few possibilities for the principal orbit type for differentiable G-actions on $\mathbb{R}^m$.

Affirmative results along this direction were first established [51] for differentiable actions of $SU(p)$ on manifolds M^m with

$$\chi(M^m) \not\equiv 0 \pmod p.$$

The technical reason for the selection of $SU(p)$, p odd prime, in that paper is the following group theoretic fact:

Proposition [51, p. 130]: *Let H be a closed subgroup of $SU(p)$, then*

$$\chi(SU(p)/H) \not\equiv 0 \pmod p$$

only when $H = SU(p)$ or H^0 is a maximal torus of $SU(p)$ and $\operatorname{ord}(H/H^0) \equiv 0 \pmod p$.

To carry out the determination of all the possibilities of principal orbit types of all possible differentiable actions of a given compact connected Lie group G on euclidean spaces is in general a very complicated and tedious job. However, the final answers after such a complicated and long analysis are surprisingly simple and usually constitute an important deep result for the study of differentiable actions of the given group G.

A systematic study of the above problem has been made in [47] for the classical groups. Since the methods and the results for $SO(n)$, $SU(n)$ and $Sp(n)$ are similar to one another, we shall only discuss the $SO(n)$ case and refer the $SU(n)$ and $Sp(n)$ cases to the original paper [§ 3 and § 4, Chapter II, 47]. The result for $SO(n)$-actions on euclidean spaces may be stated as follows:

Theorem 12. [Chapter II, 47]: *Let Φ be a differentiable action of $SO(n)$ on $\mathbb{R}^m$ with (H_Φ) its principal isotropy subgroup type. Then either*

$\dim H_\Phi = 0$ *or* H_Φ^0 *is a maximal torus, or* H_Φ^0 *is conjugate to a standard* $SO(k)$ *in* $SO(n)$.

Once the possibilities of principal orbit types for differentiable actions of a given group G on euclidean spaces is determined, it is then quite natural to see how the principal orbit type influences the action as a whole. Generally speaking, a differentiable G-space M consists of two *mutually complementary invariant subsets*, namely, the union of all principal orbits M_0 and the union of all singular orbits M_s. Suppose H is a principal isotropy subgroup. Then M_0 is an open everywhere dense submanifold and $G/H \to M_0 \to M_0/G$ is a differentiable fibration associated with the principal $(N(H)/H)$-bundle

$$\hat{\xi}: N(H)/H \to F(H, M_0) \to M_0/G.$$

Hence, it is quite clear that the principal orbit type G/H together with the associated $(N(H)/H)$-bundle $\hat{\xi}$ constitute a set of invariants of the given G-space with dominant importance. In fact, if the principal isotropy subgroups (H) are non-trivial, then the principal orbit type G/H alone already constitutes a quite strong invariant that, usually, will force the action to be quite regular. In other words, differentiable actions of compact Lie groups with "non-trivial" principal isotropy subgroups are quite regular. For example, differentiable actions of compact connected. Lie groups on homotopy spheres, euclidean spaces and disks with S^k $(k \neq 1,3)$ as principal orbit type have been satisfactorily classified in [43]. A more systematic discussion of differentiable actions of classical groups on euclidean spaces with non-trivial principal orbit types has been made in Chapter III of [47]. Again, we shall only quote those results of the $SO(n)$ case and refer to Chapter III of [47] for the corresponding cases of $SU(n)$ and $Sp(n)$ actions.

Theorem 13. *Let* Φ *be a differentiable action of* $SO(n)$ *on a smooth manifold* M *with vanishing first Pontrjagin class. If the identity component of the principal isotropy subgroups,* H_Φ^0, *of* Φ *are conjugate to a standard* $SO(\ell)$, $\ell \geq 4$, *then we have*

$$G_x^0 \sim SO(k_x) \quad \textit{for all} \quad x \in M$$

where $SO(k_x)$ *is standard in* $SO(n)$.

Theorem 14. *Let* Φ *be a differentiable action of* $SO(n)$ *on euclidean space* $\mathbb{R}^m$. *If the identity component of the principal isotropy subgroup,* H_Φ^0, *of* Φ *is non-abelian and not 3-dimensional, then*

1. $H_\Phi = H_\Phi^0$ *is conjugate to a standard* $SO(k)$, $k \geq 4$, *in* $SO(n)$,
2. *there are exactly the following* $(n-k+1)$ *orbit types:*

$$\textit{point}, \quad SO(n)/SO(n-1), SO(n)/SO(n-2), \ldots, SO(k)/SO(k),$$

3. *$H^*(F;A) \cong H^*$(point; A), where F is the fixed point set of the $SO(n)$-action and $A=\mathbb{Z}$ if n is even, $A=\mathbb{Z}_2$ if n is odd,*

4. *the orbit space $\mathbb{R}^m/\Phi$ is naturally an at least $(k-1)$-connected differentiable manifold with boundary, where the boundary points are the images of singular orbits,*

5. $\dim F = m - n\cdot(n-k)$.

Remarks. 1. Theorems of the above type seem to suggest the following general picture of differentiable actions of compact Lie groups:

For a given compact Lie group G and manifold M of a *given type*, there are but few possibilities of principal orbit types for differentiable G-actions on M. Those G-actions with non trivial principal isotropy subgroup types are usually quite "*regular*".

2. Regularity theorems of this type clearly imply some even better regularity theorems than those we quoted in C. For example, the following generalization of the theorems of C. is a direct consequence of the above theorem:

Theorem 15. *Let Φ be a differentiable action of $SO(n)$ on $\mathbb{R}^m$. If $m \leqslant n/2(n-1)-[n/2]$, then the above theorem applies and the above conclusions (i)–(v) hold for Φ. Furthermore, since $\dim F = m - n\cdot(n-k) \geqslant 0$, we have $k \geqslant n/2+1$.*

E. Differentiable Actions with Few Orbit Types

As we have already pointed out in d), the union of all principal orbits, M_0, together with the differentiable fibration $G/H \to M_0 \to M_0/G$ constitutes a set of dominant invariants for the given G-space M. It was proved in [43, 50] by the "blowing up" procedure that the above invariants already constitute a complete set of invariants in the special case that M consists only of fixed points and S^k-type of orbits ($k \neq 1, 3$). On the other hand, if one looks at the examples of Bredon [16] from the point of view of Chapter IV of [46], it is not difficult to see that one may get infinitely many different $O(n)$-actions on spheres by pasting together the same pair of $O(n)$-spaces M_0 and M_s in different ways.

Roughly speaking, the global invariants of a differentiable G-space M may be divided into the following three parts:

1. The "regular set" M_0 together with the fibration $G/H \to M_0 \to M_0/G$.

2. The "singular set" M_s together with a partition of M_s into several fibrations corresponding to different singular orbit types.

3. How M_0 and M_s *paste together* equivariantly.

In fact, the singular set M_s may consist of quite a few different orbit types and M_s should be considered as the union of several invariant subsets $M_{s_1}, M_{s_2}, \dots, M_{s_k}$ according to orbit type. Hence, we shall also consider how the $\{M_{s_i}\}$ paste together equivariantly. If the number of orbit types is small, then the above invariants are reasonably simple. The cases of two types and three types was first worked out in [Chapter IV, 46] for the purpose of classifying differentiable actions of classical groups with few orbit types [see Chapter V, 46]. Later, such results were also obtained independently by K. JÄNICH [57], and then F. HIRZEBRUCH applied them to the study of singularities and exotic spheres [40].

Clearly, the number of invariants and their complexities increase rapidly with the number of orbit types. Hence, those G-spaces with fewer orbit types are simpler and easier to study. The above observations naturally suggest "regularity conditions" that require the *number of orbit types to be reasonably small.*

The first regularity theorem along this direction is the following theorem of A. BOREL.

Theorem 16 [12, p. 190]: *Let $X \sim_{\mathbb{Z}} \mathbb{R}^n$ and let the one-point compactification of X be a n-cm$_{\mathbb{Z}}$. If the action of a compact connected Lie group G on X has only two types of orbits on X, then it has a fixed point.*

The above theorem and the examples of [46] (cf. 1. of III) suggest the following problem:

Problem 10. Let G be a compact connected non-commutative Lie group. Then there is a well-defined positive integer, $\nu(G)$, depending on G, given by "$\nu(G)$ is the *greatest lower bound* of the numbers of orbit types of all possible differentiable G-actions on euclidean spaces *without fixed points*". As usual, such a number $\nu(G)$ is easy to define but is very difficult to compute. Hence, our problem is to get reasonable "estimates" for the number $\nu(G)$. For example, the above theorem of A. BOREL shows that $\nu(G) > 2$ for all G; and the computations of C., 2 seem to suggest that $\nu(G) > \frac{1}{2} rk(G)$.

It is clear from observation (2.2) that regularity theorems of (D) automatically imply some regularity theorems on differentiable actions with few orbit types. Here, again, we only quote the $SO(n)$ case and refer the $SU(n)$ and $Sp(n)$ cases to [§ 3, Chapter III, 47].

Theorem 17. *Let Φ be a differentiable action of $SO(n)$ on $\mathbb{R}^m$. If the number of orbit types of Φ is less than* $\mathrm{rk}(SO(n)) = [n/2]$, *then all orbits must be real Stiefel manifolds and hence the conclusion (i)–(v) of Theorem 14 holds for Φ. In particular, we have that*

$$\nu(SO(n)) \geqslant [n/2].$$

Remarks. In a sense, the above theorem and the corresponding theorems for $SU(n)$ and $Sp(n)$ show that the actions of classical groups with various Stiefel manifolds as orbit types are about the only subclass of differentiable actions of classical groups that are simple and nice enough so that the study of the global pasting invariants are still feasible. Hence the above theorem somehow justifies the selection of the so-called regular $O(n)$-manifolds in [58].

2. Although the theorems of [47] stated in d) and e) were proved for actions on $\mathbb{R}^m$, in fact the proofs will go through without change for all acyclic manifolds or even-dimensional homology spheres. In the case of odd dimensional spheres, due to the fact that the EULER characteristic of odd spheres vanishes, one has to modify the proofs slightly in order to prove the parallel theorems.

3. With results of the above type, it should not be difficult to classify actions of classical groups (reasonably large) on euclidean spaces and homotopy spheres with few orbit types. For example, the cases of three types on euclidean spaces and two types on homotopy spheres are essentially done [16, 47].

F. Differentiable Actions on Compact Homogeneous Spaces

Although spheres, euclidean spaces and disks are undoubtedly the best testing spaces for the study of differentiable transformation groups, they are obviously not the limitation of the available nice testing spaces. In fact, after one has understood the behavior of differentiable transformation groups on those "best testing spaces" reasonably well, it would be then an absolute necessity to broaden the domain of testing spaces. For example, the simplicity of the topology of spheres and euclidean spaces is a good thing to help us to get some nice results and some basic understandings to *begin* with. However, it is rather dangerous to indulge in such simplicities to the exclusion of other possibilities. Homogeneous spaces are obviously the natural candidates for such a purpose; they are naturally highly symmetric, but they cover a wide range of topological types.

Let us formulate the so-called *comparison principle* for the class of compact homogeneous spaces as follows:

Let G be a compact connected Lie group with a closed subgroup H of G, and let G/H be the homogeneous space of left cosets of H. We may assume without loss of generality that $\bigcap_{g\in G} gHg^{-1}=\{e\}$. Let Z be the centralizer of H in G. Then, clearly, $G\times Z$ acts on G/H differentiably in a natural way, namely,

$$(g,z)\cdot g_1H=gg_1Hz=gg_1zH \quad \text{for all} \quad g,g_1\in G \quad \text{and} \quad z\in Z.$$

The restrictions of the above $G\times Z$ action to its subgroups is the obvious analogy to linear actions on spheres. We shall call them "natural actions". Then one natural approach to the study of differentiable actions on G/H is again to *compare the geometric behavoir of general differentiable actions with those of the "natural actions"*. Again, it seems to be rather hopeful to try to show similar regularity theorems under suitable regularity conditions. Of course, there might be considerable difficulties due to the relatively more complicated topology type of homogeneous spaces. So far, very few results have been worked out along this direction. Let us list some of them as follows:

1. Transitive Actions

Classification of transitive actions on compact homogeneous spaces is essentially classification of compact homogeneous spaces in terms of their diffeomorphic types.

a) Transitive actions on spheres (homotopy spheres) was initiated by MONTGOMERY and SAMELSON [68] and later finished by A. BOREL and PONCET [10, 91]. The results may be summarized as follows: A homogeneous homotopy sphere must be standard and besides those well-known transitive actions of spheres, one has the following cases: $S^6 = G_2/SU(3)$, $S^7 = \mathrm{Spin}(7)/G_2$, $S^{15} = \mathrm{Spin}(9)/\mathrm{Spin}(7)$.

b) Transitive actions on integral homology spheres were classified by BREDON [19]. The answer is that there is only one homogeneous integral homology sphere, the Poincaré sphere, which is not standard.

c) Transitive actions on homogeneous spaces with non-vanishing EULER characteristic were *essentially* classified by H. C. WANG [106].

d) Transitive actions on many Stiefel manifolds (real, complex, or quaternionic) have been classified in [56]. The results are that all such transitive actions must be natural in the sense we defined above.

e) One may define the *rank* of a topological space X to be the sum of the ranks of all the groups $\pi_{2k-1}(X)$ $(k=1,2,\ldots)$. In [88], ONISCIK classified transitive actions on homogeneous spaces of rank 1 and 2.

Remarks. a) In order to classify homogeneous spaces in terms of their topological types or diffeomorphic types, it is important to find explicit relationships between the infinitesimal data of the group pair (G,H) (or equivalently the Lie algebra pair $(\mathfrak{G},\mathfrak{H})$) and the topological or diffeomorphic invariants of G/H. Examples of such relationships are: (For simplicity, assume H connected)

1. $\mathrm{Rk}(G) - \mathrm{Rk}(H) = -\Sigma(-1)^k \mathrm{Rk}(\pi_k(G/H))$.

2. Let $P(X,t)$ be the POINCARÉ polynomial of X. Then it was proved in [88], that

$$P(G,t)/P(H,t) = \prod_{k=1}^{\infty} (1+t^{2k-1})^{(r_{2k-1}-r_{2k})}$$

where $r_j = \mathrm{Rk}(\pi_j(G/H))$.

3. According to BOREL-HIRZEBRUCH [13], the rational PONTRJAGIN classes can be effectively computed in terms of the infinitesimal data, and it was recently proved by NOVIKOV that rational PONTRJAGIN classes are topological invariants. It seems to me that rational PONTRJAGIN classes are going to be of increasing inportance in the study of the topology of homogeneous spaces.

b) Technically, the difficulty of classifying transitive actions on G/H increases with the corank of $H = \text{Rk}\, G - \text{Rk}\, H$.

c) In view of the classification theorem of compact Lie groups, it seems rather natural to try first on those homogeneous spaces of classical groups.

Problem 11. Let G be a classical group and H be a closed connected subgroup. Classify transitive actions on G/H. One would except that the original action is the only irreducible transitive action on G/H for almost all cases.

Problem 12. Is it true that two homeomorphic homogeneous spaces are necessarily diffeomorphic?

2. Non-transitive Actions

Although some of the results of [46, 47] automatically apply to many classical homogeneous spaces such as Stiefel manifolds, G/T, etc., to get some interesting information about non-transitive actions on them, there have been, so far, no serious and systematic study of such an obviously interesting and important subject. Hence, instead of summarizing the scattered results that appeared in [50, 53], we shall briefly mention some obvious natural problems:

Problem 12. Let K be an elementary abelian group (i.e., $K = T^r$ or $\mathbb{Z}_p^r$) and G/H be a compact homogeneous space. What is the cohomology of the fixed point set F of a differentiable K-action on G/H? Does the cohomology of F necessarily resemble the cohomology of the fixed point set of a suitable *natural action* of K?

Problem 13. Consider a differentiable action of T^r (r-dim torus) on a compact homogeneous space G/H. Is it true that there always exists at least one orbit with its dimension $\leqslant (\text{Rk}\, G - \text{Rk}\, H)$? In the case $(\text{Rk}\, G - \text{Rk}\, H) = 0$, $\chi(G/H) \neq 0$ and then it is a well-known fact that $\chi(F) = \chi(G/H) \neq 0$, hence $F \neq \emptyset$.

Problem 14. Study the geometric behavior of large subgroups of $\text{Diff}(G/H)$. To begin with, one may first investigate the geometric behaviors of actions of large simple Lie groups. For example, what are the orbit structures of differentiable $SO(m)$-actions on $SO(n)/H$ with $m \geqslant n/2$? Intuitively, it is *not* likely that $SO(n+1)$ can act differentiably and non-trivially on any homogeneous space, $SO(n)/H$, of $SO(n)$. However, it is not at all easy to show such non-trivial actions are indeed impossible.

G. Discussion and Comments

a) Because of some purely technical reasons, one knows very little about differentiable G-actions on euclidean spaces and spheres with G/T as principal orbit types. (T is the maximal torus of G). In view of the fact that all linear G-actions φ with G/T as their principal orbit types must be the sum of adjoint representations and trivial representations, i.e., $\varphi = \mathrm{Ad}_G + \ell \cdot \theta$, one is tempted to conjecture that

Conjecture. Let G be a compact connected semi-simple Lie group and T be a maximal torus of G. If Φ is a differentiable G-action on $\mathbb{R}^m$ with G/T as its principal orbit type, then

1. The fixed point set F is non-empty and connected and $f = \dim F = m - \dim G \geqslant 0$.
2. The orbit structure of Φ resembles to a great extent the orbit structure of the linear action $\varphi = \mathrm{Ad}_G + \ell\theta$. Furthermore, similar results should also be provable for differentiable actions on spheres with G/T as principal orbit type.

b) Although results of the P. A. Smith type on the cohomologies of fixed points sets no longer hold for differentiable actions of compact connected non-commutative Lie groups in general [18, 46], under suitable regularity conditions, such results are indeed provable [for example, see 46, 47, 51]. In fact, whether one can prove the existence of fixed points for actions on euclidean spaces satisfying a given regularity condition may be regarded as a test of the strength of the given conditions.

Usually, theorems on the topological or diffeomorphic invariants of the fixed point set are difficult to prove but are useful and deep results to have. Hence, it seems to be rather important to pursue results of the P. A. Smith type for *regular differentiable action* even further than just the cohomological structures of the fixed point sets. For example, it was proved in [46] that to every integral homology sphere H^k, $k \geqslant 5$, there exists a suitable homotopy sphere Σ^k such that $H^k \# \Sigma^k$ can be realized as the fixed point set of a differentiable G-action on a standard sphere. Hence, it is interesting to see

Problem 15. Which homotopy spheres can be realized as the fixed point set of a differentiable circle group action on a standard sphere? Or, more precisely, which differentiable knots of homotopy spheres can be realized as the fixed point set of differentiable actions on standard spheres?

c) As we have already remarked before, the regular set M_0 and the singular set M_s are two important mutually complementary invariant subsets for a given G-space M. The singular set M_s is the analogy to "singularities", and hence, is naturally an important invariant. Among singular orbits, fixed points are the absolutely most singular orbits; however, they may not always exist even for differentiable actions on euclidean spaces. Hence, it is very interesting to study the structure of the equivariant tubular neighborhoods of the closed submanifold formed by the points on those most singular orbits (usually there are more than one most singular orbit types) and how they influence the behavior of the action as a whole.

Problem 16. Let G be a compact connected non-commutative Lie group. Let Φ be a differentiable action of G on a euclidean space $\mathbb{R}^m$, we define

$$s(\Phi) = \min\{\dim G(x) | x \in \mathbb{R}^m\}.$$

Put

$$S(G) = \mathrm{Max}\{s(\Phi) | \text{ all possible differentiable } G\text{-actions, } \Phi, \text{ on euclidean spaces}\}.$$

Try to compute or estimate $S(G)$.

d) So far, almost all regularity theorems are stated only in the case of classical groups. However, since classical groups exhaust the set of "building blocks" for compact connected Lie groups with only five exceptions, considerable information for actions of general compact Lie groups can already be deduced from the regularity theorems for actions of classical groups.

Let $G = G_1 \times G_2$ be the product of two simple Lie groups G_1, G_2. Let $\Gamma(G,M)$, $\Gamma(G_1,M)$ and $\Gamma(G_2,M)$ be the set of differentiable equivalent classes of differentiable actions of G, G_1, and G_2 respectively. Obviously, we have the following maps of sets induced by restrictions:

$$\begin{array}{ccc} & & \Gamma(G_1,M) \\ & \overset{r_1}{\nearrow} & \uparrow \\ \Gamma(G,M) & \xrightarrow{r_1 \times r_2} & \Gamma(G_1,M) \times \Gamma(G_2,M) \\ & \underset{r_2}{\searrow} & \downarrow \\ & & \Gamma(G_2,M) \end{array}$$

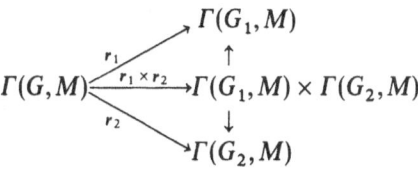

A natural approach to the study of differentiable G-actions on M is to study the map $(r_1 \times r_2)$. For example, it is interesting to know the image of $(r_1 \times r_2)$ in $\Gamma(G_1,M) \times \Gamma(G_2,M)$ and then, for each element (σ_1, σ_2) in the image of $(r_1 \times r_2)$, what is the inverse image. Problems of this type seem to be very important and fascinating. One of the simplest examples is the following:

Problem 17. Let $G = SO(m) \times SO(n)$ and $M = \mathbb{R}^{m \cdot n}$. What are those differentiable actions of G on $\mathbb{R}^{m \cdot n}$ such that the restricted actions of $SO(n)$ and $SO(m)$ both have a fixed point and their local representation around the fixed point is $m \cdot \rho_n$ and $n \cdot \rho_m$ respectively?

V. Applications

In this section, we shall summarize some geometric applications of the regularity theorems. Since most of the regularity theorems that we mentioned in IV are rather young, their geometric applications are far from being fully exploited. Intuitively, results in differentiable transformation groups should be able to find their natural applications in those geometric situations where *symmetries* play a role.

A. Degree of Symmetries of Smooth Structures

Since differentiable actions of compact Lie groups on a given smooth structure M^m may be regarded as natural generalizations of the classical $\mathbb{Z}_2$-symmetries, the maximum of the dimensions of compact subgroups in $\mathrm{Diff}(M^m)$ is obviously an interesting invariant of the smooth structure M^m that roughly measures the degree of symmetry of the smooth structure M^m. Following [52], we shall denote such a well-defined integer by $N(M^m)$ and call it the degree of symmetry of M^m. In case M^m is compact, then it is not difficult to see that $N(M^m)$ is also the maximum of the dimensions of the groups of isometries of all possible Riemmanian metrics on M^m. Hence, we may reformulate one of the well-known theorems in classical Riemannian geometry as follows:

Theorem 6 (cf. C., IV): *$N(M^m) \leqslant \frac{1}{2}m(m+1)$ and $N(M^m) = \frac{1}{2}m(m+1)$ only when M^m is the ordinary m-sphere S^m or the ordinary m-projective space P^m.*

Furthermore, if one looks more carefully at the *possible values* of $N(M^m)$ *close to* $\frac{1}{2}m(m+1)$, it is not difficult to see that there are *gaps*. We refer to a recent paper of MANN in these Proceedings for a systematic discussion of such gaps.

In general, invariants such as $N(M^m)$ are natural and easy to define but are very difficult to compute. Hence, it is rather natural that one shall first try on some specially interesting cases. For example:

a) The m-spheres S^m and the m-projective space P^m are the *most symmetric* m-manifolds, and it is well known in differential topology that there are other smooth structures on the topological m-sphere $|S^m|$ as well as the topological m-projective space $|P^m|$. Hence, it is especially interesting to compare the degrees of symmetry of the exotic m-spheres, say $N(\Sigma^m)$, with the degree of symmetry of the standard m-sphere, $N(S^m) = \frac{1}{2}m(m+1)$.

b) Let G be a semi-simple compact connected Lie group and H be a connected closed subgroup of G. It is well known that there is a canonical invariant inner product on $\mathfrak{G}$ given by the so-called *Cartan-Killing form.* Let $\mathfrak{H}$ be the Lie algebra of H and $\mathfrak{H}^\perp$ be the orthogonal complement of $\mathfrak{H}$ in $\mathfrak{G}$ (with respect to Cartan-Killing form). $\mathfrak{H}^\perp$ can be naturally considered as the tangent space of G/H at the base point and the induced inner product on $\mathfrak{H}^\perp$ naturally determines a homogeneous Riemannian metric on G/H, say σ. In a sense σ is the most natural Riemannian metric on G/H. Hence, one is tempted to ask whether the most natural Riemannian metric σ should be also the most symmetric metric. Namely, is it true that $N(G/H) = \dim(ISO(\sigma))$.

Now let us summarize the known results about a) and b) as follows:

1. Degrees of Symmetry of Homotopy Spheres

The concept of degree of symmetric was first introduced in [52], in which it was proved that

Theorem 18. *Let Σ^m be any smooth homotopy m-sphere that is not diffeomorphic to the ordinary m-sphere S^m. If $m \geqslant 40$, then we always have $N(\Sigma^m) < \frac{1}{8}m^2 + 1$.*

In another paper by WU-CHUNG HSIANG and the author [45], we prove that the above estimate is best possible when and only when Σ^m is a Kervaire sphere. To be precise, suppose $bP_{4k+2} = \mathbb{Z}_2$ and $4k+2 > 40$ and Σ^{4k+1} is the generator of bP_{4k+2}. Then

$$N(\Sigma^{4k+1}) = k(2k+1) + 1 = \tfrac{1}{8}(m^2-1) + 1.$$

It is also not difficult to see by equivariant plumbing and slightly more careful estimation that

$$(2k^2 - 3k + 1) \leqslant N(\Sigma^{4k-1}) \leqslant 2k^2 - 3k + 3$$

for $\Sigma^{4k-1} \in \{bP_{4k} - S^{4k-1}\}$.

The above results show that *exotic spheres are at least four times less symmetric than the standard spheres.* However, *those exotic spheres that bound parallelizable manifolds are in fact quite symmetric.* For homotopy spheres $\Sigma^m \notin bP_{m+1}$, one actually knows very little. Even for their existence, one has to use the PONTRJAGIN-THOM construction to identify their cosets of bP_{m+1} with elements of coker J, and no particular constructive methods for obtaining them are known so far [61]. In a recent joint paper of WU-CHUNG HSIANG and the author [116], we manage to show that

Theorem 19. *Let Σ^m $(m \geqslant 300)$ be a homotopy sphere which does not bound a π-manifold. Then*

$$N(\Sigma^m) < \tfrac{1}{16}(m+1)^2 + 3.$$

The above result shows that those homotopy spheres in the subgroup $bP_{m+1}, bP_{m+1} \supseteq \theta^m$ (i. e., those that one knows how to construct) are much more symmetric than the others.

Remarks. a) The basic reason for the first estimate of [52] is the following characterization of standard spheres in differentiable transformation groups:

Theorem 20 [43]. *A homotopy m-sphere is the standard sphere S^m $(m \geqslant 5)$ if and only if there exists a differentiable G-action with S^k $(k \neq 1, 3)$ as the principal orbit type.*

Of course, in order to deduce $N(\Sigma^m) < \frac{1}{8}m^2 + 1$, $(\Sigma^m \neq S^m)$ from the above theorem, one needs a lot more information about the regularity of large group actions. Hence, technically, such results may be considered as direct applications of the regularity theorems of [46] (cf. (C), IV).

b) One of the key steps of showing that $N(\Sigma^m) < \frac{1}{16}(m+1)^2 + 3$ for $\Sigma^m \notin bP_{m+1}$ and $m \geqslant 300$ is again a *similar characterization of the homotopy spheres of bP_{m+1} in terms of the orbit structures of certain differentiable transformation groups.* Namely,

Theorem 21 [116]. *A homotopy m-sphere Σ^m $(m \geqslant 19)$ belongs to bP_{m+1} if and only if there exists a differentiable action of $SO(n)$ $(n \geqslant 6)$, with $V_{n,2} = SO(n)/SO(n-2)$ as the principal orbit type.*

c) For $m \geqslant n \cdot k$, $(n-k) \geqslant 4$, it is not difficult to show that the fixed point set of a differentiable $SO(n)$-action on a homotopy m-sphere Σ^m with $V_{n,k} = SO(n)/SO(n-k)$ as the principal orbit type is necessarily non-empty. Hence, the subset of homotopy m-spheres in θ^m $(m \geqslant n \cdot k)$ that admit such differentiable $SO(n)$ actions forms a subgroup in θ^m. Suppose we denote it by θ^m_k. Then the above two characterizations show that

$$\theta^m_1 = \{\mathrm{Id}\} = \{S^m\} \quad \text{and} \quad \theta^m_2 = bP_{m+1}$$

respectively. Within suitable range of m, n, and k, it follows by definition that $\theta^m_k \subseteq \theta^m_{a \cdot k}$. It seems interesting to know other geometric characterization of the subgroups θ_k^m as well as the lattice structure of $\{\theta_k^m\}$. For example, it is not known whether $\theta_2^m \subseteq \theta_3^m$, etc.

Problem 18. Is it true that $\theta_k^m \subseteq \theta_{k+1}^m$? If so, what are those $k \geqslant 2$ with $\theta_k^m \nsubseteq \theta_{k+1}^m$?

Of course, one may use $SU(n)$ and $Sp(n)$ to formulate similar problems. As one can see from the proofs of [52] and [116], such problems are closely related with the degree of symmetries of elements of $\{\theta^m - bP_{m+1}\}$.

2. Degree of Symmetries of Compact Homogeneous Spaces

Unlike the case of exotic spheres, one has an obvious candidate for $N(M^m)$ in the case $M^m = G/H$ is a compact homogeneous space, namely

$$N(M^m) = \dim(ISO(\sigma))$$

where σ is the natural metric on G/H defined above. Hence, the problem here is to generalize the classical result $N(S^m) = \frac{1}{2}m(m+1)$ to other compact homogeneous spaces. It seems to me that the difficulties here are purely technical. If one knows how to study differentiable transformation groups on spaces with more complicated topological types, it should be possible to settle the above problem affirmatively. So far, there is only one trial in the case $M = V_{n,2} = SO(n)/SO(n-2)$; see [53] for details.

Once $N(G/H)$ is known as in the S^m case, then, of course, it is also interesting to compare the degrees of symmetry of G/H with those manifolds of the same homotopy type of G/H.

B. Actions of Large Compact Subgroups of Diff(M)

Let M^m be a given differentiable manifold and $\mathrm{Diff}(M^m)$ be the group of all diffeomorphisms of M^m onto itself. In the study of compact subgroups of $\mathrm{Diff}(M^m)$, it is technically easier *to fix a given compact Lie group* G and consider all possible homomorphisms of G into $\mathrm{Diff}(M^m)$, which is the same as considering all possible differentiable actions on M^m of *the given group* G. From the point of view of differentiable transformation groups, however, it is much more desirable and natural to study *all those compact subgroups of* $\mathrm{Diff}(M^m)$ *satisfying certain natural geometric conditions.* $N(M^m)$ is one of the first invariants of that nature. For deeper information, one many consider the following problem.

Problem 19. For a given manifold M with rich symmetries, what is the geometric behavior of the large compact subgroups of Diff(M), say those of dimension between $N(M)$ and $1/a\,N(M)$, ($a = 2, 3$, or 4)?

In a sense, the above problem is one of the natural "next" problems once one has successfully computed $N(M)$ for a given smooth structure M. Therefore, let us specialize the above problem into the consideration of the interesting special examples such as $\mathbb{R}^m$, S^m, P^m, $V_{n,2}=SO(n)/SO(n-2)$ and exotic spheres, Σ^m, that bound parallelizable manifolds. The cases $\mathbb{R}^m$, P^m, and S^m are parallel. We shall only mention the results about $\mathbb{R}^m$ and refer to [117] for further information.

Following the so-called comparison principal, it is natural to study first the geometric behavior of large compact *linear groups* for the sake of motivation. For example, it is not difficult to show the following results by linear representation theory

Proposition 5.1. *Let $H \subseteq SO(n)$ be a closed connected subgroup of $SO(n)$, $n \geqslant 11$. If $\dim H > \frac{1}{4}n^2$, then there exists a linear subspace $V \subseteq \mathbb{R}^m$ such that $H = SO(V) \times K$, $K \subseteq SO(V^\perp)$ and $\dim V > n/2$. (By $SO(V)$, we mean the subgroup of $SO(n)$ that leaves $V^\perp$ point-wise fixed. See Chapter II of [46] for a proof of the above proposition.)*

Proposition 5.2. *Let H be a compact connected linear transformation group on $\mathbb{R}^n$. If $\dim H > \frac{1}{8}n^2$, then there is a normal subgroup H_1 of H, such that the principal orbit of the linear action of H_1 is S^k and the fixed point set of the H_1-action is $\mathbb{R}^{n-k-1}$, $(k+1) \geqslant n/4$. (A weaker version of this proposition was proved and used in [116].)*

Again, as an application of the regularity theorems of IV, it was proved in a recent paper of the author [117] that large compact subgroups of $\mathrm{Diff}(\mathbb{R}^n)$, $\mathrm{Diff}(S^{n-1})$ or $\mathrm{Diff}(P^{n-1})$ have similar geometric properties. For example:

Theorem 22. *Let G be a compact connected subgroup of $\mathrm{Diff}(\mathbb{R}^n)$, $n \geqslant 50$, with $\dim G > \frac{1}{4}n^2$. Then $G = G_1 \times G_2$ where G_1 is isomorphic to $SO(k)$ for a suitable $k > n/2$, the fixed point set of G_1 is an acyclic submanifold of dimension $(n-k)$, X^{n-k}, and G_2 can be considered as a compact subgroup of $\mathrm{Diff}(X^{n-k})$.*

Theorem 23. *Let G be a compact connected subgroup of $\mathrm{Diff}(\mathbb{R}^n)$, $n \geqslant 50$, with $\dim G > \frac{1}{8}n^2$. Then there is a normal subgroup G_1 of G such that the principal orbit type of the G_1-action is S^k, $k > n/4$.*

Similar results for S^{n-1}, P^{n-1}, $V_{n,2}=SO(n)/SO(n-2)$ and further generalizations of the above two theorems for the $\mathbb{R}^n$ case were discussed in [117]. The results roughly confirm the belief that the geometric behavior of large subgroups of $\mathrm{Diff}(\mathbb{R}^n)$, $\mathrm{Diff}(S^{n-1})$, etc., are quite regular and *their geometric behavior models after the geometric behavior of linear actions of large subgroups rather closely*. For details, see [117].

Now, let us look at the case $M=\Sigma^m \in bP_{m+1}$. Naturally, we shall consider the $(4k+1)$-case and the $(4k-1)$-case separatly. *If* $m=4k+1$ *and* $bP_{4k+2} \approx \mathbb{Z}_2$, let Σ^{4k+1} be the $(4k+1)$-Kervaire sphere. Then $N(\Sigma^{4k+1})=k(2k+1)+1$. Of course, the first step in the study of large compact subgroups of $\mathrm{Diff}(\Sigma^{4k+1})$ would be to classify differentiable actions of compact connected Lie groups of dimension $k(2k+1)+1$ on Σ^{4k+1}. This question and some interesting geometric behavior of the differentiable action of the largest (in dimension) compact connected subgroups of $\mathrm{Diff}(\Sigma^{4k+1})$ were discussed in a joint paper of WU-CHUNG HSIANG and the author [45]. Now, using the beautiful results of BRIESKORN-HIRZEBRUCH [21, 40], the result of such a classification can be neatly restated as follows:

Theorem 24. *There are infinitely many conjugacy classes of compact connected subgroups of maximal possible dimension (i. e.,* $N(\Sigma^{4k+1}) = k(2k+1)+1$*) in* $\mathrm{Diff}(\Sigma^{4k+1})$. *The following subgroups* G_d, $d>0$ *and* $d \equiv \pm 3 \bmod (8)$, *form a complete set of representatives: According to* BRIESKORN *and* HIRZEBRUCH, *we may consider* Σ^{4k+1} *as the subset in* $\mathbb{C}^{2k+2}=\{(z_0, z_1, \ldots, z_{2k+2})\}$ *given by the following equations:*

$$z_0^d + z_1^2 + \cdots + z_{2k+2} = 0 \quad (d \equiv \pm 3 \bmod (8)),$$
$$|z_0|^2 + |z_1|^2 + \cdots + |z_{2k+2}|^2 = 1.$$

Hence, for each $d>0$ *and* $d \equiv \pm 3 \bmod (8)$, *we have a group* G_d *which is the biggest subgroup of* $U(2k+2)$ *that leaves the above set invariant, and consequently may be considered as a subgroup of* $\mathrm{Diff}(\Sigma^{4k+1})$. *It is not difficult to see that the* G_d *are all locally isomorphic to* $SO(2k+1) \times SO(2)$ *and act on* Σ^{4k+1} *with almost identical orbit structure.*

From the above result, it seems plausible that large compact subgroups of $\mathrm{Diff}(\Sigma^{4k+1})$, say with dimension bigger than $(k+1)^2$, will have similar geometric behavior to those subgroups of $\{G_d\}$ with dimension bigger than $(k+1)^2$. For example, proved in [117] was

Theorem 25. *Let* Σ^{4k+1} *be the* $(4k+1)$*-Kervaire sphere and* G *be a compact connected subgroup of* $\mathrm{Diff}(\Sigma^{4k+1})$. *If* $k \geqslant 15$ *and*

$$\dim G > (k+3)^2,$$

then there is a normal subgroup G_1 *of* G *such that* G_1 *is isomorphic to* $SO(l), \ell \geqslant 3k/4$ *and the principal orbit type of the* G_1*-action is*

$$V_{\ell,2} = SO(l)/SO(\ell-2).$$

The case of $4k-1$ is slightly more complicated. For example, the order of the group bP_{4k} is huge. However, similar results are still provable with more detailed and careful treatment. We shall refer to [117] for such discussion.

C. Differentiable Actions of Non-compact Semi-simple Lie Groups

It is well known that differentiable actions of compact Lie groups are far more regular than those of the non-compact Lie groups and the behavior of compact Lie group actions is comparatively much more well understood than that of the non-compact Lie group. Of course, one of the most basic and important cases of non-compact Lie group actions are those $\mathbb{R}^1$-actions which are naturally associated with ordinary differential equations [100]. So far, most of the important research has been concentrated in that special but basic case. However, knowledge about $\mathbb{R}^1$-actions still seems rather inadequate. Now, suppose we consider differentiable actions of semi-simple Lie groups (non-compact). Then, due to the *tight* group structure of semi-simple groups and the existence of a compact subgroup of considerable size, it is possible to bring in Lie group theory and the relatively well understood knowledge of compact differentiable transformation groups to help us. Let G be a non-compact connected semi-simple Lie group and K be a maximal compact subgroup of G. For a given differentiable manifold M, we shall denote the set of equivalence classes of differentiable G-actions on M by $\Gamma(G;M)$ and the set of equivalence classes of differentiable K-actions on M by $\Gamma(K;M)$. Naturally, the restriction induces a map

$$\rho:\Gamma(G;M)\to\Gamma(K;M).$$

Due to the fact that maximal compact subgroups of G are conjugate to each other, the map ρ does not depend on the choice of K in G. Hence the restricted K-action is naturally an invariant of the original G-action. A reasonable way to study G-actions on M is to investigate the image and the kernel of the above map ρ.

With the above point of view in mind, one finds that it is in fact possible to deduce some highly non-trivial results from the existing knowledge in Lie group theory and compact transformation groups. For example, almost as a direct consequence of the regularity theorems of § IV, one shows that the fixed point sets of differentiable actions of simple non-compact Lie groups on euclidean spaces of comparatively low dimensions are always non-empty and with acyclic cohomologies. On the other hand, one knows from the theory of symmetric spaces that the homogeneous space G/K is diffeomorphic to a euclidean space $\mathbb{R}^m$. The above observation naturally leads to the following conjecture.

Conjecture. Let G be a non-compact simple Lie group, H be a maximal compact connected subgroup of G, $m=\dim G-\dim H$. Then the fixed point set F of a given differentiable action of G on a euclidean space of dimension smaller than m is always non-empty and has acyclic cohomology with a suitable coefficient group. Moreover, any differentiable action of G on $\mathbb{R}^m$ without fixed points should be differentiably equivalent to the natural transitive action on G/H.

See [54] for results and further remarks with this point of view.

Of course, in order to study differentiable actions of non-compact Lie groups, the knowledge of compact transformation groups is helpful, but far from being adequate. One needs to understand the basic properties of orbit structures for non-compact Lie group actions. We refer to a recent paper of RICHARDSON [95] for a discussion in that direction.

D. Concluding Remarks

In differentiable transformation groups, one studies the geometric behavior of symmetries. Intuitively, symmetries of a given geometric structure are undoubtedly among the natural nice properties that should be useful and important in the study of other properties of the given geometric structure. So far, very few such potential applications of symmetries have been exploited. With more and more understanding of regularity theorems of differentiable transformation groups, it seems rather hopeful that the understanding of Riemannian manifolds with rich symmetries may increase considerably by effective exploitation of the rich symmetry.

References

1. ADAMS, F.: On the non-existence of elements of Hopf invariant one, Ann. of Math. **72,** 20—104 (1960).
2. — Vector fields on spheres, Ann. of Math. **75,** 603—632 (1962).
3. ANDERSON, D., E. BROWN, and F. PETERSON: SU-cobordism, KO-characteristic numbers and the Kervaire invariant, Ann. of Math. **83,** 54—67 (1966).
4. BING, R.: A homeomorphism between the 3-sphere and the sum of two solid horned spheres, Ann. of Math. **56,** 354—362 (1952).
5. BOCHNER, S.: Compact groups of differentiable transformations, Ann. of Math. **46,** 372—381 (1945).
6. —, and D. MONTGOMERY: Groups of differentiable and real or complex analytic transformations, Ann. of Math. **46,** 685—694 (1945).
7. — Locally compact groups of differentiable transformations, Ann. of Math. **47,** 639—653 (1946).
8. — Groups on analytic manifolds, Ann. of Math. **48,** 659—669 (1947).
9. BOREL, A.: Les bouts des espaces homogènes des groupes de Lie, Ann. of Math. **58,** 443—457 (1953).
10. — Le plan projectif de octaves et les sphères comme espaces homogènes, Comptes Rendue de l'Académie des Sciences, Paris **230,** 1378—1383 (1960).

11. — Fixed points of elementary commutative groups, Bull. of A. M. S. **65**, 322—326 (1959).
12. —, et. al.: Seminar on Transformation Groups, Ann. of Math. Studies **46**, Princeton University Press (1961).
13. —, and F. HIRZEBRUCH: Characteristic classes and homogeneous spaces I, Amer. J. of Math. **80**, 485—538 (1958); II, **81**, 351—382 (1959); III, **82**, 491—504 (1960).
14. —, and J. DE SIEBENTHAL: Sur les sous-groupes fermés de rang maximum des groupes des Lie compacts connexés, Comm. Math. Helv. **23**, 200—221 (1949).
15. —, and J. P. SERRE: Sur certain sous-groupes de Lie compacts, Comm. Math. Helv. **27**, 128—139 (1953).
16. BREDON, G.: Transformation groups on spheres with two types of orbits, Topology **3**, 115—122 (1965).
17. — Examples of differentiable group actions, Topology **3**, 103—113 (1965).
18. — Exotic actions on spheres (these Proceedings).
19. — On homogeneous cohomology spheres, Ann. of Math. **73**, 556—565 (1961).
20. — On a certain class of transformation groups, Mich. Math. J. **9**, 385—393 (1962).
21. BRIESKORN, E.: Beispiele zur Differentialtopologie von Singularitäten, Inventiones Math. **2**, 1—14 (1966).
22. — Examples of singular normal complex spaces which are topological manifolds, Proc. Nat. Acad. Sci. USA **55**, 1395—1397 (1966).
23. BROWDER, W.: Higher torsions in H-spaces, Trans. A. M. S. **108**, 353—375 (1963).
24. — Homotopy type of differentiable manifolds, Colloquium on algebraic topology, Aarhus 1962, 42—46.
25. — The Kervaire invariant of framed manifolds and its generalization (mimeo. Priceton Univ.).
26. CHEVALLEY, C.: Theory of Lie groups, Princeton Univ. Press 1946.
27. — The Betti numbers of exceptional simple Lie groups, Proc. of International Congress of Math. **2**, 21—24 (1950), Cambridge, USA.
28. CONNER, P.: Orbits of uniform dimension, Mich. J. of Math. **6**, 25—32 (1959).
29. —, and E. FLOYD: On the construction of periodic maps without fixed points, Proc. of A. M. S. **10**, 354—360 (1959).
30. — Differentiable Periodic Maps. Berlin-Göttingen-Heidelberg-New York: Springer 1964.
31. — Orbit spaces of circle groups of transformations, Ann. of Math. **67**, 90—98 (1958).
32. DYNKIN, E.: Maximal subgroups of the classical groups, Trudy Moscow Matematicheschi Obschcesta, vol. **1**, 39—166 (1952); American Mathematical Society Translations, Ser. **2**, vol. **6**, 245—378 (1957).
33. EISENHART, L. P.: Riemannian Geometry, Princeton University Press, Princeton, N. J. 1940
34. — Continuous groups of transformations, Princeton University Press, 1933.
35. FLOYD, E.: Examples of fixed point sets of periodic maps I, Ann. of Math. **55**, 167—171 (1952); II, **64**, 396—398.

36. — Fixed point sets of compact abelian Lie groups of transformations, Ann. of Math. **66**, 30—35 (1957).
37. —, and R. Richardson: An action of a finite group on an n-cell without stationary points, Bull. A. M. S. **65**, 73—76 (1959).
38. Godement, R.: Théorie des faisceaux, Actualités Scientifiques et Industrielles, 1252. Paris, 1958, Hermann ed.
39. Haefliger, A.: Differentiable embeddings of S^n in S^{n+q} for $q>2$, Ann. of Math. vol. **83**, 402—436 (1966).
40. Hirzebruch, F.: Singularities and exotic spheres, Seminaire Bourbaki **19** (1966/67).
41. — The topology of normal singularities of an algebraic surface (d'apres un article de D. Mumford) Seminaire Bourbaki **15** (1962/63).
42. Hsiang, W. C.: A note on free differentiable actions of S^1 and S^3 on homotopy spheres, Ann. of Math. **83**, 266—272 (1966).
43. —, and W. Y. Hsiang: Classification of differentiable actions on S^n, R^n and D^n with S^k as the principal orbit type, Ann. of Math. **82**, 421—433 (1965).
44. — Some differentiable actions of S^1 and S^3 on 7-spheres, Quarterly Journal of Math. (2), Oxford, vol. **15**, 371—374 (1964).
45. — On compact subgroups of diffeomorphism groups of Kervaire spheres, Ann. of Math. **85**, 359—369 (1967).
46. — Differentiable actions of compact connected classical groups I, Amer. J. Math. **89** (1967).
47. — Differentiable actions of compact connected classical groups II (to appear).
48. — Some results on differentiable actions, Bull. A. M. S. **72**, 134—137 (1966).
49. —, and R. H. Szczarba: On the tangent bundle of a Grassmann manifold, Amer. J. Math. **86**, 685—697 (1964).
50. Hsiang, W. Y.: On classification of differentiable $SO(n)$ actions on simply connected π-manifolds, Amer. J. Math. **88**, 137—153 (1966).
51. — On the principal orbit type and P. A. Smith theory of $SU(p)$ actions, Topology **6**, 125—135 (1967).
52. — On the bound of the dimensions of the isometry groups of all possible Riemannian metrics on an exotic sphere, Ann. of Math. **85**, 351—358
53. — The natural metric on $SO(n)/SO(n—2)$ is the most symmetric metric, Bull. of Math. **73**, 55—58 (1967).
54. — Remarks on differentiable actions of non-compact semisimple Lie groups on euclidean spaces (to appear in Amer. J. Math.).
55. — On compact homogeneous minimal submanifolds, Proc. of N. A. S. (USA) **56**, 5—6 (1966).
56. —, and J. C. Su: On the classification of transitive effective actions on classical homogeneous spaces (to appear in Trans. of A. M. S.).
57. Jänich, K.: Differenzierbare Mannigfaltigkeiten mit Rand als Orbiträume differenzierbarer G-Mannigfaltigkeiten ohne Rand, Topology **5**, 301—320 (1966).
58. — On the classification of $O(n)$-manifolds (to appear in these Proceedings).

59. KERVAIRE, M.: An interpretation of G. Whitehead's generalization of the Hopf invariant, Ann. of Math. **69,** 345—364 (1959).
60. — A manifold which does not admit any differentiable structure, Comment. Math. Helv. **34,** 257—270 (1960).
61. —, and J. MILNOR: Groups of homotopy spheres I, Ann. of Math. **77,** 504—537 (1963).
62. KRAMER, M.: Hauptisotropiegruppen bei endlich dimensionalen Darstellungen kompakter halbeinfacher Liegruppen (Diplomarbeit, Bonn 1966).
63. MILNOR, J.: On manifolds homeomorphic to 7-spheres, Ann. of Math. **64,** 399—405 (1956).
64. — On isolated singularities of hypersurfaces (mimeo.), Princeton, 1966.
65. MONTGOMERY, D.: Topological groups of differentiable transformations, Ann. of Math. **46,** 382—387 (1945).
66. — Analytic parameters in three-dimensional groups, Ann. of Math. **49,** 118—131 (1948).
67. — Finite dimensional groups, Ann. of Math. **52,** 591—605 (1950).
68. —, and H. SAMELSON: Transformation groups of spheres, Ann. of Math. **44,** 454—470 (1943).
69. — Groups transitive on the n-dimensional torus, Bull. A. M. S. **49,** 455—456 (1943).
70. — —, and C. T. YANG: Exceptional orbits of highest dimension, Ann. of Math. **64,** 131—141 (1956).
71. — —, and L. ZIPPIN: Singular points of compact transformation groups, Ann. of Math. **63,** 1—9 (1956).
72. —, and C. T. YANG: The existence of a slice, Ann. of Math. **65,** 108—116 (1957).
73. — — Orbits of highest dimension, Trans. A. M. S. **87,** 284—293 (1958).
74. — Differentiable actions on homotopy 7-spheres, Trans. A. M. S. **122,** 480—498 (1966).
75. — Differentiable actions on homotopy 7-spheres II, III (these Proceedings; unpublished).
76. —, and L. ZIPPIN: Topological Transformation Groups, New York Interscience (1955)
77. — A class of transformation groups in E_n, Amer. J. Math. **63,** 1—8 (1941).
78. — Existence of subgroups isomorphic to the real numbers, Ann. of Math. **53,** 298—326 (1951).
79. — Small subgroups of finite-dimensional groups, Ann. of Math. **56,** 213—241 (1952).
80. — Topological transformation groups, Ann. of Math. **41,** 778—791 (1940).
81. — Theorem on Lie groups, Bull. A. M. S. **48,** 448—452 (1942).
82. MOSTERT, P.: On compact Lie groups acting on a manifold, Ann. of Math. **65,** 447—455 (1957).
83. MOSTOW, G. D.: A new proof of E. Cartan's theorem on the topology of simple groups, Bull. A. M. S. **55,** 969—980 (1949).
84. — Equivariant embeddings in euclidean spaces, Ann. of Math. **65,** 432—446 (1957).

85. — On a conjecture of Montgomery, Ann. of Math. **65,** 513—516 (1957).
86. Myers, S., and N. Steenrod: The group of isometries of a Riemannian manifold, Ann. of Math. **40,** 400—416 (1939).
87. Oniscik, A.: Inclusion relations among transitive compact transformation groups, Trudy Moskov Mat. Obsc **11,** 199—242 (1962). (Translated in AMS Transl. vol. 50, 2nd Series).
88. — Transitive compact transformation groups, Mat. Sb. **60 (102),** 447—485 (1963). (Translated in AMS Transl. Vol. 55, Series 2).
89. Palais, R.: On the differentiability of isometries, Proc. A. M. S. **8,** 805—807 (1957).
90. — Imbedding of compact differentiable transformation groups in orthogonal representations, J. of Math. and Mech. **6,** 673—678 (1957).
91. Poncet, J.: Groupes de Lie compacts de transformations de l'espaces euclidean et les sphères comme espaces homogènes, Comment. Math. Helv. **33,** 109—120 (1959).
92. Pontrjagin, L.: Topological Groups, Princeton University Press, 1946.
93. Richardson, R.: A rigidity theorem for subalgebras of Lie and associative algebras, III. J. Math. **11,** 92—111 (1967).
94. — Some stability theorems for Lie algebras (to appear).
95. — On the variation of isotropy subalgebra (these Proceedings).
96. Samelson, H.: Topology of Lie groups, Bull. A. M. S. **58,** 2—37 (1952).
97. — Beiträge zur Topologie der Gruppenmannigfaltigkeiten, Ann. of Math **42,** 1091—1137 (1941).
98. Smale, S.: Generalized Poincaré conjecture in dimension greater than four, Ann. of Math. **74,** 391—406 (1961).
99. — On gradient dynamical systems, Ann. of Math. **74,** 199—206 (1961).
100. — Differentiable Dynamical Systems I (mimeo.), ISA Princeton, N. J.
101. Smith, P. A.: Fixed point theorems for periodic transformations, Amer. J. Math. **63,** 1—8 (1941).
102. — The topology of transformation groups, Bull. A. M. S. **44,** 497—514 (1938).
103. — Fixed points of periodic transformations, Appendix B in Lefschetz's Algebraic Topology (1942).
104. Steenrod, N.: The Topology of Fibre Bundles, Princeton University Press 1951.
105. Wang, H. C.: Two-point homogeneous spaces, Ann. of Math. **55,** 177—191 (1952).
106. — Homogeneous spaces with non-vanishing Euler characteristics, Ann. of Math. **50,** 925—953 (1949).
107. — Closed manifolds with homogeneous complex structure, Amer. J. Math. **76,** 1—32 (1954).
108. — Compact transformation groups of S^n with an $(n—1)$-dimensional orbit, Amer. J. Math. **82,** 698—748 (1960).
109. Weyl, H.: The Classical Groups, Princeton University Press 1939.
110. — Theorie der Darstellung kontinuierlicher halbeinfacher Gruppen durch lineare Transformationen I, Math. Zeit. **23,** 271—309 (1925); II, Math. Zeit. **24,** 328—395 (1926).

111. YANG, C. T.: The triangulability of the orbit space of a differentiable transformation group, Bull. A. M. S. **69**, 405—408 (1963).
112. MONTGOMERY, D., H. SAMELSON, and C. T. YANG: Groups on E^n with (n—2)-dimensional orbits, Proc. A. M. S. **7**, 719—728 (1956).
113. —, and C. T. YANG: Groups on S^n with principal orbits of dimension n—3, I, II, III. J. of Math. **4**, 507—517 (1960); **5**, 206—211 (1961).
114. KISTER, J.: Examples of periodic maps on euclidean spaces without fixed points, Bull. A. M. S. **67**, 461—474 (1961).
115. MONTGOMERY, D., and H. SAMELSON: Examples for differentiable group actions on spheres, Proc. N. S. A. (USA) **47**, 1201—1205 (1961).
116. HSIANG, W. C., and W. Y. HSIANG: Degree of symmetries of homotopy spheres (to appear) (mimeo. at Yale University).
117. HSIANG, W. Y.: Geometric behavior of large compact subgroups of Diff(M), (to appear).

Differentiable Actions on Homotopy Seven Spheres, II

DEANE MONTGOMERY and C. T. YANG*

1. Introduction

In a previous paper [1] under the same title, we studied differentiable actions of the circle group on homotopy 7-spheres and proved

Theorem A. *There is an infinite cyclic group consisting of all orientation-preserving equivariant diffeomorphism classes of differentiable actions of the circle group on oriented homotopy 7-spheres, each having an oriented 3-sphere as the fixed point set and acting freely otherwise.*

Theorem B. *There is an infinite cyclic group consisting of all orientation-preserving equivariant diffeomorphism classes of free differentiable actions of the circle group on oriented homotopy 7-spheres.*

Theorem C. *On each homotopy 7-sphere there are infinitely many differentiably distinct differentiable actions of the circle group, each having a 3-sphere as the fixed point set and acting freely otherwise.*

The proof of Theorem A is based on the following result of HAEFLIGER [2]: *There is an infinite cyclic group consisting of all isotopy classes of imbedded oriented 3-spheres in the oriented 6-sphere.* In fact, if A is an imbedded 3-sphere in the 6-sphere S^6, one can construct a differentiable action of the circle group G on a homotopy 7-sphere Σ such that 1. the fixed point set F is a 3-sphere, 2. G acts freely on $\Sigma - F$ and 3. there is a diffeomorphism h of S^6 onto the orbit space Σ/G with $h(A) = F$. Let G be oriented. The for any orientation on S^6 and A, one can use 2. and 3. to assign a definite orientation on Σ and F. Therefore for any isotopy class of an imbedded oriented 3-sphere in the oriented 6-sphere, one determines an orientation-preserving equivariant diffeomorphism class of a differentiable action of the circle group on a homotopy 7-sphere

* The second author is supported in part by the U.S. Army Research Office and by the National Science Foundation.

satisfying 1., 2. and 3. above. Since this is a one-to-one correspondence, Theorem A follows from Haefliger's theorem. Notice that, since every imbedded 3-sphere in the 6-sphere is piecewise linearly unknotted, all diffeomorphism classes in Theorem A are in the same homeomorphism class.

If α is an element of the group in Theorem A represented by a differentiable action of the circle group on an oriented homotopy 7-sphere Σ_α, we denote by $[\Sigma_\alpha]$ the orientation-preserving diffeomorphism class of Σ_α. It can be shown that $\alpha \mapsto [\Sigma_\alpha]$ is a homomorphism of the group in Theorem A *onto* the group of all orientation-preserving diffeomorphism classes of oriented homotopy 7-spheres. Hence Theorem C follows.

The proof of Theorem B is roughly as follows. Given any imbedded oriented 3-sphere A in the oriented 6-sphere one uses an appropriate surgery on A to obtain an oriented homotopy complex projective 3-space M and then construct a free differentiable action of the circle group on an oriented homotopy 7-sphere such that there is an orientation-preserving diffeomorphism of M onto the orbit space. It can be shown that this gives a one-to-one correspondence between the isotopy classes of imbedded oriented 3-spheres in the oriented 6-sphere and the orientation-preserving equivariant diffeomorphism classes of free differentiable actions of the circle group on oriented homotopy 7-spheres. Hence Theorem B follows from Haefliger's theorem.

Denote by Π the group in Theorem B. As observed by W. C. HSIANG, there is an isomorphism f of Π into the group $\mathbb{Z}$ of integers such that for any $\alpha \in \Pi$ represented by a free differentiable action of the circle group G on a homotopy 7-sphere Σ, the first rational Pontrjagin class of the orbit space Σ/G is given by

$$p_1(\Sigma/G) = (f(\alpha) + 4)a^2,$$

where a is a generator of $H^2(\Sigma/G; \mathbb{Z})$. Since rational Pontrjagin classes are topological invariants, each diffeomorphism class in Theorem B is a homeomorphism class. Homotopy complex projective 3-spaces are among those differentiable 6-manifolds considered in WALL [3]. As a consequence of Wall's result, we have

$$f(\Pi) = 24\,\mathbb{Z}.$$

It is said in [1] that on *every* homotopy 7-sphere there are infinitely many differentiably distinct free differentiable actions. But this is false. As the first purpose of the present paper, we correct this by proving the following

Theorem 1. *There are exactly six unoriented homotopy 7-spheres not diffeomorphic to one another such that on each of them there is a free*

differentiable action of the circle group. Moreover, on any homotopy 7-sphere, if there exists a free differentiable action of the circle group, then there are infinitely many topologically distinct from one another.

The proof of Theorem 1 is based on the computation of a modified Eells-Kuiper's invariant to be constructed in § 2.

Denote by $\mathbb{Z}_p$ the subgroup of the circle group of order p, where p is a prime >1. As the second purpose of the present paper, we study certain free differentiable actions of $\mathbb{Z}_p$ on homotopy 7-spheres. Making use of our modified Eells-Kuiper's invariant, we also show that there are several differentiably distinct free differentiable actions of $\mathbb{Z}_3$ on the 7-sphere; similarly it can be seen that there is a free differentiable action of $\mathbb{Z}_5$ on the 7-sphere which is not h-cobordant to any lens space, so that the action is not one of those obtained in MILNOR [7].

2. The Invariant of EELLS-KUIPER

$\mathbb{Q}$ denotes the rational field; $\mathbb{Z}$ denotes the group of integers; $\mathbb{Z}_p$ denotes the cyclic group of order p.

The invariant μ of EELLS-KUIPER [5] is defined as follows. Let M be a closed, oriented, differentiable $(4k-1)$-manifold which bounds a compact, oriented, spin, differentiable $4k$-manifold W such that

a) the boundary homomorphism

$$\partial: H_{4k}(W,M;\mathbb{Q}) \to H_{4k-1}(M;\mathbb{Q})$$

maps the fundamental class $[W,M] \in H_{4k}(W,M;\mathbb{Q})$ into the fundamental class $[M] \in H_{4k-1}(M;\mathbb{Q})$;

b) the homomorphisms

$$j^*: H^{2k}(W,M;\mathbb{Q}) \to H^{2k}(W;\mathbb{Q})$$
$$j^*: H^{4\ell}(W,M;\mathbb{Q}) \to H^{4\ell}(W;\mathbb{Q}), \qquad 0 < \ell < k,$$

induced by the inclusion map $j: W \to (W,M)$ are isomorphisms onto and

c) the homomorphism

$$i^*: H^1(W;\mathbb{Z}_2) \to H^1(M;\mathbb{Z}_2)$$

induced by the inclusion map $i: M \to W$ is onto.

Let $L_k(p_1,\ldots,p_k)$ be the k-th polynomial associated with $z^{1/2}/\tanh z^{1/2}$, $\hat{A}_k(p_1,\ldots,p_k)$ the k-th polynomial associated with $z^{1/2}/2\sinh z^{1/2}$,

$$t_k = \hat{A}_k(0,\ldots,0,1)/L_k(0,\ldots,0,1)$$

and

$$N_k(p_1,\ldots,p_{k-1}) = \hat{A}_k(p_1,\ldots,p_{k-1},0) - t_k L_k(p_1,\ldots,p_{k-1},0).$$

Let $p_\ell(W)$ be the ℓ-th rational Pontrjagin class of W, $0<\ell<k$. Then we have, by b),

$$N_k(j^{*-1}p_1(W),\dots,j^{*-1}\,p_{k-1}(W))\in H^{4k}(W,M;\mathbb{Q})$$

so that we have a rational number

$$N_k(W)=\langle N_k(j^{*-1}p_1(W),\dots,j^{*-1}p_{k-1}(W)),[W,M]\rangle.$$

Let $\tau(W)$ be the index of W and $a_k=4/(3+(-1)^k)$. Then

$$\mu(M)=(N_k(W)+t_k\tau(W))/a_k \bmod 1$$

is independent of the choice of W so that it depends only on M. We call $\mu(M)$ the Eells-Kuiper invariant of M.

Following MILNOR [4], one can generalize the invariant μ as follows. Let M and W be as above except that c) is replaced by

d) there is a preassigned spin structure s on M and there is a spin structure of s' on W such that the induced spin structure of s' on M coincides with s.

Then

$$\mu(M,s)=(N_k(W)+t_k\tau(W))/a_k \bmod 1$$

depends only on (M,s). Since c) implies d), $\mu(M,s)$ is a generalization of $\mu(M)$.

The existence of the invariant μ is based on the following integrality theorem: Let X be a closed, oriented, spin differentiable $4k$-manifold. Then the $\hat{A}$-genus of X

$$\hat{A}(X)=\langle \hat{A}_k(p_1(X),\dots,p_k(X)),[X]\rangle$$

is an integer; if k is odd, then $\hat{A}(X)$ is an even integer.

Now we are in a position to modify the invariant μ. Instead of the preceding integrality theorem, we need the following

Integrality Theorem (see, for example, [6; p. 197]). *Let X be a closed, oriented differentiable $2r$-manifold with rational Pontrjagin classes $p_j(X)$ and d an element of $H^2(X;\mathbb{Z})$ whose reduction* mod 2 *is the second Stiefel-Whitney class $w_2(X)$. Regard d as an element of $H^2(X;\mathbb{Q})$ and denote by $u^{(2r)}$ the $2r$-component of*

$$e^{d/2}\cdot\sum_{i=0}^{\infty}\hat{A}_i(p_1(X),\dots,p_i(X))\in\sum_{j=0}^{\infty}H^j(X;\mathbb{Q}),$$

and by $[X]$ the fundamental class of X. Then

$$\hat{A}(X,d/2)=\langle u^{(2r)},[X]\rangle$$

is an integer.

Notice that if $r=2k$ and X is a spin manifold, we may take $d=0$ and then we have $\hat{A}(X,0)=\hat{A}(X)$. Hence this integrality theorem is a generalization of the first part of the preceding one. However, for a spin manifold X of dimension $4k$ with odd k, this integrality theorem is not as strong as the preceding one because of a factor of 2.

Let M be a closed, oriented, differentiable $(2r-1)$-manifold such that

1. $H^1(M;\mathbb{Q})=0$ and $H^2(M;\mathbb{Z})$ is a finite group of odd order.

Let W be a compact, oriented, differentiable $2r$-manifold having M as its boundary such that

2. there is an element c of $H^2(W;\mathbb{Z})$ whose reduction mod 2 is the second Stiefel-Whitney class $w_2(W)$ of W;
3. the boundary homomorphism

$$\partial: H_{2r}(W,M;\mathbb{Q}) \to H_{2r-1}(M;\mathbb{Q})$$

maps $[W,M]$ into $[M]$;

4. the homomorphism

$$j^*: H^q(W,M;\mathbb{Q}) \to H^q(W;\mathbb{Q})$$

induced by the inclusion map is an isomorphism onto for $q=4\ell$, $0<\ell<r/2$ and for $q=r$ in case that r is even.

By (1), $H^1(M;\mathbb{Q})=H^2(M;\mathbb{Q})=0$. It follows from the cohomology sequence of (W,M) that

5. $j^*: H^2(W,M;\mathbb{Q}) \to H^2(W;\mathbb{Q})$ is an isomorphism onto.

Denote by m the order of $H^2(M;\mathbb{Z})$. By (1), m is an odd number. It follows from (2), that mc is an element of $H^2(W;\mathbb{Z})$ whose reduction mod 2 is $w_2(W)$. Since m is the order of $H^2(M;\mathbb{Z})$, mc is in the kernel of the natural homomorphism $H^2(W;\mathbb{Z}) \to H^2(M;\mathbb{Z})$. Now we regard mc as the element $mc \otimes 1 \in H^2(W;\mathbb{Z}) \otimes \mathbb{Q} = H^2(W;\mathbb{Q})$. Then, by (5), $j^{*-1}(mc) \in H^2(W,M:\mathbb{Q})$. If $r=2k-1$, we let $u^{(2r)}(W)$ be the $2r$-component of

$$e^{j^{*-1}(mc)/2} \cdot \sum_{i=0}^{k-1} \hat{A}_i(j^{*-1}p_1(W),\ldots,j^{*-1}p_i(W)) \in \sum_{j=0}^{\infty} H^j(W,M;\mathbb{Q}),$$

and define

$$v(M) = \langle u^{(2r)}(W), [W,M] \rangle \quad \bmod 1.$$

If $r=2k$, we let $u^{(2r)}(W)$ be the $2r$-component of

$$e^{j^{*-1}(mc)/2} \cdot \sum_{i=0}^{k-1} \hat{A}_i(j^{*-1}p_1(W),\ldots,j^{*-1}p_i(W)) + N_k(j^{*-1}p_1(W),\ldots,j^{*-1}p_{k-1}(W))$$

and define

$$v(M) = \langle u^{(2r)}(W), [W,M] \rangle + t_k \tau(W) \quad \bmod 1.$$

We claim that $v(M)$ is independent of the choice of W and c.

Let W' and c' be analogous to W and c. Then

$$X = W \underset{M}{\cup} (-W'),$$

is a closed, oriented, differentiable $2r$-manifold. In the commutative diagram

$$\begin{array}{ccccccc} H^1(M;\mathbb{Z}) & \to & H^2(X;\mathbb{Z}) & \to & H^2(W;\mathbb{Z}) \oplus H^2(W';\mathbb{Z}) & \to & H^2(M;\mathbb{Z}) \\ \downarrow & & \downarrow & & \downarrow & & \downarrow \\ H^1(M;\mathbb{Z}_2) & \to & H^2(X;\mathbb{Z}_2) & \to & H^2(W;\mathbb{Z}_2) \oplus H^2(W';\mathbb{Z}_2) & \to & H^2(M;\mathbb{Z}_2) \end{array}$$

where rows are MAYER-VIETORIS sequences and vertical homomorphism are reduction mod 2, we know, by (1), that $H^1(M;\mathbb{Z}) = H^1(M;\mathbb{Z}_2) = 0$ and $H^2(M;\mathbb{Z})$ is of odd order m. It follows that there is a unique $d \in H^2(X;\mathbb{Z})$ whose image in $H^2(W;\mathbb{Z}) \oplus H^2(W';\mathbb{Z})$ is $(mc, -mc')$ and whose reduction mod 2 is $w_2(X)$. Therefore, by the integrality theorem, $\hat{A}(X, d/2)$ is an integer. If $r = 2k-1$, it is easily seen that

$$\hat{A}(X, d/2) = \langle u^{(2r)}(W), [W, M] \rangle - \langle u^{(2r)}(W'), [W', M] \rangle.$$

Therefore $v(W) = v(W')$. If $r = 2k$, one may proceed as in [5] to show that $v(W) = v(W')$ again holds.

Notice that if $r = 2k$ and W is a spin manifold, then

$$v(M) = a_k \mu(M),$$

where $a_k = 4/(3 + (-1)^k)$. Hence in the special case that r is a multiple of 4, the invariant v is a generalization of the Eells-Kuiper invariant μ.

In later sections, we use only the case $r = 2k = 4$. For this case, an explicit formula for v is given as follows. Let M be a closed, oriented, differentiable 7-manifold such that $H^1(M;\mathbb{Q}) = 0$ and $H^2(M;\mathbb{Z})$ is a finite group of odd order m, and let W be a compact, oriented, differentiable 8-manifold having M as its boundary and such that (2), (3) and (4) hold. Then

$$\begin{aligned} v(M) = \langle \tfrac{1}{896}(j^{*-1} p_1(W))^2 - \tfrac{1}{192} j^{*-1} p_1(W)(j^{*-1}(mc))^2 \\ + \tfrac{1}{384}(j^{*-1}(mc))^4, [W, M] \rangle - \tfrac{1}{224} \tau(W) \quad \bmod 1. \end{aligned}$$

3. Proof of Theorem 1

By Theorem B, there is an isomorphism g of the group $\mathbb{Z}$ of integers onto the group of orientation-preserving equivariant diffeomorphism classes of free differentiable actions of the circle group G on oriented homotopy 7-spheres. For each integer i, we let $g(i)$ be represented by a free differentiable action of G on an oriented homotopy 7-sphere Σ_i. Then Σ_i/G is an oriented homotopy complex protective 3-space which can be obtained by a surgery on an imbedded oriented 3-sphere A_i in the oriented 6-sphere. Since the correspondence between i and the isotopy class of A_i is an isomorphism between $\mathbb{Z}$ and the group of isotopy classes of imbedded oriented 3-spheres in the oriented 6-sphere, it follows from [3; Theorem 5] that

$$p_1(\Sigma_i/G) = (\pm 24i + 4)a^2,$$

where a is a generator of $H^2(\Sigma_i/G; \mathbb{Z})$. We may assume

$$p_1(\Sigma_i/G) = (24i + 4)a^2$$

because, otherwise, we replace g by $i \mapsto g(-i)$.

Let W_i be the map cylinder of the projection $\Sigma_i \to \Sigma_i/G$. Then W_i is a compact differentiable 8-manifold having Σ_i as its boundary, and it can be oriented so that $\partial[W_i, \Sigma_i] = [\Sigma_i]$. Since W_i is a closed 2-cell bundle over Σ_i/G, it is easily seen that there are natural ring isomorphisms

$$H^*(W_i; \mathbb{Z}) \cong H^*(\Sigma_i/G; \mathbb{Z}),\ H^*(W_i, \Sigma_i; \mathbb{Z}) \cong H^*(CP^4, *; \mathbb{Z}),$$

where $(CP^4, *)$ is the pointed complex projective 4-space. Since $p_1(\Sigma_i/G) = (24i + 4)a^2$ and since the first Pontrjagin class of the normal bundle of Σ_i/G in W_i is a^2 (see, for example, [6]), it follows that

$$p_1(W_i) = (24i + 5)a'^2,$$

where a' is a generator of $H^2(W_i; \mathbb{Z})$. Now we can compute $\nu(\Sigma_i)$ by setting $W = W_i$ and $m = 1$. In fact, we have

$$\begin{aligned} \nu(\Sigma_i) &= \tfrac{1}{896}(24i+5)^2 - \tfrac{1}{192}(24i+5) + \tfrac{1}{384} - \tfrac{1}{224} \quad \mod 1 \\ &= \tfrac{1}{14} i(9i+2). \end{aligned}$$

Therefore we have the following table:

$i \bmod 14$	0	1	2	3	4	5	6	7	8	9	10	11	12	13
$28\nu(\Sigma_i) \mod 28$	0	22	24	6	24	22	0	14	8	10	20	10	8	14

Hence if Σ is an oriented homotopy 7-sphere on which there is a free differentiable action of the circle group, then

$$28\mu(\Sigma) \bmod 28 \ (= 28\nu(\Sigma) \bmod 28)$$

is one of

$$0,\ \pm 4,\ \pm 6,\ \pm 8,\ \pm 10,\ 14.$$

This proves that there are exactly ten such *oriented* homotopy 7-spheres or six such *nonoriented* homotopy 7-spheres.

It can be seen that there are free $\mathbb{Z}_2$ actions on each homotopy seven-sphere, and hence some of these can not be imbedded in a free circle action.

4. Applications to $\mathbb{Z}_p$ Actions

MILNOR has shown [4] how to use invariants of this kind to study involutions. A few further facts in this direction could be seen starting with HCP^3's which are more general than the standard one he uses.

However we shall not give any details on involutions but limit ourselves to a few remarks on free actions of $\mathbb{Z}_p$, p an odd prime, on homotopy spheres of dimension 7.

Let N be an HCP^3 and ξ a 2-disk bundle over N with space W and boundary $M=\partial W$. We choose ξ so that it corresponds to $p\beta$ where β is a generating 2-cocycle of N. Then M is the base space of a free $\mathbb{Z}_p$ action on a homotopy 7-sphere and $p_1(\xi)=p^2\beta^2$, $m=p$. Assume $p_1(N)=24i+4$. Then

$$p_1 W=(p^2+24i+4)\beta^2.$$

We substitute in the formula for v and make use of the cohomology ring structure. First observe

$$0\to H^2(W,M)\to H^2(W)\to H^2(M)\to 0,$$

that is

$$0\to \mathbb{Z}\to\mathbb{Z}\to\mathbb{Z}_p\to 0,$$

and similarly for H^4.

Let $\alpha=$ generator of $H^2(W,M)$. Then $\alpha\to p\beta$, $\alpha^2\to p^2\beta^2$, where we we also use β for a generator of $H^2(W)$, so that (letting generator $H^4(W,M)=a$, generator $H^8(W,M)=h$)

$$\alpha^2=pa.$$

If b is a generator of $H^4(W)$ then $a\cup b=h$ and

$$a^2=a\cup a=a\cup pb=ph.$$

$$(pa)^2=\alpha^4=p^3h.$$

We shall use the following in the formula below:

$$j^{*-1}p_1W=(p^2+24i+4)a/p(j^{*-1}p_1W)^2=(p^2+24i+4)^2h/p.$$

We can take $c=\beta$ so that

$$j^{*-1}pc=\alpha,\ (j^{*-1}pc)^2=pa(j^{*-1}pc)^4=p^3h$$

$$(j^{*-1}p_1)(j^{*-1}pc)^2=(p^2+24i+4)(a/p)pa=(p^2+24i+4)ph.$$

Using these facts, substituting gives

$$\nu(M)=(1/896)[(p^2+24i+4)^2(1(p)-4)]-(1/192)(p^2+24i+4)p+ \\ +(1/384)p^3 \qquad \bmod 1$$

$$=1/(64\cdot 42p)[3(p^2+24i+4)^2-12p-14(p^2+24i+4)p^2+7p^4] \bmod 1.$$

As an example, let $p=3$. Then

$$\nu=(1/2^7\cdot 7\cdot 9)[3(13+24i)^2-36-14\cdot 9(13+24i)+7\cdot 81] \qquad \bmod 1$$

$$=(1/2^5\cdot 7\cdot 3)[144i^2-96i-50] \qquad \bmod 1$$

$$=(1/2^4\cdot 7\cdot 3)[72i^2-48i-25] \qquad \bmod 1.$$

When $i=0$,

$$\nu=-50/672=-25/336 \qquad \bmod 1.$$

By choosing different values of i, we see that there is certainly more than one invariant, so that there is more than one free action of $\mathbb{Z}_3$ on a homotopy seven sphere. When $i=6$

$$\nu=263/336$$

so that there is more than one free $\mathbb{Z}_3$ action on S^7. MILNOR [7] has used torsion to find different actions of $\mathbb{Z}_p$ on spheres but torsion is zero for $\mathbb{Z}_3$ so this method gives some examples in addition to the ones found by MILNOR. A similar fact is true for other values of p, as can be seen by computation. The method of MILNOR is to start with a lens space and find a new space using a cobordism. This cobordism will not change ν, but computation shows there are more values of ν than there are lens spaces. Hence there are free $\mathbb{Z}_p$ actions which are not lens spaces and can not be obtained from lens spaces by h-cobordism.

References

1. MONTGOMERY, D., and C. T. YANG: Differentiable actions of homotopy seven spheres, Trans. Am. Math. Soc. **122,** 480—498 (1966).
2. HAEFLIGER, A.: Knotted $(4k-1)$-spheres in $6k$-space, Annals of Math. **75,** 452—466 (1962).
3. WALL, C. T. C.: Classification problems in differential topology V, On certain 6-manifolds, Invent. Math. **1,** 355—374 (1966).
4. MILNOR, J.: Remarks concerning spin manifolds, Differential and combinatorial topology, Symposium in honor of Morse, edited by Cairns, 55—62.

5. EELLS, J., and N. H. KUIPER: An invariant for certain smooth manifolds, Annali di Mat. **60**, 93—110 (1962).
6. HIRZEBRUCH, F.: Topological methods in algebraic geometry, third edition 1966.
7. MILNOR, J.: Some free actions of cyclic groups on spheres, Differential Analysis, Bombay Colloquium 1964 (37—42).

On the Classification of Regular $O(n)$-Manifolds in Terms of their Orbit Bundles

KLAUS JÄNICH

Introduction

If one wishes to classify or at least to understand better some large class of mathematical objects, it is often useful to study how they can be built from a comparatively small collection of more elementary objects. One possibility to do so for the class of differentiable G-manifolds is given by their decomposition into orbit bundles. More precisely: If X is a compact differentiable G-manifold (G a compact Lie group), and if $Y_{(H)} \subset X$ denotes the union of all orbits of type (H), then $Y_{(H)}$ is not only an invariant sub*set* of X, but also an invariant sub*manifold*, and therefore X is the union of a finite number of G-manifolds with just one orbit type each. We call them the "orbit bundles" of X, because $Y_{(H)}$ is in a natural way a fibre bundle with fibre G/H, group $\Gamma = N_H/H$ (see [2], pp. 157–160) and with a differentiable manifold as base space, and because its fibres are the orbits of the G-action. $Y_{(H)}$, as a G-manifold, is completely determined by its associated Γ-principal bundle.

The viewpoint of the present article is to consider the one-type-G-manifolds as those "more elementary objects". (I do not wish to obscure the fact that the classification of G-manifolds with a single orbit type can be extremely difficult itself—we are only concerned with the opposite kind of question, in some sense.) Our program then, in its most general formulation, is *to classify G-manifolds in terms of their orbit bundles.*

1. G-manifolds with two Orbit Types

To carry out this program in its full generality does not seem feasible. One will have to make some restrictions. It is natural to start with the G-manifolds which have (essentially) only two different types of orbits. This case, under some mild additional conditions, has been treated in WU-CHUNG HSIANG and WU-YI HSIANG's paper [5], Theorem 3 (resp. [6],

Theorem 4.1) and independently in my paper [7]. Here X/G becomes a differentiable manifold M with boundary, and the two orbit bundles are bundles over $M-\partial M$ and ∂M respectively. But in general they do not determine X: They can be put together in many different ways to form a closed G-manifold. The missing information is given as follows:

First, the principal bundle describing the orbit bundle over $M-\partial M$ can be naturally extended (and thereby compactified) to a bundle P over *all* of M. Then it follows from the geometrical meaning of P in X that the other bundle is given by a certain *reduction* σ of $P|\partial M$ to a smaller structure group. But the pair (P,σ) not only describes the two bundles, it also provides a "link" between them, and in fact the isomorphism classes of pairs (P,σ) are the classifying data for this particular kind of G-manifold. (See the theorem in 3.1 of [7], but beware of the false remark made in 3.2 of that paper).

2. Regular $O(n)$-Manifolds

Certain G-manifolds with *three* orbit types are also classified in [5], [6] and [7], § 6. Let for instance $G=O(n)$ and each orbit be of type $(O(n))$ (=fixed point), $(O(n-1))$ or $(O(n-2))$. Then X/G is a manifold M with boundary, the fixed point set L being a 2-codimensional submanifold of ∂M (see [6], Theorem 4.2, Prop. 4.4). If $M\cong D^4$ and $L\neq\emptyset$ and connected (hence a *knot*), then X is called a *knot-manifold.* The classification by orbit bundles led to the result, that there is a bijective correspondence between the isomorphism classes of knots and the equivariant diffeomorphism classes of knot-manifolds: For each knot there is only one possibility for the orbit bundles and the relations between them.

"Most" of the knot manifolds are exotic spheres, and very detailed results on the differentiable structures of these spheres and their relation to singularities of algebraic hypersurfaces have been obtained by HIRZEBRUCH [4].

The classifying data for a knot-manifold are very neat: Just a knot. This is highly accidental, however. Already for $O(n)$-manifolds with a fourth orbit type $(O(n-3))$, the relations between the four orbit bundles become rather complicated, although involving no new principal difficulties. The purpose of the present article is to describe these relations in a more general case, namely for what I call "regular $O(n)$-manifolds".

Definition. An $O(n)$-manifold is called *regular*, if each orbit is of the type $(O(k))$ for some $0\leqslant k\leqslant n$, ($O(k)$ being imbedded in $O(n)$ in the standard manner).

The number of different orbit types of a regular $O(n)$-manifold may even locally be up to $n+1$, but the great "regularity" of the partially ordered (in our case: ordered) set of orbit types

$$(O(n))<(O(n-1))<\cdots<(O(1))<(1)$$

minimizes other difficulties. Note also, that Cartesian product (with diagonal action), equivariant connected sum, and any invariant submanifold of regular $O(n)$-manifolds is again regular.

I will now report on the "classification of regular $O(n)$-manifolds in terms of their orbit bundles". For proofs and details I refer to [8].

3. First Step: Compactification of the Orbit Bundles

Let X be a closed regular $O(n)$-manifold. We will now carefully disassemble the orbit bundles which constitute X, in order to learn how to reverse this process.

Put $X_0=X$ and $Y_0=\{x\in X|(G_x)=(O(n))\}$. Then Y_0 is the bundle of the smallest orbits (namely the fixed point set) of X_0 and in particular is a *closed* submanifold of X_0. Now we remove the interior of a small tubular neighborhood of Y_0 from X_0. The resulting regular $O(n)$-manifold with boundary shall be denoted by $X_1=X_0 \odot Y_0$. (This removal can be described intrinsically, so that X_1 in fact does not depend on the choice of a tubular neighborhood.) X_1 does not have fixed points, each orbit in X_1 is of type $(O(n-k))$ for some $1\leqslant k\leqslant n$. We denote $\{x\in X_1|(G_x)=(O(n-1))\}$, which is the bundle of the smallest orbits in X_1, by Y_1 and define $X_2=X_1 \odot Y_1$. And so on by induction: $Y_k=\{x\in X_k|(G_x)=(O(n-k))\}$ and $X_{k+1}=X_k \odot Y_k$. Eventually, X_m will have only one orbit type, thus $Y_m=X_m$ and $X_{m+1}=\phi$.

$Y_0,\ldots,Y_m$ are compactifications of the orbit bundles of the original X.

4. $\langle k\rangle$-Manifolds

Before I can proceed, I must draw your attention to a technical but important detail, namely to the "corners" which occur upon the successive removal of tubular neighborhoods. X_0 is a smooth manifold without boundary. X_1 is a manifold with boundary, let us denote the boundary of X_1 by $\partial_0 X_1$, because it results from the removal of Y_0 from X_0. X_2 in turn is a "manifold with corners", its boundary is the union of two manifolds with boundary: $\partial_0 X_2$, resulting from the removal of Y_0, and $\partial_1 X_2$, which results from the removal of Y_1. Their common boundary is the "edge" of X_2. We call $\partial_0 X_2$ and $\partial_1 X_2$ the "faces" of X_2.

Let us now drop the subject of $O(n)$-manifolds for a moment and let me make some remarks on manifolds with corners. In the same way as a differentiable manifold with boundary is modelled on the open sets of $\{x \in R^n | x_n \geqslant 0\}$, a *differentiable manifold with corners* is modelled on

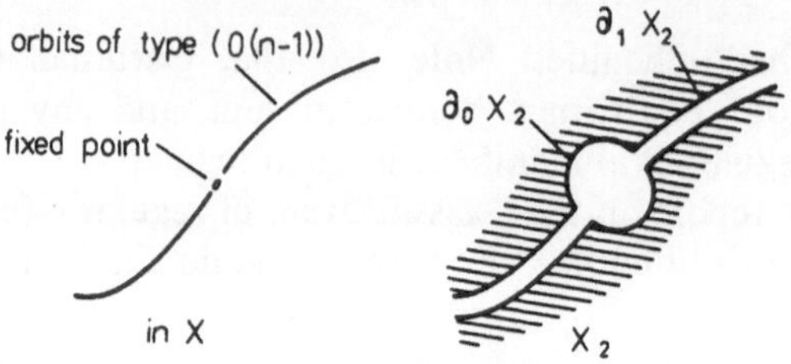

Fig. 1

the open sets of $\{x \in R^n | x_1 \geqslant 0, \ldots, x_n \geqslant 0\}$. (See A. DOUADY's "variétés a bord anguleux" [3]). The number of zeros in the n-tuple $x=(x_1, \ldots, x_n)$ is denoted by $c(x)$. On a manifold with corners X, it does not depend on the choice of the local coordinates, and so induces a map $c\colon X \to \mathbb{Z}_+$. Clearly $\{x | c(x) \neq 0\}$ is the boundary of X, and $c(x)=1$ characterizes the "good points" of the boundary, where X looks exactly like an ordinary differentiable manifold with boundary.

How do we have to define the "faces", which occurred in our example X_2? If we wish a face of a manifold with corners to be a manifold with

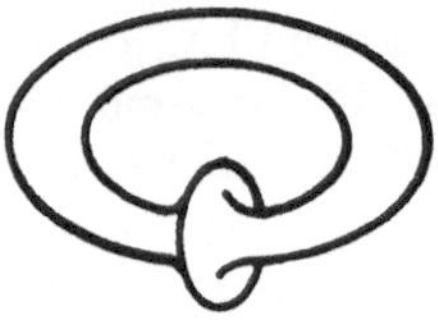

Fig. 2

corners again, then examples like the 2-disk with one corner or the solid torus with an edge don't seem to have any faces. To exclude these cases, we define:

Definition. A manifold with corners X is called a *manifold with faces*, if each $p \in X$ belongs to the closure of $c(p)$ components of $\{x | c(x)=1\}$.

Note that p belongs to at most $c(p)$ such closures anyway.

Let X be a manifold with faces. The closure of a component of $\{x | c(x)=1\}$ shall be called a *connected face*, and the union of disjoint connected faces is called a *face* of X. Such faces are in fact manifolds with faces themselves.

Having introduced "faces", we will now define something very similar to an ordinary simplex with k faces.

Definition. A manifold with faces X together with a k-tuple $(\partial_0 X, \ldots, \partial_{k-1} X)$ of faces of X is called a *$\langle k\rangle$-manifold*, if

(i): $\partial_0 X \cup \cdots \cup \partial_{k-1} X = \partial X$ and

(ii): $\partial_i X \cap \partial_j X$ is a face of $\partial_i X$ and of $\partial_j X$ for $i \neq j$.

In particular, a $\langle 0\rangle$-manifold is a manifold without boundary (of any dimension! k has nothing to do with the dimension of X), and a $\langle 1\rangle$-manifold is an ordinary manifold with boundary. Notice: $\partial_i X$ becomes a $\langle k-1\rangle$-manifold if we define $\partial_j(\partial_i X) = \partial_i X \cap \partial_j X$ for $0 \leqslant j < i$ and $\partial_i X \cap \partial_{j+1} X$ for $i < j \leqslant k-2$.

Example (and motive for the definition). If X is a regular $O(n)$-manifold, then the X_k and Y_k of § 3 are $\langle k\rangle$-manifolds, with $\partial_i X_k$ being the part of the boundary of X_k which originates from the removal of Y_i from X_i, and $\partial_i Y_k = Y_k \cap \partial_i X_k$. Moreover, the $\langle k\rangle$-structure of Y_k induces a $\langle k\rangle$-structure on the quotient $M_k = Y_k/O(n)$, and hence Y_k is a fibre bundle with fibre $O(n)/O(n-k)$ and group $\Gamma = O(k)$ over a $\langle k\rangle$-manifold M_k.

5. Regular $O(n)$-Manifolds of Type r

The operation $\odot$ is not uniquely reversible. Let for instance Σ^n, $n \geqslant 7$, be an exotic sphere. Then we always have $\Sigma^n \odot \mathrm{pt} = D^n = S^n \odot \mathrm{pt}$. It is for this reason that the compactifications $Y_0, \ldots, Y_m$ of the orbit bundles are far from being sufficient to determine the $O(n)$-manifold X. For a complete set of invariants we also have to take into account the normal bundles of the Y_k in X_k, and certain relations between all these data. To describe this smoothly, I first have to introduce one more notion.

Let again X be a regular $O(n)$-manifold, $x \in X$; let V_x be the normal space of the orbit Gx at the point x and $G_x \to \mathrm{GL}(V_x)$ the induced representation ("slice representation"). We know that $G_x \cong O(i)$ for some i.

Proposition (Wu-Yi Hsiang*). *$G_x \to \mathrm{GL}(V_x)$ is equivalent to $k\rho_i \oplus$* trivial, *where $k\rho_i$ denotes the k-fold direct sum of the standard representation ρ_i of $O(i)$ in R^i.*

k is well determined as long as $i \neq 0$, and hence we can define:

Definition. If $i \neq 0$, the number $k-i$ is called the *type* of x.

* Oral Communication.

If $X' = \{x | G_x \neq 1\}$, then the "type" defines a G-invariant map $X' \to Z$, and this map is *locally constant*. Thus the points of constant type form certain open and closed subsets of X'. The description of the classifying data, which will be given in the next section, depends on the type and hence has to be done separately for each of these subsets. Just for convenience however, I will restrict myself to the case where there is only one such subset:

Definition. Let X be a regular $O(n)$-manifold and $r \geqslant -n$. Then X is *of type* r, if each $x \in X'$ is of type r (and $X' = X$ if $r < 0$).

If we assume X/G to be connected, then the existence of just *one* point x of a type $r < 0$ implies that the whole manifold is of type r. Also, if X/G is connected, then X is of type r for some $r < 0$ if and only if the principal isotropy group is not the trivial group $\{1\}$. If the principal isotropy group *is* $\{1\}$, the manifold may be of mixed type. But the classification theorem for manifolds of type r can easely be modified to handle this case (see [8], Theorem 1′).

From now on, let X be a regular $O(n)$-manifold of type r and $m = \min(n, n+r)$. Then $X_m = Y_m$ and $X_{m+1} = \phi$. Finally, let X be endowed with an equivariant Riemannian metric.

6. The Classification Theorem

$Y_k \to M_k$ is in a canonical way a fibre bundle with fibre $O(n)/O(n-k)$ and group $O(k)$. In fact, the associated $O(k)$-principal bundle can be identified with $P_k = \{x \in Y_k | G_x = O(n-k)\} \subset Y_k$.

Now consider the *normal bundle* of Y_k in X_k. It can be described by a principal bundle Q_k over M_k with group $O(n-k+r)$; with the one exception, of course, that for $k = n$ (as it occurs for $r > 0$), the group of Q_k is not $O(r)$ but $\{1\}$. One gets the Q_k as follows:

The normal bundle E_k is an $O(n)$-vector bundle (see ATIYAH-SEGAL [1], Lecture I) over the $O(n)$-manifold Y_k. Then $(E_k | P_k)/O(k)$ is an $O(n-k)$-vector bundle over the trivial $O(n-k)$-manifold M_k; that is to say: It is a vector bundle over M_k, and each fibre is an $O(n-k)$-module. These modules however are all equivalent to $(n-k+r)\rho_{n-k}$, and using the metric we obtain an associated $O(n-k+r)$-principal bundle Q_k over M_k.

We denote the "fibered product" $P_k \oplus Q_k$ by Λ_k, which is then a principal bundle over M_k with group $O(k) \times O(n-k+r)$, resp. $O(n)$ if $k = n$. Λ_k suffices to reproduce Y_k and E_k with their structures as $O(n)$-manifold and $O(n)$-vector bundle. To rebuild X from $(\Lambda_0, \dots, \Lambda_m)$

however, we also have to use certain relations between the Λ's. These relations are the following:

Let $i<k$. Then $\partial_i \Lambda_k = \Lambda_k | \partial_i M_k$ is fibered not only over $\partial_i M_k$, but also over M_i. Moreover, the associated principal bundle of this fibration is Λ_i itself! Therefore we have $\partial_i \Lambda_k$ as an associated fibre bundle:

$$\partial_i \Lambda_k = \lambda_k^i \; {}_i\!\times \Lambda_i,$$

where ${}_i\!\times$ stands for ${}_{O(i)\times O(n-i+r)}\!\times$, and λ_k^i is the typical fibre of this bundle.

The structure of λ_k^i is quite extensive: It is an $O(k)\times O(n-k+r)$-principal bundle over a $\langle k-i-1\rangle$-manifold, endowed with an $O(i)\times O(n-i+r)$-action compatible with all this. But the point is: λ_k^i *does not depend on* X, it only depends on $n, r, i; k$. We call $\{\lambda_k^i\}_{0\leqq i<k\leqq m}$ the (n,r)-*system*. For an explicit description see [8].

Given the (n,r)-system, we can define without reference to any X:

Definition. An $(m+1)$-tuple $(\Lambda_0, \ldots, \Lambda_m)$ of $O(i)\times O(n-i+r)$-principal bundles Λ_i over $\langle i\rangle$-manifolds, with $\partial_i \Lambda_k = \lambda_k^i \; {}_i\!\times \Lambda_i$ is called an (n,r)-*chain of bordisms*. Let $C(n,r)$ denote the set of isomorphism classes of (n,r)-chains.

Notice that $\Lambda_0, \ldots, \Lambda_{k-1}$ determine the boundary $\partial_0 \Lambda_k \cup \cdots \cup \partial_{k-1} \Lambda_k$ of Λ_k. Thus to construct an (n,r)-chain means to start with a Λ_0 over a closed manifold and to solve m bordism problems.

If $S(n,r)$ denotes the set of equivariant diffeomorphism classes of regular $O(n)$-manifolds of type r, we have constructed a map $S(n,r) \to C(n,r)$.

Classification Theorem. $S(n,r)\to C(n,r)$ *is bijective.*

7. A Corollary

It requires further work to obtain from this classification theorem more explicit results on regular $O(n)$-manifolds. In particular, it is desirable (and difficult) to get relations between the topological properties of X and its orbit bundles. But I wish to mention a corollary which can be obtained more or less immediately from the theorem. Consider a G-manifold X and a closed subgroup H of G. Then $F(H,X)=\{x \mid hx=x$ for all $h\in H\}$ is a $\Gamma = N_H/H$-manifold. One asks: How much information on the G-manifold X does the Γ-manifold $F(H,X)$ contain? Which Γ-manifolds are realizable as an $F(H,X)$? If $G=O(n)$, $H=O(k)$, then $\Gamma=O(n-k)$ and $F(O(k),\ldots)$ defines a map $S(n,r)\to S(n-k,k+r)$.

Corollary. *$S(n,r) \to S(n+r,0)$ is bijective for $r \leqslant 0$.*

In particular, an $O(n)$-manifold X of type $2-n$, $n \geqslant 2$ (knot-manifolds, for instance, are of this type), is completely determined by its $O(2)$-submanifold $F(O(n-2), X)$.

It would be interesting to have a more direct, "generalizable", proof of this corollary.

References

1. ATIYAH, M. F., and G. SEGAL: Equivariant K-theory, University of Warwick, 1965.
2. BOREL, A.: Seminar on Transformation Groups, Annals of Math. Studies **46** (1960).
3. DOUADY, A.: Variétés á bords anguleux et voisinages tubulaires, Séminaire Henri Cartan **14** (1961/62).
4. HIRZEBRUCH, F.: Singularities and exotic spheres, Séminaire Bourbaki **19** (1966/67).
5. HSIANG, WU-CHUNG, and WU-YI HSIANG: Some results on differentiable actions, Bull. Am. Math. Soc. **72**, 134—137 (1966).
6. — Differentiable actions of compact connected classical groups I, Amer. J. Math. **89**, 705—786 (1967).
7. JÄNICH, K.: Differenzierbare Mannigfaltigkeiten mit Rand als Orbiträume differenzierbarer G-Mannigfaltigkeiten ohne Rand, Topology **5**, 301—320 (1966).
8. — On the classification of $O(n)$manifolds, Math. Ann. **176**, 53—76 (1968).

Involutions on Homotopy Spheres

G. R. LIVESAY and C. B. THOMAS

This survey article should be read in conjunction with that of GLEN BREDON; the difference being that he discusses a large number of particular examples, while we wish to outline the progress which is being made towards classifying involutions in general. In particular the reader wishing for a summary of the important work of MONTGOMERY and YANG on 7-spheres is referred to the article of BREDON. Our survey divides naturally into a discussion of free involutions and of involutions with a fixed point set. The subjects covered in this article reflect the author's interests. They fully recognize their ignorance of the work of many other investigators, and that this is a reflection on them only.

Fixed Point Free Involutions

At present there seem to be two approaches to this problem. The first attempts to "desuspend" an involution and leads in half the dimensions to a signature/Arf invariant obstruction. The second decomposes any pair (Σ^{2n+1}, T) into the union of two solid tori carrying the antipodal map (which we denote by A). As yet we can see no way of combining these two methods, although it is easy to formulate more or less wild conjectures.

Given a smooth (C^∞) involution (Σ^{n+1}, T) we say that T *desuspends* if there exists an invariant n-sphere S^n, smoothly imbedded in Σ^{n+1}.

Theorem 1. (BROWDER-LIVESAY, *see* [3]) (i) *If* $n \geqslant 5$ *is odd, then* (Σ^{n+1}, T) *desuspends to* $(S^n, T|S^n)$. (ii) *if* $n \equiv 2(4), n > 5$, *then* (Σ^{n+1}, T) *desuspends to* $(S^n, T|S^n)$ *if and only if a certain integer obstruction* $\sigma(\Sigma^{n+1}, T)$, *called the signature, vanishes.* (iii) *If* $n \equiv 0(4), n > 4$, *then* (Σ^{n+1}, T) *desuspends to* $(S^n, T|S^n)$ *if and only if a certain* $\mathbb{Z}_2$*-obstruction, called the Arf invariant, vanishes.*

The method of proof is to express the homotopy sphere as a union $\Sigma^{n+1} = F \cup TF$ with $F \cap TF = M^n$, and then exchange handles between the two "sides" of the invariant submanifold M^n. As usual with such arguments the main difficulties arise in the middle dimension, and account for the three cases in the theorem. It is interesting to note that the $\mathbb{Z}_2$ and $\mathbb{Z}$ obstructions seem to have been interchanged; this is because the defining bilinear form depends on T, which reverses orientation for n even. The same sort of argument leads to a theorem about the uniqueness of desuspensions, in which, as one would expect, the roles of even and odd are interchanged.

Theorem 2 (BROWDER-LIVESAY). *If* $n>4$ *is even, any two desuspensions of* (Σ^{n+1}, T) *are equivariantly concordant. In the case of* (Σ^{4k}, T) *there is an integer obstruction to equivariant concordance, and in the case of* (Σ^{4k+2}, T) *a* $\mathbb{Z}_2$*-obstruction.*

Following on from these results, S. LOPEZ DE MEDRANO [10] has proved by explicit construction that for all $n \geqslant 1$ and all integers i, there is a fixed point free differentiable involution T_i: $\Sigma_i^{4n+3} \to \Sigma_i^{4n+3}$ with $\sigma(\Sigma_i, T_i) = 8i$. In dimension three the same theorem holds with homology spheres replacing homotopy spheres. In his paper, LOPEZ DE MEDRANO remarks that an analogous construction does not work in dimension $4n+1$, so the question of the existence of an involution with non-zero Arf invariant remains open*.

Using methods independent of those in the previous paragraph one can prove:

Theorem 3. (LIVESAY-THOMAS, *see* [9]). *Let* A *denote the antipodal map. Any fixed point free involution* (Σ^{2n+1}, T) *can be obtained by identifying* $(S^n \times D^{n+1}, A)$ *and* $(D^{n+1} \times S^n, A)$ *along their boundaries by means of an A-equivariant diffeomorphism. Two diffeomorphisms which differ by one which can be equivariantly extended to either* $S^n \times D^{n+1}$ *or* $D^{n+1} \times S^n$ *define conjugate involutions.*

This theorem is proved by imbedding a copy of P^n in the quotient space Q^{2n+1} of (Σ^{2n+1}, T), and then constructing a homotopy equivalence $P^{2n+1} \to Q^{2n+1}$ which is the identity on P^n. Using results of ADAMS [1] and ATIYAH [2] one proves that the normal bundles of P^n in P^{2n+1} and

---

* I. BERSTEIN, *Bull. Amer. Math. Soc.* 74 (1968) pp. 678—682, has given examples of smooth involutions with non-zero Arf invariant on the Kervaire spheres. C. T. C. WALL, *Bull, Amer. Math. Soc.* 74 (1968) pp. 554—558, has classified free PL involutions on spheres. A similar classification has been obtained independently by S. LÓPEZ DE MEDRANO (see these Proceedings).

Many other results leading in the direction of a classification of smooth free involutions on homotopy spheres have been obtained by I. BERSTEIN and G. R. LIVESAY, W. BROWDER, C. H. GIFFEN, F. HIRZEBRUCH, K. JÄNICH, S. LÓPEZ DE MEDRANO, D. MONTGOMERY and C. T. YANG, P. ORLIK and C. ROURKE, D. SULLIVAN, and others.

Q^{2n+1} are equivalent; hence that there is an equivariant imbedding of $(S^n \times D^{n+1}, A)$ into (Σ^{2n+1}, T). That T is equivalent to A on the complementary solid torus now follows by considering the invariant diagonal sphere in the boundary, and then applying the equivariant collaring and s-cobordism theorems.

The original motivation for the last theorem was to define a composition law for involutions in such a way that the signature and Arf invariant became homomorphisms into $\mathbb{Z}$ and $\mathbb{Z}_2$ respectively.

(In [14], D. MONTGOMERY and C. T. YANG obtain free $\mathbb{Z}_2$ actions on homotopy seven spheres which exhaust all possible values of the signature. Their result is especially interesting because they obtain these actions as restrictions of free circle actions, where a group composition is defined.) Unfortunately this has not yet been done; the difficulty lies in carrying over the composition law of diffeomorphisms to their equivalence classes. It would also be of great interest to link the Livesay-Thomas and the Browder-Livesay methods. Ideally one would like to construct a list of A-equivariant diffeomorphisms of $S^n \times S^n$ such that the corresponding (Σ^{2n+1}, T)'s exhaust all conjugacy classes. A rather less ambitious program would be to construct diffeomorphisms giving rise to all possible signatures and Arf invariants; in particular one would like to realize Lopez de Medrano's examples by the method of Theorem 3. This would be a considerable advance, although the following example shows that for dimension $4n+3$ the signature does not determine the (smooth) equivalence class.

Example (MILNOR [11]). Consider (S^7, A) and let Σ_0 have order two in Θ_7. By taking the obvious involution T on $\Sigma_0 \# S^7 \# \Sigma_0$, one obtains a quotient manifold distinct from P^7. However it is clear that (S^7, T) desuspends to (S^6, A) and so has zero signature.

By considering MILNOR's exotic 7-sphere as an S^3-bundle over S^4 with an involution defined by the antipodal map on fibres, and desuspending twice, one obtains [7] an exotic involution on S^5. This involution is clearly a candidate for a non-zero Arf invariant. But as BROWDER and LIVESAY observe in their paper, the desuspension argument breaks down, because of the customary difficulties in finding a basis for $H_2(M^4)$ represented by imbedded spheres.

In dimension three one has the following result:

Theorem 4 (LIVESAY, *see* [8]). *If* (S^3, T) *is fixed point free, then* T *is conjugate to the antipodal map.*

LOPEZ DE MEDRANO has observed that if one of his homology 3-spheres is 1-connected, Theorem 4 implies that it is a counter-example to the Poincaré conjecture.

Involutions with Fixed Points

We mention here, as in the fixed point free case, only a few of the "positive" results, referring the reader to the article of BREDON, and to the many papers in the literature for examples of how badly behaved involutions with fixed points can be.

In [16], P. A. SMITH, using a remarkable theorem of R. L. WILDER concerning generalized 2-manifolds, proved that *if* T: $S^3 \to S^3$ *is of finite period, then the fixed point set of* T *is a sphere.* This result provided the stimulus for much of the further work in this area. One question which was raised, known as the Smith conjecture, was the following. *If* T: $S^3 \to S^3$ *is periodic with a circle of fixed points, can the circle be knotted*? This question remains open, at least for odd periods*. If one generalizes to periodic maps T: $S^n \to S^n$, then C. H. GIFFEN [6], has shown that a knotted S^{n-2} may occur as the fixed point set, for any $n \geqslant 4$ and any period $p > 1$. D. MONTGOMERY and H. SAMELSON, [13], showed that if the fixed point circle of T: $S^3 \to S^3$ (an involution) is unknotted, then T is equivalent to a rotation (conjugate in the group of homeomorphisms). E. E. MOISE [12], has extended this result to periodic P. L. homeomorphisms of any period. In their paper, MONTGOMERY and SAMELSON also proved that the fixed point circle of an involution T: $S^3 \to S^3$ could not be a knotted torus knot of type $(p, 2)$.

In [4], E. H. CONNELL, D. MONTGOMERY, and C. T. YANG have proved the following theorem. *Let* G *be a compact group which acts differentiably on* E^n *with a fixed point set diffeomorphic to a hyperplane with codimension at least* 3, *and with all the non-trivial orbits of the same type and of codimension at least* 5. *Then the action of* G *is differentiably equivalent to a linear one.* (In connection with this, see also [5], and [15].) They also obtain the corollary: *Let* G *be a compact group which acts differentiably on* S^n *with a fixed point set which is diffeomorphic to* S^k, $0 \leqslant k \leqslant n-3$, *and with all other orbits of the same type and of dimension* r. *Then the action of* G *is topologically equivalent to a linear action on* S^n *if* $n - r \geqslant 5$.

References

1. ADAMS, J. F.: On the groups $J(X)$-II, Topology **3**, 137—171 (1965).
2. ATIYAH, M. F.: Thom complexes, Proc. London Math. Soc. (3) **11**, 291—310 (1961).

---

* FRIEDHELM WALDHAUSEN has shown that the fixed-point circle cannot be knotted when the period is 2. If T has period $2k$, then T^k has period 2 and the same circle of fixed points as T, so Waldhausen's result holds for all even periods.

3. Browder, W., and G. R. Livesay: Fixed point free involutions on homotopy spheres, Bull. Amer. Math. Soc. **73,** 242—245 (1967).
4. Connell, E. H., D. Montgomery, and C. T. Yang: Compact groups in E^n, Ann. of Math. **80,** 94—103 (1964).
5. — — — Correction to "Compact groups in E^n", Ann. of Math. **81,** 194 (1965).
6. Giffen, C. H.: The generalized Smith conjecture, Amer. J. Math. **88,** 187—198 (1966).
7. Hirsch, M. W., and J. W. Milnor: Some curious involutions of spheres, Bull. Amer. Math. Soc. **70,** 372—377 (1964).
8. Livesay, G. R.: Fixed point free involutions on the 3-sphere, Ann. of Math. **72,** 603—611 (1960).
9. —, and C. B. Thomas: Free $\mathbb{Z}_2$ and $\mathbb{Z}_3$ actions on homotopy spheres, Topology (to appear).
10. Lopez de Medrano, S.: Involutions of homotopy spheres and homology 3-spheres, to appear.
11. Milnor, J.: Remarks concerning spin manifolds, Differential and combinatorial topology, Princeton, 1964.
12. Moise, E. E.: Periodic homeomorphisms of the 3-sphere, Illinois J. of Math. **6,** 206—225 (1962).
13. Montgomery, D., and H. Samelson: A theorem on fixed points of involutions in S^3, Canadian J. Math. **7,** 208—220 (1955).
14. —, and C. T. Yang: Differentiable actions on homotopy seven spheres, III, to appear.
15. Siebenmann, L. C.: On detecting open collars, to appear.
16. Smith, P. A.: Transformations of finite period, II, Ann. of Math. **40,** 690—711 (1939).

Involutionen auf Mannigfaltigkeiten

HEINRICH BEHNKE zum 70. Geburtstag gewidmet

F. HIRZEBRUCH

An der Konferenz über Transformationsgruppen (Tulane University, New Orleans) konnte ich leider wider Erwarten nicht teilnehmen. Der Aufforderung der Veranstalter, trotzdem einen Bericht für die Proceedings zu schreiben, komme ich gern nach. In New Orleans wollte ich über equivariant plumbing, equivariant handle body constructions und knot manifolds im Sinne von W. C. HSIANG, W. Y. HSIANG und JÄNICH vortragen. Der vorliegende Bericht hängt zwar hiermit sehr zusammen, legt jedoch das Schwergewicht auf Untersuchungen, mit denen ich mich in Berkeley (August und September 1967) beschäftigt habe. In Berkeley hatte ich zahlreiche Anregungen durch Gespräche mit D. SULLIVAN und C. T. C. WALL. Da es schwierig ist, diesen beiden Mathematikern stets an den in Frage kommenden Stellen des Berichtes zu danken, möchte ich dies hier in der Einleitung ganz herzlich tun. Die ursprünglich für New Orleans vorgesehenen Dinge kann man in den Lecture Notes von K. H. MAYER und dem Verf. [18] und in der Bonner Dissertation von D. ERLE nachlesen. Der vorliegende Bericht entspricht im wesentlichen Kolloquiumsvorträgen, die der Verfasser im Oktober 1967 in Haverford, Princeton, New York und Boston gehalten hat; der Bericht ist manchmal ausführlicher als die Vorträge, muß sich aber an manchen Stellen trotzdem auf Beweisandeutungen beschränken.

Viele Dinge dieser Arbeit können verallgemeinert werden auf G-Mannigfaltigkeiten, wo G eine kompakte Liesche Gruppe ist (vgl. ATIYAH and SINGER [3].

1. Beispiele von Involutionen

Wir betrachten die Gleichung

$$(1) \qquad z_0^{a_0} + z_1^{a_1} + \cdots + z_n^{a_n} = 0 \qquad (n \geqq 1).$$

Die Exponenten a_j sollen ganze Zahlen $\geqq 2$ sein. Wir setzen $a = (a_0, \ldots, a_n)$. Nach A. WEIL [31] ist die Anzahl der Lösungen von (1) über jedem endlichen Körper mit q Elementen gleich q^n, wenn die folgende Bedingung erfüllt ist:

$$(2)\quad \begin{cases} \text{Die Menge aller} \quad x=(x_0,\ldots,x_n)\in\mathbb{Z}^{n+1} \quad \text{mit} \\ 0<x_j<a_j \quad \text{und} \quad \sum_{j=0}^{n} \frac{x_j}{a_j}\in\mathbb{Z} \quad \text{ist leer.} \end{cases}$$

BRIESKORN ([6], [7]) hat die durch Konjunktion von (1) und der Gleichung

$$(3)\qquad \sum_{j=0}^{n} z_j \bar{z}_j = 1$$

definierte orientierte $(2n-1)$-dimensionale kompakte differenzierbare Mannigfaltigkeit $\Sigma_a^{2n-1}\subset\mathbb{C}^{n+1}$ betrachtet, die mit Kodimension 2 in der durch (3) gegebene Sphäre $\mathbb{S}^{2n+1}$ eingebettet ist. BRIESKORN hat zahlreiche interessante Resultate über die Σ_a^{2n-1} erhalten, die zum Teil im folgenden vorkommen.

Σ_a^{2n-1} ist $(n-2)$-fach zusammenhängend. Σ_a^{2n-1} ist eine rationale Homologiesphäre genau dann, wenn (2) gilt (vgl. auch [18]).

Wenn alle a_j gerade oder wenn alle a_j ungerade sind, dann führt die Involution $Tz=-z$ des $\mathbb{C}^{n+1}$ die Mannigfaltigkeit Σ_a^{2n-1} in sich über. Wir bezeichnen $T|\Sigma_a^{2n-1}$ mit T_a und erhalten Beispiele von orientierten Mannigfaltigkeiten mit orientierungserhaltenden fixpunktfreien Involutionen. Wenn alle a_j gerade sind, dann ist Σ_a^{2n-1} keine ganzzahlige Homologiesphäre [6]. Da wir hauptsächlich Involutionen auf Sphären untersuchen möchten, betrachten wir hier den Fall ungerader Exponenten.

Wenn alle a_j ungerade sind, dann ist (2) äquivalent mit der Existenz *eines* Exponenten a_i, welcher teilerfremd zu allen anderen Exponenten ist [7].

Wenn alle a_j ungerade sind, dann ist die Existenz von *zwei* Exponenten, von denen jeder teilerfremd zu allen anderen Exponenten ist, äquivalent damit, daß Σ_a^{2n-1} *eine ganzzahlige Homologiesphäre* ist ([6], [7]).

Für $n\geqq 3$ ist Σ_a^{2n-1} immer einfach-zusammenhängend. Nach SMALE [28] ist deshalb Σ_a^{2n-1} für $n\geqq 3$ eine Sphäre (d.h. homöomorph zur Standardsphäre), falls Σ_a^{2n-1} eine ganzzahlige Homologiesphäre ist.

Wir beschränken uns auf den Fall $n=2k$. Die orientierten differenzierbaren Mannigfaltigkeiten Σ_a^{4k-1} repräsentieren für $k\geqq 2$, sofern sie Sphären sind, Elemente der endlichen zyklischen Gruppe bP_{4k}, die von KERVAIRE und MILNOR [20] eingeführt wurde, und zwar ist $\Sigma_{(3,5,2,\ldots,2)}^{4k-1}$ (insgesamt $2k+1$ Exponenten) ein erzeugendes Element dieser Gruppe ([6], [17], [18]), das mit M_0^{4k-1} bezeichnet werden soll. M_0^{4k-1} ist (bis auf das Vorzeichen) die Sphäre, die sich durch "plumbing" von 8 Exemplaren des Tangentialbündels von $\mathbb{S}^{2k}$ gemäß dem Baume E_8 ergibt ([17], [18]).

Wenn (2) erfüllt ist, dann liegt jede auftretende Summe $\sum \frac{x_j}{a_j}$ entweder strikt zwischen 0 und 1 mod 2 oder zwischen 1 und 2 mod 2. Wir betrachten entsprechend für $n=2k$ die Mengen

$$\left\{x \mid x \in \mathbb{Z}^{2k+1},\quad 0<x_j<a_j,\quad 0<\sum_{j=0}^{2k} \frac{x_j}{a_j}<1 \bmod 2\right\}$$

$$\left\{x \mid x \in \mathbb{Z}^{2k+1},\quad 0<x_j<a_j,\quad 1<\sum_{j=0}^{2k} \frac{x_j}{a_j}<2 \bmod 2\right\}$$

und definieren $\tau(a_0,a_1,\ldots,a_{2k})$ als die Differenz der Anzahlen dieser beiden Mengen. Diese Definition ist auch für $k=1$ sinnvoll. Wenn Σ_a^{4k-1} eine ganzzahlige Homologiesphäre ist ($k \geqq 1$), dann ist $\tau(a_0,a_1,\ldots,a_{2k})$ durch 8 teilbar. In der Gruppe bP_{4k} gilt für $k \geqq 2$ (siehe [6], [7]; vgl. [17], [18])

(4) $$\Sigma_a^{4k-1} = (-1)^k \tfrac{1}{8} \tau(a_0,\ldots,a_{2k}) M_0^{4k-1}.$$

Es ist $\tau(3,5,2,\ldots,2)=(-1)^k \cdot 8$ (für insgesamt $2k+1$ Exponenten). Man rechnet aus, daß für $a=(a_0,\ldots,a_{2k-1},3)$ gilt:

(5) $$\begin{aligned} \tau(a_0,\ldots,a_{2k-1},3,3,3) &= -3\,\tau(a_0,\ldots,a_{2k-1},3), \\ & \qquad\qquad (k \geqq 1) \\ \tau(6j-1,18j-1,3) &= -24j(4j-1). \end{aligned}$$

Insbesondere gilt in $\mathrm{b\,P}_{4k}$

(6) $$\Sigma_{(5,17,3,\ldots,3)}^{4k-1} = 3^{k+1} M_0^{4k-1}.$$

Die Ordnungen $d(k)$ der zyklischen Gruppen bP_{4k} sind bekannt (bis auf einen Faktor 2), vgl. [20]. In [17] habe ich leider nicht auf die Schwierigkeit mit dem Faktor 2 hingewiesen. Jedenfalls teilt die Primzahl 3 keine der Ordnungen $d(k)$. Deshalb ist $\Sigma_{(5,17,3,\ldots,3)}^{4k-1}$ ein erzeugendes Element der Gruppe bP_{4k}. Diese besondere Sphäre hat die fixpunktfreie Involution $T_{(5,17,3,\ldots,3)}$. Die Standardsphäre hat die Antipodenabbildung. Eine leichte Konstruktion (vgl. BREDON [4], 8.2) liefert aus einer fixpunktfreien Involution T auf eine Sphäre Σ eine fixpunktfreie Involution T' auf der Sphäre $\Sigma + 2\Sigma'$, wo Σ' eine zu Σ gleichdimensionale Sphäre ist. Es folgt

Satz. *Jede Sphäre in* bP_{4k} $(k \geqq 2)$ *besitzt eine (orientierungserhaltende) fixpunktfreie differenzierbare Involution.*

Es entsteht die Frage, ob jedes Element von bP_{4k} in der Form Σ_a^{4k-1} mit ungeraden Exponenten a_j geschrieben werden kann.

2. Die Signatur

In diesem Paragraphen brauchen wir nicht vorauszusetzen, daß die auftretenden Mannigfaltigkeiten differenzierbar sind.

Es sei M^{4k} eine kompakte orientierte Mannigfaltigkeit mit oder ohne Rand und $T: M^{4k} \to M^{4k}$ eine orientierungstreue Involution.

Für $x, y \in H_{2k}(M^{4k}, \mathbb{R})$ werde mit $x \circ y$ die Schnittzahl von x und y bezeichnet. Es sei

$$f_T(x,y) = x \circ Ty.$$

Da $x \circ Ty = Tx \circ TTy = Tx \circ y = y \circ Tx$, ist f_T eine symmetrische Bilinearform über dem endlich-dimensionalen reellen Vektorraum $H_{2k}(M^{4k}, \mathbb{R})$ und hat als solche eine Signatur $\tau(f_T) =$ Anzahl der positiven minus Anzahl der negativen „Eigenwerte" von f_T. (Die Form f_T kann ausgeartet sein. „Eigenwerte" 0 werden bei der Bildung der Signatur nicht berücksichtigt.) Für $T =$ Identität ist $\tau(M^{4k}, T)$ die übliche Signatur $\tau(M^{4k})$ von M^{4k}, in [16] Index genannt.

Wenn T fixpunktfrei ist, dann gilt, wie leicht zu zeigen,

$$\tau(M^{4k}, T) = 2\tau(M^{4k}/T) - \tau(M^{4k}). \tag{7}$$

Die Signatur $\tau(M^{4k}, T)$ hat die folgende additive Eigenschaft, die für $T = Id$ von S. P. Novikov bemerkt und mir von D. Sullivan und C. T. C. Wall mitgeteilt wurde.

(M_1^{4k}, T_1) und (M_2^{4k}, T_2) seien kompakte orientierte Mannigfaltigkeiten mit orientierungserhaltender Involution. B_1 bzw. B_2 sei Vereinigung von Zusammenhangskomponenten des Randes von M_1 bzw. M_2 mit $T_1(B_1) = B_1$ und $T_2(B_2) = B_2$. Die Paare $(B_1, T_1|B_1)$ und $(B_2, T_2|B_2)$ seien äquivariant und orientierungstreu homöomorph. Verklebt man M_1^{4k} mit M_2^{4k} vermöge dieses Homöomorphismus, dann erhält man eine Mannigfaltigkeit

$$M^{4k} = M_1^{4k} \ominus M_2^4 k,$$

die so orientiert sei, daß auf M_1^{4k} die gegebene, auf M_2^{4k} die zur gegebenen entgegengesetzte Orientierung induziert wird. T_1, T_2 bauen sich zu einer

Involution T auf M^{4k} zusammen. Es gilt

$$(8) \qquad \tau(M^{4k},T)=\tau(M_1^{4k},T_1)-\tau(M_2^{4k},T_2).$$

Wenn B_1, B_2 Sphären (der Dimension $4k-1$) sind, dann ist (8) trivial. (8) kann allgemein durch genaueres Studium der Mayer-Vietoris-Sequenz von $(M^{4k}; M_1^{4k}, M_2^{4k})$ bewiesen werden; vgl. [3], Proposition 7.1.

3. Eine Anwendung der Transversalitätssätze

Es sei M^{4k} eine kompakte orientierte differenzierbare Mannigfaltigkeit und X eine kompakte differenzierbare Untermannigfaltigkeit. Es sei $X \cap \partial M = \phi$. Die Mannigfaltigkeit X hat also keinen Rand. X braucht nicht zusammenhängend zu sein. Wir lassen sogar zu, daß die verschiedenen Zusammenhangskomponenten von X verschiedene Dimension haben. Die Kodimensionen dieser Zusammenhangskomponenten sollen alle gerade sein; wir sagen dann, daß die Kodimension von X gerade ist. X braucht nicht orientierbar zu sein. Die Injektion $i\colon X\to M^{4k}$ besitzt eine Approximation $j\colon X\to M^{4k}$, welche zur Untermannigfaltigkeit X von M^{4k} transversal ist (vgl. [29], [22a]). Das Urbild $j^{-1}(X)$ ist eine Untermannigfaltigkeit von X, und zwar ist das Normalbündel von $j^{-1}(X)$ in X isomorph zur Beschränkung des Normalbündels von X in M^{4k} auf $j^{-1}(X)$. Der Isomorphismus wird durch j induziert. Damit ist das Normalbündel von $j^{-1}(X)$ in M^{4k} als direkte Summe eines gerade dimensionalen reellen Vektorbündels mit sich selbst kanonisch orientiert. (Sind E, F orientierte Vektorräume, dann ist $E\oplus F$ orientiert durch Zusammensetzung der Orientierung von E und der von F in dieser Reihenfolge. Ist $E=F$, dann hängt die Orientierung von $E\oplus E$ nicht von der von E ab. Ist E geradedimensional, dann bleibt die so bestimmte Orientierung von $E\oplus E$ auch bei der Summandenvertauschung $E\oplus E\to E\oplus E$ invariant.) Also ist auch $j^{-1}(X)$ kanonisch orientiert, und zwar spannen die Orientierungen von $j^{-1}(X)$ und des Normalbündels von $j^{-1}(X)$ in M^{4k} die gegebene Orientierung von M^{4k} auf. Sind j_0, j_1 zwei hinreichend nahe Approximationen von i, die beide transversal zu X sind, dann existiert eine Homotopie $J\colon X\times I\to M^{4k}$ mit $J(x,0)=j_0(x)$ und $J(x,1)=j_1(x)$, und zwar kann J transversal zur Untermannigfaltigkeit X von M^{4k} gewählt werden. $J^{-1}(X)$ etabliert dann einen orientierten Cobordismus zwischen $j_0^{-1}(X)$ und $j_1^{-1}(X)$. Das Normalbündel von $J^{-1}(X)$ in $M^{4k}\times I$ ist wieder die direkte Summe eines Vektorbündels mit sich selbst. Deshalb repräsentieren $j_0^{-1}(X)$ und $j_1^{-1}(X)$ dasselbe Element der Thomschen Cobordismus-Algebra Ω_*, siehe [29]. Dieses wohldefinierte Element von Ω_* wollen wir mit $X \circ X$ bezeichnen und als Selbstschnitt von X in M^{4k} interpretieren. Wenn $\dim X \leqq 2k$ (d. h. wenn alle Zusammenhangskomponenten von X eine Dimension $\leqq 2k$ haben), dann ist $X \circ X \in \Omega_0 = \mathbb{Z}$ und gerade die Selbstschnittzahl der Homologieklasse der Vereinigungsmenge der $2k$-dimensionalen Komponenten von X, sofern diese orientierbar sind. Die Signatur $\tau(X \circ X)$ ist wohlerklärt, da $\tau: \Omega^* \to \mathbb{Z}$ ein Homomorphismus der Cobordismusalgebra in die ganzen Zahlen ist [16]. In dem obigen Fall, wo $\dim X \leqq 2k$, ist $\tau(X \circ X)$ die erwähnte Selbstschnittzahl.

4. Ein Spezialfall des Atiyah-Bott-Singerschen Fixpunktsatzes

Es sei T eine orientierungserhaltende differenzierbare Involution $M^{4k} \to M^{4k}$. Es werde aber jetzt vorausgesetzt, daß $T|\partial M^{4k}$ fixpunktfrei ist. Die Menge Fix T der Fixpunkte von T ist eine kompakte differenzierbare Untermannigfaltigkeit von $M^{4k} - \partial M^{4k}$ gerader Kodimension. (Fix T spielt hier die Rolle von X in § 3.) Nach § 3 ist $\tau(\text{Fix}\, T \circ \text{Fix}\, T)$ erklärt!

Satz. *Wenn* $\partial M^{4k} = \phi$, *dann*

(9) $$\tau(M^{4k}, T) = \tau(\text{Fix}\, T \circ \text{Fix}\, T).$$

Der Beweis ergibt sich aus dem G-Signature-Theorem (6.12.) des noch nicht veröffentlichten Manuskriptes von Atiyah-Singer [3].

Wenn T fixpunktfrei ist, dann muß also $\tau(M^{4k}, T) = 0$ sein. Dies folgt auch aus § 2 (7), da bekanntlich die Signatur sich bei Überlagerungen orientierter *differenzierbarer* Mannigfaltigkeiten multiplikativ verhält. Da diese Eigenschaft der Signatur für topologische Mannigfaltigkeiten unbekannt ist [9], weiß man auch nicht, ob $\tau(M^{4k}, T)$ im Falle topologischer Mannigfaltigkeiten und fixpunktfreier Involutionen immer verschwindet.

Der vorstehende Satz impliziert aber sogar, daß $\tau(M^{4k}, T) = 0$, falls nur $\dim \text{Fix}\, T < 2k$ (d. h. falls alle Zusammenhangskomponenten von Fix T eine Dimension $< 2k$ haben). Dies ist in der kombinatorischen und in der topologischen Kategorie im allgemeinen falsch und kann vielleicht wieder richtig werden, wenn man über Fix T gewisse Regularitätsvoraussetzungen macht.

Wie leicht zu sehen, ist $\tau(M^{4k}, T)$ modulo 2 gleich der Euler-Poincaréschen Charakteristik $e(M^{4k})$. Dies liefert einen Satz von Conner und Floyd ([10], vgl. [3]).

Korollar. *Es sei* M^{4k} *eine kompakte orientierte differenzierbare unberandete Mannigfaltigkeit mit* $e(M^{4k})$ *ungerade und* T *eine orientierungstreue Involution von* M^{4k}. *Dann hat wenigstens eine der Komponenten von* $\operatorname{Fix} T$ *eine Dimension* $\geqq 2k$.

Falls $\partial M^{4k} \neq \phi$, dann ist (9) im allgemeinen falsch. Dieser „Fehler" kann zur Definition einer Invarianten der fixpunktfreien Involution $T|\partial M^{4k}$ benutzt werden (vgl. [3]).

Satz. *Es seien* (M_1^{4k}, T_1) *und* (M_2^{4k}, T_2) *kompakte orientierte differenzierbare Mannigfaltigkeiten mit orientierungstreuen Involutionen, für die* $T_1|\partial M_1^{4k}$ *und* $T_2|\partial M_2^{4k}$ *fixpunktfrei sind.* $(\partial M_1^{4k}, T_1|\partial M_1^{4k})$ *und* $(\partial M_2^{4k}, T_2|\partial M_2^{4k})$ *seien orientierungstreu äquivariant diffeomorph. Dann ist*

$$(10)\quad \tau(M_1^{4k}, T_1) - \tau(\operatorname{Fix} T_1 \circ \operatorname{Fix} T_1) = \tau(M_2^{4k}, T_2) - \tau(\operatorname{Fix} T_2 \circ \operatorname{Fix} T_2).$$

Beweis. Wir betrachten auf $M^{4k} = M_1^{4k} \ominus M_2^{4k}$ die Involution T, (vgl. § 2). Da $\partial M^{4k} = \phi$, gilt (9). Offensichtlich ist $\operatorname{Fix} T = \operatorname{Fix} T_1 \cup \operatorname{Fix} T_2$ und

$$\tau(\operatorname{Fix} T \circ \operatorname{Fix} T) = \tau(\operatorname{Fix} T_1 \circ \operatorname{Fix} T_1) - \tau(\operatorname{Fix} T_2 \circ \operatorname{Fix} T_2).$$

Aus dieser Gleichung und aus (8) folgt wegen (9) die zu beweisende Gleichung (10).

5. Eine Invariante für fixpunktfreie Involutionen

Es sei X^{4k-1} $(k \geqq 1)$ kompakt, orientiert, differenzierbar, unberandet. T sei eine orientierungstreue fixpunktfreie differenzierbare Involution von X. Dem Paar (X, T) soll eine rationale Zahl $\alpha(X, T)$ als „Invariante" zugeordnet werden. Wir werden sehen, daß $2\alpha(X, T)$ ganzzahlig ist.

Bisher ist kein Beispiel bekannt, in dem $\alpha(X, T)$ nicht ganzzahlig ist. (Inzwischen hat JÄNICH bewiesen, daß $\alpha(X, T)$ stets ganzzahlig ist.)

Mit mX bezeichnen wir die disjunkte Vereinigung von m Exemplaren von X. Falls $m_1 X$ eine kompakte orientierte differenzierbare Mannigfaltigkeit N_1^{4k} so berandet, daß die auf $m_1 X$ gegebene Involution sich zu einer (orientierungstreuen) Involution T_1 auf N_1^{4k} erweitern läßt, dann definieren wir

$$(11)\quad \alpha(X, T) = \frac{1}{m_1}\left(\tau(N_1^{4k}, T_1) - \tau(\operatorname{Fix} T_1 \circ \operatorname{Fix} T_1)\right).$$

Aus der Bordismus-Theorie [10] folgt, daß

$$\Omega_*(B_{Z_2}) \otimes \mathbb{Q} = \Omega_* \otimes \mathbb{Q}.$$

Da $\Omega_{4k-1}\otimes\mathbb{Q}=0$ existieren m_1, N_1^{4k}, T_1 mit den obigen Eigenschaften immer und zwar sogar so, daß $\operatorname{Fix} T_1=\phi$. Nach BURDICK [8a] ist

$$\Omega_n(B_{\mathbb{Z}_2})\cong\Omega_n+\mathfrak{N}_{n-1},$$

deshalb kann m_1 immer gleich 2 gewählt werden (mit $\operatorname{Fix} T_1=\phi$).

Es muß jetzt gezeigt werden, daß $\alpha(X,T)$ nicht von der Wahl von m_1, N_1, T_1 abhängt:

Falls m_2, N_2, T_2 mit den gleichen Eigenschaften gegeben sind, dann setzen wir $M_1=m_2 N_1$ und $M_2=m_1 N_2$ (Multiplikation mit m_2 bzw. m_1 bedeutet die disjunkte Vereinigung von m_2 bzw. m_1 Exemplaren von N_1 bzw. N_2). Da $\partial M_1=\partial M_2=m_1 m_2 X$, folgt aus § 4 (10), daß

$$m_2(\tau(N_1,T_1)-\tau(\operatorname{Fix} T_1\circ\operatorname{Fix} T_1))=m_1(\tau(N_2,T_2)-\tau(\operatorname{Fix} T_2\circ\operatorname{Fix} T_2)).$$

Damit ist $\alpha(X,T)$ wohldefiniert. $2\alpha(X,T)$ ist wegen des Resultats von BURDICK ganzzahlig.

BROWDER und LIVESAY [8] haben für eine (orientierungstreue) differenzierbare fixpunktfreie Involution T einer orientierten differenzierbaren Mannigfaltigkeit X^{4k-1}, *welche eine ganzzahlige Homologiesphäre* ist, eine Invariante $\beta(X,T)$ definiert. $\beta(X,T)$ ist eine ganze durch 8 teilbare Zahl. Ihr Verschwinden ist, jedenfalls wenn X eine topologische Sphäre mit $k\geqq 2$ ist, notwendig und hinreichend für die Existenz einer in X differenzierbar eingebetteten $\mathbb{S}^{4k-2}$ mit $T(\mathbb{S}^{4k-2})=\mathbb{S}^{4k-2}$.

Es sei $v\in H^1(X/T;\mathbb{Z}_2)$ die charakteristische Klasse der Überlagerung $p\colon X\to X/T$ und V^{4k-2} eine nicht-orientierte Untermannigfaltigkeit von X/T, welche v repräsentiert. Wir setzen $\tilde{V}=p^{-1}(V)$.

Es ist $X=A\cup TA$ mit $A\cap TA=\tilde{V}$. Die berandete Mannigfaltigkeit A ist orientiert, da X orientiert ist. $\partial A=\tilde{V}$ ist dann auch orientiert. Nach dem Alexanderschen Dualitätssatz ist für die ganzzahlige Homologie

$$(i_A,i_{TA})\colon H_{2k-1}(\tilde{V})\to H_{2k-1}(A)\oplus H_{2k-1}(TA),$$

ein Isomorphismus, wo $i_A\colon\tilde{V}\to A$, $i_{TA}\colon\tilde{V}\to TA$ die Einbettungen sind. Über $\operatorname{Kern} i_A$ betrachtet man die symmetrische Bilinearform

$$B(x,y)=x\circ Ty,\qquad x,y\in\operatorname{Kern} i_A\subset H_{2k-1}(\tilde{V}).$$

(Beachte, daß T auf $\tilde{V}$ orientierungs-umkehrend ist.)

Die Browder-Liversay-Invariante $\beta(X,T)$ ist per definitionem die Signatur von B. Vertauscht man die Rollen von A und TA, dann ändert sich β nicht, β hängt nicht von der Wahl von V ab. Orientierungswechsel von X ändert das Vorzeichen von β. Die ganzzahlige quadratische Form B hat die Determinante ± 1. Da $B(x,x)$ stets gerade ist, folgt nach einem bekannten Satz über quadratische Formen (vgl. [18], § 13), daß $\beta(X,T)\equiv 0 \bmod 8$.

Satz. *Die orientierte differenzierbare Mannigfaltigkeit* X^{4k-1} $(k \geqq 1)$ *sei eine ganzzahlige Homologiesphäre, T eine fixpunktfreie Involution von X. Dann ist*

$$\alpha(X,T) = \pm\beta(X,T).$$

Beweisandeutung: DOLD [11] konstruiert ausgehend von V eine $4k$-dimensionale Mannigfaltigkeit D mit

$$\partial D = 2(X/T) - X.$$

Lemma. *Für die Doldsche Mannigfaltigkeit D ist*

$$\tau(D) = \pm\beta(X,T).$$

Dieses Lemma stellt die Hauptschwierigkeit da. Der Beweis, der aber doch ziemlich geradlinig aus der Doldschen Konstruktion folgt, soll bei anderer Gelegenheit nachgeholt werden.

Es sei ε die Antipodenabbildung von $\mathbb{S}^{4k-1}$. Nach einer Mitteilung von WALL – man kann dies auch aus den Ergebnissen von BURDICK (siehe § 5) schließen – sind (X,T) und $(\mathbb{S}^{4k-1},\varepsilon)$ cobordant, d.h. sie sind gleich als Elemente von $\Omega_*(B_{\mathbb{Z}_2})$. Es gibt also eine Mannigfaltigkeit Y^{4k} mit fixpunktfreier Involution T', so daß

$$\partial(Y^{4k},T') = (X,T) - (\mathbb{S}^{4k-1},\varepsilon).$$

WALL hat die folgende Zahl betrachtet

$$\omega(X,T) = 2\tau(Y^{4k}/T') - \tau(Y^{4k}).$$

Lemma. $$\omega(X,T) = \alpha(X,T).$$

Beweis. Aus Y^{4k} erhält man durch Ankleben einer Zelle $\mathbb{D}^{4k}$ längs $\mathbb{S}^{4k-1}$ eine Mannigfaltigkeit $\bar{Y}^{4k}$ mit einer Involution $\bar{T}$, die genau einen Fixpunkt hat. Da $\tau(Y^{4k},T') = \tau(\bar{Y}^{4k},\bar{T})$, folgt die Behauptung wegen § 2 (7) und nach Definition von α.

Die Doldsche Konstruktion kann auch auf die berandete Mannigfaltigkeit Y^{4k} mit Involution T' angewandt werden. Man muß dazu eine Untermannigfaltigkeit V' von Y^{4k}/T' verwenden, welche die charakteristische Cohomologieklasse der Überlagerung $Y^{4k} \to Y^{4k}/T'$ repräsentiert und den Rand von Y^{4k}/T' transversal schneidet. Die Doldsche Konstruktion für Y^{4k} und T' ergibt dann eine Mannigfaltigkeit E^{4k+1}, deren Rand so aussieht

$$\partial E^{4k+1} = 2(Y^{4k}/T') - Y^{4k} - D_1 + D_2,$$

wo D_1 die Doldsche Mannigfaltigkeit für X und D_2 die für $\mathbb{S}^{4k-1}$ ist, also

$$\partial D_1 = 2(X/T) - X,$$
$$\partial D_2 = 2(\mathbb{S}^{4k-1}/\varepsilon) - \mathbb{S}^{4k-1}.$$

Ferner ist

$$\partial Y^{4k} = X - \mathbb{S}^{4k-1},$$
$$\partial(Y^{4k}/T') = X/T - \mathbb{S}^{4k-1}/\varepsilon.$$

∂E^{4k+1} ist eine $4k$-dimensionale unberandete Mannigfaltigkeit, die sich durch Zusammenkleben von berandeten Mannigfaltigkeiten längs Randkomponenten ergibt. Deshalb ist nach § 2 (8)

$$\begin{aligned}\tau(\partial E^{4k+1}) &= 2\tau(Y^{4k}/T') - \tau(Y^{4k}) - \tau(D_1) + \tau(D_2)\\ &= \omega(X,T) \pm \beta(X,T) + 0\\ &= \omega(X,T) \pm \beta(X,T).\end{aligned}$$

Da die Signatur einer Mannigfaltigkeit, die als Rand auftritt, verschwindet, ergibt sich der zu beweisende Satz.

6. Auflösung von Singularitäten

Problem. *Berechne die Invariante α für die Involution $T_{(a_0,a_1,\ldots,a_{2k})}$ von $\Sigma^{4k-1}_{(a_0,a_1,\ldots,a_{2k})}$ (alle a_j ungerade; $k \geqq 1$).*

Eine Lösung dieses Problems würde nützlich sein für die Beantwortung der Frage, welche Werte von α für Involutionen welcher Sphären aus bP_{4k} auftreten können. Diese Frage ist trotz der Ergebnisse von LOPEZ DE MEDRANO [21] über die Browder-Livesay-Invariante β offen, da dieser nur zeigt, daß β $(=\alpha)$ für jedes j den Wert $8j$ auf wenigstens einer Sphäre vorgegebener Dimension $4k-1$ (mit $k \geqq 2$) annehmen kann, aber nichts darüber aussagt, auf welcher Sphäre der Wert $8j$ vorkommt. Für $k=1$ hat LOPEZ DE MEDRANO ein entsprechendes Resultat für ganzzahlige Homologiesphären.

Das obige Problem kann vollständig gelöst werden für $k=1$. Die 3-dimensionale Mannigfaltigkeit $\Sigma^3_{(a_0,a_1,a_2)}$ ist eine ganzzahlige Homologiesphäre dann und nur dann, wenn a_0, a_1, a_2 paarweise teilerfremd sind. Sie ist keine Sphäre, wenn alle $a_j \geqq 2$. Für paarweise teilerfremde a_0, a_1, a_2 $(a_j \geqq 2)$ handelt es sich also um eine Poincarésche 3-dimensionale Mannigfaltigkeit (verschwindende erste Homologiegruppe, nicht triviale Fundamentalgruppe) und zwar ist $\Sigma^3_{(a_0,a_1,a_2)}$ die gefaserte Poincarésche Mannigfaltigkeit im Sinne von SEIFERT [27] mit drei Ausnahmefasern der Ordnungen a_0, a_1, a_2. Die Gruppe $\mathbb{S}^1 = \{t \mid t \in \mathbb{C}$ und $|t|=1\}$ operiert auf $\Sigma^3_{(a_0,a_1,a_2)}$ vermöge

$$(12) \qquad (z_0, z_1, z_2) \mapsto (t^{a_1 a_2} z_0, t^{a_0 a_2} z_1, t^{a_0 a_1} z_2).$$

Diese $\mathbb{S}^1$-Aktion hat keine Fixpunkte. Die Orbits liefern die Seifertsche Faserung (vgl. auch [25]). Sind alle a_0, a_1, a_2 ungerade, dann ergibt sich für $t=-1$ die fixpunktfreie Involution $T_{(a_0,a_1,a_2)}$.

Auf jedem orientierten Seifertschen Faserraum mit orientierter Basis und mit Ausnahmefasern ungerader Ordnung operiert $\mathbb{S}^1$ und für $t=-1$ hat man eine fixpunktfreie Involution. Unter Verwendung von Methoden von R. VON RANDOW [26] kann man in all diesen Fällen die Invariante α berechnen. Wir beschränken uns hier auf ein Beispiel und verwenden dabei die Theorie der Auflösung der Singularitäten komplexer Flächen, die mit der Arbeit von R. VON RANDOW natürlich in engem Zusammenhang steht.

Satz. *Für die ganzzahlige Homologiesphäre* $\Sigma^3_{(3,6j-1,18j-1)}$ $(j \geqq 1)$ *mit der fixpunktfreien Involution* $T=T_{(3,6j-1,18j-1)}$ *hat die Invariante* α *den Wert* $8j$.

Beweisandeutung. $\Sigma^3_{(3,6j-1,18j-1)}$ berandet eine „Mannigfaltigkeit mit Singularität" $N_j^4 \subset \mathbb{C}^3$, die wie folgt definiert ist:

$$z_0^3 + z_1^{6j-1} + z_2^{18j-1} = 0,$$

$$z_0\bar{z}_0 + z_1\bar{z}_1 + z_2\bar{z}_2 \leqq 1.$$

Der einzige singuläre Punkt von N_j^4 ist $0=(0,0,0)$. Der klassische Auflösungsprozeß bläst 0 in ein System von rationalen Kurven (2-dimensionalen Sphären) auf. Vgl. z. B. [14], [15]. Man erhält hierdurch eine Mannigfaltigkeit M_j^4 und eine eigentliche (proper) Abbildung

$$\varphi: M_j^4 \to N_j^4,$$

so daß

$$\varphi: M_j^4 - \varphi^{-1}(0) \to N_j^4 - \{0\},$$

biholomorph und $\varphi^{-1}(0)$ Vereinigung von orientierten 2-Sphären ist, die singularitätenfrei in M_j^4 eingebettet sind und deren Schnittverhalten durch einen Graphen $\mathfrak{G}_j$ angegeben wird:

Jede 2-Sphäre entspricht einem Eckpunkt von $\mathfrak{G}_j$. Wenn zwei Eckpunkte von $\mathfrak{G}_j$ nicht durch eine Kante verbunden sind, dann schneiden sich die entsprechenden 2-Sphären nicht. Wenn zwei Eckpunkte von $\mathfrak{G}_j$ durch eine Kante verbunden sind, dann schneiden sich die entsprechenden 2-Sphären transversal in genau einem Punkt mit der Schnittzahl $+1$. Der Graph $\mathfrak{G}_j$ wird bewertet, indem man jedem Eckpunkt die Selbstschnittzahl der entsprechenden 2-Sphären zuordnet. $\mathfrak{G}_j$ sieht so aus:

(13)

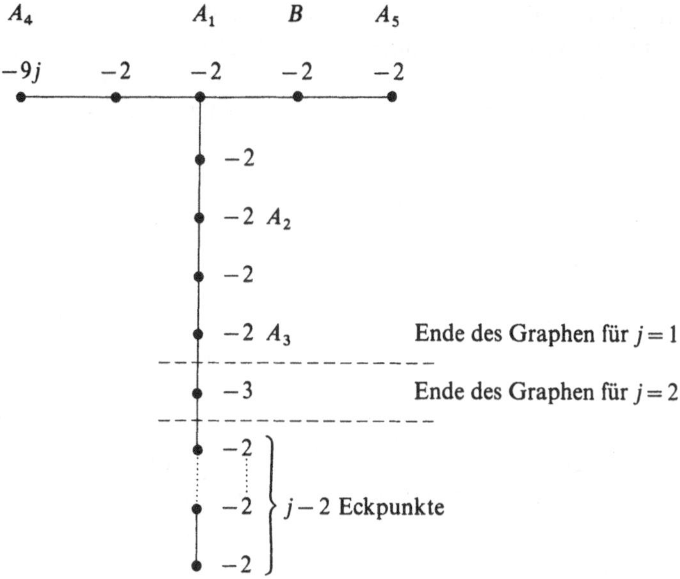

Sechs der $j+8$ Sphären haben wir einen Namen $A_1, \ldots, A_5, B$ gegeben. M_j^4 hat den Homotopietyp eines Buketts von $j+8$ zweidimensionalen Sphären. $H_2(M_j^4, \mathbb{Z})$ ist frei von Range $j+8$ mit den $j+8$ singularitätenfrei eingebetteten Sphären als Erzeugenden.

Die angegebene Auflösung ist die eindeutig bestimmte minimale Auflösung (vgl. z. B. [5], p. 81). Daraus kann man schließen, daß die durch (12) gegebene $\mathbb{S}^1$-Aktion sich von N_j^4 auf M_j^4 holomorph anheben läßt. φ wird dann äquivariant. Die $\mathbb{S}^1$-Aktion führt deshalb $\varphi^{-1}(0)$ in sich über. Außerhalb von $\varphi^{-1}(0)$ hat sie keine Fixpunkte. Da $t \in \mathbb{S}^1$ auf der Homologie von M_j^4 trivial operiert ($\mathbb{S}^1$ ist zusammenhängend), muß jede 2-Sphäre für jedes t in sich übergehen. Der Schnittpunkt zweier Sphären ist daher fix unter der $\mathbb{S}^1$-Aktion. Auf der Sphäre A_1 hat die $\mathbb{S}^1$-Aktion also 3 Fixpunkte und damit ist A_1 fix. Auf jeder anderen Sphäre haben wir mindestens 2 Fixpunkte der $\mathbb{S}^1$-Aktion. Bezüglich einer geeigneten holomorphen Karte z für die Sphäre kann man zwei Fixpunkte 0 und ∞ nennen und erreichen, daß $t \in \mathbb{S}^1$ so operiert:

$$z \mapsto t^a z \qquad (a \in \mathbb{Z}),$$

wobei a im Rahmen des Auflösungsprozesses explizit berechnet werden kann. Es zeigt sich dabei, daß für die Sphären $\neq A_1$ der Exponent a nicht verschwindet und damit genau 2 Fixpunkte vorhanden sind. Die Fixpunktmente der $\mathbb{S}^1$-Aktion ist also $A_1 \cup \{j+7 \text{ isolierte Punkte}\}$. Die

Involution T ergibt sich für $t=-1$ und kann also auch auf M_j^4 angehoben werden. Die Anhebung heiße ebenfalls T.

Auf $M_j^4-\varphi^{-1}(0)=N_j^4-\{0\}\subset\mathbb{C}^3-\{0\}$ ist T offensichtlich fixpunktfrei. $\operatorname{Fix} T$ besteht aus A_1 und den übrigen Sphären, für die der Exponent a gerade ist. Es ergibt sich, daß

(14) $\operatorname{Fix} T = A_1\cup A_2\cup A_3\cup A_4\cup A_5\cup\{j-1 \text{ isolierte Punkte}\}.$

Somit ist

$$\tau(\operatorname{Fix} T\circ\operatorname{Fix} T)=A_1\circ A_1+\cdots+A_5\circ A_5=-9j-8.$$

Da T auf der Homologie von M_j^4 trivial operiert, gilt

$$\tau(M_j^4,T)=\tau(M_j^4).$$

Da die quadratische Form von M_j^4 nach einem bekannten Satz über die Auflösung von Singularitäten negativ-definit ist, hat man

$$\tau(M_j^4)=-(j+8).$$

Es folgt

$$\begin{aligned}\alpha(\Sigma^3_{(3,6j-1,18j-1)},T)&=\tau(M_j^4,T)-\tau(\operatorname{Fix} T\circ\operatorname{Fix} T)\\&=-(j+8)-(-9j-8)=8j.\end{aligned}$$

Bemerkungen. 1. Die Formel (14) kann man sich auch mit Hilfe des folgenden Lemmas klarmachen.

Mit $\mathbb{D}^n$ wird die abgeschlossene Einheits-Vollkugel in $\mathbb{R}^n$ bezeichnet.

Lemma. *Auf $\mathbb{S}^2$ sei I die orthogonale Involution mit genau zwei Fixpunkten $x_1, x_2 (x_1=-x_2)$. Es sei E ein Bündel über $\mathbb{S}^2$ mit typischer Faser $\mathbb{D}^2$, Strukturgruppe $SO(2)$ und gerader Eulerscher Zahl, d. h., die Selbstschnittzahl des Nullschnittes als Homologieklasse der Mannigfaltigkeit E ist gerade. Dann gibt es eine Involution I_1 (bzw. I_2) von E, welche eine orthogonale Bündelabbildung ist, auf der Basis I induziert und in den Fasern E_{x_1} und E_{x_2} die Identität (bzw. die Antipodenabbildung $v\mapsto -v$) ist. Ist dagegen die Eulersche Zahl ungerade, dann gibt es eine Involution I_1 (bzw. I_2) von E, welche eine orthogonale Bündelabbildung ist, auf der Basis I induziert und in E_{x_1} (bzw. E_{x_2}) die Identität und in E_{x_2} (bzw. E_{x_1}) die Antipodenabbildung ist.*

2. Die explizite Auflösung der Singularität von $z_0^{a_0}+z_1^{a_1}+z_2^{a_2}=0$ soll bei anderer Gelegenheit dargestellt werden. – Grundsätzlich könnte auch eine Auflösung der Singularität von $z_0^{a_0}+\cdots+z_{2k}^{a_{2k}}=0$ für die Bearbeitung des zu Anfang dieses Paragraphen erwähnten Problems verwendet werden. Der Rechenaufwand scheint jedoch sehr groß zu sein.

7. Fixpunktfreie Involutionen auf den 7-Sphären

Der Graph $\mathfrak{G}_j$ von § 6 (13) kann zur Konstruktion von 7-dimensionalen Sphären benutzt werden.

Zu $\mathfrak{G}_j$ gehört eine ganzzahlige Form f_j, nämlich die über $V_j = H_2(M_j^4, \mathbb{Z})$ definierte Schnittform. Es ist $\det f_j = \pm 1$, da ∂M_j^4 eine ganzzahlige Homologiesphäre ist (vgl. [18]).

Jedem Eckpunkt P von $\mathfrak{G}_j$ ordne man ein Bündel mit typischer Faser $\mathbb{D}^4$, Strukturgruppe $SO(4)$ und Eulerscher Zahl $f_j(P,P)$ zu. (Beachte hier, daß P als Element von V_j angesehen werden kann. $f_j(P,P)$ ist die dem Eckpunkt P zugeordnete ganze Zahl; siehe (13).) Durch „plumbing“ dieser Bündel gemäß dem Graphen $\mathfrak{G}_j$ erhält man eine Mannigfaltigkeit M_j^8 vom Homotopietyp eines Buketts von $j+8$ vierdimensionalen Sphären (vgl. [18]).

Die Schnittform von M_j^8 ist gleich f_j; wenn man $H_4(M_j^8, \mathbb{Z})$ und V_j kanonisch identifiziert. Da $\det f_j = \pm 1$, ist ∂M_j^8 eine Sphäre [18] und repräsentiert ein Element der Gruppe $\Theta_7 = bP_8 = \mathbb{Z}_{28}$ (vgl. [20]). M_j^8 hängt nicht nur von j ab, sondern noch von der Wahl der den Eckpunkten von $\mathfrak{G}_j$ zugeordneten Bündel, die ja nicht durch die Eulersche Zahl allein bestimmt sind:

Die Isomorphieklassen der zu betrachtenden $SO(4)$-Bündel über $\mathbb{S}^4$ bilden die Gruppe $\pi_3(SO(4))$. Für $\xi \in \pi_3(SO(4))$ sind die ganzen Zahlen $e(\xi)$ (Eulersche Zahl) und $(p_1/2)(\xi)$ (halbe Pontrjaginsche Zahl) definiert. Es ist

$$e(\xi) \equiv \frac{p_1}{2}(\xi) \bmod 2 \tag{15}$$

und $(e, p_1/2)$ ist ein Isomorphismus von $\pi_3(SO(4))$ auf die Gruppe der Paare (a,b) ganzer Zahlen mit $a \equiv b \bmod 2$ (siehe [22]).

Für M_j^8 sind $H^4(M_j^8, \mathbb{Z})$, $H^4(M_j^8, \partial M_j^8, \mathbb{Z})$ und $H_4(M_j^8, \mathbb{Z})$ kanonisch isomorph. $H_4(M_j^8, \mathbb{Z})$ ist mit V_j zu identifizieren, und die 4-dimensionale Pontrjaginsche Klasse der Mannigfaltigkeit M_j^8 deshalb als ein Element von V_j anzusehen. Diese Pontrjaginsche Klasse ist durch 2 teilbar. Welche Elemente $w \in V_j$ können als halbe 4-dimensionale Pontrjaginsche Klassen einer Mannigfaltigkeit M_j^8 bei geeigneter Auswahl der den Eckpunkten P von $\mathfrak{G}_j$ zugeordneten Bündel auftreten? Wegen (15) muß w die Bedingung

$$f_j(w,P) \equiv f_j(P,P) \bmod 2,$$

für jeden Eckpunkt erfüllen. Das P zugeordnete Bündel ξ ist durch $f_j(w,P) = (p_1/2)(\xi)$ und $f_j(P,P) = e(\xi)$ bestimmt. Da $\det f_j = \pm 1$, sind die möglichen w genau die Elemente von V_j, welche der Bedingung

$$f(w,x) \equiv f(x,x) \bmod 2 \quad \text{für alle} \quad x \in V_j \tag{16}$$

genügen. Für ein solches w ist also durch „plumbing" eine Mannigfaltigkeit M_j^8 definiert, die wir genauer $M_j^8(w)$ nennen wollen.

Da $\det f_j = \pm 1$, liefert ein bekannter Satz über quadratische Formen ([18], § 13) die Kongruenz

$$\text{(17)} \qquad \tfrac{1}{8}(\tau(f_j) - f_j(w,w)) \in \mathbb{Z}.$$

Hier ist $\tau(f_j)$ die Signatur von f_j, in unserem speziellen Fall ist $\tau(f_j) = -(j+8)$.

Nach EELLS und KUIPER [12] bestimmt die in (17) angegebene Zahl nach Reduktion modulo 28 die differenzierbare Struktur der Sphäre $\partial M_j^8(w)$.

Die Mannigfaltigkeit $M_j^8(w)$ läßt eine Involution zu, die auf $\partial M_j^8(w)$ fixpunktfrei ist und der Involution T von M_j^4 (siehe § 6) genau nachgebildet ist. Die Beschreibung der $SO(4)$-Bündel über $\mathbb{S}^4$ mit Hilfe quaternionaler Übergangsfunktionen [22] zeigt nämlich, daß das Lemma von § 6 (Bemerkung 1.) auch im quaternionalen Fall richtig ist. (Man ersetze im Lemma $\mathbb{S}^2$ durch $\mathbb{S}^4$, $SO(2)$ durch $SO(4)$ und $\mathbb{D}^2$ durch $\mathbb{D}^4$). Dieses Lemma ermöglicht die Konstruktion von T (äquivariantes plumbing). Bis auf isolierte Fixpunkte bestimmt $\operatorname{Fix} T$ wie in § 6 die Homologieklasse

$$A_1 + \cdots + A_5 \in V_j \cong H_4(M_j^8(w), \mathbb{Z}).$$

Da T auf $H_4(M_j^8(w), \mathbb{Z})$ trivial operiert, ist die Invariante α wie in § 6 zu berechnen. Es folgt

$$\alpha(\partial M_j^8(w), T) = 8j.$$

Wir haben also eine fixpunktfreie Involution mit $\alpha = 8j$ auf allen Sphären von Θ_7, die sich durch „plumbing" gemäß dem Graph (13) darstellen lassen. Welche Sphären sind dies?

Offensichtlich hat $\operatorname{Fix} T = A_1 + \cdots + A_5$ die in (16) von w verlangte Eigenschaft. Die möglichen Werte von w sind daher gleich $\operatorname{Fix} T + 2u$ mit $u \in V_j$. Die Berechnung der in (17) angegebenen Zahl liefert

$$\text{(18)} \qquad \begin{aligned} &\tfrac{1}{8}(\tau(f_j) - f_j(\operatorname{Fix} T + 2u, \operatorname{Fix} T + 2u)) \\ &= j - \frac{f_j(\operatorname{Fix} T, u) + f_j(u,u)}{2}. \end{aligned}$$

Wir müßten also feststellen, welche ganzzahligen Werte modulo 28 in der Form (18) darstellbar sind? Wir gehen jedoch anders vor, und stellen zunächst fest, daß (vgl. § 6 (13)) durch

$$\text{(19)} \qquad \begin{aligned} &\tfrac{1}{2}(f_j(\operatorname{Fix} T, aA_1 + bB) + f_j(aA_1 + bB, aA_1 + bB)) \\ &= -a + b - a^2 + ab - b^2, \end{aligned}$$

sowohl gerade als auch ungerade Werte dargestellt werden. Bei der am Schluß von § 1 erwähnten Konstruktion (vgl. [4]), hat α für die Involution auf Σ und für die auf $\Sigma + 2\Sigma'$ den gleichen Wert. Es folgt

Satz. *Für jede Sphäre $\Sigma \in \Theta_7 = bP_8$ und jede ganze Zahl j gibt es eine differenzierbare Involution T auf Σ mit $\alpha(\Sigma, T) = 8j$. Auf jeder der 28 Sphären 7-Sphären gibt es also unendlich viele fixpunktfreie Involutionen, die paarweise differenzierbar inäquivalent sind.*

Bei der obigen Konstruktion war stets $j \geqq 1$. Orientierungswechsel beweist den Satz für $j \leqq -1$. Für $j = 0$ betrachtet man die Involution T auf der Sphäre M^7_{2h-1} ([22], [13]), die Rand des Bündels ξ über $\mathbb{S}^4$ mit Faser $\mathbb{D}^4$, Strukturgruppe $SO(4)$, mit $e(\xi) = 1$ und $(p_1/2)(\xi) = 2h - 1$ ist. Offensichtlich ist

$$\alpha(M^7_{2h-1}, T) = 1 - 1 = 0.$$

Nach [12] liefert M^7_{2h-1} sowohl gerade als auch ungerade Elemente von $\Theta_7 = \mathbb{Z}_{28}$.

8. Über S^1-Aktionen auf Sphären

Gegeben sei eine freie differenzierbare $\mathbb{S}^1$-Aktion auf einer (orientierten) Sphäre Σ der Dimension $4k-1$ ($\Sigma \in \Theta_{4k-1}$). Der Orbitraum $\Sigma/(\mathbb{S}^1)$ ist dann eine $(4k-2)$-dimensionale Mannigfaltigkeit vom Homotopietyp des komplexen projektiven Raumes $\mathbb{P}_{2k-1}(\mathbb{C})$. Die Sphäre Σ ist ein Prinzipalbündel über $\Sigma/(\mathbb{S}^1)$ mit Faser und Strukturgruppe $\mathbb{S}^1$. Dieses Bündel hat eine erste Chernsche Klasse

$$g \in H^2(\Sigma/(\mathbb{S}^1)), \mathbb{Z}).$$

Es ist

$$H^*(\Sigma/(\mathbb{S}^1), \mathbb{Z}) = \mathbb{Z}[g]/(g^{2k}).$$

Der Orbitraum werde so orientiert, daß

$$g^{2k-1}[\Sigma/(\mathbb{S}^1)] = +1.$$

Die Cohomologieklasse g kann in der orientierten Mannigfaltigkeit $\Sigma/(\mathbb{S}^1)$ durch eine orientierte Untermannigfaltigkeit X der Codimension 2 repräsentiert werden. Die Signatur $\tau(X)$ hängt nicht von der Wahl von X ab und heißt auch virtuelle Signatur (oder virtueller Index) von g

$$\tau(X) = \tau(\Sigma/(\mathbb{S}^1), g), \quad \text{vgl. [16].} \tag{20}$$

Wir definieren eine Invariante der $\mathbb{S}^1$-Aktion

$$\nu(\Sigma, (\mathbb{S}^1)) = 1 - \tau(\Sigma/(\mathbb{S}^1), g). \tag{21}$$

ν hängt nicht von der Orientierung von Σ ab. Die Invariante ν wurde von mehreren Autoren, insbesondere von BROWDER, NOVIKOV, ROTHENBERG, SULLIVAN und WALL, betrachtet. Da die rationalen Pontrjaginschen Klassen Homotopie-Invarianten mod 8 sind [2] und die L-Polynome [16] nicht die Primzahl 2 im Nenner enthalten, folgt

$$\nu(\Sigma, (\mathbb{S}^1)) \equiv 0 \bmod 8 .$$

Das ergibt sich auch aus dem folgenden Satz. Zunächst aber einige Bezeichnungen.

Dem Prinzipalbündel mit Faser und Strukturgruppe $\mathbb{S}^1 = \{t | t \in \mathbb{C}$ und $|t| = 1\}$ ist ein $\mathbb{D}^2$-Bündel assoziiert, dessen Totalraum mit B bezeichnet werde. $\mathbb{D}^2$ ist kanonisch orientiert. Da wir $\Sigma/(\mathbb{S}^1)$ in bestimmter Weise orientiert hatten, sind B und $\Sigma = \partial B$ orientiert. Wir sagen, daß Σ kompatipel orientiert ist, wenn seine Orientierung in dieser Weise aus der von $\Sigma/(\mathbb{S}^1)$ gewonnen wird.

Satz. *Die Sphäre Σ der Dimension $4k-1$ sei mit einer freien $\mathbb{S}^1$-Aktion versehen. Σ sei kompatibel orientiert. Für $-1 \in \mathbb{S}^1$ erhält man eine fixpunktfreie Involution T von Σ. Es gilt*

$$\alpha(\Sigma, T) = \nu(\Sigma, (\mathbb{S}^1)).$$

Beweis. T ist auch auf B definiert. $\operatorname{Fix} T$ ist der Nullschnitt des Bündels B, also eine Untermannigfaltigkeit von B, die mit $\Sigma/(\mathbb{S}^1)$ zu identifizieren ist. $\operatorname{Fix} T \circ \operatorname{Fix} T$ (im Sinne von § 3) ist eine Untermannigfaltigkeit von $\Sigma/(\mathbb{S}^1)$, die die Cohomologieklasse g repräsentiert, also als Mannigfaltigkeit X gewählt werden kann. Ferner ist $\tau(B, T) = 1$. Es folgt

$$\alpha(\Sigma, T) = \tau(B, T) - \tau(\operatorname{Fix} T \circ \operatorname{Fix} T) = 1 - \tau(X) = \nu(\Sigma, (\mathbb{S}^1)).$$

Bemerkung. Für $k=2$ ist die Übereinstimmung von ν und β (vgl. § 5) ein Resultat von MONTGOMERY und YANG [23]. Es ist nicht bekannt, auf welchen Sphären freie $\mathbb{S}^1$-Aktionen existieren (vgl. jedoch die Resultate von WU-CHUNG HSIANG [19]). MONTGOMERY und YANG [23] haben die 7-Sphären genau bestimmt, auf denen freie $\mathbb{S}^1$-Aktionen existieren.

Zusätze bei der Korrektur

1. Eine neue Version von [23] enthält die Gleichung $\nu = \beta$ für beliebiges k.
2. Einen vollständigen Beweis für die Gleichung $\alpha = \beta$ bringt eine Arbeit von K. JÄNICH und dem Verf. (erscheint in Proceedings Coll. Alg. Geometry, Tata Institute, Bombay).

3. Lopez de Medrano hat für ganzzahlige Homologiesphäre Σ_a^{4k-1} ($k=2$, $a=(a_0,a_1,\ldots,a_{2k})$ und alle a_j ungerade) bewiesen, daß die „normal invariants" von Σ_a^{4k-1}/T_a verschwinden und daß

$$\alpha(\Sigma_a^{4k-1}, T_a) = \tau(a_0, a_1, \ldots, a_{2k}) \bmod 16.$$

Vgl. hierzu auch P. Orlik und C. P. Rourke (Free involutions on homotopy $(4k+3)$-spheres, unveröffentlicht).

Literatur

1. Atiyah, M. F., and R. Bott: A Lefschetz fixed point formula for elliptic complexes I, Ann. of Math. **86**, 374—407 (1967).
2. —, u. F. Hirzebruch: Charakteristische Klassen und Anwendungen, Enseignement Mathématique **7**, 188—213 (1961).
3. —, and I. M. Singer: The index of elliptic operators III. Ann. of Math. **87**, 546—604 (1968).
4. Bredon, Glen E.: Exotic actions on spheres (these Proceedings).
5. Brieskorn, E.: Über die Auflösung gewisser Singularitäten von holomorphen Abbildungen, Math. Ann. **166**, 76—102 (1966).
6. — Beispiele zur Differentialtopologie von Singularitäten, Invent. Math. **2**, 1—14 (1966).
7. — Singularitäten von Hyperflächen (unveröffentlichtes Manuskript).
8. Browder, W., and G. R. Livesay: Fixed point free involutions on homotopy spheres, Bull. Amer. Math. Soc. **73**, 242—245 (1967).

8a. Burdick, R. O.: On the oriented bordism groups of $\mathbb{Z}_2$, Proc. Amer. Math. Soc. (erscheint demnächst).

9. Chern, S. S., F. Hirzebruch, and J. P. Serre: On the index of a fibred manifold, Proc. Amer. Math. Soc. **8**, 587—596 (1957).
10. Conner, P. E., and E. E. Floyd: Differentiable Periodic Maps, Ergebnisse der Mathematik und ihrer Grenzgebiete **33**. Berlin-Göttingen-Heidelberg: Springer 1964.
11. Dold, A.: Démonstration élémentaire de deux résultats du cobordisme, Séminaire de topologie et de géométrie différentielle, Paris, Mars 1959.
12. Eells, J., and N. H. Kuiper: An invariant for certain smooth manifolds, Ann. di Mat. pura ed appl. **60**, 93—110 (1963).
13. Hirsch, M. W., and J. Milnor: Some curious involutions of spheres, Bull. Amer. Math. Soc. **70**, 372—377 (1964).
14. Hirzebruch, F.: Über vierdimensionale Riemannsche Flächen mehrdeutiger analytischer Funktionen von zwei komplexen Veränderlichen, Math. Ann. **126**, 1—22 (1953).
15. — Differentiable manifolds and quadratic forms, Notes by Sebastian S. Koh, University of California, Berkeley 1962.
16. — Neue topologische Methoden in der algebraischen Geometrie, Ergebnisse der Mathematik und ihrer Grenzgebiete **9**. Berlin-Göttingen-Heidelberg: Springer 1962. Third enlarged edition, Topological methods in algebraic geometry, Berlin-Heidelberg-New York: Springer 1966.

17. — Singularities and exotic spheres, Séminaire Bourbaki, 1966/67, No 314.
18. —, u. K. H. Mayer: $0(n)$-Mannigfaltigkeiten, exotische Sphären und Singularitäten. Lecture Notes in Mathematics, Vol. **57**. Berlin-Heidelberg-New York: Springer 1968.
19. Hsiang, Wu-Chung: A note on free differentiable actions of $\mathbb{S}^1$ and $\mathbb{S}^3$ on homotopy spheres, Ann. of Math. **83**, 266—272 (1966).
20. Kervaire, M., and J. Milnor: Groups of homotopy spheres I, Ann. of Math. **77**, 504—537 (1963).
21. López de Medrano, S.: Involutions of homotopy spheres and homology 3-spheres, Bull. Amer. Math. Soc. **73**, 727—731 (1967).
22. Milnor, J.: On manifolds homeomorphic to the 7-sphere, Ann. of Math. **64**, 399—405 (1956).
22a. — Differential Topology, Notes by J. Munkres, Princeton University, 1958.
23. Montgomery, D., and C. T. Yang: Differentiable actions on homotopy seven spheres III (unveröffentlichtes Manuskript).
24. Mumford, D.: The topology of normal singularities of an algebraic surface and a criterion for simplicity, Inst. Hautes Etud. Scient. Publ. Math. **9**, Paris 1961.
25. Orlik, P., and F. Raymond: Actions of $SO(2)$ on 3-manifolds (these Proceedings).
26. v. Randow, R.: Zur Topologie von dreidimensionalen Baummannigfaltigkeiten, Bonner Math. Schriften Nr. **14**, Bonn 1962.
27. Seifert, H.: Topologie dreidimensionaler gefaserter Räume, Acta Math. **60**, 147—238 (1933).
28. Smale, S.: Generalized Poincaré's conjecture in dimensions greater than four, Ann. of Math. **74**, 391—406 (1961).
29. Thom, R.: Quelques propriétés globales des variétés différentiables, Comment. Math. Helv. **28**, 17—86 (1954).
30. Wall, C. T. C.: Free piecewise linear involutions on spheres (unveröffentlichtes Manuskript).
31. Weil, A.: Numbers of solutions of equations in finite fields, Bull. Amer. Math. Soc. **55**, 497—508 (1949).

Some Results on Involutions of Homotopy Spheres

SANTIAGO LÓPEZ DE MEDRANO

1. Introduction

This paper contains some results on fixed point free involutions of homotopy spheres, or, what is the same, on manifolds having the same homotopy type as real projective space. We will denote such an involution $T: \Sigma^n \to \Sigma^n$ simply as (T, Σ^n). All our involutions are either differentiable or piecewise linear, and, unless we mention explicitly which case we are considering, it should be understood that a result holds in both, with the appropiate interpretation of the terms used.

First, we consider the problem of finding invariant spheres. We say that (T, Σ^n) admits an invariant codimension q sphere if there is an embedded $S^{n-q} \subset \Sigma^n$ (locally flat in the piecewise linear case) such that $T(S^{n-q}) = S^{n-q}$. The case $q=1$ was studied by BROWDER and LIVESAY [2], and part of their results are included in Section 2, together with some examples. Other examples are mentioned at the end of Section 4. In Section 3, we consider the case $q=2$ when $n=4k+3$. Finally, in Section 4, we study the classification of involutions.

Complete proofs are not included, but the main ideas are sketched. Some of the details can be found in the literature [2, 4].

2. Invariant Codimension 1 Spheres

We describe now the results of BROWDER and LIVESAY concerning the problem of finding an invariant S^{n-1} for an involution (T, Σ^n). The method consists in taking an embedded invariant submanifold $W^{n-1} \subset \Sigma^n$, and trying to simplify it by surgery, keeping the property of being invariant, until it becomes a sphere. Not any invariant submanifold will do: if there is an invariant S^{n-1}, then it divides Σ^n in two parts, and T must interchange them, since it is fixed point free. Therefore we must start with an invariant submanifold that has that property.

Definition. $W^{n-1} \subset \Sigma^n$ is called a *characteristic submanifold* for (T, Σ^n) if $W = A \cap TA$ and $\Sigma = A \cup TA$, where A is a compact submanifold whose boundary is W. We will also refer to the quotient W^{n-1}/T as a characteristic submanifold for the homotopy projective space Σ^n/T.

A simple transversality argument shows that characteristic submanifolds always exist: take the covering $\Sigma \xrightarrow{p} \Sigma/T$ and its classifying map $\Sigma/T \xrightarrow{g} P^N$, N large, and make it transverse regular to $P^{N-1} \subset P^N$. Then $p^{-1} g^{-1} P^{N-1}$ is a characteristic submanifold for (T, Σ).

Now we start simplifying our characteristic submanifold by "equivariant surgery. If we denote by K_q the kernel of $i_*: H_q(W) \to H_q(A)$, we can easily see that $H_q(W) = K_q \oplus T_* K_q$. Therefore our aim will be to kill K_q inductively. Assuming $K_i = 0$ for $i < q, q < [(n-1)/2]$ and given $\alpha \in K_q$, we can represent α by an embedded sphere $S^q \subset W^{n-1}$ that bounds an embedded disc $D^{q+1} \subset A$, with $D^{q+1} \cap W = S^q$, and such that S^q and $T(S^q)$ are disjoint. If N is a tubular neighborhood of (D^{q+1}, S^q) inside (A, W) small enough that it is disjoint from $T(N)$, then we can form $A' = A - N \cup T(N)$ (corners have to be straightened in the smooth case). Then $A' \cup TA' = \Sigma$ and $A' \cap TA' = W'$ is a new characteristic submanifold that is simpler than W, in the sense that we have reduced K_q by killing α. Continuing this process, we can kill K_q for $q < [(n-1)/2]$.

When $q = [(n-1)/2]$ we run, as usual, into the difficulties of surgery in the middle dimension. The essential one arises in the case n odd and is due to the fact that α and $T_* \alpha$ may have a non-zero intersection number, in which case we cannot represent α by an embedded sphere S^q disjoint from $T(S^q)$.

The obstruction to doing this final step of the surgery is called the *Browder-Livesay* invariant of the involution, and will be denoted by $\sigma(T, \Sigma^n)$. It lies in the following groups:

$$\sigma(T, \Sigma^n) \in \begin{cases} 0 & n \text{ even}, \\ \mathbb{Z} & n = 4k+3, \\ \mathbb{Z}_2 & n = 4k+1. \end{cases}$$

When $n = 4k+3$, $\sigma(T, \Sigma^n)$ is the index (signature) of the symmetric bilinear form $x \cdot T_* y$ defined on K_{2k+1}. Notice that this form is nonsingular and even (i.e. $x \cdot T_* x$ is always an even integer) and therefore its index will be a multiple of 8 (see [5]).

When $n = 4k+1$, $\sigma(T, \Sigma^n)$ is the Arf invariant of a quadratic form $\psi: K_{2k} \to \mathbb{Z}_2$, associated with the skew-symmetric bilinear form $x \cdot T_* y$.

We can sum up the previous argument in the following

Theorem 1 (BROWDER *and* LIVESAY, [2]). *If* $n \geqslant 6$, (T, Σ^n) *admits an invariant codimension* 1 *sphere if and only if* $\sigma(T, \Sigma^n) = 0$.

Thus, if n is even, any involution (T, Σ^n) admits an invariant S^{n-1}. We now give examples of involutions that don't admit invariant codimension 1 spheres in the case $n=4k+3$. It will follow from Theorem 5 that such examples also exist for $n=4k+1$ in the piecewise linear case, but it is still not known if there are smooth examples.

Theorem 2. *For all* $i \in \mathbb{Z}$ *and* $k>0$, *there exists an involution* (T_i, Σ_i^{4k+3}) *with* $\sigma(T_i, \Sigma_i^{4k+3}) = 8i$.

The idea of the proof (see [4] for details) is to take any involution (T, Σ^{4k+3}) that admits an invariant S^{4k+2} (the antipodal map in S^{4k+3}, for example), add equivariantly $2m$ handles to this invariant sphere inside Σ^{4k+3} (the inverse of the simplifying process described in the proof of Theorem 1), thus obtaining a characteristic submanifold W, homeomorphic to the connected sum of $2m$ copies of $S^{2k+1} \times S^{2k+1}$, and then remove the complement of W and put it back in a different way.

If we denote by $\alpha_1, \ldots, \alpha_{2m}, \beta_1, \ldots, \beta_{2m}$ the standard basis for $H_{2k+1}(W)$, where $\alpha_1, \ldots, \alpha_{2m}$ is a basis for K_{2k+1}, then the matrix of the bilinear form $x \cdot T_* y$ with respect to this basis is

$$U = \begin{pmatrix} 0 \cdots 0 & 1 \\ 0 \cdots 1 & 0 \\ \cdots\cdots\cdots & \\ \cdots\cdots\cdots & \\ 1 \cdots 0 & 0 \end{pmatrix},$$

the $2m \times 2m$ matrix with 1's in the non-principal diagonal and 0's elsewhere. Now given a set of $2m$ elements

$$\alpha'_i = \Sigma\, p_{ij} \alpha_j + \Sigma\, q_{ij} \beta_j, \quad i = 1, \ldots, 2m.$$

of $H_{2k+1}(W)$ we would like to remove the complement of W in Σ^{4k+3} and put it back in such a way that we obtain a new involution (T', Σ'^{4k+3}) with the same characteristic submanifold W, and for which K_{2k+1} is the subgroup of $H_{2k+1}(W)$ freely generated by $\alpha'_1, \ldots, \alpha'_{2m}$. We can do this if

1. PQ^t is symmetric and even (i. e., it has even integers in the main diagonal),

2. $H = PUP^t - QUQ^t$ is non-singular. Then the matrix of the bilinear form $x \cdot T_* y$ with respect to the basis $\{\alpha'_i\}$ is H.

We are left, therefore, with the following algebraic problem: given a matrix H, can we find matrices P and Q that satisfy 1. and 2.? It turns out that we can do it for any matrix H that is even, symmetric and non-singular, and therefore we can realize all possible index obstructions. For example, if H is the well known 8×8 matrix with index 8, we can take

$$P=\begin{pmatrix}2&1&2&1&0&0&1&3\\2&1&2&1&0&1&1&2\\1&1&1&1&0&1&2&1\\1&1&1&1&0&0&1&1\\1&0&1&1&1&1&0&1\\0&0&2&1&1&1&0&1\\0&0&1&1&1&0&0&0\\1&0&0&1&1&0&0&0\end{pmatrix},\quad Q=\begin{pmatrix}2&1&1&0&1&1&1&2\\3&1&1&0&1&2&1&1\\2&1&0&0&1&2&1&1\\1&2&0&0&1&1&0&1\\1&1&0&1&1&2&0&1\\0&0&1&1&1&1&0&1\\0&0&1&0&1&0&0&0\\1&0&0&1&0&1&0&0\end{pmatrix}$$

Combining these, we can realize all multiples of 8.

BROWDER and LIVESAY also consider the question of uniqueness of invariant codimension 1 spheres: If S_1^{n-1} and S_2^{n-1} are invariant spheres for (T,Σ^n), are $(T|S_1^{n-1},S_1^{n-1})$ and $(T|S_2^{n-1},S_2^{n-1})$ equivalent as involutions? For this problem there are obstructions similar to those defined above, and it follows from Theorems 2 and 4 that there are examples where the index obstruction is different from 0, even in the smooth case.

3. Invariant Codimension 2 Spheres

Notice that the involutions constructed in the proof of Theorem 2 all admit an invariant S^{4k+1} contained in the characteristic submanifold W, although most of them do not admit an invariant S^{4k+2}. This fact turns out to be true in general

Theorem 3. *Every involution* (T,Σ^{4k+3}), $k>0$, *admits an invariant codimension* 2 *sphere. In the differentiable case it can be taken to be diffeomorphic to* S^{4k+1}.

There are several ways of proving this. One comes out directly from the proofs of Theorem 1 and 2: if we look at a $2k$-connected characteristic submanifold W for (T,Σ^{4k+3}), reversing the process described in the proof of Theorem 2, we can remove the complement of W and put it back in such a way that we obtain an involution with 0 Browder-Livesay invariant. This shows that we can always perform equivariant surgery on W outside Σ^{4k+3}, until we obtain a sphere. In other words, we can describe W as the result of adding handles equivariantly to an involution (T,S^{4k+2}). Since the latter has an invariant S^{4k+1}, and we can consider that the handles are attached away from it, the result follows.

This invariant S^{4k+1} will be knotted in general. The assumption that an unknotted one can always be found, implies that there are only a finite number of piecewise linear involutions in S^{4k+3}, which would contradict Theorem 2.

From Theorems 1 and 3 one can deduce that in every involution (T, Σ^{4k}), $k>1$, there is an invariant subcomplex $K \subset \Sigma^{4k}$, p.l. homeomorphic to S^{4k-2}: take K to be the suspension of an invariant S^{4k-3} contained in an invariant S^{4k-1}. But the preceding remark shows that K will not be locally flat in general, so we cannot deduce that (T, Σ^{4k}) admits an invariant codimension 2 sphere, in the sense we defined it in Section 1.

4. Classification

We now extend the results of Browder [1], Novikov [6] and Sullivan [7], regarding the classification of manifolds within a given simply connected homotopy type to the classification of involutions. To apply their ideas it is more convenient to speak in terms of homotopy projective spaces. Similar results can be proved for other manifolds with fundamental group $\mathbb{Z}_2$ and the same orientability as the projective space of the same dimension.

Our invariants will be the Browder-Livesay invariant and the normal invariant. Following Sullivan, we understand a normal invariant as a homotopy class of mappings into F/O (in the differentiable case) or F/PL (in the picewise linear case). The non-simply connected obstruction groups, defined and computed by Wall [8, 9] will play a fundamental role in the proof of our results.

We look first at the differentiable case. Let's recall Novikov's Theorem [6]: The diffeomorphism type of a simply connected smooth manifold M^n, $n \geqslant 5$, is determined, up to the addition of a homotopy sphere that bounds a π-manifold, by its normal invariant. In our case the normal invariant is not enough, since the number of possible normal invariants is finite, and we've exhibited in Section 2 an infinite number of different homotopy projective spaces. It turns out that the only other invariant we need is the Browder-Livesay invariant.

Theorem 4. *The diffeomorphism type of a smooth homotopy projective space Q^n, $n \geqslant 5$, is determined, up to the addition of a homotopy sphere that bounds a π-manifold, by its normal invariant and its Browder-Livesay invariant.*

The proof goes as follows: If Q_1^n and Q_2^n have the same normal invariant, we can obtain a cobordism V^{n+1} between them with the right normal properties to perform surgery on it. Given characteristic submanifolds W_1^{n-1} and W_2^{n-1} for Q_1^n and Q_2^n, we can find in V^{n+1} a "characteristic" cobordism U^n joining W_1^{n-1} and W_2^{n-1}.

If n is even we can always do the surgery on V^{n+1} until we get an h-cobordism, since the obstruction to doing the last step is 0 [9]. It follows that Q_1^n and Q_2^n are diffeomorphic. If n is odd and Q_1^n and Q_2^n have the same Browder-Livesay invariant, we can arrange the characteristic submanifolds until the homotopy equivalence between Q_1^n and Q_2^n induces a homotopy equivalence between them, by a relative version of Theorem 1. (This is not known for $n=5$, but the result can be shown by a different argument when $n=4k+1$.) Then, as in the case n even, we can do surgery on the characteristic cobordism U^n until it becomes an h-cobordism, and then extend this surgery to V^{n+1}. Now we are left with the problem of doing surgery in the complement of U, which is a simply connected surgery problem and can be solved in the usual way. It follows that $Q_1^n = Q_2^n \# \Sigma^n$, where Σ^n is a homotopy sphere that bounds a π-manifold.

We now say a few words about which of these invariants can be realized (cf. BROWDER [1], for the simply connected case).

If $n=4k+1$, $k>0$, all possible normal invariants can be realized. In all other cases there is an obstruction that lies in $\mathbb{Z}_2$ [9]. We will show below that in the piecewise linear case this obstruction is different from 0 in half of the cases. In the smooth case some examples with non-zero obstruction can be constructed, but results are not complete.

If $n=4k+3$ the two invariants are independent: If a normal invariant can be realized, then it can be realized together with any Browder-Livesay invariant.

If $n=4k+1$ the Browder-Livesay invariant is determined by the normal invariant, and therefore, the first one is only needed for $n=4k+3$ in Theorem 4 and the proof of Theorem 5.

We now turn to the piecewise linear case, where complete results can be obtained using results of SULLIVAN and the fact that we have the Kervaire and Milnor manifolds at our disposal, which we don't have in general in the smooth case. These manifolds are constructed by taking framed manifolds whose boundaries are homotopy spheres, that have Arf invariant 1, or index 8, respectively, and attaching a cone to their boundaries (see [3]).

SULLIVAN [7] has completely determined the mod. 2 homotopy type of F/PL, and therefore we can compute the set of normal invariants for P^n. The number of them is $2^{[n/2]}$, and as a group they are given by

$$[P^n, F/PL] \approx \mathbb{Z}_4 \oplus ([n/2]-2)\mathbb{Z}_2.$$

(Here, and in what follows, we denote by $n\mathbb{Z}_2$ the direct sum of n copies of $\mathbb{Z}_2$.)

This group structure is the one coming from the description of F/PL "localized at 2" as a Postnikov system with only one non-zero k-invariant, not the one induced by Whitney sum of F/PL-bundles (see [7]).

Let $PL(P^n)$ denote the set of p. l. homeomorphism classes of p. l. manifolds homotopy equivalent to P^n. First we show that

$$PL(P^{4k+3}) \approx PL(P^{4k+2}) \times \mathbb{Z}, \quad k>0.$$

The first component of this map is defined by taking the characteristic submanifold of the given homotopy projective space Q^{4k+3}, and doing surgery on it until it is homotopy equivalent to P^{4k+2} (cf. Section 3). The second is the Browder-Livesay invariant divided by 8. The fact that this is a monomorphism is the piecewise linear version of Theorem 4, and the fact that it is onto is a refinement of Theorem 2.

Now we can prove

Theorem 5. *For* $k>0$,

$$\begin{aligned} PL(P^{4k+1}) &\approx \mathbb{Z}_4 \oplus (2k-2)\mathbb{Z}_2, \\ PL(P^{4k+2}) &\approx \mathbb{Z}_4 \oplus (2k-2)\mathbb{Z}_2, \\ PL(P^{4k+3}) &\approx \mathbb{Z}_4 \oplus (2k-2)\mathbb{Z}_2 \oplus \mathbb{Z}, \\ PL(P^{4k+4}) &\approx \mathbb{Z}_4 \oplus (2k-1)\mathbb{Z}_2. \end{aligned}$$

The first formula follows from the fact that all normal invariants can be realized. For the second, by adding Kervaire manifolds we see that in half of the cases the realization obstruction is not 0, and that it can be identified with the component in the last $\mathbb{Z}_2$ summand of $[P^{4k+2}, F/PL]$. The third formula was proved above and it also follows that the realization obstruction is different from 0 in half of the cases. It follows immediately that this obstruction is not always 0 in $[P^{4k+4}, F/PL]$, and to see when it is 0, we notice that it is independent of the component in the last $\mathbb{Z}_2$ summand of this group, since this component can be changed by adding a Milnor manifold and this doesn't change the realization obstruction. (This obstruction is an Arf invariant, and the Arf invariant of a quadratic form obtained by reducing mod 2 a symmetric, even, non singular bilinear form over the integers is 0.) The last formula now follows.

For $n=4k+3$ the Browder-Livesay invariant is given by the $\mathbb{Z}$ component, and for $n=4k+1$ it is given by the next-to-last $\mathbb{Z}_2$ component. So only half of the piecewise linear involutions of S^{4k+1} admit invariant codimension 1 spheres.

Another proof of Theorem 5 was given by WALL in [9], using a different argument and, instead of the Browder-Livesay invariant, an invariant related to the non-simply connected surgery obstructions*.

---

* F. HIRZEBRUCH has recently shown that these two invariants are identical. One of the many applications of this result is the determination of the differentiable structure of spheres Σ_i^{4k+3} in Theorem 2. (This has been done more directly by P. ORLIK and C. P. ROURKE.) In particular, it follows that all 7-spheres admit involutions with any given Browder-Livesay invariant. (See also MONTGOMERY and YANG, *Differentiable Actions on Homotopy Seven Spheres, II*, these Proceedings.)

An essential step in [9] is provided by an explicit formula for the realization obstruction in terms of cohomology classes, due to Sullivan and Wall.

Theorem 5 shows that $PL(P^n)$, $n \geqslant 5$, can be given a group structure by adding normal and Browder-Livesay invariants. It would be interesting to find a geometric description of this operation. A step in this direction has been carried out by Livesay and Thomas. (See their paper, *Involutions on homotopy spheres*, these Proceedings.)

Note: Since this paper was first written, the subject has become extremely popular, and several new results have been obtained by M. F. Atiyah, I. Berstein, W. Browder, F. Hirzebruch, G. R. Livesay, D. Montgomery, P. Orlik, C. P. Rourke, D. Sullivan, C. T. C. Wall, C.-T. Yang and the author. They deal mainly with the smooth case and the relations between involutions and free S^1 actions.

References

1. Browder, W.: Homotopy type of differentiable manifolds, Colloq. on Alg. Top., Aarhus, 1962, 42—46.
2. —, and G. R. Livesay: Fixed point free involutions on homotopy spheres, Bull. Amer. Math. Soc. **73**, 242—245 (1967).
3. Kervaire, M., and J. Milnor: Groups of homotopy spheres I, Ann. of Math. **77**, 504—537 (1963).
4. López de Medrano, S.: Involutions of homotopy spheres and homology 3-spheres, Bull. Amer. Math. Soc. **73**, 727—731 (1967).
5. Milnor, J.: On simply connected 4-manifolds, Symposium International de Topología Algebráica, México 1958, 122—128.
6. Novikov, S. P.: Homotopy equivalent smooth manifolds I (in Russian), Izv. Akad. Nauk. USSR, Ser. Mat. **28, (2)**, 365—474 (1964).
7. Sullivan, D.: Triangulating homotopy equivalences, Princeton Ph. D. Thesis, 1965.
   See also: Triangulating homotopy equivalences, Smoothing homotopy equivalences (mimeographed notes, University of Warwick), and Triangulating and smoothing homotopy equivalences and homeomorphisms (mimeographed notes, Princeton University).
8. Wall, C. T. C.: Surgery of compact manifolds (to appear). See also: Surgery of non-simply-connected manifolds, Ann. of Math. **84**, 217—276 (1966).
9. — Free piecewise linear involutions on spheres (to appear).

Free Differentiable Actions on Homotopy Spheres

DEANE MONTGOMERY and C. T. YANG*

1. Introduction

Throughout this paper, G denotes the circle group $SO(2)$ of rotations of the euclidean plane and $\mathbb{Z}_2$ denotes the subgroup of G of order 2. $\mathbb{R}^n$ denotes the euclidean n-space, S^n denotes the unit n-sphere in $\mathbb{R}^{n+1}$ and CP^n denotes the complex projective n-space, all having the usual differentiable structure. By a *homotopy n-sphere*, abbreviated by HS^n, we mean a closed differentiable n-manifold having the homotopy type of S^n; by a *homotopy complex projective n-space*, abbreviated by HCP^n, we mean a closed differentiable $2n$-manifold having the homotopy type of CP^n.

It is well-known that for any free differentiable action of the circle group G on an HS^{2n+1}, Σ^{2n+1}, the orbit space Σ^{2n+1}/G is an HCP^n, and that for any HCP^n, M, there is, up to a differentiable equivalence, a unique free differentiable action of G on an HS^{2n+1}, Σ^{2n+1}, such that Σ^{2n+1}/G is diffeomorphic to M. (Two differentiable actions (G,X) and (G,Y) are called *differentiably equivalent* if there is a diffeomorphism $h\colon X\to Y$ such that for any $g\in G$ and $x\in X$, $h(gx)=gh(x)$.) Hence there is a one-to-one correspondence between differentiable equivalence classes of free differentiable actions of G on HS^{2n+1}'s and diffeomorphism classes of HCP^n's.

Suppose that a free differentiable action of the circle group G on an HS^{2n+1}, Σ^{2n+1}, is given, $n\geqq 3$. Then we have a free differentiable action of $\mathbb{Z}_2$ on Σ^{2n+1}, where $\mathbb{Z}_2$ is the subgroup G of order 2. As seen in BROWDER-LIVESAY [1], an index or an Arf invariant is defined for the action $(\mathbb{Z}_2, \Sigma^{2n+1})$ according as n is odd or even. Moreover, it is shown that a necessary and sufficient condition for the existence of a differentiably imbedded $\mathbb{Z}_2$-invariant HS^{2n} in Σ^{2n+1} is that this index or Arf invariant vanishes. On the other hand, if we let $M=\Sigma^{2n+1}/G$, there is a differentiable homotopy equivalence

$$f\colon M\to CP^n,$$

* The second author was supported in part by the U.S. Army Research Office and by the National Science Foundation when the work was done.

which is transverse regular on CP^{n-1}. Therefore

$$N = f^{-1}(CP^{n-1}),$$

is a differentiable submanifold of M of codimension 2. By means of framed surgery as devised in BROWDER [2] and NOVIKOV [3], it is always possible to kill the kernel of

$$f_*: \pi_i(N) \to \pi_i(CP^{n-1}),$$

for $i=0,\dots,n-2$. That means, there is a differentiable homotopy equivalence $f': M \to CP^n$ which is homotopic to f and is transverse regular on CP^{n-1} and such that, if we let $N' = f'^{-1}(CP^{n-1})$, then $f'_*: \pi_i(N') \to \pi_i(CP^{n-1})$ has a trivial kernel for $i=0,\dots,n-2$. In connection with the killing of the kernel of f_* for $i=n-1$, there is an index or an Arf invariant according as n is odd or even. In fact, a necessary and sufficient condition for the killing of the kernel of f_* for $i=0,\dots,n-2,n-1$, is that this index or Arf invariant vanishes. This index or Arf invariant turns out to be dependent only on the action (G,Σ^{2n+1}) and hence is called the index or Arf invariant of (G,Σ^{2n+1}). We note that if we succeed to kill the kernel of $f_*: \pi_i(N) \to \pi_i(CP^{n-1})$ for $i=0,\dots,n-2,n-1$, then there is a differentiably imbedded HCP^{n-1} in M and hence there is a differentiably imbedded G-invariant HS^{2n-1} in Σ^{2n+1}. As the main theorem of this paper, we shall show that the index or Arf invariant connected with the killing of the kernel of f_* for $i=n-1$ is equal to that of $(\mathbb{Z}_2,\Sigma^{2n+1})$ defined in [1]. Hence we have

Theorem A. *Let there be a free differentiable action of the circle group G on a homotopy $(2n+1)$-sphere Σ^{2n+1}, $n \geqq 3$, and let $\mathbb{Z}_2$ be the subgroup of G of order 2. Then there is a differentiably imbedded G-invariant homotopy $(2n-1)$-sphere in Σ^{2n+1} iff there is a differentiably imbedded $\mathbb{Z}_2$-invariant homotopy $2n$-sphere in Σ^{2n+1}.*

In [4], it is shown that there is an isomorphism h of the group $\mathbb{Z}$ of integers onto the group of differentiable equivalence classes of free differentiable actions of G on HS^7's such that for any $i \in Z$, if $h(i)$ is represented by a free differentiable action of G on an HS^7, Σ_i^7, then the first rational Pontrjagin class of Σ_i^7/G is given by

$$p_1(\Sigma_i^7/G) = (24i+4)\alpha^2,$$

where α is a generator of the second integral cohomology group of Σ_i^7/G and the Eells-Kuiper invariant of Σ_i^7 is given by

$$\mu(\Sigma_i^7) = \pm i(9i+2)/14 \bmod 1.$$

Using the information on $p_1(\Sigma_i^7)/G)$, we are able to shown that the index of (G,Σ_i^7) is equal to $8i$. Hence we have

Theorem B. a) *Any two free differentiable actions of the circle group G on homotopy 7-spheres Σ and Σ' respectively are differentiably equivalent iff the actions of $\mathbb{Z}_2$ on Σ and Σ' are differentiably equivalent.*

b) *Given any free differentiable action of the circle group G on a homotopy 7-sphere Σ, the following are equivalent.*

1. *The action (G,Σ) is differentiably equivalent to an orthogonal action.*
2. *The action $(\mathbb{Z}_2,\Sigma)$ is differentiably equivalent to an orthogonal action.*
3. *There exists a differentiably imbedded G-invariant 5-sphere in Σ.*
4. *There exists a differentiably imbedded $\mathbb{Z}_2$-invariant 6-sphere in Σ.*

c) *On the 7-sphere with the usual differentiable structure, there are infinitely many free differentiable actions of the circle group G such that the actions of $\mathbb{Z}_2$ are not differentiably equivalent to one another.*

Let M be an HCP^4 such that there is a stable tangential homotopy equivalence $f: M \to CP^4$. Then

$$p_1(M) = f^* p_1(CP^4),$$

where p_1 is the first rational Pontrjagin class. If M has a differentiably imbedded HCP^3, N, such that the inclusion map of N into M induces an isomorphism of $\pi_2(N)$ onto $\pi_2(M)$, then f can be so chosen that $N = f^{-1}(CP^3)$. Therefore $p_1(N) = f^* p_1(CP^3)$ so that N is diffeomorphic to CP^3 [4]. From this result, it follows that M is diffeomorphic to the connected sum of CP^4 and an HS^8 and hence is piecewise linearly homeomorphic to CP^4. However, SULLIVAN [5] constructs an HCP^4, M, such that there is a stable tangential homotopy equivalence of M into CP^4 but M is *not* piecewise linearly homeomorphic to CP^4. Therefore, if M is that HCP^4, it is impossible to have a differentiably imbedded HCP^3, N, in M such that the inclusion map of N into M induces an isomorphism of $\pi_2(N)$ onto $\pi_2(M)$. Hence we have

Theorem C. *There is a free differentiable action of the circle group G on a homotopy 9-sphere Σ^9 such that there is no differentiably imbedded $\mathbb{Z}_2$-invariant homotopy 8-sphere in Σ^9 and no differentiably imbedded G-invariant homotopy 7-sphere.*

As a by-product of our study, we also have

Theorem D. a) *On the 7-sphere with the usual differentiable structure, there is, for each integer i, a free differentiable action of the cyclic group $\mathbb{Z}_2$ of order 2 such that the Browder-Livesay index is equal to $8i$.*

b) *There is a free differentiable action of the cyclic group $\mathbb{Z}_2$ on a homotopy 9-sphere such that the Browder-Livesay Arf invariant does not vanish.*

Notice that similar results have been given by S. LOPEZ DE MEDRANO, [6] and [7].

The authors are grateful to Professor W. BROWDER and others for valuable discussions during the Conference on Transformation Groups held at Tulane University, April, 1967.

2. A Basic Construction

In this action, we give a construction which will be needed in the proof of our main theorem. It is essentially the same construction as seen in [2] and [3] except that certain modifications are made as in LEVINE [8] and HAEFLINGER [9].

All manifolds and vector bundles over manifolds will be differentiable as will be maps and imbeddings. For any manifold M, $\tau(M)$ denotes the tangent bundle of M. If M is a manifold and N is a submanifold of M, $\nu(N,M)$ denotes the normal bundle of N in M and $\xi|N$ denotes the restriction of a vector bundle ξ on M to N. By an *r-frame* on N, we mean an r-tuple of linearly independent vector fields in $\tau(M)|N$. D^n denotes the closed unit n-disk in $\mathbb{R}^n$ and $E^n = D^n - S^{n-1}$.

Let M and M' be closed ℓ-manifolds, N' a closed m-submanifold of M', $f\colon M \to M'$ a map which is transverse regular on $N', N = F^{-1}(N')$ and $h\colon S^{k-1} \to N$ an imbedding. For the sake of simplicity, we identify M with $M \times \{0\}$ by setting $x = (x,0)$ for all $x \in M$.

We now state assumptions which [8] and [9] have used to make a construction. Later these assumptions will be verified for the constructions made in this paper.

We assume

(2.1) $0 \leqq k-1 \leqq m/2$ and $5 \leqq m+2 \leqq \ell$.

(2.2) $fh\colon S^{k-1} \to N'$ is null-homotopic and in fact, $fh(S^{k-1}) = y$, a point of N'.

(2.3) $\pi_i(M-N) = 0$, $i \leqq k-1$.

(2.4) $\pi_i(M'-N') = 0$, $i \leqq \ell$.

Choose an $(\ell-m)$-frame of $\nu(N',M')$ at y and let $(v_1,\dots,v_{\ell-m})$ be its induced $(\ell-m)$-frame on $h(S^{k-1})$ by f. Then h can be extended to an imbedding of D^k into M, also denoted by h, such that $h(D^k) \subset M-N$ and v_1 points radially into $h(D^k)$.

We also assume

(2.5) The frame $(v_2,\dots,v_{\ell-m})$ can be extended to an $(\ell-m-1)$-frame on $h(D^k)$ in $\nu(h(D^k),M)$.

The obstruction to this extension is an element

$$\lambda \in \pi_{k-1}(V_{\ell-k,\ell-m-1}),$$

where $V_{\ell-k,\ell-m-1}$ is the Stiefel manifold of orthonormal $(\ell-m-1)$-frames in $\mathbb{R}^{\ell-k}$. Hence (2.5) means

(2.5′) $\lambda=0$.

Notice that when $2(k-1)<m$, $\pi_{k-1}(V_{\ell-k,\ell-k-1})=0$ so that (2.5′) holds.

Under these assumptions, HAEFLINGER [8] has shown that the map f can be extended to a map

$$F: M\times[0,1]\to M',$$

which is transverse regular on N' and such that $B=F^{-1}(N')$ is a compact $(m+1)$-manifold in $M\times[0,1]$ which is a framed cobordism from N to $N_1=B\cap(M\times\{1\})$, obtained by attaching a handle to $N\times[0,1]$ at $h(S^{k-1})$.

Let $f_1=F|M\times\{1\}$, which is also regarded as the map of M into M' given by $x\to F(x,1)$. Then f_1 is a homotopy equivalence homotopic to f and is transverse regular on CP^{n-1} and $N_1=f_1^{-1}(CP^{n-1})$ is obtained from $N=f^{-1}(CP^{n-1})$ by performing a framed surgery at $h(S^{n-1})$. Notice that HAEFLIGER [8] requires $\ell-m>2$, but LEVINE [9] shows how to handle the case $\ell-m=2$ which is the case of interest in this paper.

3. Surgery Below the Middle Dimension

Suppose that a free differentiable action of the circle group on an HS^{2n+1}, Σ^{2n+1}, is given, $n\geqq 3$, and let

$$M=\Sigma^{2n+1}/G.$$

Then there is a homotopy equivalence

$$f: M\to CP^n.$$

Let $\sigma: S^2\to M$ be an imbedding representing a generator of $\pi_2(M)$. We may assume that $f\sigma$ maps S^2 diffeomorphically onto CP^1, because there always exists such a homotopy equivalence homotopic to f. Since $\nu(\sigma(S^2),M)$ is determined by its second Stiefel-Whitney class, it is isomorphic to $\nu(CP^1,CP^n)$. Therefore we may assume that f maps a closed tubular neighborhood T of $\sigma(S^2)$ in M diffeomorphically onto a closed tubular neighborhood of CP^1 in CP^n. By the relative transversality theorem of THOM, we may also assume that f is transverse regular on CP^{n-1}. Hence

$$N=f^{-1}(CP^{n-1}),$$

is a closed submanifold of M of codimension 2, containing $\sigma(S^2)$.

Lemma 1. a) *$M-N$ is connected.*

b) *If N is connected, then $M-M$ is simply connected.*

Proof. a) is a direct consequence of the fact that N is a subset of M of codimension 2.

b) Let T be a closed tubular neighborhood of N in M and $p: \partial T \to N$ the natural circle fibration (obtained from $\nu(N,M)$). Then $p^{-1}(\sigma(S^2))$ is a 3-sphere so that for any $a \in \sigma(S^2)$, the circle $p^{-1}(a)$ is null-homotopic in ∂T. Since N is connected, then so are T and ∂T and

$$\pi_1(\partial T) \to \pi_1(T),$$

induced by the inclusion map is an isomorphism onto. Let C be the closure of $M-T$. Since $\pi_1(M)$ is trivial, it follows from van Kampen's theorem that

$$\pi_1(\partial T) \to \pi_1(C),$$

induced by the inclusion map is trivial and onto. Hence $\pi_1(C)$ is trivial and consequently $\pi_1(M-N)$ is trivial.

Unless it is stated to the contrary, the group $\mathbb{Z}$ of integers is used as the coefficient group for both homology and cohomology.

Lemma 2. *$f_*: H_i(N) \to H_i(CP^{n-1})$ is onto for all integers i. In particular, if N is connected, then $f_*: H_{2n-2}(N) \to H_{2n-2}(CP^{n-1})$ is an isomorphism onto.*

Proof. Let A' be the complement of an open tubular neighborhood of CP^{n-1} in CP^n such that $A = f^{-1}(A')$ is the complement of an open tubular neighborhood of N in M. Clearly $f^*: H^{2n}(CP^n) \to H^{2n}(M)$ is an isomorphism onto and $H^{2n}(CP^n, A') \to H^{2n}(CP^n)$ induced by the inclusion map is an isomorphism onto. Moreover, $H^{2n}(M,A) \to H^{2n}(M)$ induced by the inclusion map is onto and is an isomorphism onto when N is connected. By the isomorphism theorem of THOM,

$$H^{2n-2}(CP^{n-1}) \to H^{2n}(CP^{n-1}, A'), H^{2n-2}(N) \to H^{2n}(M,A),$$

are isomorphisms onto. It follows that

$$f^*: H^{2n-2}(CP^{n-1}) \to H^{2n-2}(N),$$

maps $H^{2n-2}(CP^{n-1})$ isomorphically onto a direct summand of $H^{2n-2}(N)$ and is an isomorphism onto when N is connected. From this result, it is easily seen that for any integer i, $f^*: H^i(CP^{n-1}) \to H^i(N)$ is an isomorphism into and its image is a direct summand of $H^i(N)$. Hence our assertion follows from duality.

Lemma 3. *If k is an integer $\leqq n$ such that $f_*: H_i(N) \to H_i(CP^{n-1})$ is one-to-one for $i < k-1$, then*

$$H_{k-1}(M-N) = 0.$$

Proof. By hypothesis and Lemma 2, $f_*: H_i(N) \to H_i(CP^{n-1})$ is an isomorphism onto for $i < k-1$. Therefore, by duality,

$$f^*: H^{2n-2-i}(CP^{n-1}) \to H^{2n-2-i}(N),$$

is an isomorphism onto for $i < k-1$ so that $H^{2n-2-i}(M) \to H^{2n-2-i}(N)$ induced by the inclusion map is an isomorphism onto for $i < k-1$. Using the exactness of the cohomology sequence of (M, N), we infer that $H_C^{2n-k+1}(M-N) = 0$. Hence our assertion follows from duality.

Lemma 4. *If k is an integer $\leqq n$ such that $f_*: \pi_i(N) \to \pi_i(CP^{n-1})$ is one-to-one for $i < k-1$, then both $f_*: \pi_i(N) \to \pi_i(CP^{n-1})$ and $f_*: H_i(N) \to H_i(CP^{n-1})$ are isomorphisms onto for $i < k-1$. Moreover, if $H: \pi_{k-1}(N) \to H_{k-1}(N)$ is the Hurewicz homomorphism, K_π is the kernel of $f_*: \pi_{k-1}(N) \to \pi_{k-1}(CP^{n-1})$ and K_H is the kernel of $f_*: H_{k-1}(N) \to H_{k-1}(CP^{n-1})$, then for $k \neq 2$, H maps K_π isomorphically onto K and for $k = 2$, H is onto and its kernel is the commutator subgroup of K_π. Furthermore, $\pi_i(M-N) = 0$ for $i \leqq k-1$.*

Proof. Clearly $f\sigma: S^2 \to CP^{n-1}$ represents a generator of $\pi_2(CP^{n-1})$ so that $f_*: \pi_2(N) \to \pi_2(CP^{n-1})$ is onto. Since $f_*: \pi_0(N) \to \pi_0(CP^{n-1})$ is onto and $\pi_i(CP^{n-1}) = 0$ for any non-negative integer $i \leqq n+1$ different from 0 and 2, it follows that $f_*: \pi_i(N) \to \pi_i(CP^{n-1})$ is onto for $i = 0, \ldots, n+1$. Hence, by hypothesis, $f_*: \pi_i(N) \to \pi_i(CP^{n-1})$ is an isomorphism onto for $i < k-1$.

Let C be the mapping cylinder of $f: N \to CP^{n-1}$. Since the homotopy sequence of (C, N) is exact and since CP^{n-1} is a deformation retract of C, we infer that $\pi_i(C, N) = 0$ for $i \leqq k-1$. Therefore $H_i(C, N) = 0$ for $i \leqq k-1$ and the Hurewicz homomorphism $H: \pi_k(C, N) \to H_k(C, N)$ is an isomorphism onto if $k \neq 2$ and is onto with the commutator subgroup of $\pi_2(C, N)$ as the kernel if $k = 2$. Since there are natural isomorphisms $\pi_k(C, N) \cong K_\pi$ and $H_k(C, N) \cong K_H$, we infer that $H: K_\pi \to K_H$ is an isomorphism onto if $k \neq 2$ and is onto with the commutator subgroup of K_π as the kernel if $k = 2$.

Since $f_*: H_i(N) \to H_i(CP^{n-1})$ is an isomorphism onto for $i < k-1$ and is onto for $i = k-1$, so is $H_i(N) \to H_i(M)$ induced by the inclusion map, so that $H_i(M-N) = 0$ for $i \leqq k-1$. By Lemma 1, $\pi_0(M-N) = 0$ and for $k > 1$, $\pi_1(M-N) = 0$. Hence, by the Hurewicz isomorphism theorem, $\pi_i(M-N) = 0$ for $i \leqq k-1$.

Now we are in a position to use the construction of the preceding section to kill the kernel of $f_*: \pi_i(N) \to \pi_i(CP^{n-1})$ for $i < n-1$.

Let us begin with the killing of the kernel of $f_*: \pi_0(N) \to \pi_0(CP^{n-1})$. If the kernel is not trivial, then N has more than one component so that there is an imbedding $h: S^0 \to N$ such that $h(S^0)$ is not contained in the same component of N. In this case we can proceed doing the construction of the preceding section with M, CP^n, CP^{n-1}, f, h in place of M, M', N', f, h respectively, as the required assumptions (2.1)–(2.5) are easily verified. Therefore we obtain a homotopy equivalence $f_1: M \to CP^n$ which is homotopic to f and is transverse regular on CP^{n-1} and such that $f_1^{-1}(CP^{n-1})$ has one less component than $N = f^{-1}(CP^{n-1})$. Repeating this construction if necessary, we can finally have a homotopy equivalence $f': M \to CP^n$ which is homotopic to f and is transverse regular on CP^{n-1} and such that $f'^{-1}(CP^{n-1})$ is connected and thus the kernel of $f'_*: \pi_0(f'^{-1}(CP^{n-1})) \to \pi_0(CP^{n-1})$ is trivial. Hence we kill the kernel of $f_*: \pi_0(N) \to \pi_0(CP^{n-1})$ by using f' to replace f.

We next consider the killing of the kernel of $f_*: \pi_1(N) \to \pi_1(CP^{n-1})$ under the assumption that the kernel of $f_*: \pi_0(N) \to \pi_0(CP^{n-1})$ is already killed. Given any element ξ of the kernel of $f_*: \pi_1(N) \to \pi_1(CP^{n-1})$, there is an imbedding $h: S^1 \to N$ representing ξ. With M, CP^n, CP^{n-1}, f, h in place of M, M', N', f, h respectively, (2.1), (2.2) and (2.4) are obvious and (2.3) follows from Lemma 1. Moreover, h can be extended to an imbedding $h: D^2 \to M$ such that $h(E^2) \subset M - N$ and v_1 points radially into $h(D^2)$, where v_1 is as in § 2. (Using v_1 to construct an imbedding of $S^1 \times [0,1]$ into M such that $S^1 \times \{0\}$ coincides with $h(S^1)$; then, using $\pi_1(M-N)=0$, attach a 2-cell to $S^1 \times \{1\}$ and eliminate self-intersections.) Since $\pi_1(V_{2n-2,1})=0$, (2.5′) (or equivalently (2.5)) holds. Hence there is a homotopy equivalence $f_1: M \to CP^n$ which is homotopic to f and is transverse regular on CP^{n-1} and such that the kernel of $f_{1*}: \pi_0(f_1^{-1}(CP^{n-1})) \to \pi_0(CP^{n-1})$ is trivial and the kernel of $f_{1*}: \pi_1(f_1^{-1}(CP^{n-1})) \to \pi_1(CP^{n-1})$ is the quotient group of the kernel of $f_*: \pi_1(N) \to \pi_1(CP^{n-1})$ by the normal subgroup generated by ξ. Repeating this construction if necessary, the kernel of $f_*: \pi_1(N) \to \pi_1(CP^{n-1})$ can be killed also.

Now we assume that k is an integer such that $1 < k-1 < n-1$ and the kernel of $f_*: \pi_i(N) \to \pi_i(CP^{n-1})$ has been killed for $i < k-1$. Again we can kill the kernel of $f_*: \pi_{k-1}(N) \to \pi_{k-1}(CP^{n-1})$ by using the same argument of the preceding paragraph. Hence we finally have a homotopy equivalence $f': M \to CP^n$ which is homotopic to f and is transverse regular on CP^{n-1} and such that the kernel of $f'_*: \pi_i(f'^{-1}(CP^{n-1})) \to \pi_i(CP^{n-1})$ is trivial for $i < n-1$.

4. The Index $I(G, \Sigma^{2n+1})$ for Odd n

Suppose that a free differentiable action of the circle group G on an HS^{2n+1}, Σ^{2n+1}, is given, $n \geqq 3$, and let

$$M = \Sigma^{2n+1}/G.$$

Let

$$f: M \to CP^n$$

be a homotopy equivalence which is transverse regular on CP^{n-1}, and let

$$N = f^{-1}(CP^{n-1}).$$

As seen in the preceding section, we may assume that the kernel of $f_*: \pi_i(N) \to \pi_i(CP^{n-1})$ has been killed for $i < n-1$. Therefore, by Lemma 4, $f_*: \pi_i(N) \to \pi_i(CP^{n-1})$ and $f_*: H_i(N) \to H_i(CP^{n-1})$ are isomorphisms onto for $i < n-1$ and the Hurewicz homomorphism H maps the kernel K_π of $f_*: \pi_{n-1}(N) \to \pi_{n-1}(CP^{n-1})$ isomorphically onto the kernel K_H of $f_*: H_{n-1}(N) \to H_{n-1}(CP^{n-1})$.

Let

$$p: \Sigma^{2n+1} \to M$$

be the projection and

$$Y = p^{-1}(N).$$

Then

$$p: Y \to N,$$

is a principal bundle of fibre G. Let G be oriented. Then for any $x \in Y$, the orbit Gx is an oriented circle in Y which bounds an oriented 2-disk D^2 in Y. Clearly $p(D^2)$ is a 2-cycle of N which represents a generator a of $H_2(N)$. Let α be a generator of $H^2(N)$ with $\langle \alpha, a \rangle = 1$. Then α^{n-1} is a generator of $H^{2n-2}(N)$. Let us assign to N the orientation represented by α^{n-1}. Then we have a bilinear form

$$F: K_H \otimes K_H \to \mathbb{Z},$$

which maps each $\xi \otimes \eta \in K_H \otimes K_H$ into the intersection number $\xi \cdot \eta$. The bilinear form F is symmetric or skew symmetric iff n is odd or even.

In this section, we assume that n is odd. Then it has a signature I. The integer I is independent of the choice of f. In fact, let f_1 be a second homotopy equivalence of M into CP^n which is transverse regular on CP^{n-1}, and $N_1 = f_1^{-1}(CP^{n-1})$. It is easily seen that f_1 is either homotopic to f or homotopic to τf, where τ is a diffeomorphism of CP^n onto itself with $\tau^* \alpha = -\alpha$. Therefore there is a homotopy $F: M \times [0,1] \to CP^n$ which is transverse regular on CP^{n-1} and such that $F^{-1}(CP^{n-1})$ is a cobordism between N and N_1. Hence we obtain the same I when f_1 is used to replace f. The integer I is called the *index* of (G, Σ^{2n+1}) and is also written $I(G, \Sigma^{2n+1})$.

Lemma 5. *For any odd integer $n > 3$, the index I vanishes iff there is a homotopy equivalence $f: M \to CP^n$ which is transverse regular on CP^{n-1} and such that, if $N = f^{-1}(CP^{n-1})$, then $f: N \to CP^{n-1}$ is a homotopy equivalence.*

Proof. If there is a homotopy equivalence $f: M \to CP^n$ such that $f: N \to CP^{n-1}$ is a homotopy equivalence, then $K_H = 0$ and hence I vanishes.

Conversely assume that I vanishes. Then we have at first a homotopy equivalence $f: M \to CP^n$ as described at the beginning of this section. Therefore K_H is a free abelian group which has a basis

$$\{\xi_1, \xi_2, \ldots, \xi_{2r-1}, \xi_{2r}\},$$

such that

$$\xi_i \cdot \xi_j = \begin{cases} 1 & \text{if } \{i,j\} = \{2t-1, 2t\} \text{ for some } t, \\ 0 & \text{otherwise.} \end{cases}$$

Let $h: S^{n-1} \to N$ be an imbedding representing ξ_1. We can use the construction of § 2 to kill both ξ_1 and ξ_2 at the same time. In fact, (2.1), (2.2) and (2.4) are obvious and (2.3) follows from Lemma 4. As seen in § 2, h can be extended to an imbedding $h: D^n \to M$ such that $h(D^n) \subset M - N$ and v_1 points radially into $h(D^n)$. Moreover, the obstruction to extending the 1-frame v_2 on $h(S^{n-1})$ to a 1-frame on $h(D^n)$ is an element $\lambda \in \pi_{n,1}(V_{n,1})$. Consider the boundary homomorphism

$$\partial: \pi_{n-1}(V_{n,1}) \to \pi_{n-2}(SO(n-1)),$$

of the fibring $SO(n) \to V_{n,1}$ with fibre $SO(n-1)$. This maps λ into the element corresponding to $v(h(S^{n-1}), N)$. Since $\xi_1 \cdot \xi_1 = 0$, $v(h(S^{n-1}), N)$ is trivial [10; Lemma 7] so that $\partial\lambda = 0$. But for this case, ∂ is known to be one-to-one. Therefore $\lambda = 0$ and hence (2.5) holds. This shows that ξ_1 and ξ_2 can be killed by the construction of § 2. Similarly $\xi_3, \xi_4, \ldots, \xi_{2r-1}, \xi_{2r}$ can be killed, two at a time. The proof is thus completed.

Observe that when $f: N \to CP^{n-1}$ is a homotopy equivalence, $Y = p^{-1}(N)$ is a G-invariant HS^{2n-1} in Σ^{2n+1}.

5. The Arf Invariant $c(G, \Sigma^{2n+2})$ for Even n

We consider the situation of the preceding section except that n is even. As before, we assume that the kernel of $f_*: \pi_i(N) \to \pi_i(CP^{n-1})$ has been killed for $i < n-1$. Then, since n is even, $K_\pi = \pi_{n-1}(N)$, $K_H = H_{n-1}(N)$ and $H: \pi_{n-1}(N) \to H_{n-1}(N)$ is an isomorphism onto. Moreover, the bilinear form

$$F: H_{n-1}(N) \otimes H_{n-1}(N) \to \mathbb{Z},$$

is skew symmetric. In order to define an Arf invariant of this bilinear form, we need a function

$$\varphi: H_{n-1}(N) \otimes \mathbb{Z}_2 \to \mathbb{Z}_2,$$

such that for any $\xi, \eta \in H_{n-1}(N) \otimes \mathbb{Z}_2$,

$$\varphi(\xi+\eta) = \varphi(\xi) + \varphi(\eta) + \xi \cdot \eta .$$

This can be done by following LEVINE [9; § 4.1].

Let $\xi \in H_{n-1}(N) \otimes \mathbb{Z}_2$ be represented by an imbedding $h: S^{n-1} \to N$. Since $fh: S^{n-1} \to CP^{n-1}$ is null-homotopic, we may assume that $fh(S^{n-1}) = y \in CP^{n-1}$. As before, we choose a 2-frame of $\nu(CP^{n-1}, CP^n)$ at y and let (v_1, v_2) be its induced frame on $h(S^{n-1})$ by f. Then h can be extended to an imbedding $h: D^n \to M$ such that $h(D^n) \subset M - N$ and v_1 points radially into $h(D^n)$. The obstruction to extending v_2 to a 1-frame on $h(D^n)$ is an element

$$\lambda(\xi) \in \pi_{n-1}(V_{n,1}) = \pi_{n-1}(S^{n-1}) \cong \mathbb{Z} .$$

Consider the natural homomorphisms in the exact sequence

$$\cdots \to \pi_{n-1}(V_{n,1}) \to \pi_{n-1}(V_{2n,n+1}) \to \pi_{n-1}(V_{2n,n}) \to \cdots .$$

The image $\mu(\xi)$ of $\lambda(\xi)$ in $\pi_{n-1}(V_{2n,n+1})$ can be constructed as follows. Let $(e_1, \ldots, e_n)$ be an n-frame in $\tau(h(D^n))$ which is also regarded as an n-frame on $h(S^{n-1})$. Then $(e_1, \ldots, e_n, v_2)$ is an $(n+1)$-frame on $h(S^{n-1})$ which, with respect to a $(2n)$-frame in $\tau(M)|h(D^n)$, define an element of $\pi_{n-1}(V_{2n,n+1})$, namely $\mu(\xi)$. As pointed out by LEVINE [9], $\mu(\xi)$ depends only on ξ. Since $\pi_{n-1}(V_{2n,n+1}) \cong \mathbb{Z}_2$, we obtain a function

$$\varphi: H_{n-1}(N) \otimes \mathbb{Z}_2 \to \mathbb{Z}_2 ,$$

such that for any $\xi \in H_{n-1}(N) \otimes \mathbb{Z}_2$, $\varphi(\xi)$ is the element of $\mathbb{Z}_2$ corresponding to $\mu(\xi)$. It can be shown that this is a desired function. (See [9] for details.) Let $\{\xi_1, \eta_1, \ldots, \xi_r, \eta_r\}$ be a symplectic basis of $H_{n-1}(N) \otimes \mathbb{Z}_2$. Then

$$c - \sum_{i=1}^{r} \varphi(\xi_i)\varphi(\eta_i),$$

is the Arf invariant associated with φ. It can be shown that c depends only on (G, Σ^{2n+1}). Therefore it is called the Arf invariant of (G, Σ^{2n+1}) and is also written $c(G, \Sigma^{2n+1})$.

As shown in [9; § 4.11], any $\xi \in H_{n-1}(N) \otimes \mathbb{Z}_2$ can be killed by a construction of § 2. Hence we have

Lemma 6. *For any even integer $n > 3$, the Arf invariant c vanishes iff there is a homotopy equivalence $f: M \to CP^n$ which is transverse regular on CP^{n-1} and such that, if $N = f^{-1}(CP^{n-1})$, then $f: N \to CP^{n-1}$ is a homotopy equivalence.*

6. Characteristic Submanifold W

In order to compare $I(G,\Sigma^{2n+1})$ and $c(G,\Sigma^{2n+1})$ with the Browder-Livesay invariants $I(\mathbb{Z}_2,\Sigma^{2n+1})$ and $c(\mathbb{Z}_2,\Sigma^{2n+1})$, we construct in this section a special characteristic submanifold W by means of which a relation can be established. The relation will be given in the next section as our main theorem.

Let
$$p: S^{2n+1} \to CP^n,$$
be the natural projection. Then there is a map $F: \Sigma^{2n+1} \to S^{2n+1}$ such that

$$\begin{array}{ccc} \Sigma^{2n+1} & \xrightarrow{F} & S^{2n+1} \\ \downarrow p & & \downarrow p \\ M & \xrightarrow{f} & CP^n \end{array}$$

is commutative. Since f is transverse regular on CP^{n-1}, F is transverse regular on S^{2n-1} and then is transverse regular on a small tubular neighborhood of S^{2n-1} in S^{2n}. On the other hand, F is transverse regular on $S^{2n}-S^{2n-1}$. Hence F is transverse regular on S^{2n} and consequently
$$W = F^{-1}(S^{2n}),$$
is a closed $\mathbb{Z}_2$-invariant $2n$-submanifold of Σ^{2n+1} so that it is a charcteristic submanifold in the sense of BROWDER-LIVESAY [1].

Let T be the generator of $\mathbb{Z}_2$ and V the closure of a component of $W-Y$. It is obvious that

Lemma 7. a) *V is a compact $(2n)$-manifold with $\partial V = Y$.*

b) *$W = V \cup T(V)$, $Y = V \cap T(V)$ and $V-Y$ is a cross-section of the action $(G,\Sigma^{2n+1}-Y)$.*

Lemma 8. a) *$\pi_i(V)=0$ for $i \leqq n-1$.*

b) *$H_0(Y) \cong \mathbb{Z}$ and $H_i(Y)=0$ for $i=1,\dots,n-2$.*

c) *The sequence*
$$0 \to H_{n-1}(Y) \xrightarrow{p_*} H_{n-1}(N) \xrightarrow{f_*} H_{n-1}(CP^{n-1}) \to 0,$$
is exact so that there is an isomorphism $p_*: H_{n-1}(Y) \cong K_H$.

d) *The homomorphism*
$$\partial: H_n(V,Y) \to H_{n-1}(Y),$$
in the homology sequence of (V,Y) is an isomorphism onto.

Proof. Whenever $i \leqslant n-1$, $\pi_i(M-N)=0$ (Lemma 4) so that $\pi_i(V-Y)=0$ (Lemma 7, b)). Hence $\pi_i(V)=0$, proving a).

b) and c) are consequences of Lemma 4 and the homomorphism of the Gysin sequence of $p: Y \to N$ into that of $p: S^{2n-1} \to CP^{n-1}$ induced by the bundle map $F: Y \to S^{2n-1}$.

In order to prove d), we have only to observe that in the commutative diagram

$$\begin{array}{ccccccccc} H_n(Y) & \longrightarrow & H_n(V) & \longrightarrow & H_n(V,Y) & \xrightarrow{\partial} & H_{n-1}(Y) & \longrightarrow & H_{n-1}(V) \\ \downarrow & & \downarrow & & \downarrow{\scriptstyle p_*} & & \downarrow{\scriptstyle p_*} & & \downarrow \\ H_n(N) & \longrightarrow & H_n(M) & \longrightarrow & H_n(M,N) & \xrightarrow{\partial} & H_{n-1}(N) & \longrightarrow & H_{n-1}(M) \end{array}$$

rows are exact, $H_n(N) \to H_n(M)$ is onto, $p_*: H_n(V,Y) \to H_n(M,N)$ is an isomorphism onto (Lemma 7, b)) and $H_{n-1}(V)=0$ (a consequence of a)).

Let $\theta: R' \to G$ be the orientation-preserving homomorphism with $\ker\theta = 2\mathbb{Z}$. Let $G_+ = \theta[0,1]$ and

$$A = G_+ V.$$

Lemma 9. a) *A is a compact $(2n+1)$-manifold with $\partial A = W$ and such that*

$$\Sigma^{2n+1} = A \cup T(A), \qquad W = A \cap T(A).$$

b) *Let K be the kernel of the homomorphism $H_n(W) \to H_n(A)$ induced by the inclusion map. Then the homomorphism $H_n(W) \to H_n(W, T(V))$ induced by the inclusion map maps K isomorphically onto $H_n(W, T(V))$.*

Proof. a) is obvious so that we have only to prove b).

Let ξ be an element of K whose image in $H_n(W,T(V))$ is 0. Then there is a singular n-cycle z of $T(V)$ representing ξ. Since the image of ξ in $H_n(A)$ is 0, there is a singular $(n+1)$-chain a of A with $\partial a = z$. However, it is not hard to see that $T(V)$ is a deformation retract of A so that we may assume that a is in $T(V) \subset W$. Hence $\xi = 0$. This shows that $K \to H_n(W,T(V))$ is one-to-one.

Let $\eta \in H_n(W,T(V))$. Since $H_n(V,Y) \cong H_n(W,T(V))$, there is a singular n-cycle b of (V,Y) representing η. Let

$$z = b + G_+(\partial b) - T(b) = -\partial(G_+ b).$$

It is clear that z is a singular n-cycle of W which represents an element ξ of K, and that the image of ξ in $H_n(W,T(V))$ is η. Hence $K \to H_n(W,T(V))$ is onto.

7. Main Theorem

Suppose that a free differentiable action of the circle group G on an HS^{2n+1}, Σ^{2n+1}, is given, $n \geqslant 3$, and $\mathbb{Z}_2$ is the subgroup of G of order 2. Then we have a free differentiable action of $\mathbb{Z}_2$ on Σ^{2n+1} so that, according to BROWDER-LIVESAY [1], an index or an Arf invariant is defined as follows. Let W be a closed $\mathbb{Z}_2$-invariant $2n$-submanifold of Σ^{2n+1} and A a compact $(2n+1)$-manifold such that $\pi_i(W)=0$ for $i<n$ and

$$\Sigma^{2n+1}=A\cup T(A), \qquad W=A\cap T(A),$$

where T is the generator of $\mathbb{Z}_2$. (For example, W and A are those given in the preceding section.) Let K be the kernel of the homomorphism $H_n(W)\to H_n(A)$ induced by the inclusion map. Let Σ^{2n+1} be so oriented that if U is an open $2n$-cell in N such that $p:p^{-1}(U)\to U$ has a cross-section U', then U' can be oriented such that $G\times U'\to GU'\subset\Sigma^{2n+1}$ and $p:U'\to N$ are orientation-preserving. Let A be oriented as Σ^{2n+1} and let W be oriented such that $\partial A=W$. Then there is a bilinear form

$$F':K\otimes K\to\mathbb{Z},$$

which maps each $\xi\otimes\eta\in K\otimes K$ into the intersection number $\xi\cdot T^*\eta$. Since $T:W\to W$ is orientation-reversing, F' is symmetric and skew symmetric iff n is odd or even. For any odd n, the signature depends only on $(\mathbb{Z}_2,\Sigma^{2n+1})$ so that it is called the index of $(\mathbb{Z}_2,\Sigma^{2n+1})$ and is denoted by $I(\mathbb{Z}_2,\Sigma^{2n+1})$. For any even n, there is a function

$$\varphi':K\otimes\mathbb{Z}_2\to\mathbb{Z}_2,$$

such that for any $\xi,\eta\in K\otimes\mathbb{Z}_2$,

$$\varphi'(\xi+\eta)=\varphi'(\xi)+\varphi'(\eta)+\xi\cdot T_*\eta.$$

Therefore we have an Arf invariant associated with φ'. Since it depends only on $(\mathbb{Z}_2,\Sigma^{2n+1})$, it is called the Arf invariant and is denoted by $c(\mathbb{Z}_2,\Sigma^{2n+1})$.

Theorem. *Suppose that a free differentiable action of the circle group G on a homotopy $(2n+1)$-sphere Σ^{2n+1} is given, $n\geqslant 3$, and let $\mathbb{Z}_2$ be the subgroup of G of order 2. Then for odd n, $I(G,\Sigma^{2n+1})=I(\mathbb{Z}_2,\Sigma^{2n+1})$ and for even n, $c(G,\Sigma^{2n+1})=c(\mathbb{Z}_2,\Sigma^{2n+1})$.*

The proof is carried out by using the characteristic submanifold W constructed in the preceding section. For that W, we shall give a natural isomorphism

$$\rho:H_H\cong K,$$

such that if we use ρ to identify K_H with K, then F coincides with F' and φ coincides with φ' and hence the theorem follows.

Lemm 10. *There is an isomorphism*

$$\rho:K_H\to K,$$

such that

$$\begin{array}{ccccccccc}
K_H & & & & \xrightarrow{\quad\rho\quad} & & & & K \\
\cap & \nwarrow & & & & & & \swarrow & \cap \\
H_{n-1}(N) & \xleftarrow{p_*} & H_{n-1}(Y) & \xleftarrow{\partial} & H_n(V,Y) & \longrightarrow & H_n(W,T(V)) & \longleftarrow & H_n(W)
\end{array}$$

is commutative.

Proof. We recall that K_H is, by definition, the kernel of

$$f_*: H_{n-1}(N) \to H_{n-1}(CP^{n-1}).$$

Then, by Lemma 8, b), p_* maps $H_{n-1}(Y)$ isomorphically onto K_H. We know that $\partial: H_n(V,Y) \to H_{n-1}(Y)$ is an isomorphism onto (Lemma 8, c)) and so are $H_n(V,Y) \to H_n(W,T(V))$ (excision theorem) and

$$K \to H_n(W,T(V))$$

(Lemma 9, b)). Hence our assertion follows.

Let $\xi \in K_H$. Then there is an immersion $h: S^{n-1} \to N$ representing ξ. (Notice that if $n>3$, we may assume that h is an imbedding.) Since $fh: S^{n-1} \to CP^{n-1}$ is null-homotopic, there is a neighborhood U of $h(S^{n-1})$ in N such that $p: p^{-1}(U) \to U$ is a trivial circle bundle. Let $h': U \to p^{-1}(U)$ be a cross-section of this bundle. Then p maps $U'=h'(U)$ diffeomorphically onto U and $h'h: S^n \to Y$ is an immersion. By Lemma 8, a), $\pi_{n-1}(V)=0$ so that $h'h$ can be extended to an immersion $h'': D^n \to V$ such that $h''(D^n)$ intersects Y transversally. Since $n \geqslant 3$, we can eliminate self-intersections in $h''(D^n)$ and hence we may assume that h'' imbeds E^n into $V-Y$.

Let us identify S^n with

$$-\partial([0,1] \times D^n) = \{0\} \times D^n - [0,1] \times S^{n-1} - \{1\} \times D^n,$$

and define

$$h^*: S^n \to W,$$

by

$$h^*(t,x) = \begin{cases} h''(x) & \text{if } (t,x) \in \{0\} \times D^n, \\ \theta(t)h''(x) & \text{if } (t,x) \in (0,1) \times S^{n-1}, \\ Th''(x) & \text{if } (t,x) \in \{1\} \times D^n, \end{cases}$$

where θ is the orientation-preserving homomorphism of $\mathbb{R}^1$ into G with kernel $2\mathbb{Z}$ and $T=\theta(1)$ is the generator of $\mathbb{Z}_2$. Clearly $h^*(S^n)$ represents $\rho(\xi)$.

Let u be a vector field in $\tau(W)$ having the following properties: 1. $u(x)=0$ for $x \in Th''(D^n)$. 2. For any $x \in h''(D^n)$, $u(x)$ is transversal to $h''(D^n)$ and finitely zero. 3. For any $x \in h''(S^{n-1})$, $u(x) \neq 0$ and is tangent to Gx in the positive direction. Using the vector field u, we obtain a deformation of W which gives an isotopy between the identity map of W and a diffeomorphism $q: W \to W$ such that for any $x \in W$, $q(x) = x$ iff $u(x) = 0$. By our choice of u, it is clear that $qh^*(S^n)$ is in $\rho(\xi)$ and

$$qh^*(S^n) \cap Tqh^*(S^n) = (h''(D^n) \cap qh''(D^n)) \cup (Th''(D^n) \cap Tqh''(D^n)),$$

is a finite set. Let

$$h''(D^n) \cap qh''(D^n) = \{x_1, \ldots, x_r\}.$$

Without loss of generality, we may assume that $qh^*(S^n)$ and $Tqh^*(S^n)$ intersect transversally at each x_i (and thus also at each Tx_i). Then the intersection number of $qh^*(S^n)$ and $Tqh^*(S^n)$ at each x_i is either 1 or -1. We shall denote by k the sum of the intersection numbers at $x_1, \ldots, x_r$.

Let

$$D = h''((1-\varepsilon)D^n),$$

where ε is a positive number which is so small that $x_1, \ldots, x_r$ are contained in the interior of D. Notice that D is an n-cell in $V - Y$ and the intersection number of qD and $-D$ is equal to k. Let us identify $\mathbb{R}^n \times \mathbb{R}^n$ with an open neighborhood of D in W such that $D = D^n \times \{0\}$. Since for each $x \in \partial D$, $u(x) \neq 0$ and is transversal to D, we may let $u(x) = (0, u_2(x)) \in \mathbb{R}^n \times \mathbb{R}^n$ with $\|u_2(x)\| = 1$. Therefore u_2 is a map of S^{n-1} into S^{n-1} and the degree of u_2 is equal to the intersection number of D and qD which is k or $-k$ according as n is odd or even.

Suppose first that n is odd. Let v be the oriented $(n-1)$-vector bundle over ∂D which is the orthogonal complement of $u|\partial D$ in $v(D, V)|\partial D$. Then

$$\begin{aligned} \xi \cdot \xi &= h(S^{n-1}) \cdot h(S^{n-1}) && \text{in } N \\ &= h''(S^{n-1}) \cdot h''(S^{n-1}) && \text{in } U' \\ &= \partial D \cdot \partial D && \text{in } v. \end{aligned}$$

As seen in the proof of [10; Lemma 7], $\partial: \pi_{n-1}(S^{n-1}) \to \pi_{n-2}(SO(n-1))$ maps the class of u_2 into the class of v and the Euler class of v is $\partial D \cdot \partial D$ times the fundamental class of ∂D. Therefore $\partial D \cdot \partial D$ in v is equal to $2k$ and hence

$$\begin{aligned} \rho(\xi) \cdot T\rho(\xi) &= qh^*(S^n) \cdot Tqh^*(S^n) \\ &= qD \cdot (-D) + TqD \cdot (-TD) \\ &= D \cdot qD + TD \cdot TqD = 2k = \xi \cdot \xi. \end{aligned}$$

From this result, we infer that for any $\xi, \eta \in K_H$,

$$\begin{aligned} \rho(\xi) \cdot T\rho(\eta) &= (\rho(\xi+\eta) \cdot T\rho(\xi+\eta) - \rho(\xi) \cdot T\rho(\xi) - \rho(\eta) \cdot T\rho(\eta))/2 \\ &= ((\xi+\eta) \cdot (\xi+\eta) - \xi \cdot \xi - \eta \cdot \eta)/2 = \xi \cdot \eta. \end{aligned}$$

Hence the bilinear forms F and F' have the same signature, proving that $I(G,\Sigma^{2n+1})=I(\mathbb{Z}_2,\Sigma^{2n+1})$.

Suppose next that n is even. Then

$$\varphi'(\rho(\xi)\bmod 2)=k\bmod 2=\gamma\bmod 2.$$

Since $u\colon S^{n-1}\to S^{n-1}$ is of degree $-k$, it follows from the definition of φ that

$$\varphi(\xi\bmod 2)=-k\bmod 2=\varphi'(\rho(\xi)\bmod 2).$$

Hence φ and φ' have the same Arf invariant, proving that $c(G,\Sigma^{2n+1})=c(\mathbb{Z}_2,\Sigma^{2n+1})$.

8. Applications

We apply the main theorem to some special cases.

Lemma 11. *Let $n=3$ and M a homotopy complex projective 3-space with $p_1(M)=(24i+4)\alpha^2$, where α is a generator of $H^2(M)$. Then the bilinear form*

$$F\colon K_H\otimes K_H\to\mathbb{Z},$$

defined by intersection is a symmetric form of signature $8i$.

We recall that f is a homotopy equivalence of M into CP^3 which is transverse regular on CP^2, $N=f^{-1}(CP^2)$ which is so oriented that $(\alpha|N)^2\in H^4(N)$ represents the orientation on N. Since $p_1(M)=(24i+4)\alpha^2$ and $p_1(\nu(N,M))=(\alpha|N)^2$, we have

$$p_1(N)=(24i+3)(\alpha|N)^2.$$

Hence, using the Hirzebruch index theorem, the index of N is equal to $8i+1$. On the other hand, K_H is the kernel of $f_*\colon H_2(N)\to H_2(CP^2)$ which is onto, and the index of CP^2 is equal to 1. It follows that the signature of F is equal to $8i$.

Now we can easily verify Theorems A – D. For Theorem A, if $n>3$, it is a consequence of Lemmas 5, 6, the Main Theorem and the Browder-Livesay theorem [1]; if $n=3$, it is a special case of Theorem B. For Theorem B, we have only to observe that, for any free differentiable action of the circle group G on an HS^7, Σ, if there is a G-invariant 5-sphere in Σ, then $I(G,\Sigma)=0$ so that, by Lemma 11 and the classification of HCP^3's [4], (G,Σ) is differentiably equivalent to an orthogonal action of the circle group G on an HS^7, Σ, if there is a G-invariant by Lemma 6, the second part of Theorem D follows. By Lemma 11 and the characterization of free differentiable actions of the circle group on homotopy 7-spheres [4], we know that, for any integer i, there is a free differentiable action of $\mathbb{Z}_2$ on a homotopy 7-sphere Σ_i such that $I(\mathbb{Z}_2, \Sigma_i) = 8i$ and the Eells-Kuiper invariant $\mu(\Sigma_i)$ is $i(9i+2)/14 \bmod 1$. Let Σ' be the homotopy 7-sphere with $\mu(\Sigma') = -1/28 \bmod 1$. Then $\Sigma_i \# 2i(9i+2)\Sigma'$ is diffeomorphic to S^7. Since the action of $\mathbb{Z}_2$ on Σ_i induces a natural free differentiable action of $\mathbb{Z}_2$ on $\Sigma_i \# 2i(9i+2)$ such that $I(\mathbb{Z}_2, \Sigma_i \# 2i(9i+2)) = I(\mathbb{Z}_2, \Sigma_i)$, the first part of Theorem D follows.

References

1. BROWDER, W., and G. R. LIVESAY: Fixed point free involutions on homotopy spheres, Bull. Amer. Math. Soc. **73**, 242—245 (1967).
2. — Homotopy type of differentiable manifolds, Colloquium on Algebraic Topology, Aarhus 1962, 42—46.
3. NOVIKOV: Homotopically equivalent smooth manifolds, Amer. Math. Soc. Translations **48**, 271—396 (1965).
4. MONTGOMERY, D., and C. T. YANG: Differentiable actions on homotopy 7-spheres II, Proceedings of the Tulane Conference on Transformation Groups.
5. SULLIVAN, D.: Triangulating homotopy equivalences, Princeton Ph. D. Thesis, 1966.
6. LOPEZ DE MEDRANO, S.: Involutions of homotopy spheres and homology 3-spheres, Bull. Amer. Math. Soc. **73**, 727—731 (1967).
7. — Some results on involutions of homotopy spheres, to appear.
8. HAEFLIGER, A.: Knotted $(4k-1)$-spheres in $6k$-space, Ann. of Math. **75**, 452—466 (1962).
9. LEVINE, J.: A classification of differentiable knots, Ann. of Math. **82**, 15—50 (1965).
10. MILNOR, J.: A procedure for killing the homotopy groups of differentiable manifolds, Symposia in Pure Math., Amer. Math. Soc. **III**, 39—55 (1961).

Some Results on Cyclic Transformation Groups on Homotopy Spheres*

J. C. Su

1. Introduction

This paper is concerned with smooth actions of a cyclic group Z_k of order k on a homotopy $(n+2)$-sphere $\tilde{\Sigma}^{n+2}$ whose fixed point set is a homotopy n-sphere Σ_1^n. The action is further assumed to be semi-free, i. e., free outside the fixed point set. An action of this type will be denoted by $(\tilde{\Sigma}^{n+2}, \Sigma_1^n, Z_k)$ and referred to simply as an "action". We assume throughout the paper that $n \geqslant 5$. It is known that if the fixed point sphere is of codimension 1, then all the homotopy spheres involved are standard and the action must indeed be linear. Thus the codimension 2 case is the first case with interest. Non-linear actions do occur, as shown by GIFFEN [3]. In contrast, the codimension 2 case is shown to be trivial for circle group actions by W. Y. HSIANG [5]. Our results generalize GIFFEN's results with somewhat different techniques.

We first observe that for any action $(\tilde{\Sigma}^{n+2}, \Sigma_1^n, Z_k)$, the orbit space $\tilde{\Sigma}^{n+2}/Z_k$ is also a homotopy sphere. Thus every action gives rise to a smooth knot $(\tilde{\Sigma}^{n+2}/Z_k, \Sigma_1^n)$ of homotopy spheres. However, not every knot can be obtained in this way (for a fixed k). In § 2, we give a criterion to characterize those knots that do come from actions and show that they in fact classify all actions (Theorem 2.3). This is analogous to a result of MONTGOMERY and YANG [8] for circle group actions.

Even though the study of actions has been reduced to the study of knots of a specific kind, we have not been able to construct knots of this kind in a systematic way. Instead, we have found some examples provided by the work of BRIESKORN and HIRZEBRUCH [4, 1]. In this way, we have managed to draw some conclusions, naturally of a very isolated and highly incomplete type. Denote by Θ^n the group of homotopy n-spheres and $\Theta^n(\partial\pi)$ the subgroup of Θ^n of those homotopy spheres which bound a π-manifold. We have for example

* This work is partially supported by NSF Grant GP-6651.

Theorem 1.1. *Let $\Sigma_1^{4m-3} \in \Theta^{4m-3}(\partial\pi)$ be a homotopy $(4m-3)$-sphere bounding a π-manifold, $\Sigma^{4m-1} \in \Theta^{4m-1}$ an arbitrary homotopy $(4m-1)$-sphere, $m \geqslant 2$, and k an odd integer $\geqslant 3$. Then there exists some homotopy $(4m-1)$-sphere $\tilde{\Sigma}^{4m-1} \in \Theta^{4m-1}$ which admits infinitely many differentiably (and topologically) distinct smooth actions of Z_k whose fixed point set is Σ_1^{4m-3} and whose orbit space is Σ^{4m-1}. If $\Sigma^{4m-1} \in \Theta^{4m-1}(\partial\pi)$, then $\tilde{\Sigma}^{4m-1}$ can be found in $\Theta^{4m-1}(\partial\pi)$. If Σ_1^{4m-3} and Σ^{4m-1} are standard spheres, then $\tilde{\Sigma}^{4m-1}$ can be taken as the standard sphere.*

Theorem 1.2. *Let $\Sigma_1^{4m-1} \in \Theta^{4m-1}(\partial\pi)$ be a homotopy $(4m-1)$-sphere bounding a π-manifold, $\Sigma^{4m+1} \in \Theta^{4m+1}$ an arbitrary homotopy $(4m+1)$-sphere, $m \geqslant 2$, and k any prime number. Then there exists some homotopy $(4m+1)$-sphere $\tilde{\Sigma}^{4m+1}$ which admits infinitely many differentiably (and topologically) distinct smooth actions of Z_k whose fixed point set is Σ_1^{4m-1} and whose orbit space is Σ^{4m+1}. If $\Sigma^{4m+1} \in \Theta^{4m+1}(\partial\pi)$ and $k \neq 2$, $\tilde{\Sigma}^{4m+1}$ can be found in $\Theta^{4m+1}(\partial\pi)$. If in addition both Σ_1^{4m-1} and Σ^{4m+1} are standard spheres, then $\tilde{\Sigma}^{4m+1}$ can be taken as the standard sphere.*

Other remarks are contained in 3.

2. A Classification Theorem

Manifolds will always be compact, C^∞, orientable, with or without boundary. But specification of orientation will not be necessary.

We use the following notation: D^n, E^n and S^n denote respectively the closed n-disk, the open n-disk, and the n-sphere. We will regard D^2 (resp. S^1) as complex numbers with norm $\leqslant 1$ (resp. $= 1$). Let k be a positive integer, Z_k the cyclic group of order k. We will use D_k^2 (resp. S_k^1) to denote D^2 (resp. S^1) together with the action of Z_k given by $z \to e^{2\pi i/k}z$. The orbit space can be identified with $D^2(S^1)$, and via this identification, the orbit map p_k: $D_k^2 \to D^2$ is given by $p_k(z) = z^k$.

Let $(\tilde{\Sigma}^{n+2}, \Sigma_1^n, Z_k)$ be an action, i.e., a smooth action of Z_k on a homotopy $(n+2)$-sphere $\tilde{\Sigma}^{n+2}$ whose fixed point set Σ_1^n is a homotopy n-sphere and the action is free outside the fixed point set. Since Σ_1^n has trivial normal bundle in $\tilde{\Sigma}^{n+2}$ and since the action is free outside Σ_1^n, there is an equivariant imbedding $\tilde{\varphi}$: $\Sigma_1^n \times D_k^2 \to \tilde{\Sigma}^{n+2}$ such that $\tilde{\varphi}(x, 0) = x$ for $x \in \Sigma_1^n$. Then $\tilde{Y} = \tilde{\Sigma}^{n+2} - \tilde{\varphi}(\Sigma_1^n \times E_k^2)$ is a manifold with a smooth free action of Z_k which preserves orientation. It follows that the orbit space $Y = \tilde{Y}/Z_k$ is a manifold. Moreover, the equivariant diffeomorphism $\tilde{\varphi}$: $\Sigma_1^n \times S_k^1 \to \partial\tilde{Y}$ induces a diffeomorphism φ: $\Sigma_1^n \times S^1 \to \partial Y$ via the diagram

$$\begin{array}{ccc} \Sigma_1^n \times S_k^1 & \xrightarrow{\tilde{\varphi}} & \partial \tilde{Y} \\ \Big\downarrow {\scriptstyle 1 \times p_k} & & \Big\downarrow \\ \Sigma_1^n \times S^1 & \xrightarrow{\varphi} & \partial Y \end{array}$$

where $\partial \tilde{Y} \to \partial Y$ is the orbit map. Clearly, the orbit space $\Sigma^{n+2} = \tilde{\Sigma}^{n+2}/Z_k$ can be identified with $\Sigma_1^n \times D^2 \bigcup_\varphi Y$, that is, the space obtained by pasting $\Sigma_1^n \times S^1 \subset \Sigma_1^n \times D^2$ onto $\partial Y \subset Y$ via φ. In this way, Σ^{n+2} becomes a manifold.

Proposition 2.1. *Σ^{n+2} is a homotopy $(n+2)$-sphere.*

Proof. We first assert that Σ^{n+2} is simply connected. This can be seen by the Van Kampen Theorem. It can also be argued geometrically as follows. Choose a base point $x_0 \in \Sigma_1^n = \Sigma_1^n \times 0 \subset \Sigma_1^n \times D^2 \subset \Sigma^{n+2}$ and let α be a loop based at x_0. We may choose α in such a way that $\alpha(t) \in \Sigma_1^n$ for all $0 < t < 1$. Take some $\varepsilon > 0$ so that $\alpha[0, \varepsilon]$ and $\alpha[1-\varepsilon, 1]$ are contained in $\Sigma_1^n \times D^2$. We may assume that over the portion $[0, \varepsilon]$ and $[1-\varepsilon, 1]$, α is of the form

$$\alpha(t) = (\alpha_1(t), (t/\varepsilon) z_1), \qquad 0 \leqslant t \leqslant \varepsilon,$$
$$\alpha(t) = (\alpha_2(t), ((1-t)/\varepsilon) z_2), \qquad 1-\varepsilon \leqslant t \leqslant 1,$$

where α_1 and α_2 are paths in Σ_1^n and $z_1, z_2 \in D^2$ are not zero. Now $\alpha[\varepsilon, 1-\varepsilon] \subset \Sigma^{n+2} - \Sigma_1^n$. Since $\tilde{\Sigma}^{n+2} - \Sigma_1^n \to \Sigma^{n+2} - \Sigma_1^n$ is a covering map, we can lift α to a path $\tilde{\alpha}$: $[\varepsilon, 1-\varepsilon] \to \tilde{\Sigma}^{n+2} - \Sigma_1^n$ covering α with $\tilde{\alpha}(\varepsilon) = (\alpha_1(\varepsilon), z_1')$, where $z_1' \in D^2$ is some point such that $(z_1')^k = z_1$. Now the end point $\tilde{\alpha}(1-\varepsilon) = (\alpha_2(1-\varepsilon), z_2')$ lies over $\alpha(1-\varepsilon)$ so that $(z_2')^k = z_2$. If we just extend $\tilde{\alpha}$ by defining

$$\tilde{\alpha}(t) = \left(\alpha_1(t), \left(\frac{t}{\varepsilon}\right)^{1/k} z_1'\right), \qquad 0 \leqslant t \leqslant \varepsilon,$$

$$\tilde{\alpha}(t) = \left(\alpha_2(t), \left(\frac{1-t}{\varepsilon}\right)^{1/k} z_2'\right), \qquad 1-t \leqslant t \leqslant 1,$$

then we get a loop $\tilde{\alpha}$ in $\tilde{\Sigma}^{n+2}$ based at x_0 covering α, showing that $\pi_1(\tilde{\Sigma}^{n+2}, x_0) \to \pi_1(\Sigma^{n+2}, x_0)$ is an epimorphism. Hence Σ^{n+2} is simply connected. It remains to show that Σ^{n+2} is an integral homology $(n+2)$-sphere. From the pair $(\tilde{\Sigma}^{n+2}, \tilde{Y})$, it is easily computed that $H_*(\tilde{Y}) = H_*(S^1)$ (integral coefficients) and that $\tilde{\varphi}_*: H_1(\Sigma_1^n \times S_k^1) \to H_1(\tilde{Y})$ is an isomorphism. In particular, Z_k acts trivially on $H_*(\tilde{Y})$. From the spectral sequence of the covering $\tilde{Y} \to Y$, one then computes easily that $H_*(Y) = H_*(S^1)$ and that $\varphi_*: H_1(\Sigma_1^n \times S^1) \to H_1(Y)$ is an isomorphism. Looking at the pair (Σ^{n+2}, Y) once more, one concludes that Σ^{n+2} is an integral homology $(n+2)$-sphere.

Thus for every action $(\tilde{\Sigma}^{n+2}, \Sigma_1^n, Z_k)$, the orbit space pair $(\tilde{\Sigma}^{n+2}/Z_k, \Sigma_1^n)$ is a pair of homotopy spheres, i.e., a smooth knot. Naturally, what we want to do next is to reconstruct the action from this knot. So now we start from an arbitrary smooth knot $(\Sigma^{n+2}, \Sigma_1^n)$ of homotopy spheres. Since Σ_1^n has trivial normal bundle in Σ^{n+2}, there is an imbedding $\varphi: \Sigma_1^n \times D^2 \to \Sigma^{n+2}$ such that $\varphi(x,0)=x$ for $x \in \Sigma_1^n$. Let $Y = \Sigma^{n+2} - \varphi(\Sigma_1^n \times E^2)$. Then Y is a compact $(n+2)$-manifold. As before, the homology sequence of the pair (Σ^{n+2}, Y) shows that $H_*(Y) = H_*(S^1)$ and that $\varphi_*: H_1(\Sigma_1^n \times S^1) \to H_1(Y)$ is an isomorphism. Let $\rho: \pi_1(Y) \to H_1(Y)$ be the Hurewicz homomorphism and $\nu: H_1(Y) \to H_1(Y)/kH_1(Y) \simeq Z_k$ the natural reduction. We can then consider the regular covering space $\tilde{Y} \to Y$ corresponding to the normal subgroup $\ker \nu\rho \subset \pi_1(Y)$. Then $\tilde{Y}$ is a $(n+2)$-manifold having a natural smooth free action of Z_k on it. Moreover, ∂Y is equivariantly diffeomorphic to $\Sigma_1^n \times S_k^1$ and we can choose an equivariant diffeomorphism $\tilde{\varphi}: \Sigma_1^n \times S_k^1 \to \partial \tilde{Y}$ such that the diagram

$$\begin{array}{ccc} \Sigma_1^n \times S_k^1 & \xrightarrow{\tilde{\varphi}} & \partial \tilde{Y} \\ \downarrow{\scriptstyle 1 \times p_k} & & \downarrow \\ \Sigma_1^n \times S^1 & \xrightarrow{\varphi} & \partial Y \end{array}$$

is commutative. Now form the space

$$\tilde{\Sigma}^{n+2} = \Sigma_1^n \times D_k^2 \cup_{\tilde{\varphi}} \tilde{Y}.$$

Clearly, $\tilde{\Sigma}^{n+2}$ is an $(n+2)$-manifold with a smooth action of Z_k on it. The action has Σ_1^n as fixed point set and is free outside Σ_1^n, and the orbit pair $(\tilde{\Sigma}^{n+2}/Z_k, \Sigma_1^n)$ is diffeomorphic to $(\Sigma^{n+2}, \Sigma_1^n)$. It is not hard to see that the equivariant diffeomorphism class of $(\tilde{\Sigma}^{n+2}, \Sigma_1^n, Z_k)$ depends only on the diffeomorphism class of $(\Sigma^{n+2}, \Sigma_1^n)$. To see whether $\tilde{\Sigma}^{n+2}$ is a homotopy sphere, we need to check that $\tilde{\Sigma}^{n+2}$ is simply connected and has the homology of a $(n+2)$-sphere. In general, neither one of these statements holds. We first discuss the fundamental group $\pi_1(\tilde{\Sigma}^{n+2})$. Let $\alpha \in \pi_1(\Sigma_1^n \times S_k^1) \simeq Z$ (the group of integers) be a generator and $\langle \alpha \rangle \subset \pi_1(\tilde{Y})$ the normal subgroup of $\pi_1(\tilde{Y})$ generated by α, where α is regarded as in $\pi_1(\tilde{Y})$ via the inclusion $\varphi_*: \pi_1(\Sigma_1^n \times S_k^1) \to \pi_1(\tilde{Y})$. Then by van Kempen's theorem, we have

$$\pi_1(\tilde{\Sigma}^{n+2}) \simeq \pi_1(\tilde{Y})/\langle \alpha \rangle.$$

This can also be described as follows. The covering map $\tilde{Y} \to Y$ identifies $\pi_1(\tilde{Y})$ with $N = \ker \nu\rho \subset \pi_1(Y)$. Under this identification, α is identified with β^k, where $\beta \in \pi_1(\Sigma_1^n \times S^1) \subset \pi_1(Y)$ is a generator. Let $\langle \beta^k \rangle \subset N$ be the normal subgroup in N generated by β^k. Then we have $\pi_1(\tilde{\Sigma}^{n+2}) \simeq N/\langle \beta^k \rangle$. Thus the first condition we must impose on the knot $(\Sigma^{n+2}, \Sigma_1^n)$ is that

$$(A_k, 1): N/\langle \beta^k \rangle = 0.$$

Notice that this condition is always true if $\pi_1(\Sigma^{n+2}-\Sigma_1^n)$ is infinite cyclic.

Next we consider the homology $H_*(\tilde{\Sigma}^{n+2})$. The question here is to see if $H_*(\tilde{Y})=H_*(S^1)$. Let $\tilde{Y}_0\to Y$ be the regular covering space corresponding to the commutator subgroup of $\pi_1(Y)$. The group of covering transformations is then the group of integers Z. In other words, $H_*(\tilde{Y}_0)$ is a module over $J(Z)$, the integral group ring of Z. Let $\beta\in Z$ be a generator, $\sigma_k=\Sigma_{i=0}^{k-1}\beta^i\in J(Z)$. We can consider the homomorphism $\sigma_k:\tilde{H}_*(\tilde{Y}_0)\to\tilde{H}_*(\tilde{Y}_0)$, the multiplication by σ_k on the reduced homology of $\tilde{Y}_0$. We have

Lemma 2.2. *$H_*(\tilde{Y})=H_*(S^1)$ if and only if σ_k: $\tilde{H}_*(\tilde{Y}_0)\to\tilde{H}_*(\tilde{Y}_0)$ is an isomorphism.*

Proof. Let $Z'\subset Z$ be the subgroup generated by $\beta^k\in Z$. Then we can identify $\tilde{Y}$ as the orbit space $\tilde{Y}_0/Z'$. In the spectral sequence of the covering $\tilde{Y}_0\to\tilde{Y}$, we have

$$E^2_{p,q}=H_p(Z',H_q(\tilde{Y}_0)),\qquad E^\infty\Rightarrow H_*(\tilde{Y}).$$

Recall that $H_p(Z,G)=0$ for any $J(Z)$-module G when $p\geqslant 2$. The spectral sequence is easily seen to be trivial and reduces to a series of short exact sequences

$$0\to E^2_{0,i}\to H_i(\tilde{Y})\to E^2_{1,i-1}\to 0\quad\text{for}\quad i\geqslant 1.$$

Now $E^2_{0,i}=\operatorname{coker}(1-\beta^k)$, $E^2_{1,i}=\ker(1-\beta^k)$ for $i\geqslant 1$. From this we see that $H_*(\tilde{Y})=H_*(S^1)$ if and only if $1-\beta^k$ is an isomorphism on $\tilde{H}_*(\tilde{Y}_0)$. Apply the same argument to the covering $\tilde{Y}_0\to Y$. Since we know that $H_*(Y)=H_*(S^1)$, we conclude that $1-\beta$ is an isomorphism on $\tilde{H}_*(\tilde{Y}_0)$. This proves Lemma 2.2.

Back now to our construction, we see the necessary and sufficient condition that the manifold $\tilde{\Sigma}^{n+2}$ is a homotopy sphere is that

$$(A_k):\quad\begin{cases}(A_k,1): & N/\langle\beta^k\rangle=0,\\(A_k,2): & \sigma_k:\tilde{H}_*(\tilde{Y}_0)\to\tilde{H}_*(\tilde{Y}_0)\ \textit{is an isomorphism.}\end{cases}$$

Thus every knot $(\Sigma^{n+2},\Sigma_1^n)$ satisfying condition (A_k) is the orbit pair of some action. It is straightforward to verify that the knot then indeed determines the action. We have therefore the following theorem.

Theorem 2.3. *For any integer $k>0$, the assignment $(\tilde{\Sigma}^{n+2},\Sigma_1^n,Z_k)\to(\tilde{\Sigma}^{n+2}/Z_k,\Sigma_1^n)$ is a one-to-one correspondence between the set of equivariant diffeomorphism classes of actions and the set of diffeomorphism classes of smooth knots of homotopy spheres satisfying condition (A_k).*

Remark. Let $(\Sigma^{n+2},\Sigma_1^n)$ and $(\Sigma'^{n+2},\Sigma_1'^n)$ be two knots corresponding to actions $(\tilde{\Sigma}^{n+2},\Sigma_1^n,Z_k)$ and $(\tilde{\Sigma}'^{n+2},\Sigma_1'^n,Z_k)$ respectively. The equivariant relative connected sum $(\tilde{\Sigma}^{n+2},\Sigma_1^n,Z_k)\#(\tilde{\Sigma}'^{n+2},\Sigma_1'^n,Z_k)$ is then a new action $(\tilde{\Sigma}^{n+2}\#\tilde{\Sigma}'^{n+2},\Sigma_1^n\#\Sigma_1'^n,Z_k)$ whose orbit is just the relative connected sum $(\Sigma^{n+2},\Sigma_1^n)\#(\Sigma'^{n+2},\Sigma_1'^n)$. In particular, for a fixed k, the set of knots satisfying condition (A_k) is closed under the operation of relative connected sum.

Although Theorem 2.3 classified all possible actions, its application is very limited. For if we start from a knot $(\Sigma^{n+2},\Sigma_1^n)$ satisfying condition (A_k), we still do not know in the corresponding $(\tilde{\Sigma}^{n+2},\Sigma_1^n,Z_k)$ what homotopy sphere $\tilde{\Sigma}^{n+2}$ might be. Here we offer a partial answer to this question.

Proposition 2.4. *Let k be an odd integer, $(\tilde{\Sigma}^{n+2},\Sigma_1^n,Z_k)$ an action with orbit knot $(\Sigma^{n+2},\Sigma_1^n)$. If Σ^{n+2} bounds a π-manifold, so does $\tilde{\Sigma}^{n+2}$.*

Proof. Let W^{n+3} be a π-manifold such that $\partial W^{n+3}=\Sigma^{n+2}$. Since $n\geqslant 5$, we may always assume that W^{n+3} is at least 2-connected. It is known [6] that Σ_1^n bounds a parallelizeable manifold $'V^{n+1}$ in Σ^{n+2}. Using a normal field of Σ^{n+2} in W^{n+3}, we can push $'V^{n+1}$ into the interior of W^{n+3}, thereby obtain a diffeomorphic copy of $'V^{n+2}$ lying near Σ^{n+2}. Joining the boundary of this copy to Σ_1^n by normal trajectors, we see that Σ_1^n bounds a parallelizeable manifold V^{n+1} in W^{n+3} such that $V^{n+1}\cap\partial W^{n+3}=\partial V^{n+1}=\Sigma_1^n$. Since V^{n+1} has trivial normal bundle in W^{n+3}, we can extend the inclusion to an imbedding $\psi\colon V^{n+1}\times D^2\to W^{n+3}$. Now let

$$X=W^{n+3}-\psi(V^{n+1}\times E^2).$$

Then X is a manifold with boundary

$$\partial X=\Sigma^{n+2}-\psi(\Sigma_1^n\times E^2)\textstyle\bigcup_\psi V^{n+1}\times S^1$$

and

$$W=V^{n+1}\times D^2\textstyle\bigcup_\varphi X,$$

where $\varphi=\psi|V^{n+1}\times S^1$.

Let $j\colon D^2\to V^{n+1}\times D^2$ be an inclusion $z\to(x,z)$ for some $x\in V^{n+1}$. From the diagram

$$\begin{array}{ccccccc}
0=H_2(W^{n+3})\to & H_2(W^{n+3},X) & \xrightarrow[\cong]{\partial} & H_1(X) & \to H_1(W^{n+3})\to 0 \\
& \cong\uparrow & & \uparrow\varphi_* & \\
& H_2(V^{n+1}\times D^2,V^{n+1}\times S^1) & \longrightarrow & H_1(V^{n+1}\times S^1) & \\
& \cong\uparrow & & \uparrow j_* & \\
& H_2(D^2,S^1) & \xrightarrow[\cong]{\partial} & H_1(S^1) &
\end{array}$$

we conclude that $H_1(X) \simeq Z$. Let $\pi: \tilde{X} \to X$ be the regular covering space corresponding to the kernel of $\pi_1(X) \xrightarrow{\rho} H_1(X) \to H_1(X)/k H_1(X) \simeq Z_k$. The imbedding $\varphi: V^{n+1} \times S^1 \to X$ induces an equivariant imbedding $\tilde{\varphi}: V^{n+1} \times S^1_k \to \tilde{X}$ such taht the diagram

$$\begin{array}{ccc} V^{n+1} \times S^1_k & \xrightarrow{\tilde{\varphi}} & \tilde{X} \\ \downarrow{\scriptstyle 1 \times p_k} & & \downarrow \\ V^{n+1} \times S^1 & \xrightarrow{\varphi} & X \end{array}$$

is commutative. Now form

$$\tilde{W}^{n+3} = V^{n+1} \times D^2_k \cup_{\tilde{\varphi}} \tilde{X}.$$

From the way the action $(\tilde{\Sigma}^{n+2}, \Sigma^n_1, Z_k)$ is reconstructed from $(\Sigma^{n+2}, \Sigma^n_1)$, we can see easily that $\partial \tilde{W}^{n+3} = \tilde{\Sigma}^{n+2}$. It remains to show that $\tilde{W}^{n+3}$ is a π-manifold, or what is the same, a parallelizeable manifold. Now W^{n+3} is parallelizeable, so there exists a framing $(d_1, \ldots, d_{n+3})$ of the tangent bundle of X which can be extended to a framing of the tangent bundle of W^{n+3}. This can also be phrased as follows. Let $(v_1, \ldots, v_{n+1})$ be an arbitrary framing of the tangent bundle of V^{n+1} and (e_1, e_2) the standard framing of the tangent bundle of D^2. If $(v,z) \in V^{n+1} \times S^1$, $z=(x,y)$. Then $(v_1, \ldots, v_{n+1}, v_{n+2})$ is a framing of the tangent bundle of $V^{n+1} \times S^1$, where $v_{n+2}(v,z) = y e_1(z) - x e_2(z)$. Therefore under $\varphi: V^{n+1} \times S^1 \to X$, $\varphi_*(v_i)$, $i=1,2,\ldots,n+2$, is defined. Let $v_{n+3}(v,z) = x e_1(z) + y e_2(z)$ be the outward normal field of $V^{n+1} \times S^1$ in $V^{n+1} \times D^2$. We extend φ_* by letting $\varphi_*(v_{n+3})$ be the inward normal field of ∂X in X. In this way we obtain a map $\Theta: V^{n+1} \times S^1 \to O(n+3)$ given by

$$\begin{aligned} &\varphi_*(v_i(v,z)) = \Sigma_j \Theta_{ij}(v,z) d_j(\varphi(v,z)), \quad 1 \leq i \leq n+1, \\ &\varphi_*(e_1(v,z)) = \Sigma_j \Theta_{n+2,j}(v,z) d_j(\varphi(v,z)), \\ &\varphi_*(e_2(v,z)) = \Sigma_j \Theta_{n+3,j}(v,z) d_j(\varphi(v,z)). \end{aligned}$$

Saying that $(d_1, \ldots, d_{n+3})$ is extendable to W^{n+3} simply means that Θ can be extended to a map $\Theta_1: V^{n+1} \times D^2 \to O(n+3)$. Now $\pi: \tilde{X} \to X$ is a covering space. Hence there exists a framing $(\tilde{d}_1, \ldots, \tilde{d}_{n+3})$ of the tangent bundle of $\tilde{X}$ such that $\pi_*(\tilde{d}_i(\tilde{x})) = d_i(\pi(\tilde{x}))$, $\tilde{x} \in \tilde{X}$. On $V^{n+1} \times D^2_k$, we use the framing $(v_1, \ldots, v_{n+1}, (1/k) e_1, (1/k) e_2)$ of its tangent bundle. We have again $\tilde{\varphi}_*(v_1), \ldots, \tilde{\varphi}_*(v_{n+1})$, $\tilde{\varphi}_*((1/k) v_{n+2})$ defined and again we let $\tilde{\varphi}_*((1/k) v_{n+3})$ be the inward normal of $\partial \tilde{X}$ in $\tilde{X}$. As before, we obtain a map $\tilde{\Theta}: V^{n+1} \times S^1_k \to O(n+3)$ given by

$$\begin{aligned} &\tilde{\varphi}_*(v_i(v,z)) = \Sigma_j \tilde{\Theta}_{ij}(v,z) \tilde{d}_j(\tilde{\varphi}(v,z)), \quad 1 \leq i \leq n+1, \\ &\tilde{\varphi}_*((1/k) e_1(v,z)) = \Sigma_j \tilde{\Theta}_{n+2,j}(v,z) \tilde{d}_j(\tilde{\varphi}(v,z)), \\ &\tilde{\varphi}_*((1/k) e_2(v,z)) = \Sigma_j \tilde{\Theta}_{n+3,j}(v,z) \tilde{d}_j(\tilde{\varphi}(v,z)). \end{aligned}$$

It is then easily computed that

$$\tilde{\Theta} = \alpha \cdot [\Theta \circ (1 \times p_k)]$$

($\cdot$ is multiplication in $O(n+3)$, $\circ$ is composition of maps), where $\alpha: V^{n+1} \times S_k^1 \to O(n+3)$ is the composition of the projection $V^{n+1} \times S_k^1 \to S_k^1$ followed by the map α': $S_k^1 \to O(n+3)$ given by

$$\alpha'(e^{2\pi i\theta}) = \left[\begin{array}{ccc|cc} 1 & & & & \\ & 1 & & & \\ & & \ddots & 0 & \\ & & & 1 & \\ \hline & 0 & & \mathrm{Cos}(k-1)\theta & -\mathrm{Sin}(k-1)\theta \\ & & & \mathrm{Sin}(k-1)\theta & \mathrm{Cos}(k-1)\theta \end{array}\right].$$

Clearly, $\Theta \circ (1 \times p_k)$ extends to $V^{n+1} \times D_k^2$. If k is odd, it is well known that α' represents the zero element in $\pi_1(O(n+3)) = Z_2$ and therefore α can be extended to $V^{n+1} \times D_k^2$. This says $\tilde{\Theta}$ can be extended to $V^{n+1} \times D_k^2$, that is $\tilde{W}^{n+3}$ is parallelizeable.

If we look more closely, then what we have actually proved is more than we needed. Namely we have

1. $\tilde{\Sigma}^{n+2}$ bounds a π-manifold $\tilde{W}^{n+3}$.
2. The action of Z_k can be extended to $\tilde{W}^{n+3}$.
3. The fixed point set of Z_k on $\tilde{W}^{n+3}$ is a parallelizeable manifold V^{n+1} of $\tilde{W}^{n+3}$ such that $V^{n+1} \cap \partial\tilde{W}^{n+3} = \partial V^{n+1} = \Sigma_1^n$, and Z_k acts freely outside V^{n+1}.
4. $\tilde{W}^{n+3}$ has a framing of its tangent bundle which is equivariant in $\tilde{W}^{n+3} - V^{n+1}$.

Let us call an action $(\tilde{\Sigma}^{n+2}, \Sigma_1^n, Z_k)$ "extendable" if conditions 1. to 4. are satisfied. Then Proposition 2.4 can be strengthened to

Proposition 2.5. *Let k be an odd integer, $(\tilde{\Sigma}^{n+2}, \Sigma_1^n, Z_k)$ an action with orbit knot $(\Sigma^{n+2}, \Sigma_1^n)$. Then Σ^{n+2} bounds a π-manifold if and only if the action is extendable.*

3. Examples of Actions

We proceed to give some examples of non-trivial smooth knots satisfying condition (A_k). These examples rely on the explicit description of homotopy spheres given by Brieskorn and Hirzebruch [1, 4]. Following the notation of [4], for a sequence $a = (a_0, \ldots, a_n)$, $n \geq 3$, of positive integers $\geqslant 2$, define

$$V_a = \{(z_0, \dots, z_n) \in C^{n+1} \mid z_0^{a_0} + \dots + z_n^{a_n} = 1\},$$
$$V_a^0 = \{(z_0, \dots, z_n) \in C^{n+1} \mid z_0^{a_0} + \dots + z_n^{a_n} = 0\},$$

and

$$\Sigma_a^{2n-1} = V_a^0 \cap S^{2n+1}.$$

This is a homotopy $(2n-1)$-sphere if the sequence a satisfies certain conditions. One associates to the sequence $a = (a_0, \dots, a_n)$ its graph $\Gamma(a)$ by taking one vertex of each a_i. Two distinct vertices a_i, a_j are joined if and only if $(a_i, a_j) > 1$, where (a_i, a_j) denotes the greatest common divisor. Then Σ_a^{2n-1} is a homotopy sphere if and only if $\Gamma(a)$ satisfies the following:

Condition (B):

1. $\Gamma(a)$ has at least two isolated points, or,
2. $\Gamma(a)$ has one isolated point and at least one connected component K with an odd number of vertices such that $(a_i, a_j) = 2$ for all $a_i, a_j \in K$ and $i \neq j$.

As we shall make use of this condition, we indicate briefly its proof given in [1]. Up to homotopy type, the compliment $Y_a = S^{2n+1} - \Sigma_a^{2n-1}$ can be regarded as the total space of a fibre bundle over S^1 with typical fibre V_a. The fundamental group of S^1 however acts on $H_*(V_a)$ in a non-trivial way as follows. First of all, the only non-vanishing homology of V_a (except $H_0(V_a)$) is $H_n(V_a)$. Let Z_{a_i} be the cyclic group of order a_i generated by w_i, $J_a = J\left(\prod_{i=0}^{n} Z_{a_i}\right) = \bigotimes_{i=0}^{n} J(Z_{a_i})$ the integral group ring of the group $\prod_{i=0}^{n} Z_{a_i}$, and $I_a \subset J_a$ the ideal generated by the elements $1 + w_i + \dots + w_i^{a_i - 1}$, $i = 0, \dots, n$. Then $H_n(V_a) = J_a / I_a$ and a generator of $\pi_1(S^1) = Z$ acts on it as multiplication by $w = w_0 w_1, \dots, w_n$. From the spectral sequence of the fibring $V_a \to Y_a \to S^1$, one concludes that Σ_a^{2n-1} is a homotopy sphere if and only if $1 - w$ is an isomorphism on $H_n(V_a)$. Let $\Delta_a(t) = \det(t - w)$ be the characteristic polynomial of w. It is shown in [1] that

$$\Delta_a(t) = \prod_{0 < j_i < a_i} (t - \xi_0^{j_0} \dots \xi_n^{j_n}),$$

where $\xi_i = e^{2\pi i / a_i}$. By analizing the cyclotomic polynomials involved in $\Delta_a(t)$, Brieskorn showed that the statement $\Delta_a(1) = \pm 1$ is equivalent to the statement that $\Gamma(a)$ satisfies condition B.

Now let k be a fixed integer $\geqslant 0$ and $a = (a_0, \dots, a_n)$ a sequence such that Σ_a^{2n-1} is a homotopy sphere. We have then naturally a knot $(S^{2n+1}, \Sigma_a^{2n-1})$ and we wish to see when it satisfies condition (A_k). First of all, it is known [1] that $\pi_1(S^{2n+1} - \Sigma_a^{2n-1})$ is infinite cyclic. Hence condition $(A_k, 1)$ is always true as remarked before. Therefore we need only worry about the homology condition $(A_k, 2)$. Let $d_i = (a_i, k)$, $b_i = a_i/d_i$. We define the reduced sequence b as the sequence which consists of those b_i for which $b_i > 1$.

Proposition 3.1. *Let k be a fixed integer $\geqslant 0$ and $a = (a_0, \ldots, a_n)$ a sequence such that Σ_a^{2n-1} is a homotopy sphere. Then the knot $(S^{2n+1}, \Sigma_a^{2n-1})$ satisfies condition (A_k) if and only if the following is true: If in the reduced sequence any number of b_i such that $b_i < a_i$ is deleted, the graph of the remaining sequence still satisfies condition (B).*

Proof. The universal covering space of $Y_a = S^{2n+1} - \Sigma_a^{2n-1}$ can be identified with $R \times V_a$ and $H_*(R \times V_a) = H_*(V_a)$ as a $J(Z)$-module described as above. It follows that condition (A_k) is equivalent to the fact that $1 + w + \cdots + w^{k-1}$ is an isomorphism on $H_n(V_a)$, or that $1 - w^k$ is an isomorphism because we know $1 - w$ is already an isomorphism. Certainly, we have

$$\Delta_a^k(t) = \det(t - w^k) = \prod_{0 < j_1 < a_1} (t - \xi_0^{kj_0} \ldots \xi_n^{kj}).$$

Our question is to decide when $\Delta_a^k(1) = \pm 1$. Let $\eta_i = \xi_i^k$. We have

$$\Delta_a^k(t) = \prod_{0 < j_1 < a_1} (t - \eta_0^{j_0} \ldots \eta_n^{j_n}),$$

and η_i is a primitive b_i-th root of unity. Suppose that we have deleted a number of $b_i < a_i$ and $b = (b_{i_1}, \ldots, b_{i_l})$ are the remaining vertices. The polynomial associated with b is

$$\Delta_0(t) = \prod_{0 < j_{i_\ell} < b_{i_\ell}} (t - \eta_{i_1}^{j_{i_1}} \ldots \eta_{i_\ell}^{j_l}).$$

This polynomial is certainly a factor of $\Delta_a^k(t)$. Conversely, a typical term $(t - \eta_0^{j_0} \ldots \eta_n^{j_n})$ is a factor of $\Delta_b(t)$ for some b. In fact, let $j_{i_1}, \ldots, j_{i_m}$ be powers such that j_{i_ℓ} is not a multiple of $b_{i_\ell}, \ell = 1, \ldots, m$. If j_i is not one of them, that is, if j_i is a multiple of b_i. Then of course $b_i < a_i$ because $j_i < a_i$. We can write $\eta_{i_\ell}^{j_{i_\ell}} = \eta_{i_\ell}^{p_{i_\ell}}$ for some $0 < p_{i_\ell} < b_{i_\ell}$. Then

$$t - \eta_0^{j_0}, \ldots, \eta_n^{j_n} = t - \eta_{i_1}^{p_{i_1}} \ldots \eta_{i_n}^{p_i}$$

is a factor of $\Delta_b(t)$. It follows that $\Delta_a^k(t)$ is a product of $\Delta_b(t)$ for various sequences b obtained by deleting vertices $b_i < a_i$. Since each $\Delta_b(t)$ is a polynomial with integral coefficients, $\Delta_a^k(1) = \pm 1$ if and only if $\Delta_b(1) = \pm 1$ for all possible b. That is precisely what we asserted.

For example, for the exotic Kervaire sphere $\Sigma^{8m+1} = \Sigma(3, 2, \ldots, 2)$ $(4m + 1$ 2's), the knot $(S^{8m+3}, \Sigma^{8m+1})$ fails to satisfy condition (A_k) for any even integer k. Many other "non-realizeable" knots can be obtained in similar fashion.

On the basis of Proposition 3.1, we can now prove the theorems stated in § 1.

Proof of Theorem 1.1. Given $\Sigma_1^{4m-3} \in \Theta^{4m-3}(\partial\pi)$, m odd $\geqslant 2$. Then $\Theta^{4m-3}(\partial\pi) = Z_2[2]$, and according to [1, Satz 2], Σ_1^{4m-3} can be represented by $\Sigma_d = \Sigma(d, 2, \ldots, 2)$ $(2m-1$ 2's), where $d = \pm 1 \bmod 8$ if Σ_1^{4m-3} is the standard sphere and $d = \pm 3 \bmod 8$ if Σ_1^{4m-3} is the exotic sphere. Take for instance the exotic sphere. For any odd integer k, it is easily seen that we can find infinitely many values of d such that $d = +3 \bmod 8$ and $(d,k) = 1$. By Proposition 3.1, (S^{4m-1}, Σ_d) satisfies condition A_k for each such d. Moreover, they are all distinct knots. For it has been shown [1, Lemma 6] that $\pi_{2m-1}(S^{4m-1}, \Sigma_d)$ is free abelian of rank $d-1$. Now if $\Sigma^{4m-1} \in \Theta^{4m-1}$ is given, we can replace S^{4m-1} by Σ^{4m-1} by changing the differentiable structure of $S^{4m-1} - \Sigma_d$, i. e., taking the connected sum of S^{4m-1} and Σ^{4m-1} by removing a cell from $S^{4m-1} - \Sigma_d$. This does not change the homeomorphism type of $S^{4m-1} - \Sigma_d$ and therefore condition (A_k) is not effected. Each $(\Sigma^{4m-1}, \Sigma_d)$ gives a different action $(\tilde{\Sigma}_d^{4m-1}, \Sigma_d, Z_k)$ for some $\tilde{\Sigma}_d^{4m-1} \in \Theta^{4m+1}$. Since Θ^{4m-1} is a finite group [7], there is some $\tilde{\Sigma}^{4m-1} \in \Theta^{4m-1}$ such that $\tilde{\Sigma}^{4m-1} = \tilde{\Sigma}_d^{4m-1}$ for infinitely many values of d. This is the first part of Theorem 1.1. If Σ^{4m-1} is chosen in $\Theta^{4m-1}(\partial\pi)$, $\tilde{\Sigma}^{4m-1}$ is also in $\Theta^{4m-1}(\partial\pi)$ by Proposition 2.4. Finally, if Σ_1^{4m-3} is the standard sphere, e. g., $d = 1 \bmod 8$ and Σ^{4m-1} is also standard, we simply take equivalent relative connected sums of each $(\tilde{\Sigma}_d^{4m-1}, \Sigma_d, Z_k)$ with itself enough number of times, say r times, where $r =$ order of $\Theta^{4m-1}(\partial\pi)$. We get an action $(S^{4m-1}, S^{4m-3}, Z_k)_d$ for each d for which the orbit space S^{4m-1}/Z_k is still the standard sphere and these are still distinct actions. In fact, $\pi_{2m-1}(S^{4m-1} - S^{4m-1})$, for a given d, is free abelian of rank $r(d-1)$.

The proof of Theorem 1.2 is almost the same. Given $\Sigma_1^{4m-1} \in \Theta^{4m-1}(\partial\pi)$, we can represent it by $\Sigma_\ell = \Sigma(3, 6\ell - 1, 2, \ldots, 2)$ $(2m-1$ 2's) for some ℓ. Two distinct values ℓ and ℓ' represent the same sphere if $\ell - \ell' = 0$ mod. order of $\Theta^{4m-1}(\partial\pi)$, but the knots (S^{4m-1}, Σ_ℓ) will not be the same ($\pi_{2m}(S^{4m+1} - \Sigma_\ell)$ is free abelian of rank $2(6\ell - 2)$.). If k is prime, it is easily checked that condition A_k will be satisfied. The rest of the theorem goes the same way as before.

Remark. The values of k allowed in Theorem 1.2 can be broadened. k can be any integer whose prime factorization contains, among three prime factors 2, 3 and 5, at most one of them. This is because it suffices to prove Theorem 1.2 for the generator $\Sigma_1^{4m-1} = \Sigma(3, 5, 2, \ldots, 2)$. This is still not the best possible, but we do not pursue the matter further.

We conclude this paper with two more particular examples, one giving actions on a homotopy sphere which does not bound a π-manifold, and the other giving actions on a homotopy sphere which does bound a π-manifold but for which the action cannot be extended over to any π-manifold it bounds in the sense of § 2. This is done quite easily by taking advantage of the very special fact that $\Theta^{13}(\partial\pi)=0$ and $\Theta^{13}=Z_3$. Let $\Sigma^{11}\in\Theta^{11}(\partial\pi)$. According to Theorem 1.2, there exists some $\tilde{\Sigma}^{13}\in\Theta^{13}$ and infinitely many involutions $(\tilde{\Sigma}^{13},\Sigma^{11},Z_2)$ such that $\tilde{\Sigma}^{13}/Z_2=S^{13}$. By taking connected sums, we may assume that $\tilde{\Sigma}^{13}=S^{13}$. Now if we take the connected sum of the orbit pair (S^{13},Σ^{11}) with a generator Σ^{13} of Θ^{13} on the compliment of Σ^{11}, then $\tilde{\Sigma}^{13}$ is evidently replaced by $\tilde{\Sigma}^{13}\#\Sigma^{13}\#\Sigma^{13}$ since the involution preserves orientation. Since we do not distinguish orientation, $\Sigma^{13}\#\Sigma^{13}$ is diffeomorphic to Σ^{13}. Thus we have

Theorem 3.2. *Let $\Sigma^{11}\in\Theta^{11}(\partial\pi)$ and Σ^{13} be the exotic non-boundary homotopy 13-sphere. Then there are infinitely many distinct smooth involutions on Σ^{13} with Σ^{11} as fixed point set.*

If we do the same thing with the action of Z_3 instead of involution, then $\tilde{\Sigma}^{13}$ is replaced by $\tilde{\Sigma}^{13}\#\Sigma^{13}\#\Sigma^{13}\#\Sigma^{13}$, which is S^{13}, while the orbit sphere is $S^{13}\#\Sigma^{13}=\Sigma^{13}$. Since the orbit sphere does not bound, the action of Z_3 on S^{13} is not extendable by Proposition 2.5. Thus we have

Theorem 3.3. *Let $\Sigma^{11}\in\Theta^{11}(\partial\pi)$ and S^{13} be the standard 13-sphere. Then there are infinitely many distinct smooth actions of Z_3 on S^{13} with Σ^{11} as fixed point set, and none of these actions is extendable.*

References

1. BRIESKORN, E.: Beispiele zur Differentialtopologie von Singularitäten, Inventiones mathematicae, **2,** 1—14 (1966).
2. BROWN, E. H., and F. P. PETERSON: The Kervaire invariant of $(8k+2)$-manifolds, Bull. Amer. Math. Soc., **70,** 670—675 (1964).
3. GIFFEN, C. H.: The generalized Smith conjecture, Amer. Journ. of Math., **88,** 187—198 (1966).
4. HIRZEBRUCH, F.: Singularities and exotic spheres, Séminaire Bourbaki, 1966—67.
5. HSIANG, W. Y.: On the unknottedness of the fixed point set of differentiable circle group actions on spheres – P. A. Smith conjecture, Bull. Amer. Math. Soc., **70,** 678—680 (1964).
6. KERVAIRE, M. A.: On higher dimensional knots, Differential and combinatorial topology, Princeton Mathematical Series 27, Princeton, University Press, 105—119 (1965).
7. —, and J. MILNOR: Groups of homotopy spheres, I, Ann. of Math., **77,** 504—537 (1963).
8. MONTGOMERY, D., and C. T. YANG: Differentiable transformation groups on homotopy spheres, Michigan J. Math. **14,** 33—46 (1967).

Examples of Free Involutions on 3-Manifolds

PETER ORLIK

Let Σ denote an irreducible homology 3-sphere. The actions of $SO(2)$ on Σ are given in [2]. The involution imbedded in such an action is free if and only if all the α_j are odd. In [1] MEDRANO constructs examples of free involutions on certain Σ, but it is not known whether these are imbedded in $SO(2)$ actions. More generally:

Q: Is every free involution of an irreducible homology 3-sphere imbedded in some $SO(2)$ action?

By LIVESAY, the answer is yes for S^3. The examples below show that the conditions of Q are necessary.

a) Let Σ be a homology 3-sphere which it not simply connected. The connected sum $\Sigma \# \Sigma$ is again a homology sphere. It admits a free involution which is the antipodal map on the 2-sphere of the # operation and exchanges the summands. By [2], however, such a connected sum never admits an effective $SO(2)$ action.

b) Let F denote a 2-torus of genus 2 and M the 3-manifold obtained from $F \times [-1, 1]$ by identifying $(x, -1)$ with $(h(x), 1)$, where $h: F \to F$ is a self-homeomorphism described below.

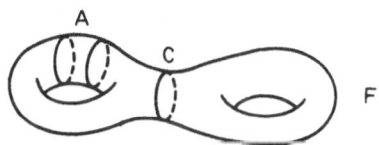

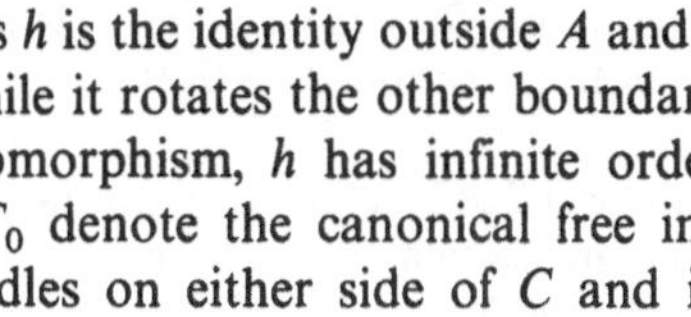

Let A denote the annular region on the figure and h the "Lickorish twist" of that region. Thus h is the identity outside A and on one of the boundary circles of A, while it rotates the other boundary circle once around. As an auto-homeomorphism, h has infinite order in the homeotopy group $\Lambda(F)$. Let T_0 denote the canonical free involution of F which exchanges the handles on either side of C and is the antipodal map on C. Note that hT_0 has no fixed points.

Now define an orientation preserving free involution, T, on M by $T(x, y) = (T_0(x), -y)$. It is clearly free everywhere but at the identification, where it is free by the above remark.

By STALLINGS [4] M is an orientable, closed, irreducible 3-manifold. In [3, Corollary 1] it is proved, in a slightly more general setting, that if a 3-manifold with effective $SO(2)$ action can be fibered over S^1, then (apart from some specified genus-one-torus bundles) the identifying homeomorphism has finite order. Thus M admits no $SO(2)$ action.

This shows that we cannot ask Q for manifolds with infinite homology.

c) Consider the Seifert manifold $M=\{-1;(o,0,0,0);(2,1);(2,1);(2,1)\}$ as described in [2]. Note that $H_1(M;Z)=Z_2+Z_2$. This manifold is homeomorphic to the Seifert manifold M' obtained as follows. Take a Klein bottle with one hole. Over this, put a circle bundle which has the property that the orientation of the fiber is reversed when describing an orientation reversing curve in the base space. The result is an orientable manifold with a torus boundary. Close it up with a solid torus with $b=-2$ as in [2].

In the notation of [3], $M'=\{-2;(n_2,2,0,0)\}$. The fibration by circles according to M' does not admit an action of $SO(2)$, but mapping each point of M' to the diametrically opposite point on its fiber is a continuous, orientation preserving free involution. The only $SO(2)$ action this manifold admits is described by the invariants of M (see [2]), and the involution imbedded in that action is not free.

d) Finally, it should be mentioned that a counterexample to Q cannot be found by taking a homology sphere with $SO(2)$ action such that, say, α_1 is even (and hence the imbedded involution T has the corresponding exceptional orbit e_1 as fixed point set) and alter T in v_1 to make it free there. Note that v_1 is a trivial disc bundle ξ_1 over e_1 and T on the boundary, t_1, is the antipodal map on the fibers of the circle bundle associated to ξ_1. Looking in the universal cover of v_1 it is clear that any free involution of v_1 must preserve the fibers of ξ_1, which is impossible.

References

1. LOPEZ DE MEDRANO, S.: Involutions of homotopy spheres and homology 3-spheres, Bull. Amer. Math. Soc. **73**, 727—731 (1967).
2. ORLIK, P., and F. RAYMOND: Actions of $SO(2)$ on 3-manifolds, these Proceedings.
3. — On 3-manifolds with local $SO(2)$ action to appear in Quar. J. Math. Oxford).
4. STALLINGS, J.: On fibering certain 3-manifolds, Topology of 3-manifolds, Prentice Hall 95—100 (1962).

Cyclic Group Actions on Homotopy Spheres

C. N. LEE

F. HIRZEBRUCH asked whether the homotopy spheres $\Sigma^{2n-1} \notin bP_{2n}$ admit any group action. The least odd number $2n-1$ for which $bP_{2n} \neq \Theta_{2n-1}$ is 9. The purpose of this note is to show: *If G is a cyclic group of order 5, every homotopy sphere $\Sigma^9 \in \Theta_9$ admits infinitely many free differentiable actions of G which are differentiably distinct from each other.* This is an immediate consequence of the following two general facts:

1. *If G is a cyclic group of order r relatively prime to the order q of the group Θ_{2n-1}, then every homotopy sphere $\Sigma \in \Theta_{2n-1}$ admits a free differentiable action of G.*

Proof. Let S^{2n-1} be the standard sphere. There is the standard orthogonal free action of G on S^{2n-1} with the lens space $L = L(r, 1, \ldots, 1)$ as its orbit space. Let p be an integer (possibly negative) such that $pr \equiv 1 \mod q$. Now given any homotopy sphere Σ^{2n-1}, take the connected sum $X = L \# p\Sigma$ where $p\Sigma$ is the connected sum of $|p|$ copies of Σ (with the orientation reversed if $p<0$). Clearly the universal covering space of X is diffeomorphic with $S^{2n-1} \# pr\Sigma$ which is in turn diffeomorphic with Σ by the choice of p. This gives us the desired action of G on Σ.

2. (J. MILNOR) *If G is a cyclic group of order $6p \pm 1$ $(p \geqslant 1)$ and if a homotopy sphere Σ^{2n-1} $(n \geqslant 3)$ admits a free differentiable action of G, then Σ^{2n-1} admits infinitely many such actions which are differentiably distinct from each other.*

This follows from the same argument as used by MILNOR in order to prove a similar statement for the standard spheres and the standard orthogonal free actions of G in his paper "Some Free Actions of Cyclic Groups on Spheres", which appeared in Proc. Bombay Colloq. 37–42, 1964.

Now our claim about the homotopy spheres in Θ_9 follows from the fact that the order of Θ_9 is 8. (See MILNOR-KERVAIRE, "Group of Homotopy Spheres: I", Ann. of Math. 504–537 (1963).)

Non-Existence of Free Differentiable Actions of S^1 and $\mathbb{Z}_2$ on Homotopy Spheres

RONNIE LEE

In this note, we shall find a necessary condition for a homotopy sphere to admit a free S^1 differentiable action, and will give an example of an odd dimensional homotopy sphere which does not admit any such action. A similar result is obtained by applying the same method to the case of free $\mathbb{Z}_2$ actions.

By a spin manifold, we mean a manifold with first and second Stiefel-Whitney classes equal to zero.

Proposition 1. *A necessary condition for a homotopy sphere Σ^{4k+1} to admit a free S^1 differentiable action is that it bounds a spin manifold.*

Proof. It is a well-known fact that if Σ^{2n+1} admits a free action, we will have the following commutative diagram:

$$\begin{array}{ccc} S^1 & & S^1 \\ \downarrow & & \downarrow \\ \Sigma^{2n+1} & \xrightarrow{b(f)} & S^{2n+1} \\ \xi \downarrow \pi & & \downarrow \eta \\ \Sigma^{2n+1}/S^1 & \xrightarrow{f} & CP^n \end{array}$$

where f is a homotopy equivalence, covered by a bundle map $b(f)$.

Let $D(\xi)$ be the disk bundle associated with the bundle ξ: $S^1 \to \Sigma^{2n+1} \xrightarrow{\pi} \Sigma^{2n+1}/S^1$, then the boundary of $D(\xi)$ will be Σ^{2n+1} and to show the proposition, we only have to show that $D(\xi)$ is a spin manifold when $n=2k$.

As $\tau(D(\xi))=\pi^*(\xi \oplus \tau(\Sigma^{2n+1}/S^1))$, we have

$$\begin{aligned} W(D(\xi)) &= W(\tau(D(\xi))) \\ &= W(\pi^*(\xi \oplus \tau(\Sigma^{2n+1}/S^1))) \\ &= \pi^*(W(\xi) \cup W(\tau(\Sigma^{2n+1}/S^1))) \\ &= \pi^*(f^* W(\xi) \cup W(\tau(\Sigma^{2n+1}/S^1))). \end{aligned}$$

And since Stiefel-Whitney classes are homotopy invariants for a manifold, we have

$$W(\tau(\Sigma^{2n+1}/S^1)) = f^* W(\tau(CP^n)),$$
$$W(D(\xi)) = \pi^* f^*(W(\eta) \cup W(\tau(CP^n)))$$
$$= \pi^* f^*((1+\alpha)(1+\alpha)^{n+1}))$$

where α is the generator of $H^2(CP^n)$. When $n=2k$, we have $W_1(D(\xi)) = W_2(D(\xi)) = 0$, and hence the proposition.

It has been shown by MILNOR (as in [1]) that there is a homotopy q sphere Σ^q which does not bound a spin manifold, and by the above proposition, this homotopy sphere will not admit any free S^1 differentiable action.

Corollary 1. *There exists a homotopy q sphere which does not admit any free S^1 differentiable action.*

Apply the same method to the $\mathbb{Z}_2$ case; it is easy to show the following:

Proposition 2. *A necessary condition for a homotopy sphere Σ^{4k+2} to admit a free $\mathbb{Z}_2$ differentiable action is that Σ^{4k+2} bound a spin manifold.*

Corollary 2. *There exists a homotopy sphere Σ^{10} which does not admit any free $\mathbb{Z}_2$ differentiable action. However, $2[\Sigma^{10}]=0$ in Θ_{10}.*

The last assertion follows from the fact that $\Theta_{10} = \mathbb{Z}_6$, and the image of the natural map $\Theta_{10} \to \Omega_{10}^{\text{spin}}$ is $\mathbb{Z}_2$.

Reference

1. MILNOR, J.: Remarks concerning Spin manifolds, Differential and combinatory topology, a symposium in honor of Marston Morse, Princeton University Press, Princeton, N. J., 1965.

On Differentiable Actions of Compact Lie Groups on Compact Manifolds

EUGENIO CALABI*

The present note is devoted to a new proof of a theorem by R. S. PALAIS and T. E. STEWART [1, 2] on the title subject and restated below, after the preliminary definitions.

Definition 1. Let G be a Lie group and X a differential manifold. A differentiable action F of G in X is a differentiable map $F: G \times X \to X$ satisfying the following relations:

1. $F(g_1, F(g_2, x)) = F(g_1 g_2, x)$,
$$(g_1, g_2 \in G,\ x \in X);$$
2. $F(1, x) = x \quad (1 = \text{identity of } G,\ x \in X)$.

Definition 2. Two actions F_1, F_2 of a Lie group G on a differential manifold X are called *(differentiably) equivalent*, if there exists a (differentiable) automorphism $\sigma: X \to X$ such that for all $g \in G$, and $x \in X$
$$F_2(g, \sigma(x)) = \sigma(F_1(g, x)).$$

Definition 3. Let G be a Lie group and M, X two differential manifolds. A differentiable family of actions of G on X parametrized by M is a differentiable map $\Phi: M \times G \times X \to X$ such that, for each $t \in M$ the restriction $\Phi_t: G \times X \to X$ defined by $\Phi_t(g, x) = \Phi(t, g, x)$ is a differentiable action of G.

Definition 4. Let $\Phi: M \times G \times X \to X$ be a differentiable family of actions of G on X parametrized by M and let t_0 be a point in M. This family is called *locally trivial at* t_0, if there exists an open neighborhood U of t_0 in M and a differentiable map $\Psi: U \times X \to X$ such that

1. Ψ is a *differentiable flow* on X parametrized by U with base point t_0, i.e.

a) For each $t \in U$ the map $\Psi_t: X \to X$ defined by $\Psi_t(x) = \Psi(t, x)$ is a differentiable automorphism of X, and

* Supported in part under NSF Contract GP 4509.

b) for $t=t_0$, Ψ_t is the identity map on X;

2. Ψ *trivializes* Φ, i.e. for each $t\in U$ and each $g\in G$

$$\Phi(t,g,\Psi(t,x)) = \Psi(t,\Phi(t_0,g,x)).$$

A differentiable family Φ of group actions of G on X is called *locally trivial*, if and only if it is so at each point t of the parameter space M.

The statement of the theorem of PALAIS-STEWART is the following

Theorem 1. *A differentiable family Φ of actions of a compact Lie group G on a compact manifold X without boundary, parametrized by M, is locally trivial at each $t\in M$.*

Proof. Let γ_0 be a Riemannian metric on X. For each $t\in M$ and each $g\in G$ let $\Phi_{t,g}$ be the automorphism of X defined by $\Phi_{t,g}(x)=\Phi(t,g,x)$ and $\Phi^*_{t,g}(\gamma_0)$ the transformation of the metric γ_0 induced contragrediently by $\Phi_{t,g}$ and denote by $d\mu$ the normalized Haar measure on G. Then for each $t\in M$ we obtain a Riemannian metric

$$\gamma_t = \int_G \Phi^*_{t,g}(\gamma_0)\,d\mu$$

satisfying the two following properties: a) γ_t is invariant under the action Φ_t of G on X; b) γ_t depends differentiably on t.

We now apply some results from the potential theory in the manifold X with any Riemannian metric γ_t:

1. The Laplace-Beltrami operator $\Delta_t = -\nabla_t\cdot\nabla$ with respect to γ_t, acting on scalar functions is self-adjoint and has a non-negative, discrete spectrum. This fact, which is equivalent to the compactness of the Green's integral operator G_t on X (with respect to γ_t), means that, for any positive real number A, the invariant space of Δ_t corresponding to the segment $[0,A]$ in the spectrum of Δ_t is finite dimensional. Furthermore, the eigenfunctions corresponding to the point spectrum, which is contained in $[0,\infty]$, include a complete orthonormal basis for the space of functions f under any of the Hilbert norms $\|f\|_{k,t}$ with

$$(\|f\|_{k,t})^2 = (f,f)_{\gamma_t} + (\Delta_t^k f,f)_{\gamma_t}, \quad k=1,2,\ldots,$$

where Δ_t^k denotes Δ_t iterated k times and the volume element in the integrals is with respect to the metric γ_t. From this and from the Sobolev inequalities, one can deduce the second result we need; it can be stated as follows.

2. Given any $t\in M$, there exists a sufficiently large, positive real number A (depending on t), such that an orthogonal basis for the eigenfunctions corresponding to the eigenvalues of Δ_t in the interval $[0,A]$ (say $u_1,u_2,\ldots,u_N$) defines a differentiable imbedding of X in $\mathbb{R}^N$.

Finally the continuity with respect to t of the Green's operator G_t corresponding to Δ_t implies the following (see [3]).

3. The spectrum of Δ_t depends continuously on t. More precisely, if a positive A is not in the spectrum of Δ_{t_0} for any given $t_0 \in M$, there exists a neighborhood U_0 of t_0 in M such that;

a) A is not in the spectrum of Δ_t for any $t \in U_0$;

b) for each differentiable arc Γ imbedded in U_0 and passing through t_0, there exists a family of differentiable functions $v_j : \Gamma \times X, \to \mathbb{R}$ $(j = 1, \ldots, N)$ with the property that for each t the N functions $v_{j,t} : X \to \mathbb{R}$, $v_{j,t}(x) = v_j(t,x)$, are a basis for the eigenfunctions of Δ_t corresponding to all the eigenvalues in $[0, A]$.

Now let t_0 be any given point in M, and let U be a neighborhood of t_0 in M with the following properties:

a) U is in a local coordinate system with coordinates $\tau = (\tau_1, \ldots, \tau_m)$ with $\tau_\mu(t_0) = 0$ and range $\sum_{\mu=1}^{m} \tau_\mu^2 \leqslant \delta$ for some $\delta > 0$.

b) There exists an $A > 0$ which is not in the spectrum of Δ_t for any $t \in U$ and one can choose N continuous functions $v_j : U \times X \to \mathbb{R}$ $(1 \leqslant j \leqslant N)$ whose restriction to $\Gamma_{t_1} \times X$, where Γ_{t_1} is any diametral arc in U defined by $\tau(t) = \lambda \tau(t_1), (-1 \leqslant \lambda \leqslant 1)$, is differentiable, and such that their restrictions $v_{j,t} : X \to \mathbb{R}$ are an orthonormal basis for the eigenfunctions of Δ_t corresponding to all eigenvalues in $[0, A]$ under the Hilbert norm defined by γ_t.

c) For each $t \in U$ the N functions $v_{j,t}$ $(1 \leqslant j \leqslant N)$ define a family of differentiable imbeddings of X in $\mathbb{R}^N$ parametrized by t.

It is clear that, for each $t \in U$, the imbedding $(v_{j,t})$ of X into $\mathbb{R}^N$ is invariant (under the orthogonal group) under the action Φ_t of G on X, since the metric γ_t is invariant, and hence each eigenspace of Δ_t furnishes an orthogonal representation of G. It is well known that the set of equivalence classes of these representations is finite and that each is stable under deformations. Therefore there exists a map $\gamma : U \to O(N)$ such that the family $(v'_{j,t})_{t \in U}$ of imbeddings $(v'_{j,t}) = \lambda(t) \cdot (v_{j,t})$ has each of the images $v'_{j,t}(X)$ in $\mathbb{R}^N$ invariant under a fixed representation $\chi : G \to O(N)$ of G, the latter being equivariant with the action Φ_t of G on X. Consider now the (radial) contraction h of U to $\tau^{-1}(0) = t_0$, defined by $\tau_\mu(h(\sigma, t)) = \sigma \tau_\mu(t)$, $0 \leqslant \sigma \leqslant 1$, and for each $t \in U$, the restriction of the family $(v'_{j,t})$ to the orbit of t in U under h; this is a one-parameter family of imbeddings of X in $\mathbb{R}^N$. It is evident that the normal variation if this family with respect to the parameter σ is equivariant with the variation of the action Φ_t of G on X, since it is invariant under $\chi(G) \subset O(N)$. Integration of this normal variation of $v'_{j,t}$ with respect to σ finally establishes, for each $t \in U$ the family $(\Psi_t)_{t \in U}$, $\Psi_t : X \to X$, of diffeomorphisms of X onto itself that establish the equivalence of the action Φ_t of G with Φ_{t_0}. This completes the proof of the theorem.

The above theorem has an immediate generalization to compact manifolds X with boundary (e.g. consider the "double" $\tilde{X}$ of X and repeat the proof using the extension of actions of G by the reflection of $\tilde{X}$ across the boundary of X). The stability of differentiable actions of G on X fails, if G is not compact [1, 2]. On the other hand if G is compact and X is not compact, it seems likely that the conclusion still holds; in order to prove it, one should try to consider the family of orbit spaces under the actions of G and prove that such a family, considered as a family of singular real varieties, is locally trivial. If such be the case, the conjectured conclusion on X would follow by considering an open covering of the orbit space by open subdomains with compact closure and boundary smooth in the generic orbits and transversal to the singular ones, in such a way that the counter-images in X of these compact subsets are submanifolds with smooth boundary.

References

1. PALAIS, R. S., and T. E. STEWART: Deformations of compact, differentiable transformation groups, Amer. Journal of Math., **82**, 935—937 (1960).
2. — Equivalence of nearby differentiable actions of a compact group, A. M. S. Bull., **67**, 362—364 (1961).
3. BROWDER, F. E.: Families of linear operators depending upon a parameter, Amer. J. of Math., **87**, 752—758 (1965).

Examples of Differentiable Involutions*

Hsu-Tung Ku

In this note we shall construct, from some sphere bundles over spheres, examples of differentiable actions of $\mathbb{Z}_2$ on $S^n \times S^m$. There are many examples of exotic differentiable actions of compact Lie groups on homotopy spheres [4] but few examples of actions on $S^n \times S^m$. Our construction enables us to produce many exotic involutions on $S^n \times S^m$ with interesting fixed point sets. We also have examples to show that there are infinitely many differentiably (and topologically) distinct involutions on $S^n \times S^m$ with fixed point sets of the same homotopy type but with no homeomorphism between them. These are the first such examples known.

We use the notation $\mathscr{B} = \{B, p, X, S^r, SO(r+1)\}$ for a (differentiable) sphere bundle over X with total space B, fibre S^r, structure group $SO(r+1)$ and projection $p: B \to X$. We always assume that X is a closed manifold. In case $X = S^q$, it is well-known that the set of equivalence classes of bundles over S^q, with structure group $SO(r+1)$ is in $1-1$ correspondence with the homotopy group $\pi_{q-1}(SO(r+1))$ [13, p. 99]. Suppose there exist integers q, r such that $\pi_{q-1}(SO(r+1)) \approx \mathbb{Z} \oplus \mathbb{Z}$, or $\pi_{q-1}(SO(r+1)) \approx \mathbb{Z}$, and let m, n be any integers. Then we have the corresponding non-equivalent sphere bundles over S^q which we shall denote by

$$\mathscr{B}_{m,n}^{(q,r)} = \{B_{m,n}^{(q,r)}, p, S^q, S^r, SO(r+1)\},$$

or

$$\mathscr{B}_{n}^{(q,r)} = \{B_{n}^{(q,r)}, p, S^q, S^r, SO(r+1)\}.$$

Since $\pi_3(SO(4)) \approx \pi_7(SO(8)) \approx \pi_{11}(SO(12)) \approx \mathbb{Z} \oplus \mathbb{Z}$, $\pi_3(SO(r)) \approx \mathbb{Z} (r \geqslant 5)$, $\pi_7(SO(r)) \approx \mathbb{Z} (r \geqslant 9)$, $\pi_{11}(SO(r)) \approx \mathbb{Z} (r \geqslant 13)$ [3, p. 428], [13], it follows that the bundles $\mathscr{B}_{m,n}^{(4,3)}$, $\mathscr{B}_{n}^{(4,r)} (r \geqslant 4)$, $\mathscr{B}_{m,n}^{(8,7)}$, $\mathscr{B}_{n}^{(8,r)} (r \geqslant 8)$, $\mathscr{B}_{m,n}^{(12,11)}$, and $\mathscr{B}_{n}^{(12,r)} (r \geqslant 12)$ are well defined. Moreover the total spaces $\mathscr{B}_{m,n}^{(4,3)}$, $\mathscr{B}_{n}^{(4,r)}$, $\mathscr{B}_{m,n}^{(8,7)}$ and $\mathscr{B}_{n}^{(8,r)}$ can be given natural differentiable structures [9, 10]. The non-trivial cohomology groups with integral coefficients $\mathbb{Z}$ of these manifolds are easily computed by using Gysin exact sequences [9]. The

* This work was supported by National Science Foundation grant GP 7952 X.

results are as follows:

$$H^4(B^{(4,3)}_{m,n}) \approx H^8(B^{(8,7)}_{m,n}) \approx \mathbb{Z}_n,\ H^4(B^{(4,4)}_n) \approx H^8(B^{(8,8)}_n) \approx \mathbb{Z} \oplus \mathbb{Z},$$
$$H^4(B^{(4,r)}_n) \approx H^r(B^{(4,r)}_n) \approx \mathbb{Z}(r \geqslant 5),$$
$$H^8(B^{(8,r)}_n) \approx H^r(B^{(8,r)}_n) \approx \mathbb{Z}(r \geqslant 9).$$

The main construction of this paper is the following: *Sphere bundle construction.* Given a differentiable sphere bundle

$$\mathscr{B} = \{B, p, X, S^r, SO(r+1)\},$$

let η^{r+1} be the $(r+1)$-dimensional (real) vector bundle over X associated with $\mathscr{B}$, and ξ^k be a k-dimensional vector bundle over X such that the Whitney sum $\eta^{r+1} \oplus \xi^k$ is a trivial $(k+r+1)$-dimensional vector bundle. Thus the total space of the unit sphere bundle of $\eta^{r+1} \oplus \xi^k$ is diffeomorphic to $X \times S^{r+k}$. In case $X = S^q$, k can be any integer greater than or equal to $r+1$ [2, p. 26–27]. The following standard construction will always give a differentiable action of $\mathbb{Z}_2$ on $X \times S^{r+k}$ with fixed point set $F = B$. Let $\mathbb{Z}_2$ always act trivially on η^{r+1}, and $\mathbb{Z}_2$ act on ξ^k by reversing vectors. (More generally, this construction also applies to the following case: Let G be a compact subgroup of $O(k)$ and $C(G)$ the centralizer of G in $O(k)$. Assume that there exists a reduction of the associated principal bundle $\hat{\xi}^k$ of ξ^k to a principal $C(G)$-bundle ξ'^k. Let $E(\xi)$ denote the total space of ξ. Then $E(\xi) = E(\hat{\xi}^k) \times_{O(k)} \mathbb{R}^k = E(\xi'^k) \times_{C(G)} \mathbb{R}^k$, and the action of G on $E(\xi)$ is induced by the action defined by $g(x,y) = (x, gy)$, where the action of G on $\mathbb{R}^k$ is via the standard linear action of $O(k)$ on $\mathbb{R}^k$).

The sphere bundle construction shows that the total space of any sphere bundle over S^q can be the fixed point set of some action of $\mathbb{Z}_2$ on $S^q \times S^m$ for suitable m. Now let us specify some interesting examples.

Theorem 1. *There exist actions of $\mathbb{Z}_2$ on $S^4 \times S^{3+k}$ $(k \geqslant 4)$ (resp. $\Sigma^8 \times S^{7+k}$ for any $\Sigma^8 \in \theta_8$, $k \geqslant 8$) with fixed point sets some exotic homotopy 7-sphere $\Sigma^7 \in \theta_7$ (resp. $\Sigma^{15} \in \theta_{15}$).*

Proof. Eells-Kuiper [5] have shown that there exist precisely 15 exotic homotopy 7-spheres which can be the total space of some S^3-bundle over S^4 with structure group $SO(4)$, and 4095 exotic homotopy spheres which are diffeomorphic to the total space of some S^7-bundles over a base space homeomorphic to S^8. Since $\theta_8 \approx \mathbb{Z}_2$ [6], the results follow from the sphere bundle construction.

Theorem 2 [9, 10]. *Let β_4, β_8 be a generator of $H^4(B^{(4,r)})$, $H^4(B^{(8,r)})$, respectively. Then*

1. *if* $m \equiv \pm m' \bmod (12, n)$, $B^{(4,3)}_{m,n}$ *and* $B^{(4,3)}_{m,\pm n}$ *have the same homotopy type, where* $(12, n)$ *means g.c.d., and*

$$p_1(B^{(4,3)}_{m+12i,n}) = \pm(4m+48i)\beta_4 \bmod n;$$

2. *for* $r \geqslant 4$, $B^{(4,r)}_{\pm n+24i}$ $(i=0, \pm 1, \pm 2, \ldots)$ *are of the same homotopy type and*

$$p_1(B^{(4,r)}_{n+24i}) = \pm(2n+48i)\beta_4;$$

3. *if* $m \equiv m' \bmod (120, n)$, $B^{(8,7)}_{m,n}$ *and* $B^{(8,7)}_{m',\pm n}$ *have the same homotopy type and*

$$p_2(B^{(8,7)}_{m+120i,n}) = \pm(12m+1440i)\beta_8 \bmod n;$$

4. *for* $r \geqslant 8$, $B^{(8,r)}_{\pm n+240i}$ $(i=0, \pm 1, \pm 2, \ldots)$ *are of the same homotopy type and*

$$p_2(B^{(8,r)}_{n+240i}) = \pm(6n+1440i)\beta_8,$$

where p_i *denotes the i-th Pontrjagin class as usual.*

S. P. NOVIKOV has shown that the rational Pontrjagin classes are topological invariants. Hence Theorem 2 together with our construction gives the following theorem:

Theorem 3. 1. *There are an infinite number of distinct actions of* $\mathbb{Z}_2$ *on* $S^4 \times S^{3+k} (k \geqslant 4)$ *with fixed point set* $B^{(4,3)}_{m+12i,n}$ *for any* $m, n, i \in \mathbb{Z}$. *Recall that* $H^4(B^{(4,3)}_{\pm m \pm 12i, \pm n}) \approx \mathbb{Z}_n$ *(if* $n \geqslant 0$*).*

2. *For* $r \geqslant 4$, *and any fixed integer* $n, 0 \leqslant n < 24, S^4 \times S^{3+k} (k \geqslant 4)$ *admits infinitely many actions of* $\mathbb{Z}_2$ *with the fixed point set* $B^{(4,r)}_{n \pm 24i}$. *All these sets are of the same homotopy type, but there are no two of which that are homeomorphic.*

3. *There are an infinite number of distinct actions of* $\mathbb{Z}_2$ *on* $S^8 \times S^{7+k} (k \geqslant 8)$ *with fixed points sets* $B^{(8,7)}_{m+120i,n}$ *for any* $m, n, i \in \mathbb{Z}$.

4. *For* $r \geqslant 8$, *and any fixed integer* $n, 0 \leqslant n < 240, S^8 \times S^{7+k} (k \geqslant 8)$ *admits infinitely many actions of* $\mathbb{Z}_2$ *such that the actions are distinguished by the second Pontrjagin classes of the fixed point sets* $B^{(8,r)}_{\pm n \pm 240i}$, $i \in \mathbb{Z}$.

In particular $B^{(q,r)}_{0,0}$, $B^{(q,r)}_0$ are $S^q \times S^r$. Thus we have

Corollary 4. *For* $r \geqslant 3$ *(resp.* $r \geqslant 7$*), there exist infinitely many distinct actions of* $\mathbb{Z}_2$ *on* $S^4 \times S^{3+k}$ *(resp.* $S^8 \times S^{7+k}$*) with fixed point sets infinitely many* $(4+r)$*- (resp.* $(8+r)$*-)dimensional manifolds of the same homotopy type as* $S^4 \times S^r$ *(resp.* $S^8 \times S^r$*), whose* p_1 *(resp.* p_2*) is divisible by* 48 *(resp.* 1440*) and no two of which are homeomorphic.*

Remark 5. As $\pi_{11}(SO(12)) \approx \mathbb{Z} \oplus \mathbb{Z}$, $\pi_{11}(SO(r)) = \mathbb{Z}$, $r \geqslant 13$, the spaces $B^{(12,11)}_{m,n}$, $B^{(12,r)}_n$ can be the fixed point sets of the actions of $\mathbb{Z}_2$ on

$S^{12}\times S^{11+k}(k\geqslant 12)$. More generally, we can consider the bundle $\mathscr{B}^{(4m,r)}$, $m\geqslant 3$, $r\geqslant 4m-1$. But the computation of $p_m(B^{(4m,r)})$ is difficult.

Next, we shall exhibit examples which have fixed point set the real, complex, or quaternion second Stiefel manifolds. We denote the Stiefel manifolds by:

$$V_{n,q}=SO(n)/SO(n-q)(1\leqslant q\leqslant n),\quad W_{n,q}=SU(n)/SU(n-q)$$

and

$$X_{n,q}=Sp(n)/Sp(n-q).$$

Theorem 6. 1. *There exists an action of* $\mathbb{Z}_2$ *on* $S^n\times S^m(m\geqslant 2n-1)$ *with fixed point set* $V_{n+1,2}$ *which is not of the same homotopy type of* $S^n\times S^{n-1}$ *if* $n+1\neq 2,4,$ *or* 8.

2. *There exists an action of* $\mathbb{Z}_2$ *on* $S^{2n+1}\times S^m(m\geqslant 4n-1)$ *with fixed point set* $W_{n+1,2}$ *which is not of the same homotopy type of* $S^{2n+1}\times S^{2n-1}$ *if* $n+1\neq 2,4,$ *or* 8.

3. *There exists an action of* $\mathbb{Z}_2$ *on* $S^{4n+3}\times S^m(m\geqslant 8n-1)$ *with fixed point set* $X_{n+1,2}$ *which is not of the same homotopy type of* $S^{4n+3}\times S^{4n-1}$ *if* $n+1\neq 2,4,$ *or* 8.

Proof. Since $SO(n+1)/SO(n)=S^n$, $SU(n+1)/SU(n)=S^{2n+1}$ and $Sp(n+1)/Sp(n)=S^{4n+3}$, we have the bundles:

$$\{V_{n+1,2},p,S^n,S^{n-1},SO(n)\},\quad \{W_{n+1,2},p,S^{2n+1},S^{2n-1},U(n)\},$$

and

$$\{X_{n+1,2},p,S^{4n+3},S^{4n-1},Sp(n)\}.$$

The sphere bundle construction is easily applied to these cases because $U(n)\subset SO(2n)$, and $Sp(n)\subset SO(4n)$. From [7] we know that $V_{n+1,2}$ is of the same homotopy type as $S^n\times S^{n-1}$ if and only if $\pi_{2n+1}(S^{n+1})$ has an element with Hopf invariant one. But J. F. Adams has shown that if $\pi_{2n+1}(S^{n+1})$ contains an element whose Hopf invariant is one, then $n+1=2,4$ or 8 [1], [2, p. 137]. Similar proofs apply to the other cases.

Theorem 7. *There are infinitely many actions of* $\mathbb{Z}_2$ *on* $S^2\times S^{1+k}(k\geqslant 2)$ *with fixed point sets the lens spaces* $L(n,1)$ *and* $S^2\times S^1$.

Proof. Since $\pi_1(SO(2))=\mathbb{Z}$, we have bundles $\mathscr{B}_n^{(2,1)}$. But for $n>0$, $B_n^{(2,1)}=S^3/\mathbb{Z}_n=L(n,1)$, and $B_0^{(2,1)}=S^2\times S^1$ [12, p. 135].

Theorem 8. *There are infinitely many actions of* $\mathbb{Z}_2$ *on* $S^4\times S^{2+k}(k\geqslant 3)$ *such that the actions are distinguished by the third homotopy group* π_3 *of the fixed point sets.*

Proof. As $\pi_3(SO(3))\approx\mathbb{Z}$, we have the corresponding infinitely many nonequivalent bundles $\mathscr{B}_n^{(4,2)}$. By [13, p. 137], we can see that $B_0^{(4,2)}$

$=S^4\times S^2$, $B_1^{(4,2)}=CP^3$, and $\pi_3(B_n^{(4,2)})=\mathbb{Z}_n(n>0)$. The proof is complete. Notice that $\pi_7(SO(5))\approx\mathbb{Z}$ [11], hence we have bundles $\mathscr{B}_n^{(8,4)}$, one of which is QP^3. We conjecture that the same results hold for this case. The $\mathbb{Z}_2$ action with fixed point set $B_1^{(4,2)}$ is obtained independently by J. C. SU [12]. His example yields some other information.

The sphere bundle construction also gives us an example of the action of $\mathbb{Z}_2$ on $X\times D^{r+k+1}$ with fixed point set F, the total space of the unit disk bundle of η^{r+1}, which is homotopy equivalent to $X\times D^{r+k+1}$. In particular, if we take $X=S^q$, then we have the following theorem:

Theorem 9. *There are infinitely many distinct actions of $\mathbb{Z}_2$ on an integral cohomology q-sphere M ($q=2,4$ or 8), say $M=S^q\times D^{r+k+1}$ ($r=1, r\geqslant 3$ or 7 resp.), of which the fixed point sets are integral cohomology q-spheres but with no homeomorphism between them.*

Finally, I want to give another kind of example by using the same construction. In [6], W. C. HSIANG showed that there are an infinite number of non-homeomorphic homotopy quaternionic projective n-spaces, say $HQP^n(i)$, for each $i\in\mathbb{Z}$, $n\geqslant 2$, and there exist sphere bundles $\{HCP^{2n+1}(i), p, HQP^n(i), S^2\}$, $i\in\mathbb{Z}$, where $HCP^{2n+1}(i)$ are non-homeomorphic homotopy complex projective $(2n+1)$-spaces. Thus we have the following examples:

Example 10. For any integer $i\in\mathbb{Z}$, $n\geqslant 2$; $HQP^n(i)\times S^m$ (for suitable m) admits at least one action of $\mathbb{Z}_2$ with fixed point set $HCP^{2n+1}(i)$.
There are still some well-known sphere bundles, say,
$\{V_{n+k+1,n+1}, p, V_{n+k+1,n}, S^k\}$ etc. We can certainly continue this argument ad infinitum, but we shall content ourselves with this.

Added in Proof

BOREL, HIRZEBRUCH and BOTT have shown that for any integers m and n with $n\geqslant 4k$, there exists an n-dimensional vector bundle η over S^{4k} with

$$p_k(\eta)=(2k-1)!\,\{\text{g.l.d.}(k+1,2)\}\,m\alpha,$$

where α is the standard generator of $H^{4k}(S^{4k},\mathbb{Z})$. (See MILNOR's Lectures on Characteristic Classes, Princeton University, 1957.) This is related to the question of Remark 5. Hence Corollary 4 and Theorem 9 have obvious generalizations.

References

1. ADAMS, J. F.: On the non-existence of elements of Hopf invariant one, Ann. of Math. **72,** 20—104 (1960).
2. ATIYAH, M. F.: *K*-theory, Harvard University, Cambridge 1964.
3. BOREL, A.: Topology of Lie groups and characteristic classes, Bull. Amer. Math. Soc. **61,** 397—432 (1955).
4. BREDON, G. E.: Exotic actions on spheres, these Proceedings.
5. EELLS, J., and N. H. KUIPER: An invariant for certain smooth manifolds, Annali di Math. **60,** 93—110 (1962).
6. HSIANG, W. C.: A note on free differentiable actions of S^1 and S^3 on homotopy spheres, Ann. of Math. **83,** 266—272 (1966).
7. JAMES, I. M., and J. H. C. WHITEHEAD: The homotopy theory of sphere bundles over spheres I, Proc. Lond. Math. Soc. (3) **4,** 196—218 (1954).
8. KERVAIRE, M., and J. MILNOR: Groups of homotopy spheres I, Ann. of Math. **77,** 504—537 (1963).
9. TAMURA, I.: On Pontrjagin classes and homotopy types of manifolds, J. Math. Soc. Japan **9,** 250—262 (1957).
10. — Homeomorphy classification of total spaces of sphere bundles over spheres, J. Math. Soc. Japan **10,** 29—43 (1958).
11. SERRE, J.-P.: Quelques calculs de groupes d'homotopie, C. R. Acad. Sci. Paris **236,** 2475—2477 (1953).
12. SU, J. C.: An example, these Proceedings.
13. STEENROD, N.: The Topology of Fibre Bundles, Princeton Univ. Press, Princeton, 1957.

3-Dimensional G-Manifolds with 2-Dimensional Orbits

WALTER D. NEUMANN

In [3] PAUL S. MOSTERT classifies all topological actions of compact connected Lie groups on connected $(n+1)$-dimensional manifolds which have n-dimensional orbits. If one starts from the assumption of differentiability a classification is very much easier, but our results show that the list given by MOSTERT (loc. cit.) for the compact case with $n=2$ has some omissions. In this note we therefore give a completed list for this case, and a brief indication of how it may be obtained when the simplifying assumption of differentiability is made.

In the following let G be a compact connected Lie group acting differentiably on a compact connected $(n+1)$-dimensional manifold M with at least one n-dimensional orbit. Let the principal orbit type be (H).

It follows that M/G is a compact connected 1-manifold, so it is homeomorphic either to the circle S^1 or to the interval $I=[0,1]$.

In the former case (H) is the only orbit type, so M is isomorphic to a fibre bundle over S^1 with fibre G/H and structure group $\Gamma=N(H)/H$. Every such bundle gives a G-manifold with the given properties, and they may be classified (up to equivariant diffeomorphism) by the components of Γ.

In the case $M/G\cong I$ we have two non-principal orbit types (U_1) and (U_2), which may be equal, corresponding to the orbits over the endpoints 0 and 1 of I. All other orbits are principal of type (H). Further we may assume $H\subset U_i$ $(i=1,2)$ and then U_i/H is an r_i-sphere $(i=1,2)$. One has an orthogonal action of U_i on $\mathbb{R}^{r_i+1}$ which is transitive on the unit sphere $S^{r_i}\subset\mathbb{R}^{r_i+1}$ $(i=1,2)$. If D^{r_i+1} is the closed unit ball in $\mathbb{R}^{r_i+1}$ then M is isomorphic to the G-manifold obtained by sticking $G\times_{U_1}D^{r_1+1}$ to $G\times_{U_2}D^{r_2+1}$ by an equivariant diffeomorphism between their boundaries, which are isomorphic to G/H. This construction always yields a G-manifold with the desired properties, and they may be classified by the components of the double coset space $N_1\backslash N(H)/N_2$, where $N_i=N(H)\cap N(U_i)$ for $i=1,2$.

We now assume that M is 3-dimensional and G acts effectively. Since G acts effectively and transitively on any 2-dimensional principal orbit, either $G=T^2$ and $H=(e)$ or $G=SO(3)$ and $H=O(2)$ or $SO(2)$.

First consider $G=T^2=S^1\times S^1$. For any two relatively prime integers p and q let $U(p,q)=\{(e^{2\pi ipt},e^{2\pi iqt})|t\in R\}$; these are just the subgroups of T^2 which are isomorphic to S^1. The possible T^2-orbit structures are:

$S1=\{(e)\}$; $S2=\{(e),(\mathbb{Z}_2\times(1))\}$;
$S3=\{(e),(\mathbb{Z}_2\times(1)),((1)\times\mathbb{Z}_2)\}$; $S4=\{(e),(\mathbb{Z}_2\times(1)),(U(1,0))\}$;
$S5=\{(e),(\mathbb{Z}_2\times(1)),(U(0,1))\}$; $S6=\{(e),(U(1,0))\}$;
$S7(p,q)=\{(e),(U(1,0)),(U(p,q))\}$, ($p,q\geqslant 0$ and relatively prime);

and all orbit structures obtainable from these by automorphisms of T^2.

If $G=SO(3)$ the possible orbit structures are:

$S8=\{(O(2))\}$; $S9=\{(SO(2))\}$;
$S10=\{(SO(2)),(O(2))\}$; $S11=\{(SO(2)),(SO(3))\}$;
$S12=\{(SO(2)),(O(2)),(SO(3))\}$.

For $S9$ $N(SO(2))/SO(2)=\mathbb{Z}_2$, so there are two possibilities for M. In all other cases M is uniquely defined by the orbit structure.

In explanation to the table: Kl, Mb, $\mathbb{P}^n$, T^n are respectively the Klein bottle, closed Möbius band, projective n-space, and n-torus. A is the manifold $Mb\times S^1\cup S^1\times Mb$ intersecting canonically in $S^1\times S^1$, and B denotes the non-trivial bundle with base S^1 and fibre S^2. $L(q,p)$ ($p,q\geqslant 0$ and relatively prime) denotes the lens space. It is shown in [1] that a manifold with orbit structure $S7(p,q)$ is $L(q,p)$; and an effective T^2-action on B with orbit structure $S5$ is easily found. In all other cases the manifolds in the table are given directly by the constructions described above.

Table

Manifold	$H_1(M)$	$H_2(M)$	$H_3(M)$	Orbit structure	
				T^2	$SO(3)$
T^3	$\mathbb{Z}\oplus\mathbb{Z}\oplus\mathbb{Z}$	$\mathbb{Z}\oplus\mathbb{Z}\oplus\mathbb{Z}$	$\mathbb{Z}$	$S1$	
$Kl\times S^1$	$\mathbb{Z}\oplus\mathbb{Z}\oplus\mathbb{Z}_2$	$\mathbb{Z}\oplus\mathbb{Z}_2$	0	$S2$	
A	$\mathbb{Z}\oplus\mathbb{Z}$	$\mathbb{Z}\oplus\mathbb{Z}_2$	0	$S3$	
$\mathbb{P}^2\times S^1$	$\mathbb{Z}\oplus\mathbb{Z}_2$	$\mathbb{Z}_2$	0	$S4$	$S8$
B	$\mathbb{Z}$	$\mathbb{Z}_2$	0	$S5$	$S9$
$S^2\times S^1$	$\mathbb{Z}$	$\mathbb{Z}$	$\mathbb{Z}$	$S6$	$S9$
$L(q,p)$	$\mathbb{Z}_q$	0	$\mathbb{Z}$	$S7(p,q)$	
$S^3=L(1,0)$	0	0	$\mathbb{Z}$	$S7(0,1)$	$S11$
$\mathbb{P}^3=L(2,1)$	$\mathbb{Z}_2$	0	$\mathbb{Z}$	$S7(1,2)$	$S12$
$\mathbb{P}^3\#\mathbb{P}^3$	$\mathbb{Z}_2\oplus\mathbb{Z}_2$	0	$\mathbb{Z}$		$S10$

The fundamental groups of the manifolds in the table are abelian except for

$$\pi_1(Kl \times S^1) = \langle a, b, c \mid a^2 = b^2, ac = ca, bc = cb \rangle,$$

$\pi_1(A) = \langle a, b \mid a^2 b = b a^2, b^2 a = a b^2 \rangle$, and $\pi_1(\mathbb{P}^3 \# \mathbb{P}^3) = \langle a, b \mid a^2 = b^2 = 1 \rangle$.

We remark finally that $L(q,p)$ and $L(q,p')$ are diffeomorphic if and only if $p' \equiv \pm p^\varepsilon \bmod q$, where p^ε is p or the inverse of p modulo q ([2]). It follows that two orbit structures of type $S7$ are related by an automorphism of T^2 if and only if the corresponding lens spaces are diffeomorphic.

Call the G-manifolds M and M' equivalent if there is an automorphism φ of G and a diffeomorphism $\psi: M \to M'$ such that $\psi(gx) = \varphi(g)\psi(x)$ for all $g \in G$ and $x \in M$. Then the above results show that a compact connected 3-manifolds admits at most one effective differentiable T^2-action up to equivalence.

References

1. JÄNICH, K.: Differenzierbare Mannigfaltigkeiten mit Rand als Orbiträume differenzierbarer G-Mannigfaltigkeiten ohne Rand, Topology **5**, 301—320 (1966).
2. MILNOR, J.: Whitehead torsion, Bull. Amer. Math. Soc. **72**, 358—426 (1966).
3. MOSTERT, P. S.: On a compact Lie group acting on a manifold, Ann. Math. **65**, 447—455 (1957); Errata, Ann. Math. **66**, 589 (1957).

Some Problems in Differentiable Transformation Groups

WU-CHUNG HSIANG* and WU-YI HSIANG

Introduction

We collect here some problems in the realm of differentiable transformation groups which have occurred to us during our collaboration in the past three years. Instead of just listing them as a sequence of questions, we try to put them together with the motivations and comments in a more organized way. It is our hope that such an effort will, indeed, make the problems more natural and inviting. We also suggest the readers to read this article together with another survey article written by the second named author for further motivation and the connections with known results (these same proceedings).

A general guiding principle. In differentiable transformation groups, one primarily studies the classifications and the geometric behavior of Lie subgroups of Diff(M) of a given smooth manifold M. Since *the existence of some interesting differentiable actions* on a smooth manifold M is itself a non-trivial fact which is *not* universally enjoyed by general manifolds, it seems rather logical to consider first the transformation groups on manifolds with sufficiently rich natural symmetries, such as euclidean spaces, spheres, homogeneous spaces, etc. For sufficiently symmetric manifolds, a natural approach to the study of transformation groups is simply *to compare the behavior of general differentiable actions with the behavior of certain "natural actions"*. Most of the known results, as well as the problems in transformation groups, *may be regarded as specifications of the above general principle of comparison.* Naturally, we shall follow this viewpoint of comparison in formulating the problems.

* The first named author was partially supported by NSF Grant NSF GP 6520 and an ALFRED P. SLOAN Fellowship.

1. Differentiable Actions of Compact Lie Groups on Euclidean Spaces, Spheres and Disks

Due to the existence of natural linear actions on euclidean spaces, spheres and disks, it is quite fair to say that they are the best testing spaces in the study of differentiable transformation groups. So far, most of the important results in transformation groups are concentrated in this *special but fundamental case.* Hence, we shall begin with the problems concerning differentiable compact Lie group actions on R^n, S^n and D^n.

a) Geometric Invariants of the Linear Actions

Following the general principle of comparison, the study of differentiable transformation groups on R^n, S^n, and D^n consists mainly in the comparison of the geometric behavior of differentiable actions to that of the linear actions. Hence, *the first step should be the study of the geometric invariants of the linear actions.* Traditionally, the linear actions of compact Lie groups are studied in terms of *the character functions* of the representations of *a fixed group.* However, very little about geometric invariants of linear actions has been understood so far. See § 2 of [14] for some simple results along this direction. Since the basic natural questions of this type are many, we include only a few of them as examples:

Problem 1. How can one express the number of orbit types of a linear G-action φ in terms of the character function χ_φ? Or, is it possible to have at least some reasonably accurate estimate of the number of orbit types of a linear G-action of φ in terms of the roots of G and the weights of φ?

Problem 2. Classify the (effective) linear actions with the number of orbit types $< \frac{1}{2}$ (the rank of the group G).

Problem 3. For a given compact connected Lie group G, classify the effective linear G-actions with their principal isotropy subgroups not equal to $\{e\}$.

Remark. A list of such classifications for simple Lie groups, and some partial results for non-simple Lie groups was worked out in Chapter I of [10]. These were very important for the proofs of the regularity theorems of [10].

Problem 4. Classify the (effective) linear actions of G on R^n with

$$\dim G > r \cdot n,$$

where r is a fixed number bigger than 5.

Problem 5. Classify the linear actions on R^n with codim of principal orbits $< \frac{1}{2}\sqrt{n}$.

Let φ be a linear action of a compact Lie group G on R^n and $\mathfrak{D}(\varphi)$ be the ring of invariant polynomial functions. It is well known that $\mathfrak{D}(\varphi)$ is finitely generated. Suppose $\{P_i(x); i=1,2,\ldots,k\}$ is a set of generators of $\mathfrak{D}(\varphi)$. Then it is not difficult to see that $\{P_i(x)\}$, considered as a set of functions of the orbit space R^n/G in the natural way, separates points of R^n/G, and hence, provides an embedding of R^n/G into an affine variety in R^k. However, the image of R^n/G is only a piece of the affine variety defined by $\mathfrak{D}(\varphi)$ and some basic inequalities among $\{P_i(x)\}$. They are the *basic invariant inequalities* which are natural generalizations of the usual Schwartz inequality.

Problem 6. What are the relationships between the basic invariant inequalities of φ and x and the basic inequalities of $\varphi \oplus \psi$, $\varphi \otimes \psi$ etc.?

Problem 7. Suppose that there exist finite generators of $\mathfrak{D}(\varphi)$ *without algebraic relations*. Is the orbit space of the unit sphere, S^{n-1}/G, necessarily contractible?

b) Regularities of Differentiable Actions of Compact Lie Groups on Euclidean Spaces, Spheres and Disks

Although the classes of differentiable actions on R^n, S^n or D^n are in every sense much wider than the classes of linear actions, it is generally believed that differentiable actions should resemble the linear actions at least under some suitable regularity conditions. To be sure, the "degree" of resemblance varies according to how strong the regularity condition is. The proofs of such resemblances always reveal further understanding about the behavior of differentiable transformation groups. Hence, one type of natural problem in differentiable transformation groups on R^n, S^n or D^n is to investigate differentiable actions satisfying certain geometric regularity conditions and try to see how far they resemble the linear actions. Again, problems of the above type are many, and we shall only list some of them as examples.

Problem 8. Let G be a compact Lie group (not necessary connected) and Φ be a differentiable action of G on a euclidean space R^n. Is the orbit space R^n/G contractible? For further discussion and related problems, see [5].

Problem 9. Classify differentiable actions of compact connected Lie groups on homotopy spheres Σ^m with 2 or 3 orbit types.

Problem 10. Classify differentiable actions of compact connected Lie groups on homotopy spheres Σ^m with the codimension of the principal orbit 1, 2, or 3 etc.

(See [3, 9, 27] for some of the known information about Problem 9 and 10.)

For the linear actions on euclidean spaces, the *origin* has an outstanding position as "a fixed center". However, it has been shown in § 2, Chapter I of [9] that for every compact connected *non-abelian* Lie group G, there exists a G-action *without fixed point* on certain high dimensional euclidean spaces. Hence, it would be interesting to see, *under what conditions* the *analogy of origin—fixed points—exists* for differentiable actions on euclidean spaces or disks. To be precise, let us introduce the following definitions:

Definitions. Let G be a compact connected simple Lie group. We set

$m(G)$ = the smallest possible dimension m such that there is an effective differentiable G-action on R^m without fixed points.

$t(G)$ = the smallest possible number of orbit types t such that there exists an effective differentiable G-action on certain euclidean space with t types of orbits but without fixed points.

d_n = l.u.b $\{\dim G \mid G$ is a compact connected simple subgroup of $\mathrm{Diff}(R^n)$ and acts on R^n without fixed points$\}$.

$\mathrm{cod}(n)$ = the smallest possible codimension r such that there exists a differentiable action on R^n with codimension r principal orbits but without fixed points.

Problem 11. Estimate $m(G)$ and $t(G)$ in terms of the roots and the rank of G, and estimate d_n and $\mathrm{cod}(n)$ as a function of n.

Remarks. 1. A theorem of A. BOREL [Chapter XIV of 1] showed that $t(G) \geqslant 3$ for all G and the construction of § 2, Chapter I of [9] provides a rather big upper bound to $t(G)$ in terms of the structure of G. It seems to us that it is possible to show that $t(G)$ at least is bigger than $\frac{3}{4}\mathrm{rk}\, G$. If G is a classical group, the above fact has been proved in [10].

2. It seems to us that $m(G)$ should be at least bigger than $\dim G$ and d_n should be at most smaller than n.

3. It should not be difficult to show that $\mathrm{cod}(n)$ is at least bigger than $\frac{1}{2}\sqrt{n}$. Furthermore, of interest would be:

Problem 12. Determine the orbit structures of the following classes of differentiable actions on euclidean spaces:

a) Differentiable actions of compact connected Lie groups on R^n (n large), with the codimension of the principal orbits less than $\frac{1}{2}\sqrt{n}$.

b) Differentiable actions of a compact connected Lie group G on euclidean spaces with the number of orbit types less than $\frac{1}{2}\mathrm{rk}(G)$.

Problem 13. For large n, it is not difficult to classify "large subgroups" of $SO(n)$, i.e., those connected subgroups H with $\dim H > \frac{1}{4}n^2$ or even, $\dim H > 5n$. Then, it would be interesting and useful if we could show that a "large compact connected subgroup" of $\mathrm{Diff}(R^n)$ must be isomorphic to a certain "large subgroup" of $SO(n)$, and moreover, with similar orbit structure.

Remarks. 1. In fact, following the argument of [15], it is not difficult to prove the following theorem (see § 5 of [14]).

Theorem. *Let G be a compact connected subgroup of* $\mathrm{Diff}(R^n)$, $n \geq 40$ *and* $\dim G > \frac{1}{4}n^2$. *Then* $G = SO(k) \times G'$ *for certain* $k > n/2$ *and the action of* $SO(k)$ *on* R^n *has* S^{k-1} *as its principal orbit type.*

2. Although we always formulate the problems only for R^n, similar (or parallel) problems can clearly be formulated for S^n and D^n and their solutions will be similar to one another.

Let $H \subseteq SO(n)$ be a connected closed subgroup of $SO(n)$. Then, Schur's lemma provides an effective tool for determining the centralizer of H in $SO(n)$. Now, suppose $G \subseteq \mathrm{Diff}(R^n)$ is a compact connected subgroup of $\mathrm{Diff}(R^n)$. We denote the group of all equivalent diffeomorphisms with respect to G by $\mathrm{Diff}(R^n; G)$. Then the study of the large compact subgroups of $\mathrm{Diff}(R^n; G)$ is a natural intermediate step for Problem 13.

Problem 14. Suppose that $G \subseteq \mathrm{Diff}(R^n)$ is a compact connected Lie subgroup with $F(G) \neq \phi$, and $\varphi\colon G \to SO(n)$ is the local representation of G around a fixed point $x \in F(G)$. Is it true that

$$\dim K \leq \dim \text{ of the centralizer of } \varphi(G) \text{ in } SO(n)$$

for all compact subgroup K in $\mathrm{Diff}(R^n; G)$?

For the linear actions of a compact connected Lie group G, i.e., the linear representations of G, the so called *weight systems*, are their *most convenient complete invariants*. Furthermore, they satisfy rather restrictive properties, such as

1. they are symmetric under the action of Weyl group $W(G)$,
2. $2(w,\alpha)/(\alpha,\alpha)$ is an integer for every weight w and root α, etc.

Now, for a given differentiable action Φ of G on an euclidean space R^n, we may define the concept of weight system for Φ as follows. By cohomology theory, it is well known that the fixed point set, $F(T)$, of the restricted action $\Phi|T$ is non-empty and acyclic. Hence the local representation of T around a point $x \in F(T)$ is constant, i.e. independent of the choice of x, and can be used to define the weight system of Φ.

Problem 15. Does the weight system of a differentiable action Φ of a compact connected Lie group G possess the same nice properties of the weight system of a linear action? In fact, it is not difficult to show that the weight system of a differentiable G-action is also symmetric under $W(G)$. However, it is more interesting to know whether the weight system of Φ also has such properties as the integrality of $2(w,\alpha)/(\alpha,\alpha)$, and to know the relationships among multiplicities of various weights, etc.

According to the differentiable slice theorem [19], the orbit type of $G(x)$ and the slice representation φ_x of G_x form a complete set of local invariants for differentiable G-spaces. Hence, it is useful to know whether the slice representations φ_x and φ_y of two points x, y on *orbits of the same type* are equivalant or not. It is easy to observe that for *linear* G-actions, the slice representations φ_x and φ_y of two points x, y on orbits of the same type are *always equivalent*. Hence, we have

Problem 16. Let Φ be a differentiable G-action on R^n. Is it true that the slice representations φ_x, φ_y of two points on orbits of the *same type* are always equivalent?

Remark. If G_x is connected, then it is not difficult to prove the above problem affirmatively.

For a semi-simple compact connected Lie group G with not too many simple normal cofactors, it is possible to determine all the possibilities of principal orbit types for *linear* G-actions.

(For example, see Chapter I of [10].)

Problem 17. Determine all the possibilities of principal orbit types for differentiable G-actions, where G is a semi-simple compact connected Lie group. Is it true that the possibilities of principal orbit types for differentiable G-actions are essentially the same as those already realized by the linear actions?

Remark. See [10] for some known results of the above problem in the case of classical groups.

c) Equivariant Embeddibility

According to Mostow and Palais [23, 24], any separable, metrizable G-space of finite orbit type and finite dimension admits an equivariant embedding in a certain euclidean space with suitable linear G-action. Although it is a beautiful general theory, there are almost no controls over the *orbit structure* and the *dimension* of *the ambient euclidean space*. In fact, with such generality in their formulation, it is rather unreasonable to expect any control. Consequently, it is almost impossible to apply their general results to get deep results for interesting *special* cases. Thus, it seems to be preferable for application to have some *refined* equivariant embedding theorems for the interesting special classes. For example

Problem 18. Let Φ be a differentiable $SO(n)$ action on a homotopy sphere Σ^m with $SO(k)$ $(k \geq 3)$ as the principal isotropy subgroup. Is it possible to embed the $SO(n)$-space (Σ^m, Φ) equivariantly into a *linear* $SO(n)$-*space* R^{m+q}, *again with* $SO(k)$ *as principal isotropy subgroups*? If possible, is there any estimate of codimension q in terms of m, n and k?

Remarks. 1. Surely, we may also formulate parallel problems by changing the group to $SU(n)$, $Sp(n)$ or by changing the total space to euclidean spaces, discs, etc.

2. The case that the principal orbit type is S^k $(k \neq 1, 3)$ was proved to be embeddable with codimension one in [8]. The case of $SO(n)$-action $(n \geq 5)$ on homotopy spheres Σ^m with $SO(n-2)$ as principal isotropy subgroups was proved to be imbeddable with codimension 4. We refer to [8, 13] for details.

3. It seems to us that the codimension q would be approximately bounded by the dimension of the orbit space.

Problem 19. Let Φ be a differentiable G-action on a homotopy m-sphere (or R^m, D^m, etc.) with G/T (T is a maximal torus) as the principal orbit type. Is the G-space equivariantly embeddable in a linear G-space R^{m+q}, again with G/T as the principal orbit type?

2. Degrees of Symmetry of Homotopy Spheres and Manifolds of the Same Homotopy Type of Homogeneous Spaces

Let M be a given smooth structure and $\mathrm{Diff}(M)$ be the group of all diffeomorphisms of M onto itself. The upper bound $N(M)$ of the dimensions of all the compact subgroups of $\mathrm{Diff}(M)$ measures, in some crude sense, the degree of symmetry of the smooth structure M. In general, though $N(M)$ is a natural invariant of M with obvious importance, it is usually very difficult to compute or even to estimate. In view of the known results of [11, 13, 15, 16] (see also § 5 of [14]), it seems rather natural and inviting to study the following problems:

Problem 20. Is it true that $N(\Sigma_1^m)$ and $N(\Sigma_2^m)$ are more or less of the same magnitude for $\Sigma_1^m, \Sigma_2^m \notin bP_{m+1}$ but $(\Sigma_1^m - \Sigma_2^m) \in bP_{m+1}$?

Problem 21. Are there homotopy spheres Σ^m with $N(\Sigma^m)=0$? If not, what is $\nu(m)=\mathrm{Min}\{N(\Sigma^m), \Sigma^m \in \Theta^m\}$?

Let $M=G/H$ be a homogeneous space of compact Lie group G. It would be interesting to see whether or not the *natural homogeneous Riemmanian metrics* are among the most symmetric metrics. To be precise, for a compact homogeneous space M, we may set $N'(M)=$ l.u.b. $\{\dim G;\ G$ a compact *transitive* subgroup of $\mathrm{Diff}(M)\}$. Then, clearly $N'(M) \leqslant N(M)$, and one is tempted to guess that $N'(M)=N(M)$, *at least for most of the compact homogeneous spaces.* The cases $M=S^m$ and RP^m are trivial, and the case $M=V_{n,2}=SO(n)/SO(n-2)$ was proved affirmatively in [16]. The following are some interesting cases:

Problem 22. Let M be the underlying smooth structure of a compact symmetric space, and G be the biggest transitive group on M. Is $N(M)=\dim G$?

Problem 23. Let G be a semi-simple compact connected Lie group and T be a maximal torus of G. Is it true that $N(G/T)=\dim G$?

Problem 24. Let M be a homogeneous space of a compact Lie group. Is it true that $N(M)>N(M')$ for all other M' with the same homotopy type of M?

Remarks. 1. So far, only the cases of homotopy spheres and homotopy second Stiefel manifolds have been worked out affirmatively [11, 13, 15, 16]. For example, one can easily construct many differentiable manifolds of the same homotopy type of CP^n or QP^n but with different rational Pontrjagin classes (see [17]). It would be very interesting to know how to estimate the change of the degree of symmetry in terms of the change of rational Pontrjagin classes.

2. An affirmative answer to the above problem will confirm the belief that homogeneous differentiable structures are indeed more symmetric than other smooth structures of the same homotopy type.

Problem 25. One can define the number $N(M)$ for M with additional structures, e.g. almost complex structures, spin structures, holomorphic structures, etc. How do they behave? In particular, it would be interesting to compare the degrees of symmetry of *different* structures over the *same smooth structure* and then to compute their upper bound.

3. Differentiable Actions on Homogeneous Spaces

So far, most of the known results in differentiable transformation groups are about actions on spheres, euclidean spaces or disks. We share the prevailing conviction that the study of differentiable actions on these "best testing spaces" is *probably* still the most important topic in transformation groups. However, it seems to us that it will be helpful to the understanding of the general picture if one broadens the set of testing spaces to include more nice spaces such as homogeneous spaces. For a homogeneous space G/H of a compact Lie group G, $G\times$(centralizer of H in G) acts naturally on G/H via translations, and hence any restriction of the above natural action of $G\times Z(H;G)$ to its subgroup K can be considered as the analogy of orthogonal actions on spheres $S^m=0(m+1)/0(m)$, and we shall call them *the natural actions.* Then, it is quite natural to formulate problems for differentiable actions on homogeneous spaces according to the principle of comparison. Here are some examples:

Problem 26. Does the cohomology (with suitable coefficients) of the fixed point set of a Z_p action on G/H (p prime or $Z_0=S^1$) necessarily resembles the cohomology of the fixed point set of *the natural* action of a suitable Z_p subgroup of $G\times Z(H;G)$?

Remark. If the above problem can be answered in the affirmative for some homogeneous spaces, G/H, besides those known cases, it will be a useful and meaningful generalization of the classical P. A. Smith theory. Higher Stiefel manifolds and compact symmetric spaces will probably be good candidates to try.

Problem 27. Let G be a simple compact connected Lie group and T be a maximal torus of G. It is not difficult to see that the principal isotropy subgroup of every non-transitive *natural* action (i.e., acting as translations) is trivial. Hence, it is natural to ask whether the principal isotropy subgroup of any non-transitive action is at most a finite group.

Remark. If the above problem can be proved affirmatively, then it follows easily that $N(G/T)=\dim G$ for a simple compact Lie group G.

Problem 28. Classify the transitive actions on a certain class of compact homogeneous spaces.

Remarks. 1. For the case of spheres, see [2, 21, 25]; for the case of homology spheres, see [Chapter XV of 1]. See also [28] for the case of homogeneous spaces with non-vanishing Euler characteristics and [17] for the case of Stiefel manifolds.

2. Technically, it seems important to develop some decomposition theorem and non-decomposition theorem for homogeneous spaces similar to those appearing in [17].

Problem 29. Is it true that two homeomorphic compact homogeneous spaces M_1, M_2 are necessarily diffeomorphic? Or does the above statement hold with only a few exceptions?

Problem 30. Classify homogeneous spaces according to their topological or diffeomorphic invariants. For example: What are the homogeneous spaces with certain specific type of cohomology rings? What are homogeneous spaces with almost complex structures, etc.?

4. Differentiable Actions of Semi-simple Non-compact Lie Groups

So far, one knows very little about differentiable actions of non-compact Lie groups. We include here a few problems about differentiable actions of *semi-simple* non-compact Lie groups which may be viewed as an extension of the study of actions of compact Lie groups.

Problem 31. Let G be a simple non-compact Lie group and K be a maximal compact subgroup of G. Is it true that every differentiable G-action on R^m with $m < \dim(G/K)$ necessarily has fixed points? Furthermore, what is the orbit structure of such actions?

Problem 32. Again consider differentiable G-actions on R^m with $m < \dim(G/K)$. How does one characterize the linear actions in terms of their geometric behavior?

Remarks. 1. For further discussion of the above two problems, see a paper of the second named author to appear in Amer. J. Math.

2. In fact, the orbit structure of linear actions of non-compact Lie groups are themselves very interesting and yet very little understood.

Problem 33. What is the geometric characterization to the fact that a smooth action of a non-compact Lie group G on a non compact manifold M can be realized as the isometry group of a suitable Riemannian metric on M?

References

1. BOREL, A. et. al.: Seminar on transformation groups, Annals of Math. Studies **46**, Princeton University Press (1961).
2. — Le plan projectif des octaves et les spheres comme espaces homogènes, Completes Rendue de l'Academie des Sciences, Paris **230**, 1378—1383 (1960).
3. BREDON, G.: Transformation groups on spheres with two types of orbits, Topology **3**, 115—122 (1965).
4. — Examples of differentiable group actions, Topology **3**, 103—113 (1965).
5. CONNER, P.: Retraction properties of the orbit spaces of a compact topological transformation group, Duke Math. J. **27**, 341—357 (1960).
6. HIRZEBRUCH, F.: Séminaire Bourbaki 1966.
7. HSIANG, W. C.: A note on free differentiable actions of S^1 and S^3 on homotopy spheres, Ann. of Math. **83**, 266—272 (1966).
8. —, and W. Y. HSIANG: Classification of differentiable actions on S^n, R^n and D^n with S^k as the principal orbit type, Ann. of Math. **82**, 421—433 (1965).
9. — — Differentiable Actions of compact connected classical groups I, Amer. J. Math. **89**, 705—786 (1967).
10. — Differentiable actions of compact connected classical groups II (in preparation).
11. — On compact subgroup of the diffeomorphism groups of Kervair spheres. Ann. of Math. **85**, 359—369 (1967).
12. — Some differentiable actions of S^1 and S^3 on 11-spheres, Quarterly J. Math. (2) Oxford **15**, 371—374 (1964).
13. — The degree of symmetry of homotopy spheres, to appear in Ann. of Math.
14. HSIANG, W. Y.: A survey article in this issue.
15. — On the bound of the dimensions of the isometry groups of all possible riemannian metrics on an exotic sphere, Ann. of Math. **85**, 351—358 (1967).
16. — The natural metric on $SO(n)/SO(n-2)$ is the most symmetric metric, Bull. AMS **73**, 55—58 (1967).
17. —, and J. C. SU: (to appear in Trans. AMS).
18. JÄNICH, K.: Differenzierbare Mannigfaltigkeiten mit Rand als Orbiträume differenzierbarer G-Mannigfaltigkeiten ohne Rand, Topology, **5**, 301—320 (1966).
19. MONTGOMERY, D., and C. T. YANG: The existence of slice, Ann. of Math. **65**, 108—116 (1957).
20. — Free differentiable actions on homotopy spheres, these Proceedings.
21. —, and H. SAMELSON: Transformation groups of spheres, Ann. of Math. **44**, 454—470 (1943).
22. —, and L. ZIPPIN: Topological transformation groups, Interscience, 1955.
23. MOSTOW, G. D.: Equivariant embedding in Euclidean space, Ann. of Math. **65**, 432—446 (1957).
24. PALAIS, R.: Embedding of compact differentiable transformation groups in orthogonal representations, J. of Math. and Mech. **6**, 673—678 (1957).

25. PONCET, J.: Groupes die Lie compacts de transformations de l'espace euclidean et les spheres comme espaces homogènes, Comment. Math. Helv. **33,** 109—120 (1959).
26. SMITH, P. A.: Fixed points of periodic transformations, Appendix B in Lefschetz Algebraic Topology, 1948, New York.
27. WANG, H. C.: Compact transformation groups of S^n with an $(n—1)$-dimensional orbit, Amer. J. Math. **82,** 698—748 (1960).
28. — Homogeneous spaces with non-vanishing Euler characteristics, Ann. Math. **50,** 925—953 (1949).

Problems

1. Classify the differentiable actions of S^1 (for example) on S^n which have fixed point set the standard S^r and are free outside the fixed point set. (See e. g., J.-C. SU, *Some results on cyclic transformation groups*, these Proceedings.) – G. E. BREDON.

2. Which homotopy spheres admit (infinitely many) free actions of the circle, or of Z_p? There are infinitely many such actions for the standard sphere by the work of MONTGOMERY and YANG and by the HSIANG brothers for $n \geqslant 7$. It is also known that the Kervaire sphere admits free circle group actions. There are more isolated results of both a positive and negative nature. – G. E. BREDON and C. N. LEE.

3. Does there exist a homotopy sphere which does not admit any affective (not necessarily free) action of S^1? (Such a sphere cannot bound a π-manifold. There are spheres in dimensions 9, 10, 13, 15, 17, and 18, not bounding π-manifolds, but which admit S^1 actions. The exotic 8-sphere is not known to admit such an action.) – G. BREDON.

4. In dimensions $8k+1$, there exist homotopy spheres which do not bound spin-manifolds (also $8k+2$). It has been shown by R. LEE that these cannot admit free S^1 actions or even free involutions. Do they admit non-free S^1 actions? (At least one of those in each dimension $8k+2$ does admit such an action, but this construction cannot be used in the odd case.) – G. E. BREDON.

5. It has been shown by W.-Y. HSIANG that the dimension of a compact Lie group acting effectively on an exotic sphere Σ^m is less than $(1/8)m^2+1$. The action of $SO(4k+1)\times S^1$ on the Kervaire sphere Σ^{8k+1} achieves the bound $(1/8)m^2+7/8$. What is the upper bound for the dimension of a torus T^r acting effectively on an exotic sphere Σ^m? A conjecture is $r \leqslant [(m+1)/4]+1$. – L. N. MANN.

6. One can ask the same question for Z_2-tori $(Z_2)^s$. – L. N. MANN.

7. Are the free involutions by actions of $-I \in SO(2n+1)$ on M_k^{4n+1} desuspendible? ($M_k^{4n+1} = \Sigma(2k+1, 2, \ldots, 2)$ in Hirzebruch's notation.) – G. E. BREDON.

8. SULLIVAN has found an example of an 8-manifold M^8 which is tangentially equivalent to CP^4 but not homeomorphic to CP^4. There is a natural principal S^1-fibration $\Sigma^9 \to M^8$, where Σ^9 is the Kervaire sphere. Restricting one's attention to the involution $Z_2 \subset S^1$ acting on Σ^9, is this involution desuspendable? – G. E. BREDON.

9. Let A denote the antipodal map of S^n into itself. Any fixed point free involution on Σ^{2n+1} can be obtained from two copies of $(S^n \times D^{n+1}, A)$

by identifying the boundaries by means of an equivariant diffeomorphism. If two such diffeomorphisms differ by one which extends equivariantly to either $S^n \times D^{n+1}$ or $D^{n+1} \times S^n$, the homotopy spheres are equivariantly diffeomorphic. The proof of this result hinges on two assertions: 1. $\mathrm{Wh}(Z_2)=0$ and 2. $\tilde{K}(RP(n)) \cong \tilde{J}(RP(n))$. A similar argument holds for fixed point free Z_3 actions on Σ^{4n+3}. There are then the following two questions:

a) Does the set isomorphism described above enable one to define an addition for involutions? If so, is the Browder-Livesay signature a homomorphism into Z for dimensions $4n+3$? Recent work by LOPEZ DE MEDRANO (see his paper, these Proceedings) suggest an affirmative answer to both cases.

b) What can one say for Z_p, p a prime $\geqslant 5$. The difficulty here is that the projection $\tilde{K}(L^{2n+1}(p)) \to \tilde{J}(L^{2n+1}(p))$ has a non-zero kernel, and the argument used for involutions breaks down. – G. R. LIVESAY and C. B. THOMAS.

10. There is the following theorem: *Let Y^7 be a compact Poincaré complex such that $\tilde{Y} \sim S^7$ and $\pi_1 Y$ has cohomology of period 4. If $\pi_1 Y$ has odd order, there exists a disc bundle over $(Y^7-\text{point})$ with spherical Thom class. If G has even order, there is an obstruction in $H^3(\pi_1 : Z_2)$.* It is reasonable to suppose that in the first case the bundle can be closed over Y, after which we are in a position to do framed surgery on a preimage of Y in S^{n+7}. In the presence of π_1, there is an obstruction to surgery in the middle dimensions, and hence an obstruction to obtaining a manifold homotopy equivalent to Y, which lies in a group related to $\mathrm{Wh}(\pi_1)$. In the p. l. category, the theorem quoted can be extended to higher dimensions, since the coefficients in the obstruction cohomology are periodic. Problem: show that surgery is possible for π_1 non-abelian of order $2p$. The obstruction in the theorem would then be the only one to produce a fixed point free group action on S^{8k-1}. The obstruction would then have to be non-zero by an old result of J. MILNOR. – C. B. THOMAS.

11. When can a (differentiable) involution on S^n (with or without fixed points) be extended to a (differentiable) involution on D^{n+1}, where $\partial D^{n+1} = S^n$? What will be the relations between $F(Z_2, S^n)$ and $F(Z_2, D^{n+1})$? In a few particular cases, the answers are known. – J. PAK.

12. Can the exotic spheres which are not in bP_{2n} be written down in a similar manner as those in bP_2 ? Do they admit group actions? (Some are known to admit group actions. See, e. g., BREDON'S *Exotic Actions on Spheres*, these Proceedings.) – F. HIRZEBRUCH.

13. Study the invariant Riemannian metric on $\Sigma^7(3,5,2,2,2)$ induced by the embedding in S^9. What are the eigenvalues of the Laplacian? – F. HIRZEBRUCH.

PART II

Algebraic Topological and Other Techniques

Equivariant Homology Theories

C. N. LEE

The bordism theory, originally invented by THOM, has been generalized to a (generalized) homology theory by ATIYAH, and independently by CONNER and FLOYD. Meanwhile, a similar phenomenon occured to the K-theory, originally invented by GROTHENDIECK, which was responsible for solutions of many delicate problems in differential topology. Since then an (abstract) form of such homology theory was formulated and studied by a number of authors, notably G. W. WHITEHEAD, E. BROWN, A. DOLD. These theories are defined on a suitable category of topological spaces and continuous maps.

It is natural to ask if one can formulate equivariant analogues of these theories on a suitable category of topological spaces with a group action and equivariant maps, and apply them to problems in differential transformation group theory. In fact, ATIYAH and SEGAL did exactly that for the K-theory [3] in 1965. On the other hand, CONNER and FLOYD studied an equivariant analogue of the original Thom bordism groups and brought forth many applications. Very recently, BREDON studied an equivariant cohomology theory defined on the category of CW complexes with a discrete group action [4], but with somewhat different purpose.

The purpose of this note is to consider the abstract equivariant homology and cohomology theories defined on a large category of spaces with a compact Lie group action. No details of proofs will be given here. The layout of the contents is as follows: Section 1 contains basic definitions. We introduce a canonical equivariant theory induced by a given nonequivariant theory. We conclude the section with a remark on the representability of the canonical equivariant cohomology theory. In Section 2, we give two examples of equivariant theory, namely the G-bordism homology theory, and the K_G-cohomology theory as defined by ATIYAH-SEGAL. We show that these two theories compare favorably with their respective canonical theories. In Section 3, the CARTAN-LERAY type of spectral sequence is obtained. This is achieved by considering a space originally defined by E. DYER for the similar purpose [6], and also used by ATIYAH-SEGAL [3]. This part is actually an equivariant analogue of a joint work (unpublished) with F. RAYMOND.

1. Equivariant Homology Theories

Denote by $\mathscr{C}_G$ a category of topological pairs (X,A) with an action of a topological group G, and equivariant maps between such pairs. Objects and morphisms in $\mathscr{C}_G$ shall be refered to as G-pairs and G-maps respectively. For simplicity, we shall assume that G is a compact Lie group and the pairs (X,A) are HAUSDORFF such that the second space A is a closed subspace of X. If (X,A) and (Y,B) are G-pairs then we define $(X,A)\times(Y,B)=(X\times Y, A\times Y\cup X\times B)$ with the diagonal G-action, i.e., $g\cdot(x,y)=(gx,gy)$, $g\in G$, $x\in X$, $y\in Y$. This is the direct product in the category $\mathscr{C}_G$. As usual, single spaces will be regarded as a pair with the second space empty. Let I denote the unit interval with the trivial G-action. A G-map $\phi:(X,A)\times I\to(Y,B)$ is called a *G-homotopy* (between the two maps ϕ_0 and ϕ_1, $\phi_0\simeq\phi_1$ by notation).

Definition. An (equivariant) homology theory h_*^G on $\mathscr{C}_G$ is a sequence $\{h_n^G\}$, $n\in\mathbb{Z}$, of covariant functors $h_n^G:\mathscr{C}_G\to Ab$ (the category of abelian groups), together with natural transformations $\partial:h_n^G(X,A)\to h_{n-1}^G(A)$ satisfying;

Axiom (Homotopy). *If ϕ_0 and ϕ_1 are G-maps which are G-homotopic, then $h_n^G(\phi_0)=h_n^G(\phi_1)$, all $n\in\mathbb{Z}$.*

Axiom (Exact Sequence). *For every $(X,A)\in\mathscr{C}_G$, the following sequence is exact,*

$$\cdots\to h_n^G(A)\to h_n^G(X)\to h_n^G(X,A)\xrightarrow{\partial} h_{n-1}^G(A)\to\cdots$$

where the unlabelled homomorphisms are those induced by the obvious inclusion maps.

Axiom (Excision). *If $(X,A)\in\mathscr{C}_G$ and U is an open subset of X such that $\bar{U}\subset\operatorname{int}.A$ and the inclusion map $i:(X-U,A-U)\subset(X,A)$ is in $\mathscr{C}_G$, then the induced homomorphism $h_n(i)$ is an isomorphism for all n.*

These constitute the least number of requirements imposed on any homology theory. However, we shall impose more conditions as needed in the future.

An *(equivariant) cohomology theory* h_G^* on $\mathscr{C}_G$ is defined in the dual fashion.

We shall define a *canonical equivariant homology (respectively, cohomology) theory associated with a given non-equivariant homology (respectively, cohomology) theory.* Thus, let h_* be a homology theory defined on an admissible category $\mathscr{C}$ of topological pairs without any group action, and maps between such pairs. Choose once and for all a CW complex E which is contractible and admits a free action of G. If $(X,A)\in\mathscr{C}_G$ then $E\times(X,A)\in\mathscr{C}_G$, and on it the action of G is free.

Definition. $h_*^{[G]}(X,A)=h_*(E\times X)/G,(E\times A)/G)$.

To make $h_*^{[G]}$ a homology theory on $\mathscr{C}_G$, we must define boundary homomorphisms $\partial: h_n^{[G]}(X,A)\to h_{n-1}^{[G]}(A)$ and the induced homomorphism $f^*=h_*^{[G]}(f)$ of G-maps f in $\mathscr{C}_G$. But these are all defined in the obvious fashion. Now, we claim,

Lemma. *$(h_*^{[G]},\partial)$ is an (equivariant) homology theory on $\mathscr{C}_G$.*

This is not difficult to prove.

In practice, the two homology theories h_*^G (arising naturally) and $h_*^{[G]}$ (defined artifically) compare favorably. See examples in the next section.

The corresponding definitions and the claims for cohomology theory are also valid.

As E. BROWN has shown in [5], some (nonequivariant) cohomology theories are equivalent to those defined by means of spectra. Such a cohomology theory is said to be *homotopy-representable*. It may be possible to prove an equivariant analogue of the Brown's representability theorem. However, we will be satisfied here only with the following observation.

Remark. *If a cohomology theory h^* on $\mathscr{C}$ is homotopy-representable, then so is its associated equivariant cohomology theory $h_{[G]}^*$ on locally compact G-pairs of $\mathscr{C}_G$.*

Proof. Let $B=\{B_n\}$, $n\in\mathbb{Z}$, be an Ω-spectrum for h^*. Define $B^E=\{B_n^E\}$, $n\in\mathbb{Z}$, where each B_n^E is the function space with the compact-open topology. Define the action of G on each B_n^E by: $f\in B_n^E$, $g\in G\Rightarrow(g\cdot f)(e)=f(g^{-1}\cdot e)$, $e\in E$. It is then not difficult to show that the action of G on B_n^E is well defined, and that $h_{[G]}^n(X,A)\approx[X/A;B_n^E]_G$, the G-homotopy classes of the G-maps preserving the canonical base points.

2. Examples

a) G-bordism Theory

CONNER-FLOYD defined in [6], the G-bordism groups $\Omega_n(G)$ of oriented G-manifolds with free G-actions. They defined also the singular bordism groups $\Omega_n(X,A)$ for arbitrary topological pairs (X,A) which formed a homology theory. Unifying these, one can define the singular G-bordism groups $\Omega_n^G(X,A)$ for G-space pairs (X,A) which will form an equivariant homology theory.

First recall that an oriented G-manifold is an oriented manifold M^n with an orientation preserving free differential action of G. A singular oriented G-manifold in a G-pair (X,A) is a pair (M^n,f) consisting of an oriented G-manifold M^n and a G-map $f:(M^n,\partial M^n)\to(X,A)$. Note that the G-action on (X,A) is not required to be free. Two such pairs (M_1^n,f_1) and (M_2^n,f_2) are *G-bordant to each other* if there is an oriented G-manifold N^{n+1} with a free G-action and a G-map $F:N^{n+1}\to X$ such that (1) ∂N^{n+1} contains $M_1^n\cup(-M_2^n)$ as a regular G-submanifold with the induced orientation, and (2) $F|M_i^n=f_i\,(i=1,2)$ and $F(\partial N^{n+1}-(M_1^n\cup M_2^n))\subset A$. Now $\Omega_n^G(X,A)$ is defined to be the abelian group of all G-bordant classes of singular oriented G-manifolds of dimension n in (X,A), where the group operation is given by disjoint union as usual. Note that $\Omega_n^G(\text{point})=\Omega_n(G)$ and $\Omega_n^{\{e\}}(X,A)=\Omega_n(X,A)$ in the notation of CONNER-FLOYD [6], and that each Ω_n^G is certainly a covariant functor defined on $\mathscr{C}_G$ with values in Ab. Define the boundary operator $\partial:\Omega_n^G(X,A)\to\Omega_{n-1}^G(A)$ by restriction. The following two assertions are generalizations of corresponding theorems of CONNER-FLOYD [6].

Lemma. *$\{\Omega_*^G,\partial\}$ is a homology theory on $\mathscr{C}_G$.*

Proof for $G=\{e\}$ is given in CONNER-FLOYD [6]. There is no essential difficulty in modifying the proof and hence we leave it to the reader.

Theorem. *There is a natural equivalence*

$$\Omega_*^G\approx\Omega_*^{[G]}.$$

We shall outline a proof. First choose an orientation on the manifold G so that its restrictions on connected components of G are permuted by the left translations by the elements of G. There are two such orientations on G.

Let (X,A) be a G-space pair. We shall define a map $\phi_{X,A}:\Omega_n^G(X,A)\to\Omega_{n-t}^{[G]}(X,A)$, where $t=\dim G$. Let (M^n,f) represent an element of $\Omega_n^G(X,A)$. Since $M^n\to M^n/G$ is a principal G-bundle, there is a G-map $\phi:M^n\to E$ unique up to G-homotopy as is well known. Thus we get a G-map: $M^n,\partial M^n\to E\times(X,A)$ defined by $y\mapsto(\phi(y),f(y))$, $y\in M^n$. This induces a map $u:M^n/G\to E\times X/G$. Orient the manifold M^n/G coherently with the orientations of M^n and G. Then $(M^n/G,u)$ represents an element which is defined to be $\phi_{X,A}([M^n,f])$. It is not difficult to see that $\phi_{X,A}([M^n,f])$ depends only on the oriented G-cobordism class of (M^n,f) so that $\phi_{X,A}$ is well defined. In fact, by varying the object $(X,A)\in\mathscr{C}_G$, we can see that the $\phi_{X,A}$ define a natural transformation $\phi:\Omega_*^G\to\Omega_*^{[G]}$ between the two homology theories. We now wish to show that ϕ is an equivalence. For this, it would be sufficient to show $\phi_X:\Omega_n^G(X)\xrightarrow{\approx}\Omega_{n-t}^{[G]}(X)$ for any G-space X. To show that ϕ_X is surjective, let $u:N^{n-t}\to(E\times X)/G$

represent an element of $\Omega_{n-t}^{[G]}(X)$. Let M^n be the "pull-back" by u of the bundle $E\times X\to(E\times X)/G$. There is the commutative diagram

$$\begin{array}{ccc} M^n & \overset{\tilde{u}}{\cdots\to} & E\times X \\ \downarrow & & \downarrow \\ N^{n-t}=M^n/G & \xrightarrow{u} & E\times X/G \end{array}$$

where u is the induced G-map. Orient M^n coherently with the orientations of N^{n-t} and G. Following the map $\tilde{u}$ by the projection map, we get a G-map $v:M^n\to X$. The pair (M^n,v) represents an element of $\Omega_n^G(X)$ and clearly $\phi_X([M^n,v])=[N^{n-t},u]$.

The injectivity of ϕ_X is somewhat more complicated but similar to the proof of the surjectivity.

One may define various other equivariant bordism homology theories, such as by considering complex manifolds with an analytic group action. An immediately related one would be that of unoriented singular G-manifolds with free actions. One might be able to define an equivariant bordism theory of singular G-manifolds with effective group actions instead of free actions. The coefficient groups of such a bordism theory have been recently studied by CONNER, FLOYD, and BOARDMAN in connection with differential transformation group theory.

b) K^*-theory

For a detailed exposition, we refer to the lecture notes by ATIYAH-SEGAL [3]. For the sake of the completeness, we shall include here a quick review of the K_G-theory as an example of an equivariant cohomology theory and relations between the two cohomology functors K_G^* and $K_{[G]}^*$. Following ATIYAH, we define $K_G(X)$ to be the GROTHENDIECK ring of all G-vector bundles over the compact G-space X. By a G-vector bundle, we mean a (complex) vector bundle E with a G-action such that each element $g\in G$ is a linear transformation: $E_x\to E_{gx}$, $x\in X$. $K_G^n(X,A)$ is defined (for negative integers n) in the usual way, and then via the equivariant Bott periodicity theorem, one completes the definition of the equivariant cohomology theory K_G^* on the category $\mathscr{C}_G^{\mathrm{cpt}}$ of compact G-pairs.

The coefficient ring $K_G(\mathrm{pt})=R(G)$ is the ring of all virtual (complex) representations of G. For any G-space $X\in\mathscr{C}_G^{\mathrm{cpt}}$, $K_G^*(X)$ is an $R(G)$-module. Let $\varepsilon:R(G)\to\mathbb{Z}$ be the ring-homomorphism defined by $\varepsilon(V)=\dim_{\mathbb{C}}V$, where V is a representation space of G. Then the kernel of ε is an ideal $I(G)$ of $R(G)$. Define $\hat{K}_G^*(X)$ to be the completion of $K_G^*(X)$ by $I(G)$. Next, we shall define $\hat{K}_{[G]}^*(X)$. First, choose a sequence of compact spaces E_k with a free G-action, $k=1,2,\ldots$, such that E_k is k-connected and there is a G-map $\varepsilon_k: E_k \to E_{k+1}$ for each k. Define $K^*_{[G]}(X)$ to be the inverse limit of

$$\cdots \leftarrow \hat{K}^*_{[G]}(E_k \times X) \overset{\tilde{\phi}^*}{\leftarrow} \hat{K}^*_{[G]}(E_{k+1} \times X) \leftarrow \cdots$$

where $\phi(y,x) = (\varepsilon_k(y), x)$, $(y,x) \in E_k \times X$.

Theorem (ATIYAH-HIRZEBRUCH-SEGAL). *For* $X \in \mathscr{C}_G^{\mathrm{cpt}}$,

$$\hat{K}^*_G(X) \approx \hat{K}^*_{[G]}(X)$$

provided $K^*_G(X)$ *is a finitely generated* $R(G)$*-module.*

Special cases were proved earlier by ATIYAH and HIRZEBRUCH [1, 2]. The general case was proved by SEGAL in [3].

3. Spectral Sequences

Let h^G_* be a homology theory on $\mathscr{C}_G$ satisfying,

Axiom (compact supports). For each $\alpha \in h^G_n(X,A)$, *there is a compact* G*-subspace* C *of* X *such that* $\alpha \in \operatorname{Im}(h^G_n(C, C \cap A) \to h^G_n(X,A))$.

The G-bordism theories satisfy this axiom.

Let $f:(X,A) \to (Y,B)$ be a G-map. Assume:

1. $f(X) = Y$ *and* $f^{-1}(B) = A$.
2. (Y,B) *is a pair admitting a locally finite triangulation.*
3. G *acts on* Y *trivially.*

Choose a triangulation of (Y,B) and let $h^G_n(f)$ denote the local (homology) coefficient system defined by $h^G_n(f)(\sigma) = h^G_n(f^{-1}(\mathrm{st}.\sigma))$ for each simplex σ of Y where $\mathrm{st}.\sigma$ = the star of σ in Y. With these assumptions and notations we have,

Theorem (Homology). *There is a homology spectral sequence* $E(f) = \{E^r_{ij}\}$ *such that*

$$E^2_{pq} \approx H_p(Y, B; h^G_q(f))$$

and E^∞ *is the graded group of a suitable filtration of* $h^G_*(X,A)$.

To outline the proof, let α be a closed G-covering of X, that is, a closed covering of X such that each set of α is a G-subset of X, and furthermore, the interiors of the sets of α form an open covering of X. Then $\alpha|A = \{u \cap A \mid u \in \alpha\}$ is a closed G-covering of A. Let (X_α, A_α) be the simplicial pair (and also its geometric realization) of the nerve of $(\alpha, \alpha|A)$. For each simplex σ of X_α, let $\mathrm{car}_\alpha(\sigma)$ be the intersection of the sets forming the vertices of σ. Following DYER [7], define $\underline{X}_\alpha = \bigcup_{\sigma \in X_\alpha} \mathrm{car}_\alpha(\sigma) \times \sigma$ with the topology as a subspace of $X \times X_\alpha$, and similarly for $\underline{A}_\alpha$. We regard (X_α, A_α) as a G-pair with the trivial action of G, and hence $X \times (X_\alpha, A_\alpha) = (X \times X_\alpha, X \times A_\alpha)$ is a G-pair with the diagonal action. Then $(\underline{X}_\alpha, \underline{A}_{\alpha|A})$ is a G-subspace pair of $X \times (X_\alpha, A_\alpha)$.

Now assume that α is numerable in the sense of A. DOLD. That is equivalent to the assumption that there is a partition of unity, $\{\lambda_u | u \in \alpha\}$. Since G is compact, we may average the functions λ_u to get a new partition of unity whose functions are constant on each orbit, and hence the associated barycentric map $(X,A) \longrightarrow (X_\alpha, A_\alpha)$ is a G-map. Let $\pi_1 : (\underline{X}_\alpha, \underline{A}_{\alpha|A}) \to (X,A)$ be the projection map restricted to the indicated subspaces.

Lemma. *$\pi_1 : (\underline{X}_\alpha, \underline{A}_\alpha) \to (X,A)$ is a G-homotopy equivalence.*

Proof. Define ϕ: $(X,A) \to (\underline{X}_\alpha, \underline{A}_\alpha)$ by $\underline{\phi}_\alpha(x) = (x, \phi_\alpha(x))$, $x \in X$, where ϕ_α is barycentric map defined preceeding this lemma. This map is well defined. In fact, it can be shown that ϕ_α is a G-homotopy inverse to π_1, whereby completing the proof.

Now we set up the exact couple which will induce the desired spectral sequence. We shall freely use the notation in HU [8, Chapter 8]. Let π_2: $(\underline{X}_\alpha, \underline{A}_{\alpha|A}) \to (X_\alpha, A_\alpha)$ be the projection map restricted to the indicated subspaces. Denote $\underline{X}^p_\alpha = \pi_2^{-1}(X^p_\alpha) \cup \underline{A}_{\alpha|A}$, where X^i_α is the i-th skeleton of the complex X_α. Let

$$E_{p,q} = h^G_{p+q}(\underline{X}^p_\alpha, \underline{X}^{p-1}_\alpha),$$

$$D_{p,q} = h^G_{p+q}(\underline{X}^p_\alpha, \underline{A}_\alpha),$$

and, let i: $h^G_{p+q}(\underline{X}^p_\alpha, \underline{A}_\alpha) \to h^G_{p+q}(\underline{X}^{p+1}_\alpha, \underline{A}_\alpha)$ and j: $h^G_{p+q}(\underline{X}^p_\alpha, \underline{A}_\alpha) \to h^G_{p+q}(\underline{X}^p_\alpha, \underline{X}^{p-1}_\alpha)$ be the homomorphisms induced by the obvious inclusion maps. Finally, let $k : \bar{h}^G_{p+q}(\underline{X}^p_\alpha, \underline{X}^{p-1}_\alpha) \to h^G_{p+q-1}(\underline{X}^{p-1}_\alpha, \underline{A}_\alpha)$ be the boundary homomorphisms. Putting these together, we have the exact couple:

$$\begin{array}{ccc} D & \xrightarrow{\;i\;} & D \\ & {}_{k}\nwarrow \quad \swarrow_{j} & \\ & E & \end{array}$$

Now it follows from computation that

$$E^2_{p,q} \approx H_p(X_\alpha, A_\alpha : h^G_q(\pi_2)),$$

where $h^G_q(\pi_2)$ denotes the homology local coefficient system defined by

$$h^G_q(\pi_2)(\sigma) = h^G_q(\operatorname{car}_\alpha(\sigma)), \quad \sigma \in X_\alpha;$$

and

$$E^\infty_{p,q} \Rightarrow h^G_{p,q}(\underline{X}_\alpha, \underline{A}_\alpha) \approx h^G_{p+q}(X,A).$$

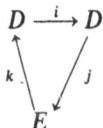

This gives us the spectral sequence associated with a closed G-covering of a G-pair (X,A). Now the spectral sequence claimed in the theorem is obtained simply by taking the covering consisting of the inverse sets $f^{-1}(\text{st. } v)$ of the closed stars of all vertices of X_α.

There is a cohomology analogue of this theorem. Let h_G^* be a cohomology theory defined on $\mathscr{C}_G$, satisfying the direct product axiom: If $X = \bigcup X_i$ is the disjoint union of open subsets belonging to $\mathscr{C}_G$, then the induced homomorphisms $h_G^p(X) \to h_G^p(X_i)$ represent $h_G(X)$ as the direct product of the $h_G^p(X_i)$. Retaining the notation set-up above, we have the cohomology local system $h_G^*(f)$ defined by $h_G^*(f)(\sigma) = h_G^*(f^{-1}\text{ st. }\sigma)$, for simplicies σ of Y, and we have

Theorem (Cohomology). *There is a cohomology spectral sequence $E(f) = \{E_Y^{pq}\}$ such that*

$$E_2^{pq} \approx H^p(Y, B; h_G^q(f)),$$

and if $\dim Y < \infty$, *then* $E(f) \Rightarrow h_G^*(X, A)$.

Remark. Both theorems can be generalized to a larger category.

References

1. ATIYAH, M.: Characters and cohomology of finite groups, Publ. Math., Inst. Hautes Etudes Scient. 1961.
2. —, and F. HIRZEBRUCH: Vector bundles and homogeneous spaces, Proc. of Symp. in Pure Math. **3**, Amer. Math. Soc., 1961.
3. —, and G. SEGAL: Equivariant *K*-theory, Lecture notes, Warwick, 1965.
4. BREDON, G.: Equivariant cohomology *K*-theories, Lecture Notes in Math. 32, Berlin-Heidelberg: Springer 1967.
5. BROWN, E. Jr.: Cohomology theories, Ann. of Math. **75**, 467—484 (1962).
6. CONNER, P., and E. FLOYD: Differentiable periodic maps, Berlin-Heidelberg: Springer 1964.
7. DYER, E.: Regular mappings and dimensions, Ann. of Math. **67**, 119—149 (1957).
8. HU, S. T.: Homotopy theory, Academic Press, 1958.

Cohomological Aspects of Transformation Groups

GLEN E. BREDON*

In this article I shall review the applications of cohomological methods to the study of periodic transformations and circle group actions and shall also obtain some new results in this direction. In order to keep the article within reasonable length we will restrict our attention to one general "method", that of BOREL, and to one type of problem, that of the implications of the global cohomology structure of a space for that of the fixed point set.

Let us fix our notation. p will denote a prime or zero, and is regarded as being fixed throughout. G denotes the cyclic group of order p for $p \neq 0$ and the circle group for $p=0$. K denotes the prime field of characteristic p and, except for explicit mention to the contrary, K is the coefficient domain for cohomology. X is a space on which G operates and F denotes the stationary point set of G on X. The cohomology theory used will be the sheaf theoretic cohomology [5].

Generally the theorems of the type we shall consider can be proved under either of the following two conditions on X:

(A) *X is paracompact and* $\dim_K X < \infty$.

(B) *X is compact.*

In (A) $\dim_K$ refers to the cohomological dimension as defined in [5; Chapter II, 15.7]. It is well-known that some assumption such as (A) or (B) is necessary to obtain any implications about F, since the examples of KLEE, and more recently of J. E. WEST, show that most any type of fixed point set F can occur when X is a HILBERT space.

We shall assume throughout that at least one of (A) or (B) is satisfied by X. Condition (A) is somewhat easier to deal with than (B) and it certainly suffices for most practical applications. However, certain results in case (B) for infinite dimensional X have proven to be of value in the theory of compact semigroups so that we shall deal here with both cases.

The type of results we shall deal with generally state that if X has a certain type of cohomology, then so does F. Let us use the notation

* The author is supported by an ALFRED P. SLOAN fellowship and by National Science Foundation grant GP 3990.

$X \sim Y$ to mean $H^*(X;K) \approx H^*(Y;K)$ *as a ring*. The simplest type of space is an acyclic space (i.e. $\sim$ point) and the theorem here is

$$X \sim \text{point} \Rightarrow F \sim \text{point}$$

and is due to P. A. SMITH (with somewhat stronger conditions on X). The next simplest space is a sphere and the celebrated Smith theorem states that

$$X \sim S^n \Rightarrow F \sim S^r \quad \text{for some} \quad r \leqslant n.$$

(Note that the coefficients used and the restrictions on the group acting are vital for these results. See [4] for counterexamples when these conditions are lifted.)

The next simplest space, as far as the cohomology *ring* structure is concerned is a space whose ring is generated by one element (and hence is a truncated polynomial ring; recall (A) and (B)). These include the real, complex and quaternionic projective spaces (P^n, CP^n and QP^n) and the CAYLEY projective plane as well as other more exotic spaces. The first theorem regarding these is also due to SMITH [8] who proved, in the case $p=2$, that

$$X = P^n \Rightarrow F \sim P^r + P^{n-r-1} \quad \text{or} \quad F = \phi$$

(+ denotes disjoint union). The method used was to lift the action to a $Z_2 \oplus Z_2$ action on S^n and study the fixed points of elements of this group. In [12], J. C. SU used the same method to obtain the result for more general $X \sim P^n$ $(p=2)$ and also for $p=0$ and $X \sim CP^n$. The general case in which $H^*(X)$ is a truncated polynomial ring was investigated by the author in [2]. The methods also apply to cohomology lens spaces (also treated by SU in [12]). The results on these cases are remarkably complete, with the theorems and examples in quite close proximity. This work will be reviewed below.

The next simplest case is that of the product of two spheres. It was investigated by SU in [11]. This case is quite rich in the sense that several different types of phenomena occur. We shall review some of SU's results and shall also prove some new theorems and obtain new examples which nearly close the gap between theorems and examples. However, there still remain some interesting unsolved problems in this case, and we shall comment on them below.

The case of involutions on a product of several n-spheres (n fixed) will also be studied. When F has trivial cohomology below dimension n the results are remarkably complete. The case $n=1$ was previously studied by CONNER [6] using covering space methods.

By studying actions on mapping cones it is possible to obtain some information about the non-existence of certain equivariant maps. Some important results in this direction will be discussed below.

1. Basic Facts

Let E_G^N be an N-universal bundle for G with base space $B_G^N = E_G^N/G$ (the classifying space). X_G^N denotes the twisted product

$$X_G^N = X \times_G E_G^N = (X \times E_G^N)/G$$

where G acts diagonally on $X \times E_G^N$. We may take $E_G^N = S^N$ when N is odd and S^{N+1} when N is even. Then we have the bundle maps $E_G^N \to E_G^{N+1}$ (inclusion) and consequently we have the canonical maps $B_G^N \to B_G^{N+1}$ and $X_G^N \to X_G^{N+1}$. These induce cohomology isomorphisms in degrees less than N and we *define* (coefficients in K):

$$H^*(B_G) = \varprojlim H^*(B_G^N),$$

$$H^*(X_G) = \varprojlim H^*(X_G^N).$$

This procedure is used for certain technical reasons and should be kept in mind. However, we shall generally write B_G or X_G when we mean B_G^N or X_G^N, N large. Recall that

$$p=2 \;\Rightarrow\; H^*(B_G) = K[t], \qquad \deg t = 1,$$
$$p=0 \;\Rightarrow\; H^*(B_G) = K[t], \qquad \deg t = 2,$$
$$p \neq 0,2 \;\Rightarrow\; H^*(B_G) = \Lambda(s) \otimes K[t], \; \deg s = 1, \; \deg t = 2, \; t = \beta(s),$$

where β is the mod p Bockstein homomorphism. This notation will be retained throughout.

There is the natural projection

$$\pi\colon X_G \to B_G$$

which is a bundle map with fiber X and group G (see [5; Chap. IV, § 9]). The study of the Leray spectral sequence of π will be of primary importance. This spectral sequence has

$$E_2^{p,q} = H^p(B_G; \mathscr{H}^q(X)) \;\Rightarrow\; H^{p+q}(X_G)$$

where $\mathscr{H}^q(X)$ is a locally constant sheaf (i.e. a bundle of coefficients) with stalk $H^q(X)$ and "group" G. [In case (A) the local constancy of this sheaf is a delicate matter which depends on the fact that B_G^N is locally contractible; see *loc. cit.*] When G operates trivially on $H^*(X)$ we have

$$E_2^{p,q} = H^p(B_G) \otimes H^q(X)$$

(valid since K is a field and B_G^N is compact). The cup product in the spectral sequence of π (see [5; Chapter IV, 6.5]) induces a module structure for the spectral sequence over $H^*(B_G) = E_2^{*,0}$ (when X is connected). On E_∞ this is compatible with $\pi^*\colon H^*(B_G) \to H^*(X_G)$ followed by the cup product in $H^*(X_G)$.

We need the following fact:

Lemma 1.1. *Let $\mathscr{A}$ be a locally constant sheaf of K-modules on B_G. Then multiplication by $t: H^r(B_G;\mathscr{A}) \to H^{r+d}(B_G;\mathscr{A})$ ($d=\deg t=1$ for $p=2$ and 2 otherwise) is an isomorphism for $r>0$ and an epimorphism for $r=0$. If $\mathscr{A}$ is constant then this is also an isomorphism for $r=0$.*

Proof. If $\mathscr{A}$ is constant then this follows from the universal coefficient formula $H^*(B_G;\mathscr{A}) \approx H^*(B_G)\otimes\mathscr{A}$. For $p=0$ any locally constant sheaf is constant since $\pi_1(B_G)=0$.

In general a locally constant sheaf on B_G can be regarded as a $G=\pi_1(B_G)$-module. Let T generate G and let $\tau=1-T$ as an endomorphism of $\mathscr{A}$. Consider the sequence

$$\mathscr{A}\supset\tau\mathscr{A}\supset\tau^2\mathscr{A}\supset\cdots\supset\tau^p\mathscr{A}=0$$

of subsheaves of $\mathscr{A}$. Each quotient $\tau^i\mathscr{A}/\tau^{i+1}\mathscr{A}$ is *constant*, since $\tau=1-T$ is trivial on it. Thus the lemma holds for these. The lemma then follows in general by an induction (i.e. if true for $\tau^{i+1}\mathscr{A}$ and $\tau^i\mathscr{A}/\tau^{i+1}\mathscr{A}$, then the 5-lemma implies that it is true for $\tau^i\mathscr{A}$).

It can be shown that for $p\neq 0$ the cohomology groups $H^*(B_G;\mathscr{A})$ are the cohomology groups of the cochain complex

$$0\longrightarrow\mathscr{A}\xrightarrow{\tau}\mathscr{A}\xrightarrow{\sigma}\mathscr{A}\xrightarrow{\tau}\mathscr{A}\xrightarrow{\sigma}\cdots$$

(with the first $\mathscr{A}$ in degree zero) where $\sigma=1+T+\cdots+T^{p-1}$, and the multiplication is also known since $H^*(B_G;\mathscr{A})$ is just the cohomology of the cyclic group G with coefficients in the G-module $\mathscr{A}$.

Proposition 1.2. *Suppose that $H^i(X)=0$ for $i>n$. Then the product by $t\in H^d(B_G)$:*

$$t\cup(\cdot): H^r(X_G)\to H^{r+d}(X_G)$$

is an isomorphism for $r>n$ and an epimorphism for $r=n$. If G acts trivially on $H^*(X)$, then this is an isomorphism for $r\geqslant n$ ($r\geqslant n-1$ for $p=0$).

Proof. $t\cup(\cdot)$ may be regarded as a spectral sequence endomorphism of bidegree $(d,0)$:

$$t\cup(\cdot): E_k^{r,q}\to E_k^{r+d,q}.$$

On $E_2^{r,q}=H^r(B_G;\mathscr{H}^q(X))$ it is an isomorphism (resp. epi) for $r>0$ (resp. $r=0$) by (1.1). The result follows from a straightforward spectral sequence argument. The refinement for $p=0$ results from the fact that $H^1(B_G)=0$ in that case.

The inclusion $F\subset X$ induces an inclusion $j: F_G\subset X_G$. Note that $F_G=F\times B_G$. A basic tool will be the exact cohomology sequence of the pair (X_G, F_G). Since X is paracompact it follows that X_G (i.e. each X_G^N) is paracompact. Thus

$$H^*(X_G, F_G) \approx H^*_\psi(X_G - F_G)$$

where ψ is the family of all closed sets of X_G which do not meet F_G. Also note $X_G - F_G = (X-F)_G$ maps into $(X-F)/G$ with acylic "fibers". Thus the Vietoris-Begle theorem [5; Chapter IV, 6.4] implies that

$$H^*_\psi(X_G - F_G) \approx H^*_\varphi((X-F)/G)$$

where φ is the family of all closed sets of X/G which do not meet $F/G = F$.

Thus the cohomology sequence of (X_G, F_G) becomes

$$\cdots \longrightarrow H^i(X_G) \xrightarrow{j^*} H^i(F_G) \xrightarrow{\delta^*} H^{i+1}_\varphi((X-F)/G) \longrightarrow \cdots$$

Note that all of these homomorphisms commute (up to sign for δ^*) with the $H^*(B_G)$-module structures. Also note that by the Künneth formula [5; Chapter IV, 8.3], which is valid since B_G^N is locally contractible and compact, we have

$$H^*(F_G) = H^*(B_G) \otimes H^*(F).$$

Now the $H^*(B_G)$-module structure of $H^*(X_G, F_G)$ induces one on $H^*_\varphi((X-F)/G)$. Following this through we see that this is $\pi^*\colon H^*(B_G) \to H^*((X-F)_G) \approx H^*((X-F)/G)$ followed by the cup product

$$H^*((X-F)/G) \otimes H^*_\varphi((X-F)/G) \to H^*_\varphi((X-F)/G).$$

The $H^*(B_G)$-module product will always be denoted by juxtaposition.

Now in case (A) we have that $\dim_K (X-F)/G$ is finite. This follows from the fact that the dimension of a locally paracompact space has a local character [5; Chapter II, 15.8]. For $p=0$ it also uses the existence of slices and a little extra work. We need something similar to this in case (B). The relevant fact is that $\varphi = c$ (compact supports) in case (B) and that for a locally compact space Y and elements $\alpha \in H^k(Y)$, $\beta \in H^n_c(Y)$ with $k>0$ we have $\alpha^i \beta = 0$ for sufficiently large i [5; Chapter II, ex. 53]. Thus in either case (A) or (B) we have that any given element of $H^*_\varphi((X-F)/G)$ is annihilated by a sufficiently high power of $t \in H^*(B_G)$. Using this we can prove the following basic fact:

Theorem 1.3. *If* $H^i(X) = 0$ *for* $i > n$, *then* $H^i(F) = 0$ *for* $i > n$.

Proof. Let $a \in H^i(F)$. For r large $t^r \delta^*(1 \otimes a) = 0$ so that $t^r \otimes a \in \operatorname{image} j^*$. However j^* preserves (or decreases) the filtration degree (of the spectral sequences of the maps into B_G). All elements of $H^*(X_G)$ have filtration $\leqslant n$ and $t^r \otimes a$ has filtration degree $i = \operatorname{degree} a$. Thus $i \leqslant n$.

It should be noted that already in (1.3) the use of coefficients in K is essential, at least for $p \neq 0$. For example, there are involutions on disks whose fixed point sets have nontrivial cohomology in positive degrees (necessarily odd torsion); see [4].

Now we can also prove, in either case (A) or (B):

Theorem 1.4. *If* $H^i(X)=0$ *for* $i>n$, *then* $H^i_\varphi((X-F)/G)=0$ *for* $i>n$ (*for* $i\geqslant n$ *if* $p=0$).

Proof. From (1.3), (1.2), the exact sequence and the 5-lemma, it follows easily that

$$t\cup(\cdot): H^i_\varphi((X-F)/G) \to H^{i+d}_\varphi((X-F)/G)$$

is an isomorphism for $i>n$ ($i\geqslant n$ if $p=0$). But as noted above every element is killed by a power of t so that the result follows.

We now have that if $H^i(X)=0$ for $i>n$ then

$$j^*: H^k(X_G) \to H^k(F_G)$$

is an *isomorphism* for $k>n$. (Having this for *sufficiently large* k is sufficient for our purposes.)

The equality $H^k(F_G)\approx\oplus(H^i(B_G)\otimes H^{k-i}(F))$ shows that $\dim H^k(F_G) = \sum_{i=0}^{k} \dim H^i(F)$. Thus if $H^i(X)=0$ for $i>n$, we have $\dim H^k(F_G) = \dim H^*(F)$ (i.e. $\Sigma \dim H^i(F)$) for $k>n$. But also $\dim H^k(F_G) = \dim H^k(X_G) \leqslant \dim H^*(X)$ by an elementary spectral sequence argument. Thus

$$\dim H^*(F)\leqslant\dim H^*(X). \tag{1.5}$$

A more refined argument using the fact that j^* preserves filtration shows that

$$\sum_{k\geqslant i}\dim H^k(F) \leqslant \sum_{k\geqslant i}\dim H^k(X) \tag{1.6}$$

and a somewhat more extended argument shows that $\dim H^i_\varphi((X-F)/G)$ can be added to the left hand side of this inequality. There are other refinements of this inequality; see [13] for some of these.

The equality of (1.5) occurs iff the spectral sequence of $\pi: X_G\to B_G$ degenerates and $\dim H^0(B_G;\mathscr{H}^q(X))=\dim H^q(X)$ for each q. The latter happens iff the coefficient sheaf is constant; that is, when G acts trivially on $H^q(X)$. Multiplicative properties of the spectral sequence show that these conditions are met iff the edge homomorphism $H^q(X_G)\to E_2^{0,q} \to H^q(X)$ is onto. This homomorphism is just restriction

$$i^*: H^q(X_G) \to H^q(X)$$

where $i: X\to X_G$ is inclusion as some fiber of π. When i^* is onto, X is said to be totally non-homologous to zero in X_G. Thus we have that the following three conditions are equivalent *when* $\dim H^*(X)<\infty$:

1. X is totally non-homologous to zero in X_G.

2. G acts trivially on $H^*(X)$ and the spectral sequence of $\pi: X_G \to B_G$ degenerates.

3. $\dim H^*(X) = \dim H^*(F)$.

In case these conditions hold, one has a very firm grasp of the situation, mainly because of the Leray-Hirsch theorem which, in this case, states the following.

Theorem 1.7. *Let $A \subset H^*(X_G)$ be a cohomology extension of the fiber (i.e. a graded submodule mapping isomorphically via i^* onto $H^*(X)$). Then the natural map*

$$H^*(B_G) \otimes A \to H^*(X_G)$$

is an isomorphism of $H^*(B_G)$-modules.

Since we are using a field as coefficients, the existence of A is equivalent to X being totally non-homologous to zero. The theorem, however, is valid in more general circumstances. A proof of this theorem for singular cohomology can be found in [9]. It is not hard to adapt it to our situation. In the case of interest to us ($\dim H^*(X) < \infty$ and coefficients in a field), an easy proof is available though a dimension count (see [2]).

Since $H^*_\varphi((X-F)/G) \to H^*(X_G)$ is an $H^*(B_G)$-module homomorphism, multiplication by high powers of t shows that it must be zero when $H^*(X_G)$ is *free* over $H^*(B_G)$. Thus

$$(1.8) \qquad 0 \longrightarrow H^i(X_G) \xrightarrow{j^*} H^i(F_G) \xrightarrow{\delta^*} H^{i+1}_\varphi((X-F)/G) \longrightarrow 0$$

is *exact* when X is totally non-homologous to zero in X_G.

If X is not totally non-homologous to zero but if G acts trivially on $H^*(X)$ the spectral sequence of $\pi: X_G \to B_G$ still provides a great deal of information. When G acts non-trivially on $H^*(X)$ the situation is much more difficult and very few results have been obtained in this case, most of them rather crude. (See, however, Section 9.) Nevertheless there is one general result which can be quite useful here. We shall formulate it for $p=2$ only. The reader can easily generalize it to p odd but there is no analogue for $p=0$. This is

Theorem 1.9. *Let $p=2$ and denote the involution by $T: X \to X$. There is a natural map*

$$Q: H^n(X; Z_2) \to H^{2n}(X_G; Z_2)$$

(not a homomorphism) such that

$$i^* Q(a) = a \cup T^* a \in H^{2n}(X)$$

and such that

$$j^* Q(a) = \Sigma_i t^{n-i} \otimes \mathrm{Sq}^i(a|F)$$

in $H^*(F_G) = H^*(B_G) \otimes H^*(F)$.

Proof. Consider the product $X \times X$ with the involution T taking (x,y) to (y,x). The map

$$h: X \to X \times X,$$
$$h(x) = (x, Tx)$$

is equivariant and defines an induced map $h_G: X_G \to (X \times X)_G$ such that

$$\begin{array}{ccccc}
X & \xrightarrow{i} & X_G & \xleftarrow{j} & F_G = B_G \times F \\
\downarrow{\scriptstyle h} & & \downarrow{\scriptstyle h_G} & & \downarrow{\scriptstyle 1 \times k} \\
X \times X & \xrightarrow{i} & (X \times X)_G & \xleftarrow{j} & \Delta_G = B_G \times \Delta
\end{array}$$

commutes. Here $\Delta \approx X$ is the diagonal and k is inclusion $F \subset X \approx \Delta$. Now in [10] a natural map

$$P: H^n(X) \to H^{2n}((X \times X)_G)$$

is defined such that $i^* P(a) = a \times a$ and $j^* P(a) = \Sigma\, t^{n-i} \otimes \mathrm{Sq}^i(a) \in H^*(\Delta_G) = H^*(B_G) \otimes H^*(\Delta)$. [This is defined for cell complexes, but extends by naturality to the Čech cohomology which coincides here with the sheaf cohomology.]

Put

$$Q(a) = h_G^* P(a) \in H^{2n}(X_G).$$

Then

$$i^* Q(a) = i^* h_G^* P(a) = h^* i^* P(a) = h^*(a \times a) = a \cup T^* a$$

since $h = 1 \times T$. Also

$$\begin{aligned} j^* Q(a) &= j^* h_G^* P(a) = (1 \otimes k^*) j^* P(a) = (1 \otimes k^*)\, \Sigma\, t^{n-i} \otimes \mathrm{Sq}^i(a) \\ &= \Sigma\, t^{n-i} \otimes k^*(\mathrm{Sq}^i(a)) = \Sigma\, t^{n-i} \otimes \mathrm{Sq}^i(k^* a) = \Sigma\, t^{n-i} \otimes \mathrm{Sq}^i(a|F) \end{aligned}$$

as claimed.

We remark that since $i^*: H^*(X_G) \to H^*(X)$ is given by the edge homomorphism of the spectral sequence it follows that the element of $E_2^{0,2n} = H^0(B_G, \mathscr{H}^{2n}(X))$ representing $a \cup T^* a$ is a permanent cocycle.

In case X is totally non-homologous to zero $T^* = 1$ and j^* is a monomorphism and we obtain from (1.9) the following

Corollary 1.10. *If $p = 2$ and X is totally non-homologous to zero then for any $a \in H^*(X)$ with $a^2 \neq 0$ we have $a|F \neq 0$ in $H^*(F)$.*

[The corresponding result for odd p requires $a^p \neq 0$.]

The following corollary of (1.9) is an existence theorem for fixed point sets of high dimension.

Corollary 1.11. *Let T be an involution on X $(p = 2)$. Suppose that $H^i(X; Z_2) = 0$ for $i > 2n$ and that T^* is the identity on $H^{2n}(X; Z_2)$ (eg. X*

a compact 2n-manifold). Suppose that $a \cup T^* a \neq 0$ *for some* $a \in H^n(X; Z_2)$. *Then* $a|F \in H^n(F)$ *is not zero.*

Proof. Since $Q(a) \in H^{2n}(X_G)$ represents $a \cup T^* a$ we have that

$$1 \otimes (a \cup T^* a) \in H^0(B_G) \otimes H^{2n}(X)$$

is a permanent cocycle. Consequently $t^k \otimes (a \cup T^* a)$ is also a permanent cocycle for all k and since it has highest fiber degree it cannot be killed in the spectral sequence. This implies that

$$0 \neq t^k Q(a) \in H^{2n+k}(X_G) \quad \text{all} \quad k \geqslant 0.$$

Since j^* is an isomorphism in high degrees we must have

$$j^* Q(a) \neq 0$$

and by (1.9) this implies that $a|F \neq 0$.

This result has the following corollary which is a strengthening of a result of CONNER and FLOYD (see [7; 27.4]).

Corollary 1.12. *Let T be an involution on* $X(p=2)$. *Assume that* $H^*(X; Z_2)$ *is finitely generated and vanishes in degrees larger than* $2n$. *Assume further that* $H^{2n}(X; Z_2) \approx Z_2$ *and that the cup product pairing* $H^n(X) \otimes H^n(X) \to H^{2n}(X) \approx Z_2$ *is non-singular. If* $\dim H^n(X; Z_2)$ *is odd then* $F \neq \phi$ *and* $H^n(X) \to H^n(F)$ *is non-zero.*

Proof. Consider the bilinear form $\langle a, b \rangle = a \cup T^* b$ on $H^n(X)$. This is symmetric and non-singular. Moreover, since $H^n(X)$ has odd dimension, there must exist a non-isotropic vector $a \in H^n(X)$; that is, $a \cup T^* a \neq 0$. By (1.11), $a|F \neq 0$ and this completes the proof.

We end this section with some further remarks and a notational convention. Note that if x is a *fixed point* then we have the inclusion $B_G \approx \{x\} \times_G E_G \to X \times_G E_G = X_G$ which is a cross-section for the fibration π. This induces the homomorphism

$$\eta_x: H^*(X_G) \to H^*(B_G)$$

(which depends on the component of F in which x lies) and gives the splitting

$$H^*(X_G) = \operatorname{Ker} \eta_x \otimes \operatorname{Im} \pi^*.$$

This is useful in some arguments since many elements of $H^*(X_G)$ of interest to us may be assumed to be in $\operatorname{Ker} \eta_x$ (with x given). This property ($\in \operatorname{Ker} \eta_x$) is preserved by cohomology operations and constitutes a type of normalization. Suppose, for example, that we have $\alpha \in H^n(X_G)$ and consider $j^*(\alpha)$. Let $F_0 \subset F$ be the component containing x and restrict $j^*(\alpha)$ to $(F_0)_G$ giving an element of $H^*(B_G)\otimes H^*(F_0)$. Let $A\otimes 1$ be the component of $j^*(\alpha)$ in $H^n(B_G)\otimes H^0(F_0)$. Then it is clear that $A=\eta_x(\alpha)$, so that $\alpha\in\operatorname{Ker}\eta_x$ iff the term of $j^*(\alpha)$ involving $H^0(F_0)$ vanishes.

2. Generalizations

Up to now we have considered only cohomology with coefficients in K, the prime field of characteristic p. It is well known that for $p\neq 0$ there can be no significant results for coefficients in the integers or in fields of characteristic different from p. However, the use of integer coefficients can sometimes be of value to prove results for mod p cohomology. The spectral sequence of $X_G\to B_G$ with integral coefficients is defined and reduction mod p maps it into the spectral sequence considered in section one. As is well-known

$$\begin{cases} H^i(B_G;Z)=0 & \text{for } i \text{ odd and} \\ H^i(B_G;Z) \xrightarrow{\approx} H^i(B_G;Z_p) & \text{for } i>0 \text{ even.}\end{cases}$$

This fact can sometimes be used to prove that certain differentials vanish in the mod p spectral sequence. For examples of the use of this method see [13].

In the case $(p=0)$ of circle actions, results on integral coefficients can often be obtained. Rather than using integral coefficients in the proofs, however, the following procedure is usually employed. First one must assume that there are only a finite number of orbit types; otherwise very little is known. This is known to always be the case when X is a compact manifold [1; Chapter IV]. One also must assume that X has finitely generated integral cohomology from which it can be deduced that F also has this property [1; p. 61]. Then one uses the information obtained from rational coefficients and that from mod p coefficients (applied to the large cyclic p-groups in the circle) together with the universal coefficient formula to deduce information about integral cohomology.

We can also generalize the remarks in Section one by considering more general support families on X. One may take any family θ of supports on X which is closed with respect to saturation by orbits (i.e. θ is the inverse image of a support family on X/G) and which satisfies either the analogue of (A) that θ be paracompactifying and $\dim_\theta X<\infty$ or that of (B) that X be locally compact and $\theta=c$.

However, both of these cases may be reduced to the one of closed supports by the trick of embedding X in a space with one ideal point. For the locally compact $(\theta=c)$ case this is done with the one-point compactification and with the other case it is done with a generalization of this known as a "one-point paracompactification" (the topology depends on θ in the same way that the topology of the one point compactification is defined from c). Thus, as regards supports, the case of closed supports suffices.

Notwithstanding this, it is sometimes convenient to use more general supports. This is particularly true when one considers the subspace $X-F$ and the supports θ consisting of those subsets of $X-F$ which are closed in X. Here we have the spectral sequence of $X_G-F_G \to B_G$ which takes the form

$$E_2^{p,q}=H^p(B_G;\mathscr{H}_\theta^q(X-F)) \Rightarrow H_\varphi^{p+q}((X-F)/G)$$

(with φ as in Section one). In case (A) or (B) is satisfied this spectral sequence converges to zero in sufficiently high degrees. There is also the natural map of this spectral sequence into that of $X_G \to B_G$. These facts can be important in the investigation of non-trivial differentials in the spectral sequence of $X_G \to B_G$ (when this spectral sequence is non-trivial).

3. Poincaré Duality

Assume that X satisfies Poincaré duality over K. This means that X has finitely generated cohomology over K and that there is an integer n such that $H^i(X;K)=0$ for $i>n$, $H^n(X;K)\approx K$ and the cup product pairing

$$H^i(X;K)\otimes H^{n-i}(X;K) \to H^n(X;K)\approx K$$

is non-singular.

It was conjectured by J. C. SU that if G acts on such an X, and if F_0 is a component of the fixed point set F, then F_0 also satisfies Poincaré duality.

If $m=\max\{i|H^i(F_0;K)\neq 0\}$ then dimensional parity is said to hold if $n-m$ is even.

Another conjecture is that dimensional parity holds when $p\neq 2$ (and X satisfies Poincaré duality).

These conjectures are well-known to be true when X is a compact orientable manifold (or orientable cohomology manifold over K) since the local Smith theorem then implies that F_0 is a compact orientable cohomology manifold over K and Poincaré duality is known to hold for such spaces (see [5; Chapter V]). Thus the interest in the conjectures arises from other cases.

The conjectures have been proved only in certain special cases and in the general case when X *is totally non-homologous to zero in* X_G. The latter result is given in [2]. The proof is given there only for the case $p\neq 0,2$ since the other cases are similar but easier.

Since this is one of the few results of this generality known we believe that the proof is of some interest. Thus, as an illustration of method, we shall prove the case $p=2$.

First let $A\subset H^*(X_G)$ be a cohomology extension of the fiber, so that by (1.7) a K-basis of A is an $H^*(B_G)$-basis of $H^*(X_G)$. Also let $x\in F_0$ be chosen once and for all and regard $H^*(F_0)$ as an internal direct summand of $H^*(F)$ (for notational purposes).

We note, for later use, the following fact. Let $\Sigma t^i \alpha_i$ be a homogeneous element of $H^*(X_G)$ with $\alpha_i\in A$ and with $0\neq a_0=i^*(\alpha_0)$; $\deg a_0<n$. Then we can find an element $a\in H^*(X)$ of *positive degree* such that $a a_0\neq 0$. By assumption, a is represented by an element $\alpha\in H^*(X_G)$ (that is $i^*(\alpha)=a$) and it may be assumed that $\eta_x(\alpha)=0$. [This is true since the projection $1-\pi^*\eta_x$ onto $\ker\eta_x$ does not alter the image under i^* of elements of positive degree, since $i^*\pi^*$ kills elements of positive degree.] Then $i^*(\alpha\Sigma t^i\alpha_i)=i^*(\alpha\alpha_0)=a a_0\neq 0$ so that

$$\alpha\Sigma t^i\alpha_i\neq 0.$$

We must show that $H^m(F_0)$ has dimension one and that for $0\neq b\in H^i(F_0)$, $0<i<m$, there is a $b'\in H^*(F_0)$ of *positive degree* such that $b b'\neq 0$ for then we can continue to multiply nontrivially until we reach degree m.

To this end let $0\neq b\in H^*(F_0)$ have positive degree. Since j^* is onto in high degrees we can find an integer k and elements $\alpha_i\in A$ such that

$$t^k\otimes b=j^*(\Sigma t^i\alpha_i). \tag{3.1}$$

Since powers of t can be cancelled from such equations we may assume that $0\neq a_0=i^*(\alpha_0)$.

If $\deg a_0<n$ in (3.1) then, as shown above, we can find some $\alpha\in\operatorname{Ker}\eta_x$ with $\alpha\Sigma t^i\alpha_i\neq 0$. Then put $j^*(\alpha)=\Sigma t^i\otimes b_i$, where $b_i|F_0$ is either zero or of positive degree (since $\eta_x(\alpha)=0$), and note that since j^* is a monomorphism we have

$$0\neq j^*(\alpha\Sigma t^i\alpha_i)=\Sigma(t^i\otimes b_i)(t^k\otimes b)=\Sigma t^{i+k}\otimes b_i b.$$

Thus $b_i b\neq 0$ for some i.

This conclusion is impossible when b has degree m. Thus $\deg b=m$ implies $\deg a_0=n$ in (3.1). It follows that $H^m(F_0)$ has dimension one, for if b, b' are both non-zero elements of $H^m(F_0)$, we could write

$$t^k\otimes b=j^*(\Sigma t^i\alpha_i);\quad t^k\otimes b'=j^*(\Sigma t^i\alpha_i'),\ \alpha_i,\alpha_i'\in A$$

(note $k=n-m$) and $\alpha_0=\alpha_0'$ since there is a *unique* element of A with degree n. Then

$$t^k\otimes(b-b')=j^*(t(\alpha_1-\alpha_1')+\cdots),$$

which is contrary to the general facts just obtained, unless $b-b'=0$.

Now fix b to be the non-zero element of $H^m(F_0)$ and let b' be some given element of $H^*(F_0)$ of degree less than m. Again b satisfies (3.1) with $\deg a_0=n$ and b' satisfies an analogous equation

$$t^r\otimes b'=j^*(\Sigma t^i\alpha_i')$$

where $\alpha_i'\in A$. If $\deg\alpha_0'<n$ then b' can be multiplied, non-trivially, by an element of positive degree, as shown above. Thus we may assume that $\deg\alpha_0'=n$ and thus that

$$\alpha_0=\alpha_0'.$$

Then

$$0\neq t^k\otimes b-t^r\otimes b'=j^*(\Sigma t^i\alpha_i'')$$

where $\alpha_i''=\alpha_i-\alpha_i'\in A$ has degree less than n. As shown above there is an $\alpha\in H^*(X_G)$ with $\eta_x(\alpha)=0$ and $\alpha\Sigma t^i\alpha_i''\neq 0$. Let

$$j^*(\alpha)=\Sigma t^i\otimes b_i$$

($b_i|F_0=0$ or of positive degree). Then

$$0\neq j^*(\alpha\Sigma t^i\alpha_i'')=\Sigma(t^i\otimes b_i)(t^k\otimes b-t^r\otimes b')=-\Sigma t^{i+r}\otimes b_ib'$$

since $b_ib=0$ by degree. Thus $b_ib'\neq 0$ for some i and this finishes the proof.

4. Projective Spaces

Let us use the notation $X\sim P^h(n)$ to signify that the cohomology ring of X is

$$H^*(X;K)\approx K[a]/a^{h+1};\quad \deg a=n$$

(a truncated polynomial ring on an element of degree n and height h). As always, p is understood as given here. Thus $RP^h\sim P^h(1)$, $CP^h\sim P^h(2)$, $QP^h\sim P^h(4)$ and the CAYLEY plane is a $P^2(8)$. It is well-known that for $p=2$, $h>1$ these are the only possibilities for the cohomology ring. If $p\neq 2$ there are examples of X with $X\sim P^2(n)$ for all even n (not manifolds) [2].

Note that acyclic spaces and spheres are contained in this class of spaces ($h=0$ and $h=1$ respectively). These spaces are the easiest to deal with because of the fact that, for G acting on X, X is always totally non-homologous to zero in X_G *unless* F *is empty*. To see this one first notes that G acts trivially on $H^*(X)$ since, for example, Z_p has no automorphisms of period p (as a Z_p-module). Next, if $X\sim P^h(n)$ with $a\in H^n(X)$ a generator then $1\otimes a\in H^0(B_G)\otimes H^n(X)=E_2^{0,n}$ is transgressive. Moreover, if $F \neq \phi$ then $\pi: X_G \to B_G$ has a cross-section so that the transgression must be zero. Thus all differentials vanish on $1 \otimes a$ and using the ring structure it follows that all differentials vanish everywhere in the spectral sequence of π.

The basic result for projective-like spaces is the following:

Theorem 4.1. [2]. *Suppose $X \sim P^h(n)$ and let F_0 be a (non-empty) component of F. Then $F_0 \sim P^k(m)$ for some $m \leqslant n$. If $p \neq 0$ then the number of components of F is at most p. The sum of the numbers $k+1$ over the components of equals $h+1$.*

We shall indicate the proof. The last statement merely expresses the equality

$$\Sigma \dim H^i(F) = \Sigma \dim H^i(X).$$

By the Smith theorems we may assume $h \geqslant 2$.

Also note that if $p \neq 2$ then $H^i(X)$ vanishes in odd degrees and it can be deduced that this is also true for $H^i(F)$; see [2] for details.

Let a generate $H^n(X)$, let $x \in F_0$ and let $\alpha \in H^n(X_G)$ represent a with $\eta_x(\alpha)=0$. Let $j_0: B_G \times F_0 = (F_0)_G \to X_G$ be the inclusion and write

$$j_0^*(\alpha) = 1 \otimes b_0 + t \otimes b_1 + t^2 \otimes b_2 + \cdots + t^k \otimes b_k \tag{4.2}$$

where $b_i \in H^*(F_0)$ and $b_k \neq 0$. Since $\eta_x(\alpha)=0$ and $x \in F_0$ it follows easily that the b_i have *positive* degree (or are zero). (Terms in s do not appear since all degrees are even when $p \neq 2$).

Now j_0^* is just j^* followed by projection of $H^*(F_0)$ onto its direct summand $H^*((F_0)_G)$ and hence it is *onto in high degrees.* In the first place this implies that $H^*(F_0)$ is generated, as a K-algebra, by $1, b_k, b_{k-1}, \ldots, b_0$ (in order of ascending degree). In the second place this implies that for sufficiently large m we can write

$$t^{k+m} \otimes b_k = j_0^*(A_0 + A_1 \alpha + A_2 \alpha^2 + \cdots + A_h \alpha^h) \tag{4.3}$$

since $1, \alpha, \alpha^2, \ldots, \alpha^h$ are an $H^*(B_G)$-basis of $H^*(X_G)$. Here $A_i \in H^*(B_G)$.

Now use (4.2) to expand the right hand side of (4.3). Comparing terms of degree zero *in* $H^*(F_0)$ we see that $A_0=0$. Then comparing terms of lowest degree ($=\deg b_k$) *in* $H^*(F_0)$ we see that $A_1 = t^m$.

From (4.2) we have

$$j_0^*(t^m \alpha) - t^{m+k} \otimes b_k = t^m \otimes b_0 + \cdots + t^{m+k-1} \otimes b_{k-1}$$

and from (4.3) we have

$$j_0^*(t^m \alpha) - t^{m+k} \otimes b_k = -j_0^*(A_2 \alpha^2 + \cdots + A_h \alpha^h).$$

Thus

$$t^m \otimes b_0 + \cdots + t^{m+k-1} \otimes b_{k-1} = -j_0^*(A_2 \alpha^2 + \cdots + A_h \alpha^h).$$

Consider the expansion of the right hand side of this equation using (4.2) and compare coefficients of t^i. This shows that $b_0, \ldots, b_{k-1}$ are all decomposable (in terms of $b_0, \ldots, b_k$). In particular the lowest degree term b_{k-1} must be a power of b_k and b_{k-2} is a polynomial in b_{k-1} and b_k and hence is a power of b_k, and so on.

It follows that 1 and b_k generate $H^*(F_0)$ as a K-algebra and hence that $F_0 \sim P^r(m)$ for some r and for $m = \deg b_k \leqslant n$ (by (4.2)).

The statement on the number of components is obtained by looking at the part of $j^*(\alpha)$ which has degree zero in $H^*(F)$. Suppose this is $t^n \otimes c$, $c \in H^0(F)$. Since j^* is onto in high degrees we see that $1, c, c^2, \ldots$ generate $H^0(F)$ and since $c^p = c$ it follows that $\dim H^0(F) \leqslant p$ which proves the contention.

We shall state a few more results from [2] which strengthen the basic result (4.1). For $p = 2$ the following gives complete information since examples exist for all possibilities. These are given in [2].

Theorem 4.4. [2]. *Let $p = 2$ and $X \sim P^h(n)$, $h \geqslant 2$. Then either F consists of two components F_1 and F_2 with $F_i \sim P^{h_i}(n)$ and $h_1 + h_2 = h - 1$ or F is connected and $F \sim P^h(n/2)$ or $F \sim P^h(n)$.*

One should note here that n must be 1, 2, 4 or 8. Also in the case $F \sim P^h(n)$ one can see that inclusion $F \subset X$ induces an isomorphism in cohomology. When X is a manifold this case arises only for the trivial action.

The proof is accomplished by comparing $j^*(\mathrm{Sq}^i \alpha)$ with $\mathrm{Sq}^i j^*(\alpha)$ where α is as in the proof of (4.1). The details may be found in [2].

The following improves the estimate on the number of components given in (4.1). Examples can be given to show that it is best possible on quaternionic projective spaces.

Theorem 4.5. [2]. *Let $p \neq 0,2$ and $X \sim P^h(4)$. Assume either that $h \geqslant p$ or that $X = QP^h$. Then F has at most $(p+1)/2$ components.*

The proof makes use of the Steenrod first power operation. There is an example of a $P^2(4)$, not QP^2, for $p = 3$ with F consisting of three points [2].

Similar methods apply to $p = 5$ and X the Cayley projective plane, to show that F has at most 2 components. There is such an example with $F = \mathrm{pt} + S^6$.

We regard the following result as one of the more interesting facts along these lines:

Theorem 4.6. [2]. *Let $p \neq 0,2$ and $X \sim P^h(4)$. Assume either that $h \geqslant p$ or that $X = QP^h$ and $h \geqslant p-2$. Then there exists at most one* *component F_0 of F with $F_0 \sim P^k(4)$, $k \geqslant 1$. The other components are all of type $P^j(2)$ or are acyclic.*

The proof is based on a close examination of the effect of the first two Steenrod power operations. The situation with $p=2$ is quite different, see (4.4). For $p=0$ the theorem can also be proved without hypothesis on h provided one assumes that $X = QP^h$.

It appears to us that this result should hold true, when $X = QP^h$, without hypothesis on h, but we don't know how to prove it for $h < p-2$.

5. Lens Spaces

Let p be odd in this section. We write $X \sim L^{2n+1}$ to signify that

$$H^*(X;K) \approx \Lambda(a) \otimes K[b]/b^{n+1}$$

where $\deg a = 1$ and $b = \beta_p(a)$ where β_p is the mod p Bockstein homomorphism. This is the cohomology ring of a lens space of dimension $2n+1$ and fundamental group Z_p.

J. C. SU showed in [12] that such cohomology rings are inherited by components of fixed point sets of Z_p-actions. It was remarked in [2] that the present cohomological methods provide an easy proof of this result but the proof was not given. Here we shall indicate this proof as an illustration of the methods. The proof will also apply to a related type of space: We write $X \sim L^{2n}$ to signify that

$$H^*(X;K) \approx (\Lambda(a) \otimes K[b])/(b^{n+1}, ab^n)$$

where $\deg a = 1$ and $b = \beta_p(a)$. An example of such a space is a lens space with a point removed.

Theorem 5.1. *Let $X \sim L^{2n+1}$ and let $F_1, \ldots, F_k$ be the components of F. Then $F_i \sim L^{2r_i+1}$ for some integers r_i and $\Sigma(r_i+1) = n+1$ (or F is empty). Moreover, the number k of components of F does not exceed p.*

Similarly:

Theorem 5.2. *Let $X \sim L^{2n}$ and let $F_1, \ldots, F_k$ be components of F. Then, after a reindexing of the F_i, we have $F_1 \sim L^{2r_1}$ and $F_i \sim L^{2r_i+1}$ for $i > 1$. Moreover, $k \leqslant p$ and $\Sigma(r_i+1) = n+1$. F cannot be empty in this case.*

Proof. The last statement of (5.2) follows from Floyd's euler characteristic formula $\chi(X) \equiv \chi(F) (\mathrm{mod}\, p)$. Both theorems will be proved together.

Let a generate $H^1(X)$ and let $b = \beta_p(a)$ as above. Assuming F to be non-empty we see that a transgresses to zero and, since the BOCKSTEIN commutes with transgression, b also transgresses to zero. Since a and b generate $H^*(X)$ it follows that X is totally non-homologous to zero in X_G. The equality $\Sigma(r_i+1)=n+1$ merely results from the equality $\dim H^*(F)=\dim H^*(X)$.

Let $\alpha\in H^1(X_G)$ represent a such that $\eta_x(\alpha)=0$ for some given $x\in F$ and let $\beta=\beta_p(\alpha)$. Then β represents b and $\eta_x(\beta)=0$. Let F_0 be the component of F containing x (which is arbitrary) and let $j_0:(F_0)_G\subset X_G$ be the inclusion. Then

$$j_0^*(\alpha)=1\otimes z\in H^0(B_G)\otimes H^1(F_0).$$

The term in $H^1(B_G)\otimes H^0(F_0)$ vanishes because of the normalization $\eta_x(\alpha)=0$. Then

$$j_0^*(\beta)=\beta_p(j_0^*(\alpha))=\beta_p(1\otimes z)=1\otimes y$$

and z and its Bockstein y generate $H^*(F_0)$ since j_0^* must be onto in high degrees.

Clearly a cohomology ring generated by an element $z\in H^1$ and its Bockstein $y\in H^2$ must be of the type L^i for some i.

We note that the number of components of F cannot exceed p by exactly the same proof as used in (4.1).

To finished the proof we must show that the number m of components of F of type L^i for i *even* is zero or one in cases (5.1) and (5.2) respectively. To do this note that L^i has euler characteristic zero or one according as i is odd or even. Also we have Floyd's formula

$$\chi(X)\equiv\chi(F)\ (\mathrm{mod}\,p).$$

The left hand side is zero in case (5.1) and one in case (5.2). The right hand side is just m. But $m\leqslant p$ since the number of components of F does not exceed p. Thus in case (5.2) the above congruence is the equality $m=1$ and in case (5.1) we have $m=0$ or $m=p$. If $m=p$ in case (5.1) then a dimension count shows that $\dim H^*(F)$ is odd contrary to the fact that $\dim H^*(X)$ is even. Thus $m=0$ in case (5.1) and $m=1$ in case (5.2) as was to be shown. (Note that in case (5.1) the fact $m=0$ also follows from the Poincaré duality or the dimensional parity theorems mentioned in section 3, but this method does not apply to case (5.2).)

6. Products of Spheres, I

In this section and the following two we consider the case in which $X\sim S^n\times S^m$. This would seem to be the case next in difficulty to that of projective spaces. It is also a case which is quite rich in the possible types of fixed point sets which can occur and thus it is eminently worthy of study. The first noteworthy work on this case was that of J. C. SU in [11] following some remarks of SWAN in [13].

This is the first case in which X need not be totally non-homologous to zero in X_G when $F \neq \phi$. This may happen in two ways: G may act non-trivially on $H^n(X)$ if $n=m$; or G may act trivially but the spectral sequence may be non-trivial. (Both cannot happen when $F \neq \phi$ as can be readily verified.) We take up these two cases in the next two sections.

In this section we assume X to be totally non-homologous to zero in X_G. Most results in this case follow directly from the Poincaré duality theorem of Section 3. It is clear from this theorem that the following cases are the only possibilities for F:

1. $F \sim S^q \times S^r$,
2. $H^*(F)$ generated by u, v in degree q with $u^2 \neq 0$, $v^2 \neq 0$ and $uv=0$ (e. g. $p=2$ and F a Klein bottle).
3. $F \sim P^3(q)$.
4. $F \sim \mathrm{pt} + P^2(q)$.
5. $F \sim S^q + S^r$ (q and/or r can be zero).

A direct proof of this was given by SU in [11]. In case 1 it is not hard to deduce from the spectral sequence that if $n \leqslant m$ then q and r can be chosen so that $q \leqslant n$, $r \leqslant m$ and, when $p \neq 2$, such that $n-q$ and $m-r$ are even. Examples in this case are obvious.

As was shown in [11] case 2 reduces to case 1 unless $p=2$. Let us indicate the proof of this. Assume that $p \neq 2$, $n \leqslant m$, $\dim H^q(F)=2$, and $\dim H^{2q}(F)=1$. If q is odd then all squares are zero and thus $F \sim S^q \times S^q$. Thus we may assume that q is even. Then since $\chi(X) \equiv \chi(F) = 4 \pmod p$ we see that n and m are even. Let $\alpha, \beta \in H^*(X_G)$ represent generators of $H^*(X)$ in degrees n and m with $\eta_x(\alpha)=0=\eta_x(\beta)$ for $x \in F$. Put $j^*(\alpha) = A \otimes w + B \otimes u$ and $j^*(\beta) = C \otimes w + D \otimes v$ where $u, v \in H^q(F)$, $w \in H^{2q}(F)$ and $A, B, C, D \in \bar{H}^*(B_G)$. Clearly $B \neq 0 \neq D$ and u, v span $H^q(F)$ since j^* is onto in high degrees. Since $0 \neq j^*(\alpha\beta) = BD \otimes uv$ we have $uv \neq 0$, and since α^2 depends on α, β over $H^*(B_G)$ and $j^*(\alpha^2) = B^2 \otimes u^2$ we must have $u^2=0$. Let $v^2 = d(uv)$, $d \in Z_p$. Then u and $v_1 = v-(d/2)u$ span $H^q(F)$ and $u^2 = 0 = v_1^2$ which shows that $F \sim S^q \times S^q$ as was to be shown.

We shall now construct examples of case 2 for $p=2$. Let $n=1,2,4,8$ and let η be the HOPF n-plane bundle over S^n. Let ε be a trivial line bundle on S^n and let ν be any k-plane bundle on S^n such that $\eta \oplus \nu$ is trivial. (This exists for all $k \geqslant n$.) Let T be the involution on $\eta \oplus \varepsilon \oplus \nu$ which is trivial on $\eta \oplus \varepsilon$ and reverses vectors in ν. The unit sphere bundle $S(\eta \oplus \varepsilon \oplus \nu) \approx S^n \times S^{n+k}$ is invariant under T and has $F = S(\eta \oplus \varepsilon)$

which is just the connected sum $P^2(n)\# -P^2(n)$ where $P^2(n)$ stands for the real, complex, quaternionic or CAYLEY projective plane according as $n=1,2,4$, or 8. [Note that $P^2(n)\# -P^2(n)$ falls in case 2 for $p=2$, where $q=n$, but that for *odd p* and $n\neq 1$ this space is $\sim S^n\times S^n$.]

An example of case 3 with $q=1$ is given by complex conjugation on $SU(3)\sim S^3\times S^5$. The fixed point set is $SO(3)=P^3$. Here X differs from $S^3\times S^5$, since Sq^2 is non-zero, and we asked in [2] whether or not such an example is possible on $S^3\times S^5$. In private correspondence, J. C. SU provided me with such an example which we shall now give here. Let $n\geqslant 2$ and let $m\geqslant n$. Consider the tangent bundle τ of S^n and let ε be a trivial $(m-n+1)$-plane bundle on S^n. Let T be an involution on S^n fixing S^2. Also extend T to τ by the differential and to ε by (-1) in the fibers. Then T acts on the WHITNEY sum $\tau\oplus\varepsilon$ which is a trivial $(m+1)$-plane bundle on S^n and the fixed set is the tangent space to S^2. T also acts on the unit tangent space $X\approx S^n\times S^m$ and has $F=P^3$, the unit tangent space to S^2.

In connection with case 3 we should mention that in [13] SWAN claims to prove that if $G=Z_p$ acts on $X\sim S^n\times S^m$ with n and m even, with G acting trivially on $H^*(X;Z)$, and with F connected, then $H^*(F)\approx H^*(S^q\times S^r)$ as a *group*. Moreover q and r are even and if $q\neq r$ then $F\sim S^q\times S^r$. His proof, however, does not eliminate the possibility of case 3 that $F\sim P^3(q)$, q even. We don't know whether or not this can occur in these circumstances. [Added in proof: J. C. SU gives an example of this elsewhere in these Proceedings.]

An example of 4 is the involution $A\to A^{-1}$ on $SU(3)\sim S^3\times S^5$ which has $F=\mathrm{pt.}+CP^2$. Also the canonical involution on the symmetric space $SU(3)/SO(3)\sim S^2\times S^3$ (since $p=2$) has $F=\mathrm{pt.}+P^2$. However, we do not know of any examples of 4 for actual products of spheres.

The remainder of this section will be spent on case 5. Note that if $q=r$ then $F=S^0\times S^q$ which is not of interest. Thus the basic question is whether q and r can differ. First we shall give some examples.

An example, given by SU [11], is the involution $A\to A^{-1}$ on $U(2)\approx S^1\times S^3$ which has $F\approx S^0+S^2$. Another example is the canonical involution on the symmetric space $\mathrm{Sp}(2)/U(2)\sim S^2\times S^4$ (but $\mathrm{Sq}^2\neq 0$) where it can be seen that $F=S^0+S^2$.

For $p\neq 2$ consider an action of G on CP^2 with fixed point set $S^2+\mathrm{pt}$. Let $X=CP^2\# -CP^2\sim S^2\times S^2$ (since $p\neq 2$). Taking the connected sum at a point of the S^2 component of the fixed point sets one obtains a G action on X with $F=S^2\# -S^2+\mathrm{pt}+\mathrm{pt}\approx S^2+S^0$.

We shall now construct our main class of examples of case 5 in the case, $p=2$, of involutions. Let T be an orthogonal involution on S^n with two fixed points x_0 and x_1. Suppose we are given a map $\varphi: S^n\to SO(m+1)$ such that $\varphi(Tx)=\varphi(x)^{-1}$ for all $x\in S^n$ and

$$\varphi(x_0)=\begin{pmatrix} I_{q+1} & 0 \\ 0 & -I_{m-q}\end{pmatrix},\quad \varphi(x_1)=\begin{pmatrix} I_{r+1} & 0 \\ 0 & -I_{m-r}\end{pmatrix}$$

(where $-1\leqslant r\leqslant q\leqslant m$). Then we define an involution T on $S^n\times S^m$ by $T(x,y)=(Tx,\varphi(x)\cdot y)$. Clearly $F=S^q+S^r$. (Note that when $r=-1$, $F=S^q$.) Now we shall show how to construct such maps φ. Let Φ be the well-known function defined by

$$\Phi(n)=\#\{k|0<k\leqslant n \quad \text{and} \quad k\equiv 0,1,2 \text{ or } 4 \ (\text{mod}\, 8)\}.$$

We claim that if $2^{\Phi(n)}|q-r$ then such a map φ exists. (We conjecture that this is necessary and sufficient.) By inclusion of orthogonal groups it clearly suffices to treat the case $r=-1$ and $q=m$. That is, we wish to find $\varphi: S^n\to SO(m+1)$ with $\varphi(Tx)=\varphi(x)^{-1}$ and $\varphi(x_0)=I$, $\varphi(x_1)=-I$, when $2^{\Phi(n)}|m+1$. It also suffices to treat the case $m+1=2^{\Phi(n)}$. For $n=1$ and $m+1=2$ define φ on a semicircle in S^1 from x_0 to x_1 to give any path from I to $-I$ in $SO(2)$ and then extend φ to the other semicircle by the equation $\varphi(Tx)=\varphi(x)^{-1}$. Suppose we have defined $\varphi: S^n\to SO(2^{\Phi(n)})$. If $n\neq 0,1,3,7 \pmod 8$ (where $\Phi(n)=\Phi(n+1)$) then $\pi_n(SO(2^{\Phi(n)}))=0$ and φ can be extended to one hemisphere of S^{n+1} arbitrarily and to the other by equivariance. In the other cases let $B: S^n\to S^n$ be the reflection through any plane containing x_0 and x_1. This reverses orientation and commutes with T. Define $\psi: S^n\to SO(2\cdot 2^{\Phi(n)})$ $=SO(2^{\Phi(n+1)})$ by

$$\psi(x)=\begin{pmatrix} \varphi(x) & 0 \\ 0 & \varphi(Bx)\end{pmatrix}.$$

Then $\psi(Tx)=\psi(x)^{-1}$, $\psi(x_0)=I$ and $\psi(x_1)=-I$. Moreover, ψ is clearly inessential and thus the above remarks show that ψ may be extended to S^{n+1}. The result follows by induction.

The following is a result in the other direction. The hypothesis $n<r$ is made for convenience only, and the other cases ($r\geqslant 0$) can be treated by the same method and with the same conclusions.

Theorem 6.1. *Let $p=2$ and let $X\sim S^n\times S^m$ with $n\leqslant m$. Suppose $F\sim S^r+S^q$ with $n<r<q$. Then $q\leqslant m$. Moreover, if $2^k<n$ or if $2^k=n$ and $\mathrm{Sq}^{2^k}: H^m(X)\to H^{m+n}(X)$ is trivial (e.g. if $X=S^n\times S^m$) then $2^{k+1}|(q-r)$.*

Proof. Let a and b be generators of $H^*(X)$ in degrees n and m respectively. Let x be a point in the "S^r-component" of F. Let $c\in H^0(F)$ be the generator for the S^q-component, and let d and e be generators of $H^*(F)$ in degrees r and q respectively. Thus $cd=0$ and $ce=e$. Let α and β in $H^*(X_G)$ represent a and b such that $\eta_x(\alpha)=0=\eta_x(\beta)$. If $q>m$ then

$j^*(\alpha)$ and $j^*(\beta)$ do not involve e. But then $j^*(\alpha\beta)$ also does not involve e and j^* could not be onto in high degrees. Thus $q \leqslant m$.

Now $j^*(\alpha)$ cannot involve d or e, by degree, and since $j^*(\alpha) \neq 0$ and $\alpha \in \ker \eta_x$ we must have

$$j^*(\alpha) = t^n \otimes c.$$

The coefficients of d and e in $j^*(\beta)$ must be non-zero for otherwise j^* will not be onto in high degrees. Thus

$$j^*(\beta) = At^m \otimes c + t^{m-r} \otimes d + t^{m-q} \otimes e$$

where $A = 0$ or 1.

By assumption we must have that $\mathrm{Sq}^i(\beta)$ is dependent upon α and β over $H^*(B_G)$ for all $i \leqslant 2^k$. But

$$\begin{aligned} j^*(\mathrm{Sq}^i\beta) &= \mathrm{Sq}^i j^*\beta \\ &= A\binom{m}{i} t^{m+i} \otimes c + \binom{m-r}{i} t^{m-r+i} \otimes d + \binom{m-q}{i} t^{m-q+i} \otimes e \end{aligned}$$

and clearly this implies that

$$\binom{m-r}{i} \equiv \binom{m-q}{i} \pmod 2 \quad \text{for} \quad 1 \leqslant i \leqslant 2^k.$$

Letting i run through powers of 2 shows that

$$m - r \equiv m - q \pmod{2^{k+1}},$$

that is $2^{k+1} | q - r$. (See [10] for the relevant facts concerning binomial coefficients mod 2.)

It seems likely that this theorem could be improved so that the conclusion $2^{k+1} | (q-r)$ is replaced by $2^{\Phi(n)} | (q-r)$ as in the examples. However, we don't know how to do this. [See the note added in proof at the end of the paper.]

7. Products of Spheres, II

In this section we examine the case of $X \sim S^n \times S^n$ in which G acts non-trivially on $H^n(X; K)$. Our attention will be concentrated on the case $X = S^n \times S^n$ and we shall not hesitate to make such an assumption if it serves to simplify the discussion.

For $X = S^n \times S^n$ only the cases $p = 2,3$ can occur for it is a well-known crystallographic fact that $Z \oplus Z$ admits automorphisms of period k only for $k = 1,2,3,4,6$. [For an easy proof let A be an integral

2×2 matrix with $A^k=I$. Let λ and μ be the eigenvalues of A. Then $\lambda\mu=\det A=\pm 1$, $\lambda+\mu=tr\ A$ is an integer and λ and μ are k-th roots of 1. It follows that $\lambda\mu=1$ (or $k=2$) and that $\lambda+\mu=-2,-1,0,1$ or 2. These cases yield, respectively, $k=2,3,4,6$ and 1 (since A, being of finite period, can be diagonalized over the complex numbers).]

The case $p=3$ and $X=S^n\times S^n$ is still further restricted as we now show. Let i_1, i_2 be the inclusions $S^n\times(*)\subset S^n\times S^n$ and $(*)\times S^n\subset S^n\times S^n$ and let π_1,π_2 be the projections. Let $T\colon S^n\times S^n\to S^n\times S^n$ be of period three and consider T_* on $H_n(S^n\times S^n;Z)$. Let $\alpha_1,\alpha_2\in H_n(S^n\times S^n)$ be the basis given by i_1 and i_2 respectively and suppose

$$\begin{cases} T_*(\alpha_1)=a\alpha_1+b\alpha_2,\\ T_*(\alpha_2)=c\alpha_1+d\alpha_2. \end{cases}$$

We have:

$$\begin{cases} ad-bc=\det T_*=1,\\ a+d=\operatorname{tr} T_*=-1. \end{cases} \tag{7.1}$$

Now consider the maps $\pi_i T\colon S^n\times S^n\to S^n$ for $i=1,2$. Computation shows that these have bidegrees (a,c) and (b,d) respectively. The Hopf construction yields maps $S^{2n+1}\to S^{n+1}$ with Hopf invariants ac and bd respectively. Thus for $n\neq 1,3,7$ ac and bd must be even (actually zero for n even). This is clearly inconsistant with (7.1).

For $n=1,3,7$ such examples do exist and are given by Su in [11]. One such construction is to regard $S^n\times S^n$ as the subset of $S^n\times S^n\times S^n$ consisting of those (x,y,z) with $xyz=1$ (where S^1,S^3, and S^7 have the usual multiplications). (Note that associativity in the Cayley numbers is valid for such triples.) Then merely permute the coordinates $(x,y,z)\to(y,z,x)$. F consists of those (x,x,x) with $x^3=1$. Thus $F=\text{pt}+S^{n-1}$.

If $G=Z_3$ acts on $X=S^n\times S^n$ $(n=1,3,7)$ and is non-trivial on $H^n(X;Z_3)$ then it can be shown that $F\sim\text{pt}+S^r$ for some *even* $r<n$. In fact $\dim H^*(F)\leqslant 3$ and $\chi(F)\equiv 0\ (\text{mod}\ 3)$ so that it need only be shown that F is non-empty and disconnected. The proof can be found in [11]. (Although Su only claims that $r\leqslant 2n$, the spectral sequence argument he uses in showing F is disconnected also shows that $r<n$.) We conjecture, in fact, that r must equal $n-1$.

We shall now turn to the case $p=2$. Here the involution T resembles, cohomologically, the interchange of factors of $S^n\times S^n$, in order that T^* be non-trivial on $H^n(S^n\times S^n;Z_2)$. Thus one would expect F to resemble the diagonal S^n. We shall show that this is indeed the case.

Theorem 7.2. *Let T be an involution on $X\sim S^n\times S^n$ such that $T^*\neq I$ on $H^n(X;Z_2)$. Then $F\sim S^n$ and the restriction $H^n(X)\to H^n(F)$ is onto.*

Moreover, if $X=S^n\times S^n$, $n\neq 1,3,7$, *and if* $\pi: X\to S^n$ *is the projection on either factor then* $(\pi|F)^*: H^n(S^n)\to H^n(F)$ *is an isomorphism.*

Proof. First it is clear that $F\sim S^r$ since $\chi(F)\equiv\chi(X)\equiv 0\ (\mathrm{mod}\ 2)$ and $\dim H^*(F)<4$. Let $x\in H^n(X)$ with $y=T^*x\neq x$. Then x,y generate $H^n(X)$ and necessarily $x^2=y^2=0$ (since coefficients are in Z_2). By (1.11) $x|F\neq 0$ so that $r=n$. If $X=S^n\times S^n$ then let $A=\begin{pmatrix} a & b\\ c & d\end{pmatrix}$ be the matrix of T^* on $H^n(S^n\times S^n; Z)$ with respect to the usual basis. By the HOPF invariant argument given above we see that ac and bd are even when $n\neq 1,3,7$. Also $a+d=Tr\,A=0,\pm 2$. If a is odd then d is odd and b,c must be even, in which case $A\equiv I\ (\mathrm{mod}\ 2)$, contrary to assumption. Thus a and d are even and $A\equiv\begin{pmatrix} 0 & 1\\ 1 & 0\end{pmatrix}\ (\mathrm{mod}\ 2)$. This shows that the elements x and y in the first part of the proof can be taken to be the usual basis elements and the last statement of the theorem follows.

The last statement is not valid in the cases $n=1,3,7$ as the involution $(z,w)\to(z^{-1},zw)$ shows. [Of course this is merely a result of the fact that $S^n\times S^n$ admits "more" homeomorphisms for $n=1,3,7$ than for other n and it is easily seen that one may change the representation of $S^n\times S^n$ as a product so that the last statement of the theorem is true in these cases also.]

8. Products of Spheres, III

In this section we consider the case in which G acts trivially on $H^*(X)$ but the spectral sequence of $\pi: X_G\to B_G$ is *non-trivial*. As before, $X\sim S^n\times S^m$ with $n\leqslant m$.

In the spectral sequence we have

$$E_2^{p,q}=H^p(B_G)\otimes H^q(X)$$

and it is clear that when $F\neq\phi$ there must be a non-zero differential from fiber degree m to fiber degree n. Let $a\in H^n(X)$ and $b\in H^m(X)$ be generators. Then $E_2=E_r$ for $r=m-n+1$ and we must have (when $F\neq\phi$)

$$d_r(1\otimes b)=A\otimes a \quad\text{where}\quad 0\neq A\in H^r(B_G).$$

We have

$$0=d_r((1\otimes b)^2)=(A\otimes a)(1\otimes b)+(-1)^m(1\otimes b)(A\otimes a)=(1+(-1)^m)(A\otimes ab)$$

which shows that for $p\neq 2$ and $F\neq\phi$ we must have that $m=\deg b$ is odd. Clearly n is even if $p=0$. If $p\neq 0,2$ and if X has the *integral* cohomology of $S^n \times S^m$ (it suffices that X has finitely generated integral cohomology) then the integral spectral sequence shows that n must be even. All these remarks are contained in [11].

It now follows easily that $F \sim S^q$ for some q. Moreover, for $p \neq 2$, $\chi(F) \equiv 0 \pmod p$ and it follows that q is odd.

Our best class of examples of this type of action has already been constructed in Section 6. There we constructed involutions on $S^n \times S^m$ with $F = S^q$ for any $q \leqslant m$ with $q+1$ divisible by $2^{\Phi(n)}$. In these examples note that F is contained in the factor S^m and hence is inessentially embedded in X for $q \neq m$.

It is worthwhile to also give some other types of examples. A further class of examples can be constructed as follows. Let $n = 1,2,4$ or 8 and let η be the HOPF n-plane bundle over S^n. Let $-\eta$ be the inverse of η (i. e. the n-plane bundle whose characteristic class in $\pi_{n-1}(O(n))$ is minus that of η). Let ε be a trivial $(m-2n+1)$-plane bundle $(m \geqslant 2n-1)$. If $p \neq 2$ let m be odd. Also let $p=2$ if $n=8$. Then there is an action of G by bundle maps on $-\eta \oplus \varepsilon$ leaving fixed only the base space S^n. Together with the trivial action on η, this gives an action on the trivial $(m+1)$-plane bundle $\eta \oplus -\eta \oplus \varepsilon$ whose fixed set is η. Passing to the unit sphere bundle we obtain an action on $S^n \times S^m$ $(m \geqslant 2n-1$, m odd if $p \neq 2$, $n = 1,2,4,8$ and $p=2$ if $n=8)$ with fixed set $F \approx S^{2n-1}$. In these examples the projection $F \to S^n \times S^m \to S^n$ is the HOPF map so that F is essentially embedded in X.

Other constructions can be made based on the HOPF maps such as the following example from [11]. Let $h\colon S^{2n-1} \to S^n$ be the HOPF map $(n = 2,4,8)$ and let X be the union of the mapping cylinders of the sphere bundle maps

$$S^n \times S^{m-n} \xleftarrow{h \times 1} S^{2n-1} \times S^{m-n} \xrightarrow{\text{proj.}} S^{2n-1}.$$

Then $X \sim S^n \times S^m$ (over the integers) when $n < m$ and X is a manifold. (However, it can be seen that $\mathrm{Sq}^n\colon H^m(X) \to H^{m+n}(X)$ is non-zero and hence X does not have the homotopy type of $S^n \times S^m$.) Let G act freely on S^{m-n} and act trivially on S^n and S^{2n-1}. (Thus if $p \neq 2$ assume that m is odd.) Then G acts on $X \sim S^n \times S^m$ with $F \approx S^{2n-1}$ (the base of the right hand fibration). By using certain actions on S^n and S^{2n-1} from [2], one can alter this example so that $F \approx S^{n-1}$ (with $p=2$ if $n=2$).

For $p \neq 2$ the latter construction can be carried out with any map $h\colon S^{2n-1} \to S^n$ of HOPF invariant *two*. Necessarily n is even and m is odd. The resulting space satisfies $X \sim S^n \times S^m$, but is not a manifold of course, and it can be arranged so that $F = S^{2r-1}$ for any $1 \leqslant r \leqslant n$. We shall see presently that the case $p=2$ is much more restrictive, as the examples given have indicated.

We shall now restrict our attention to the case $p=2$. It should be expected that a result similar to (6.1) also holds in this case and in fact we shall prove such a theorem below. The proof of (6.1) does not work in this case, however, and the present situation seems to be much more difficult, at least with the present methods. It is also necessary to make stronger assumptions on the space involved. The proof depends on the assumption that X is a compact manifold (or generally a compact separable metric cohomology manifold over Z_2). To expedite the proof we will, in fact, assume that the involution is differentiable on a differentiable manifold X.

Theorem 8.1. *Let X be a compact differentiable manifold with $X \sim S^n \times S^m$ where $n<m$ (and $p=2$). Suppose T is a differentiable involution on X with fixed point set $F \sim S^q$ for $q+1 \geqslant 2n$. Then $q \leqslant m$ and if $2^k<n$ we must have $2^{k+1} | (q+1)$.*

We will indicate the proof since it is the only non-trivial proof we know of in the case of a non-trivial spectral sequence of $X_G \to B_G$. Note that the conclusion of the theorem is weaker than that of (6.1) when $n=2^k$. This is probably just a deficiency of the proof. The same is probably also true of the assumption $q+1 \geqslant 2n$.

For convenience in the proof we shall *assume* that $q<m-1$. The other case can easily be dealt with by the same methods.

Let $a \in H^n(X)$ and $b \in H^m(X)$ be generators. In the spectral sequence of X_G we have

$$d_{m-n+1}(1 \otimes b) = t^{m-n+1} \otimes a$$

as shown above. Schematically this can be represented by Diagram 1.

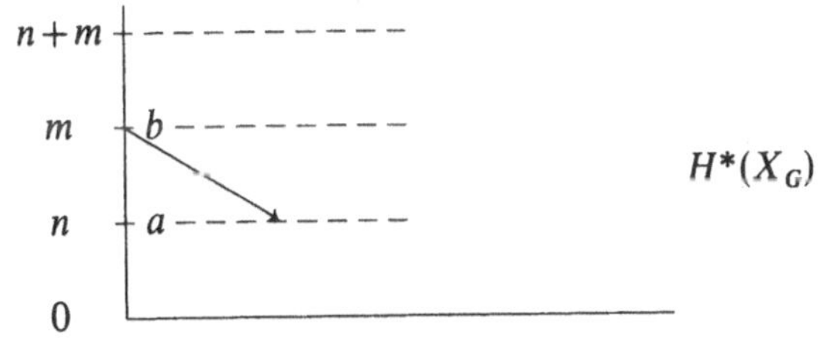

Diagram 1

Now consider $X \bmod F$, that is $X-F$ with compact supports. This has cohomology Z_2 in dimensions $n, m, n+m$ and $q+1$ by the exact cohomology sequence. Also the spectral sequence of $(X-F)_G \to B_G$

must kill all elements in high degrees. There are two possibilities for this to happen. In one case the spectral sequence looks like Diagram 2.

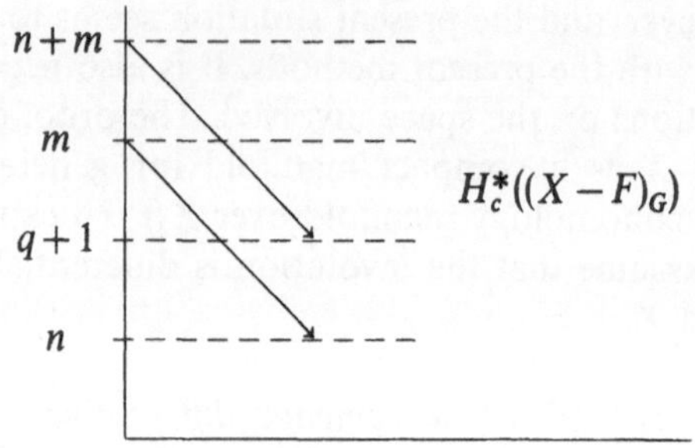

Diagram 2

For convenience we will also use a and b for the generators in degrees n and m respectively of $H_c^*(X-F)$. Then

$$d_k(1\otimes ab)=d_k(1\otimes a)\cdot(1\otimes b)+(1\otimes a)d_k(1\otimes b)=(1\otimes a)d_k(1\otimes b).$$

This is zero for $k\leqslant m-n+1$; hence $q+1\leqslant 2n$. Thus, by assumption, we have $q+1=2n$ and for $k=m-n+1$ we have

$$0\neq d_k(1\otimes ab)=(1\otimes a)(t^{m-n+1}\otimes a)=(t^{m-n+1}\otimes a^2).$$

It follows that $a^2\neq 0$ so that $n=1,2,4,8$ and $q+1=2,4,8,16$ respectively.

The other possibility for the spectral sequence of $(X-F)_G\rightarrow B_G$ with compact supports is pictured in Diagram 3.

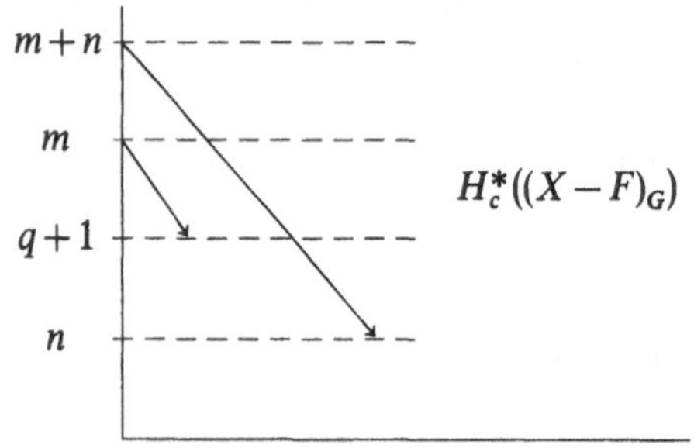

Diagram 3

Let α be the non-zero element of $H_c^n((X-F)_G)\approx Z_2$. Then from Diagram 3 we see that

$$(8.2)\qquad \begin{cases} t^m\alpha\neq 0,\\ t^{m+1}\alpha=0.\end{cases}$$

Consider $\delta^*: H^q(F_G) \to H_c^{q+1}((X-F)_G)$. Let y generate $H^q(F)$ and put

$$\gamma = \delta^*(1 \otimes y).$$

Note that $\mathrm{Sq}^i \gamma = 0$ for $i > 0$. Considering $j^*: H^*(X_G) \to H^*(F_G)$ we see that $t^i \otimes y \in \ker \delta^*$ iff $i \geqslant m+n-q$. Thus from Diagram 3 we see that

$$t^{m-q}\gamma = t^{m-n+1}\alpha.$$

Applying Sq^k to this equation and using (8.2) we see that

$$\binom{m-q}{k} \equiv \sum_{j=0}^{k} \binom{m-n+1}{j} a_{k-j} \pmod 2 \quad \text{for} \quad k<n \tag{8.3}$$

where the a_i are defined by

$$a_i = \begin{cases} 0 & \text{if} \quad \mathrm{Sq}^i \alpha = 0, \\ 1 & \text{if} \quad \mathrm{Sq}^i \alpha = t^i \alpha. \end{cases}$$

Now we consider $X-F$ with *closed* supports. By Poincaré duality (mod 2) we see that $H^i(X-F)$ is Z_2 for $i = 0, n, m+n-q-1, m$ and is zero otherwise. We consider the spectral sequence with closed supports of $(X-F)_G \to B_G$. All elements of high degree must be killed in this spectral sequence and, in fact, it is easy to see, using the cup product of the spectral sequences of $(X-F)_G \to B_G$ with closed and compact supports into that with compact supports, that this spectral sequence has the form of Diagram 4.

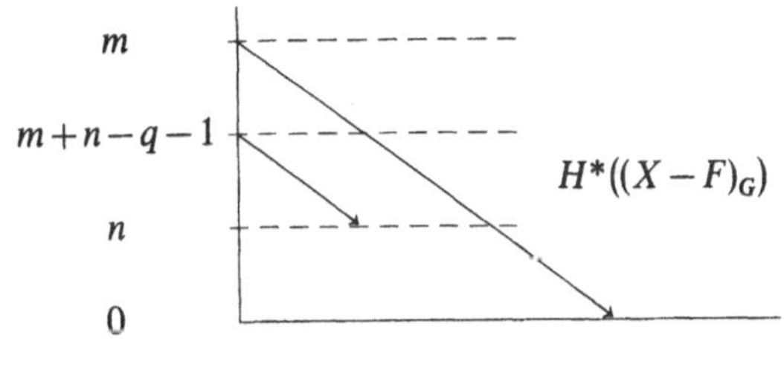

Diagram 4

Let α' be the image of α in $H^n(X_G)$. Then $\mathrm{Sq}^i \alpha'$ is zero or $t^i \alpha'$ according as $a_i = 0$ or 1. We consider the spectral sequence of $X_G \to B_G$ modulo $(X-F)_G$ (see [5; p. 140]). Since $H^i(X, X-F) = Z_2$ for $i = m+n$, $m+n-q$ and is zero otherwise, this spectral sequence shows that there is an element $x \in H^{m+n-q}(X_G, (X-F)_G)$ such that

$$\mathrm{Sq}^i x \quad \text{is either} \quad 0 \quad \text{or} \quad t^i x \quad \text{for} \quad i<q \tag{8.4}$$

(hence for $i<n$). Let h^* be the canonical homomorphism

$$h^*: H^*(X_G,(X-F)_G) \to H^*(X_G)$$

and consider $h^*(x)$. Then

$$h^*(x) = A\,t^{m-q}\alpha' + B\,t^{n+m-q} \qquad (A,B=0,1)$$

(since these are the only elements in $H^{m+n-q}(X_G)$). From Diagram 4 we see that $t^{m+1} \in \operatorname{Im} h^*$ which can only result if $B=1$. However $t^{n+m-q} \notin \operatorname{Im} h^*$ since t^{n+m-q} does not map to zero in $H^*((X-F)_G)$ (by Diagram 4 since $q>n$). Thus we must have

(8.5) $$h^*(x) = t^{m-q}\alpha' + t^{n+m-q}.$$

We also note, from Diagram 1, that $t^j\alpha'$ differs from 0 and t^{n+j} for $j \leqslant m-n$. That is $t^i(t^{m-q}\alpha')$ is independent of $t^i t^{n+m-q}$ for $i<n$ (since $q+1 \geqslant 2n$). Thus applying Sq^k $(k<n)$ to (8.5) and using (8.4) we see that we must have

(8.6) $$\binom{n+m-q}{k} \equiv \Sigma_{j=0}^{k} \binom{m-q}{j} a_{k-j} (\operatorname{mod} 2) \quad \text{for} \quad k<n.$$

Now (8.6) may be regarded as a recursion formula for the a_i (up to a_{n-1}). Thus, using the identity

$$\binom{a+b}{k} = \Sigma_0^k \binom{a}{j}\binom{b}{k-j}$$

we conclude that

(8.7) $$a_i = \binom{n}{i} \quad \text{for} \quad i<n.$$

Substituting this in (8.3) we obtain the congruence

$$\binom{m-q}{k} \equiv \Sigma_0^k \binom{m-n+1}{j}\binom{n}{k-j} = \binom{m+1}{k}$$

for $k<n$. By letting k run through powers of two, less than n, and using the fact that $\binom{a}{2^i}$ (mod 2) is the coefficient of 2^i in the dyadic expansion of a we conclude easily that $q+1 \equiv 0 (\operatorname{mod} 2^{i+1})$ if $k=2^i<n$. This completes the proof.

We conjecture that, for $X=S^n \times S^m$, $q+1$ must actually be divisible by $2^{\Phi(n)}$. It would be interesting if K-theory could be used to prove such a result. It would also be interesting of the condition $q+1 \geqslant 2n$ of (8.1) turned out to be necessary for $X=S^n \times S^m$.

9. Products of Several n-Spheres

In this section we study involutions on spaces $X \sim S^n \times S^n \times \cdots \times S^n$ (k factors). (Thus $p=2$ and coefficients are in Z_2.) If $H^i(F)=0$ for $0<i<n$ we shall obtain quite complete information. The case in which $X=S^1 \times \cdots \times S^1$ was studied by CONNER in [6] by the use of covering space methods. CONNER's method generalizes to the case in which X is a finite dimensional $K(\pi,1)$-space. Cohomological methods are not applicable to this generalization to $K(\pi,1)$-spaces but, on the other hand, the covering space method is not available when $n>1$.

Our first result shows that $\dim H^*(F)$ depends only on the action of T^* on $H^n(X)$ (where T is the given involution).

Theorem 9.1. *Let T be an involution on $X \sim S^n \times \cdots \times S^n$ (k-times). Let r be the dimension of the subspace of $H^n(X;Z_2)$ consisting of the elements fixed by T^*. Then if $F \neq \phi$ we have*

$$\dim H^*(F;Z_2)=2^r,$$

and the spectral sequence of $\pi: X_G \to B_G$ degenerates (i.e. all differentials are zero).

Before giving the proof we need some elementary algebraic lemmas.

Lemma 9.2. *Let V be a Z_2-vector space and let $T: V \to V$ be an automorphism of period two. Put $\tau=1-T=1+T$. Let $x_1, \ldots, x_p, z_1, \ldots, z_q$ be a basis of $\ker\tau$ such that $z_1, \ldots, z_q$ is a basis of $\operatorname{Im}\tau$. Write $z_i=y_i+Ty_i$ in any way. Then $x_1, \ldots, x_p, y_1, Ty_1, \ldots, y_q, Ty_q$ is a basis of V.*

The proof is accomplished by showing inductively that

$$x_1, \ldots, x_p, y_1, Ty_1, \ldots, y_r, Ty_r$$

are linearly independent and then noting that $\dim V = \dim \ker\tau + \dim \operatorname{Im}\tau$. The details are left to the reader, as are the proofs of the following two corollaries:

Corollary 9.3. *With the notation of* (9.2) *we can write V as a direct sum*

$$V=U \oplus TU \oplus W$$

where $W \subset \ker\tau$.

Corollary 9.4. *With the notation of* (9.2) *suppose we are given a set of vectors which span V, is invariant (as a set) under T, and contains vectors of $\ker\tau$ which map to a spanning set in $\ker\tau/\operatorname{Im}\tau$. Then from this set we may extract vectors $x_1, \ldots, x_p$, $y_1, \ldots, y_q$ such that $Tx_i=x_i$ and $x_1, \ldots, x_p$, $y_1, \ldots, y_q$, $Ty_1, \ldots, Ty_q$ from a basis of V.*

We shall now prove Theorem 9.1. By (9.3) we can write $H^j(X) = U_j \oplus T^* U_j \oplus W_j$ for any given j. Then $H^*(B_G; \mathscr{H}^j(X))$ is the cohomology of $G = Z_2$ with coefficients in the G-module $H^j(X)$. Thus in the spectral sequence of $X_G \to B_G$ we have

(9.5) $E_2^{i,j} = H^i(Z_2; (U_j \oplus T^* U_j) \oplus W_j) = H^i(Z_2; U_j \oplus T^* U_j) \oplus H^i(Z_2; W_j)$.

For $i > 0$ the term $H^i(Z_2; U_j \oplus T^* U_j)$ vanishes while the term $H^i(Z_2; W_j)$ is isomorphic to W_j. For $i = 0$ the first term in (9.5) is the diagonal $\{(u, T^* u) | u \in U_j\}$ and will have no importance in our discussion.

Note that operation by $t \in H^1(B_G) = H^1(Z_2; Z_2)$ annihilates the first summand of (9.5) and is an isomorphism $H^i(Z_2; W_j) \to H^{i+1}(Z_2; W_j)$ on the second for $i \geqslant 0$.

Now by (9.4) we can find elements

$$a_1, \ldots, a_\mu, b_1, \ldots, b_\nu$$

of $H^n(X)$ such that $T^* a_i = a_i$ and the a_i, b_i and $T^* b_i$ from a basis of $H^n(X)$. Note that $r = \mu + \nu$ in (9.1).

Let $x \in F$, assuming $F \neq \phi$. Then $1 \otimes a_i \in E_2^{0,n}$ is transgressive and must transgress to zero. Thus there are elements $\alpha_i \in H^n(X_G)$ representing a_i and with $\eta_x(\alpha_i) = 0$. We also put

$$\beta_i = Q(b_i) \in H^{2n}(X_G)$$

which represents $b_i \cup T^* b_i$. (Q is as in (1.9).) Note that both $1 \otimes a_i$ and $1 \otimes (b_i \cup T^* b_i)$ are permanent cocycles in the spectral sequence. Now for any integer j we see from (9.4) that the subspace W_j of (9.5) may, and shall, be assumed to have, as a basis, the elements which are cup products of several a_i with several $(b_i \cup T^* b_i)$. It follows that the summand $H^0(Z_2; W_j)$ of $E_2^{0,j}$ consists of permanent cocycles. The other summand of $E_2^{0,j}$ also consists of permanent cocycles by an easy inductive argument on j recalling that t annihilates this summand.

Thus we have shown that the spectral sequence degenerates. It follows that $\dim H^*(F) = \dim H^N(X_G) = \sum_j \dim E_2^{N-j,j}$ for N large, which is just $\sum_j \dim W_j$. For $j = \ell n$ the dimension of W_j is $\sum_i \binom{\nu}{i} \binom{\mu}{\ell - 2i}$ which is the coefficient of x^ℓ in $(1+x^2)^\nu (1+x)^\mu$. Summing over ℓ we obtain $\dim H^*(F) = 2^\nu 2^\mu = 2^r$ as claimed.

If we assume that $H^i(F; Z_2) = 0$ for $0 < i < n$ we obtain the following complete information about the cohomology structure over Z_2 of F:

Theorem 9.6. *With the hypotheses of* (9.1) *assume further that* $H^i(F, Z_2) = 0$ *for* $0 < i < n$. *Then* F *consists of* 2^ℓ *components each of which has the* mod 2 *cohomology ring of* $S^n \times \cdots \times S^n$ *(*$(r - \ell)$*-times) for some* $0 \leqslant \ell \leqslant \mu$. *Here* $\mu = \dim(\ker \tau / \operatorname{Im} \tau)$, *where* $\tau = 1 - T^*$ *on* $H^n(X; Z_2)$.

If n is odd, then $r-\ell$ is the dimension of the image in $H^n(X;Z_2)$ of the subgroup of $H^n(X;Z_4)$ consisting of the elements invariant under T^* on $H^n(X;Z_4)$.

Proof. We shall use the notation and facts established during the proof of (9.1). Let $W=\oplus W_j\subset H^*(X)$ and let $\mathscr{W}=\oplus \mathscr{W}_j\subset H^*(X_G)$ be the cohomology extension of W which has as a Z_2-basis the monomials in the α_i and $Q(b_i)$ (including $1\in H^0(X_G)$). It is clear that the map

$$H^*(B_G)\otimes \mathscr{W}\to H^*(X_G) \tag{9.7}$$

is a monomorphism and is an isomorphism in high degrees. Moreover, j^* must be a monomorphism on the image of (9.7) to $H^*(F_G)$. Thus the situation is similar to the case in which X is totally non-homologous to zero with the exception that now $\mathscr{W}$ is an extension of only part of $H^*(X)$ and that (9.7) is not onto in low degrees. Note that the homogeneous elements of $\mathscr{W}$ of positive degree (i. e. except for $1\in\mathscr{W}$) are all in $\ker\eta_x$.

Now suppose that $\alpha\in\mathscr{W}_n\subset H^n(X_G)$ (i. e. α is a linear combination of the α_i). Consider α^2 and note that since $\eta_x(\alpha)=0$ and $i^*(\alpha)^2=0$ there are only two possibilities: $\alpha^2=0$ or $\alpha^2=t^n\gamma$ for some $\gamma\in H^n(X_G)$. If $\alpha^2=t^n\gamma$ we must have that $\eta_x(\gamma)=0$ and, moreover, we can alter γ by an element annihilated by t and hence we can assume that $\gamma\in\mathscr{W}_n$. Now suppose, in the case $\alpha^2=t^n\gamma$, that

$$\begin{cases} j^*(\alpha)=1\otimes a+t^n\otimes c, \\ j^*(\gamma)=1\otimes d+t^n\otimes c' \end{cases}$$

where $a,d\in H^n(F)$ and $c,c'\in H^0(F)$. Then we have $j^*(\alpha^2)=t^{2n}\otimes c^2 =t^{2n}\otimes c$ (since $a^2=i^*(\alpha)^2|F=0$). Thus we have $d=0$ and $c=c'$. Then

$$\begin{cases} j^*(\alpha-\gamma)=1\otimes a, \\ j^*(\gamma)=t^n\otimes c. \end{cases}$$

Since $a^2=0$ we have $j^*(\mathrm{Sq}^n(\alpha-\gamma))=1\otimes a^2=0$ so that $\alpha-\gamma\in\ker\mathrm{Sq}^n$ and clearly $\gamma\notin\ker\mathrm{Sq}^n$. It follows that we may alter the given basis $\{a_i\}$ of W_n so that for some integer $\ell\leqslant\mu$,

$$j^*(\alpha_i)=\begin{cases} t^n\otimes c_i & \text{for} \quad 1\leqslant i\leqslant\ell, \\ 1\otimes a_i' & \text{for} \quad \ell<i\leqslant\mu \end{cases} \tag{9.8}$$

where $a_i'=a_i|F\in H^n(F)$ and $c_i\in H^0(F)$. Also note that

$$\alpha_i^2=\begin{cases} t^n\alpha_i & \text{for} \quad 1\leqslant i\leqslant\ell, \\ 0 & \text{for} \quad \ell<i\leqslant\mu. \end{cases} \tag{9.9}$$

Of course we also have,

$$(9.10) \qquad j^*(\beta_i) = j^*(Q(b_i)) = t^n \otimes b_i'$$

where $b_i' = b_i | F \in H^n(F)$.

It follows from (9.8), (9.10) and the fact that $\mathscr{W}$ generates $H^*(X_G)$ over $H^*(B_G)$, that $1, c_1, \ldots, c_\ell$ generate $H^0(F)$ as an *algebra*. (Presently we shall see that they are an algebra *basis*.) It also follows that for any component F_0 of F the $a_i | F_0 (i > \ell)$ and $b_i | F_0$ and 1 generate $H^*(F_0)$ as an algebra. Moreover, $a_i^2 = 0 = b_i^2$ so that all squares of elements of $H^*(F_0)$ of positive degree vanish.

Now let $c \in H^0(F)$ be the element which distinguishes F_0 (i.e. $c = 1$ on F_0 and $c = 0$ on other components). Note that F_0 is an arbitrary component and need not contain the base point x. We can write $c = P(c_1, \ldots, c_\ell)$ a square-free polynomial in the c_i. Define γ to be that element of $H^*(X_G)$ obtained by replacing c_i by α_i in P and multiplying each monomial by an appropriate power of t to bring the total degree up to ℓn. Then clearly

$$j^*(\gamma) = t^{\ell n} \otimes c .$$

Also let $\alpha = \alpha_{\ell+1} \cup \ldots \cup \alpha_\mu$ so that

$$j^*(\alpha) = 1 \otimes a' \quad \text{where} \quad a' = (a_{\ell+1} \cup \ldots \cup a_\mu) | F .$$

Finally put $\beta = \beta_1 \cup \ldots \cup \beta_\nu$ so that

$$j^*(\beta) = t^{n\nu} \otimes b' \quad \text{where} \quad b' = (b_1 \cup \ldots \cup b_\nu) | F .$$

Now it is clear that $\gamma \alpha \beta \neq 0$ since it is some power of t times an element of $H^*(X_G)$ which represents in $H^*(X)$ an element which is a monomial in the a_i and $(b_i \cup T^* b_i)$ with no repetitions. Moreover $\gamma \alpha \beta$ is in the image of $H^*(B_G) \otimes \mathscr{W} \to H^*(X_G)$ so that

$$j^*(\gamma \alpha \beta) \neq 0 .$$

But

$$j^*(\gamma \alpha \beta) = t^{n(\ell + \nu)} \otimes c a' b'$$

and $c a' b'$ is just $a' b' | F_0$, that is

$$0 \neq c a' b' = a_{\ell+1} \cup \ldots \cup a_\mu \cup b_1 \cup \ldots \cup b_\nu | F_0 .$$

Thus, for any component F_0 of F, $H^*(F_0)$ is the exterior algebra on the restrictions to F_0 of $a_{\ell+1}, \ldots, a_\mu, b_1, \ldots, b_\nu$ so that

$$F_0 \sim S^n \times \cdots \times S^n \qquad (r - \ell \text{ factors}).$$

Since $\dim H^*(F) = 2^r$, F must have 2^ℓ-components. (Note that it follows that 1 and the square-free monomials in the c_i are a basis for $H^0(F)$.)

It remains to prove the last statement of the theorem. To do this, one notes that the cohomology sequence associated with $0 \to Z_2 \to Z_4 \to Z_2 \to 0$ shows that $H^i(X, Z_4) = 0$ for $0 < i < n$. Thus the spectral sequences mod 2 and mod 4 give rise to Diagram 5 (n odd).

$$\begin{array}{ccccc}
H^n(B_G; Z_4) & \xrightarrow{0} & H^n(B_G; Z_2) & & \\
\pi^* \downarrow \uparrow \eta_x & & \pi^* \downarrow \uparrow \eta_x & & \\
H^n(X_G; Z_4) & \longrightarrow & H^n(X_G; Z_2) & \xrightarrow{Sq^1} & H^{n+1}(X_G; Z_2) \\
\downarrow & & \downarrow i^* & & \\
H^n(X; Z_4)^G & \xrightarrow{\rho} & H^n(X; Z_2)^G & &
\end{array}$$

Diagram 5

The columns are split exact and the second row is exact. A diagram chase shows that i^* maps $\ker \eta_x \cap \ker \mathrm{Sq}^1$ isomorphically onto $\mathrm{Im}\,\rho$. Now to find $\mathrm{Im}\,\rho$ we note first that $b_i + T^* b_i$ is in $\mathrm{Im}\,\rho$ since $H^n(X; Z_4) \to H^n(X; Z_2)$ is onto (Sq^1 being zero). Let $a = \Sigma \lambda_i a_i \in W_n$ and put $\alpha = \Sigma \lambda_i \alpha_i$. Then $a \in \mathrm{Im}\,\rho$ iff $\mathrm{Sq}^1(\alpha) = 0$ which holds iff $\mathrm{Sq}^1(j^*(\alpha)) = 0$. But $\mathrm{Sq}^1(j^*(\alpha)) = t^{n+1} \otimes \sum_{i=1}^{\ell} \lambda_i c_i$ since n is odd, and this is zero iff a is in the span of $a_{\ell+1}, \ldots, a_\mu$, since the c_i are independent. Thus $a_{\ell+1}, \ldots, a_\mu, b_1 + T^* b_1, \ldots, b_\nu + T^* b_\nu$ form a basis of $\mathrm{Im}\,\rho$ whence $\dim \mathrm{Im}\,\rho = \mu + \nu - \ell = r - \ell$ as claimed.

Corollary of the proof 9.11. *With the hypotheses of* (9.6), *there exists a basis* $a_1, \ldots, a_\mu, b_1, T^* b_1, \ldots, b_\nu, T^* b_\nu$ *of* $H^n(X)$ *with* $T^* a_i = a_i$ *and an integer* ℓ, $0 \leq \ell \leq \mu$, *such that for each component* F_0 *of* F *the restrictions of* $a_{\ell+1}, \ldots, a_\mu, b_1, b_2, \ldots, b_\nu$ *to* F_0 *from an exterior algebra basis for* $H^*(F_0)$ *and the restrictions of* $a_1, \ldots, a_\ell$ *to* F_0 *are zero.*

Remark 9.12. If T is an involution on $X \sim S^n \times \cdots \times S^n$ (k-times) with k even and if $\ker \tau$ on $H^n(X, Z_2)$ has dimension $k/2$ then it can be seen by the foregoing methods that $F \neq \phi$ and $H^i(F) = 0$ for $0 < i < n$. Thus $F \sim S^n \times \cdots \times S^n$ ($k/2$-times) in this case. Compare (7.2).

10. Involutions on Mapping Cones

Following the notation of [3] let $S^n(r)$ be the space S^n together with a linear involution with fixed point set of codimension r (i. e. a reflection in E^{n+1} through a subspace of dimension $n - r + 1$). We are interested in the existence or non-existence of equivariant maps $f\colon S^n(r) \to S^k(t)$. (We assume $n > r$ and $k > t$.) In the work reported on in [3] we obtain a considerable amount of information on this subject, but the proofs, when published, will depend on some facts in special cases which can be obtained via the present methods. We shall report on these special cases here. The method used will be to study the resulting involution on the mapping cone C_f. Clearly, for $n>k$, C_f has (mod 2) cohomology Z_2 in degrees k and $(n+1)$. The fixed point set F on C_f is the mapping cone of $f^G: S^{n-r} \to S^{k-t}$. Thus F has cohomology Z_2 in degrees $n-r+1$ and $k-t$ unless $n-r=k-t$ and f^G has odd degree.

The cohomological method yields results when Sq^{n-k+1} is non-zero on C_f. This situation was studied in [2] where the following theorem was proved:

Theorem 10.1. *Suppose T is an involution on X, that $H^i(X;Z_2) \approx Z_2$ for $i=0,m,m+k$ and is zero otherwise, and that $H^i(F;Z_2) \approx Z_2$ if $i=0,q,q+\ell$ and is zero otherwise. Assume that $\mathrm{Sq}^k: H^m(X) \to H^{m+k}(X)$ is non-zero and that $k \neq \ell$. If $\ell<k$ then $2\ell=k$ and $\mathrm{Sq}^\ell: H^q(F) \to H^{q+\ell}(F)$ is non-zero. If $\ell>k$ then $\ell \equiv k (\mathrm{mod}\, 2k)$ and Sq^ℓ is zero in F. Moreover $q+\ell \leqslant m$. The possibility that $\ell=0$ and $H^q(F) \approx Z_2 \oplus Z_2$ is also impossible.*

We shall not repeat the proof here, but shall formulate the applications of this result to equivariant maps.

Theorem 10.2. *Let $f: S^{m+1}(r) \to S^m(t)$ be equivariant and suppose that f represents the non-zero element of $\pi_1 = \lim_n \pi_{n+1}(S^n) \approx Z_2$. Then either $r-t \equiv 0 (\mathrm{mod}\, 4)$ or $r-t=1$. In the latter case $f^G: S^{m-t} \to S^{m-t}$ has odd degree.*

Proof. The mapping cone $X=C_f$ of f satisfies the hypotheses of (10.1) with $k=2$ and the hypotheses on F are satisfied with $\ell=|r-t-2|$ except in the case $r-t=1$ and $\deg f^G$ is odd. Thus there are two cases, $\ell=1$ and $\ell \equiv 2$ (mod 4). If $\ell=1$, then either $r-t=1$ and $\deg f^G$ is odd since Sq^1 must be non-zero on F, or $r-t=3$. If $r-t=3$ then $m+1-r<m-t$ and $f^G: S^{m-r+t} \to S^{m-t}$ is inessential contrary to Sq^1 being non-zero on F. The case $\ell \equiv 2$ (mod 4) is equivalent to $r-t \equiv 0(4)$ and this finishes the proof.

A similar proof applies to the cases $k=4$ and 8 and yields the following two theorems:

Theorem 10.3. *Let $f: S^{m+3}(r) \to S^m(t)$ be equivariant and suppose that f represents an odd multiple of the generator of $\pi_3 \approx Z_{24}$. Then either $r-t \equiv 0 (\mathrm{mod}\, 8)$ or $r-t=2$. In the latter case $f^G: S^{m-t+1} \to S^{m-t}$ is essential (stably).*

Theorem 10.4. *Let $f: S^{m+7}(r) \to S^m(t)$ be equivariant and suppose that f represents an odd multiple of the generator of $\pi_7 \approx Z_{240}$. Then* *either* $r-t\equiv 0 \pmod{16}$ *or* $r-t=4$. *In the latter case* $f^G: S^{m-t+3}\to S^{m-t}$ *represents an odd multiple of the generator of* $\pi_3 \approx Z_{24}$.

11. Problems and Conjectures

In this section we shall state a few open problems and conjectures pertaining to the type of results considered in this article.

Problem 11.1. Is there an involution on some $S^n\times S^m$ with $F=CP^3$ or with $F=\mathrm{pt}+CP^2$? Can this happen when n and m are even and the action on $H^*(S^n\times S^m;Z)$ is trivial? (See Section 6.)

Problem 11.2. What can be said about the fixed point set of an involution on the Stiefel manifold $V_{n,2}$ for $n>3$ and odd? (The cohomology ring, mod 2, of $V_{n,2}$ consists of a class in degree $n-2$, its Bockstein and their product.)

Problem 11.3. Suppose T is an involution on a compact globally symmetric space X and suppose that T has (at least) an isolated fixed point. How closely must F resemble the fixed point set of the symmetry on X?

Problem 11.4. If $X\sim S^n\times S^m$ and $F\sim S^r+S^q$ (possibly $r=-1$), what can be said about the possible values of $q-r$ in the cases $p\neq 2$? (See Sections 6 and 8.)

Conjecture 11.5. If Z_3 acts on $S^n\times S^n$, $n=3,7$, and acts non-trivially on $H^n(S^n\times S^n;Z_3)$ then $F\sim \mathrm{pt}+S^{n-1}$ (see Section 7).

Conjecture 11.6. If T is an involution on $S^n\times S^m$ with $n<m$ and $F\sim S^r+S^q$ with $r<q$ then $2^{\Phi(n)}|(q-r)$. This includes the case $r=-1$ (see Sections 6 and 8).

Conjecture 11.7. If $H^*(X;K)$ is generated as a ring by k elements then, for any component F_0 of F, $H^*(F_0;K)$ is generated by k elements at most.

Conjecture 11.8. (J. C. Su). If X satisfies Poincaré duality, then so does each component of F (see Section 3). It may be possible to prove this by cohomological methods if the spectral sequence of $X_G\to B_G$ is trivial (but the action on $H^*(X)$ is non-trivial) or, more generally, if the top class in $H^*(X)$ is non-homologous to zero in X_G. On the other hand the full conjecture may be provable by geometric methods in the case of simplicial actions on a Poincaré complex.

Note added in proof

Concerning Problem 11.1, J. C. SU has given an example with $F = CP^3$ elsewhere in these Proceedings. In a recent paper entitled "Representations at fixed points of smooth actions of compact groups", I have shown that $F = \text{pt} + CP^2$ cannot occur with the possible exception of $X = S^3 \times S^4$. Similarly, $F = \text{pt} + QP^2$ cannot occur except in six possible cases. $F = \text{pt} + \text{Cayley plane}$ cannot occur at all.

Concerning Conjecture 11.5, I have shown (loc. cit.) that, for $n = 7$, $F = \text{pt} + S^6$ or F is 3 points. Thus the conjecture has been reduced to the question of whether or not F can consist of exactly 3 points.

I have also proved conjecture 11.6 except for the case $r = -1$ (loc. cit.).

References

1. BOREL, A. et. al.: Seminar on Transformation Groups, Ann. of Math. Studies **46,** Princeton (1960).
2. BREDON, G. E.: The cohomology ring structure of a fixed point set, Ann. of Math. **80,** 524—537 (1964).
3. — Equivariant homotopy (these Proceedings).
4. — Exotic actions on spheres (these Proceedings).
5. — Sheaf Theory, McGraw-Hill, New York (1967).
6. CONNER, P. E.: Transformation groups on a $K(\pi, 1)$. II., Mich. Math. J., **6,** 413—417 (1959).
7. —, and E. E. FLOYD: Differentiable Periodic Maps, Academic Press, New York (1964).
8. SMITH, P. A.: New results and old problems in finite transformation groups, Bull. Amer. Math. Soc. **66,** 401—415 (1960).
9. SPANIER, E. H.: Algebraic Topology, McGraw-Hill, New York (1966).
10. STEENROD, N. E., and D. B. A. EPSTEIN: Cohomology Operations, Ann. of Math. Studies **50,** Princeton (1962).
11. SU, J. C.: Periodic transformations on the product of two spheres, Trans. Amer. Math. Soc. **112,** 369—380 (1964).
12. — Transformation groups on cohomology projective spaces, Trans. Amer. Math. Soc. **106,** 305—318 (1963).
13. SWAN, R. G.: A new method in fixed point theory, Comm. Math. Helv. **34,** 1—16 (1960).

Equivariant Homotopy

GLEN E. BREDON*

This report concerns the homotopy classification of equivariant maps between spheres with involutions. Some of the results we shall describe are also mentioned in the research announcement [3]. For the main part the proofs of the results stated here are much too complicated by details to be given here. A full treatment of the subject will eventually appear in another publication.

Let us begin by considering the simplest case, that of antipodal maps. The classical theorem of BORSUK and ULAM states that if $f: S^n \to S^m$ is equivariant with respect to the antipodal map (i.e. if $f(-x) = -f(x)$), then $n \leqslant m$. The best known and easiest proof of this fact is obtained by considering the induced map $\hat{f}: P^n \to P^m$ on the orbit spaces since it is easily deduced that $\hat{f}^*: H^*(P^m; Z_2) \to H^*(P^n; Z_2)$ is an epimorphism.

Another proof of this fact may be obtained by the use of the P. A. SMITH theory. From this point of view it is technically simpler to accomplish the proof if the involutions in question have fixed points. But this situation is easily achieved merely by suspending f to obtain Sf: $S^{n+1} \to S^{m+1}$ where both involutions now have two fixed points. Actually it is desirable to suspend twice so as to obtain connected fixed point sets (circles). Note that the induced map between fixed point sets is then the identity.

Before indicating the Smith theory proof let us introduce some convenient notation. We let $S^n(r)$ denote the space-with-involution whose underlying space is S^n and whose involution is given by the matrix

$$\begin{pmatrix} -I_r & 0 \\ 0 & I_{n-r+1} \end{pmatrix}.$$

Note that the fixed point set of $S^n(r)$ is S^{n-r} so that the argument "r" refers to the codimension of the fixed point set. Also note that this codimension is unchanged by suspension.

---

* The author is supported in part by an ALFRED P. SLOAN fellowship and by National Science Foundation grant GP 3990.

Some further convenient notation is the use of X^G to denote the stationary point set of a G-space X (for us, $G = Z_2$) and f^G for the induced map $X^G \to Y^G$ from an (equivariant) map $f: X \to Y$.

To return to the question at hand, let $X = S^{r+k}(r)$ and $Y = S^{t+k}(t)$ ($r = n+1$ and $t = m+1$ in the former notation). Then we are given a map (equivariant)

$$f: X \to Y$$

such that the induced map $f^G: S^k = X^G = Y^G = S^k$ is the identity. We may also assume $k \geqslant 1$. Then Smith theory provides the following commutative diagram

$$\begin{array}{ccccccccccc}
H^k(Y^G) & \xrightarrow{\approx} & {}_\sigma H^{k+1}(Y) & \xrightarrow{\approx} & {}_\sigma H^{k+2}(Y) & \xrightarrow{\approx} & \cdots & \xrightarrow{\approx} & {}_\sigma H^{k+t}(Y) & \xleftarrow{\approx} & H^{k+t}(Y) \\
\approx \big\downarrow f^{G*} & & \big\downarrow & & \big\downarrow & & & & \big\downarrow & & 0 \big\downarrow f^* \\
H^k(X^G) & \xrightarrow{\approx} & {}_\sigma H^{k+1}(X) & \xrightarrow{\approx} & {}_\sigma H^{k+2}(X) & \xrightarrow{\approx} & \cdots & \xrightarrow{\approx} & {}_\sigma H^{k+t}(X) & \longleftarrow & H^{k+t}(X)
\end{array}$$

with coefficients in Z_2. (For background on this see [4] and [5]. The dual proof in homology works equally well.) The isomorphisms on the bottom row follow from the assumption that $t < r$ as does the fact that $f^* = 0$. But the diagram is clearly self-contradictory. This proof gives more information than we have claimed since it clearly shows, more generally, that if

$$(1) \qquad f: S^{r+k}(r) \to S^{t+k}(t)$$

and if $r > t$ then $f^G: S^k \to S^k$ has *even degree* (i.e. $(f^G)^* = 0$ on $H^k(S^k; Z_2)$).

The question arises as to what extent this latter result is best possible. One can, in fact, achieve any even degree for f^G.

Although such maps can be written down quite explicitly, we shall give a suggestive general procedure for constructing them. First, we note that there is a map $g: S^{t+k}(t) \to S^{t+k}(t)$ which has degree -1 (forgetting the action) and such that g^G has degree $+1$. (If t is odd, then the involution itself is such a map g.) Second, note that $S^{t+k-1}(t) \subset S^{t+k}(t)$ (the latter is the suspension of the former) and that the quotient space is the one point union $S^{t+k}(t) \vee S^{t+k}(t)$. Consider the resulting composition

$$h: S^{t+k}(t) \longrightarrow S^{t+k}(t) \vee S^{t+k}(t) \xrightarrow{1 \vee g} S^{t+k}(t).$$

This map h has degree zero and h^G has degree 2. Now consider $S^{t+k}(t) \subset S^{t+k+1}(t+1)$. This divides S^{t+1+k} into two hemispheres *which are interchanged by the involution.* Thus h may be extended to one hemisphere (as a map) and then to the other hemisphere by equivariance. This defines a map

$$f: S^{t+1+k}(t+1) \to S^{t+k}(t)$$

and $f^G = h^G$ has degree two. Such maps of any even degree can now be constructed by using the group structure (see below).

We have found maps

$$f: S^{r+k}(r) \to S^{t+k}(t)$$

with f^G of any even degree *when* $r=t+1$. It is natural to ask whether this can be done when $r=t+2$. The construction outlined above will provide such maps with f^G of degree *four* (and hence of any degree divisible by four). It turns out however that this is the best that can be done. Unfortunately the proof of this fact, and of similar ones to follow, is much to difficult to give here and we will confine ourselves to a very sketchy discussion of it. A general result of this nature is given by the following theorem:

Theorem A. *Let* $f: S^{r+k}(r) \to S^{t+k}(t)$ *and put* $d=r-t$. *Then* $\deg f^G$ *is divisible by* $2^{\Phi(d-1)+1}$, *where* $\Phi(n)$ *is the number of integers* i *with* $0<i\leqslant n$ *and* $i\equiv 0,1,2,$ *or* 4 (mod 8). *This result is best possible when* $d\not\equiv 0 \pmod 4$. *For* $d=4$, 8, *or* 12 *we have that* $\deg f^G$ *is divisible by* $2^{\Phi(d-1)+2}$ $(=16, 32,$ *and* 256 *respectively), which is also best possible.*

We conjecture that for $d\equiv 0 \pmod 4$, $\deg f^G$ is divisible by $2^{\Phi(d-1)+2}$ (which would be best possible). [Added in proof: P. LANDWEBER has found a proof of this conjecture using operations in equivariant K-theory.]

A small part of Theorem A was also stated, in a different form, in our research announcement [2], which contains more detailed information in the cases $d\leqslant 8$.

So far we have dealt only with the case in which the fixed point sets have the same dimension. Let us now consider equivariant maps

$$f: S^{r+k+j}(r) \to S^{t+k}(t)$$

so that

$$f^G: S^{k+j} \to S^k.$$

We are able to obtain results here only for $j=1,2,3$ and also only for the stable class $\{f^G\}\in\pi_j=\lim\limits_{i}\pi_{j+i}(S^i)$ of f^G. As in Theorem A the method of Smith theory is completely inadequate for this situation. Our computations have yielded the following information:

Theorem B. *Let* $f: S^{r+k+j}(r) \to S^{t+k}(t)$ *and put* $d=r+j-t$.

If $j=1$ *and* $d\geqslant 4$ *then* $\{f^G\}=0\in\pi_1\approx Z_2$.
If $j=2$ *and* $d\geqslant 7$ *then* $\{f^G\}=0\in\pi_2\approx Z_2$.

$$\textit{If } j=3 \textit{ and } d\begin{Bmatrix} =8 \\ =9 \\ \geqslant 10\end{Bmatrix} \textit{ then } \{f^G\}\in\begin{Bmatrix} 2\pi_3 \\ 4\pi_3 \\ 8\pi_3\end{Bmatrix}\subset\pi_3\approx Z_{24}.$$

This result is best possible (stably) in the sense, for example, that if $j=1$ and $d<4$ then any value of $\{f^G\}\in\pi_1$ is possible.

There is an elementary general result along the lines of Theorem B which says that if $d \leqslant 2j$ then any element of π_j can be achieved by $\{f^G\}$. In fact, if $g: S^n \to S^m$ is any map then

$$g \wedge g: S^n \wedge S^n \to S^m \wedge S^m$$

is equivariant with respect to switching factors and hence is a map $f: S^{2n}(n) \to S^{2m}(m)$ with $f^G = g$. Here $j = n - m$ and $d = 2n - 2m = 2j$ and any smaller value of d is obtained by restriction of f to a subspace. As Theorem B shows, this result is not best possible. It does seem reasonable to us, nevertheless, that in some asymptotic sense this result is best possible.

Up to now we have discussed the question of which maps f^G can arise from a map $f: S^n(r) \to S^k(t)$. We shall now ask which maps $\bar{f}: S^n \to S^k$ can arise from a map $f: S^n(r) \to S^k(t)$ by forgetting equivariance. This question can be rather thoroughly answered and we shall list a few examples here:

Theorem C. *Suppose* $f: S^{n+j}(t+k) \to S^n(t)$ *so that the map* $\bar{f}: S^{n+j} \to S^n$ *on the underlying spaces defines the class* $\{\bar{f}\} \in \pi_j$. *Then*

$$\begin{array}{lll}
j=1, \ k \neq 1 & \text{and } k \not\equiv 0(4) & \Rightarrow \{\bar{f}\} = 0, \\
j=2, \ k \neq 2 & \text{and } k \not\equiv 0,1(4) & \Rightarrow \{\bar{f}\} = 0, \\
j=3 \left\{ \begin{array}{l} k \text{ even} \\ k \text{ odd} \left\{ \begin{array}{l} \\ \\ \end{array} \right. \end{array} \right. & \begin{array}{l} \neq 2 \text{ and } k \not\equiv 0(8) \\ \neq 3 \text{ and } k \not\equiv 1(4) \\ \text{otherwise} \end{array} & \begin{array}{l} \Rightarrow \{\bar{f}\} \in 2\pi_3, \\ \Rightarrow \{\bar{f}\} = 0, \\ \Rightarrow \{\bar{f}\} \in 12\pi_3, \end{array} \\
j=6, \ k \neq 4 & \text{and } k \not\equiv 0,2,3(8) & \Rightarrow \{\bar{f}\} = 0.
\end{array}$$

This result is best possible in the sense that, for example, if $j = 1$ and $k = 1$ or $k \equiv 0(4)$ then such maps f exist (when n and t are sufficiently large; see [3]) with $\{\bar{f}\} \neq 0$.

Similar results are known to us for $j \leqslant 9$ at least and can probably be deduced in the range $j \leqslant 13$ without great trouble.

The exceptional cases ($j = 1$, $k = 1$, $j = 2$, $k = 2$; $j = 3$, $k = 2,3$; $j = 6$, $k = 4$) arise because of the possibility of non-trivial f^G. If one required that f^G be inessential then the same theorem would hold with the absence of these exceptions.

Let us state as a corollary the most interesting special case of Theorem C, that in which $t = 0$ (i.e. no action on S^n) and $t + k = n + j + 1$ (i.e. the antipodal action on S^{n+j}). This can be reformulated as follows. Consider the diagram

$$\begin{array}{ccc}
S^{n+j} & \xrightarrow{\ f\ } & S^n \\
& \searrow & \nearrow \\
& P^{n+j} &
\end{array}$$

Corollary. *If f factors as shown above, then*

$$\begin{array}{lll} j=1 \quad \text{and} \quad n\not\equiv 2(4) & \Rightarrow & \{f\}=0, \\ j=2 \quad \text{and} \quad n\not\equiv 1,2(4) & \Rightarrow & \{f\}=0, \\ j=3\begin{cases} n \text{ even} \quad \not\equiv 4(8) \\ n \text{ odd} \begin{cases} \not\equiv 1(4) \\ \equiv 1(4) \end{cases} \end{cases} & \begin{array}{l} \Rightarrow \\ \Rightarrow \\ \Rightarrow \end{array} & \begin{array}{l} \{f\}\in 2\pi_3, \\ \{f\}=0, \\ \{f\}\in 12\pi_3, \end{array} \\ j=6 \quad \text{and} \quad n\not\equiv 1,3,4(8) & \Rightarrow & \{f\}=0. \end{array}$$

Actually this is a "stable" result and in the stable category it is best possible. As stated, however, some of the possibilities for f consistent with the theorem would not actually exist. We don't know the best possible results in the non-stable case.

Again similar results are known for $j\leqslant 13$. The case $j=1$ of this corollary was originally proved by J. H. C. WHITEHEAD in 1941 and was rediscovered by CONNER and FLOYD in 1962. The cases $j=2,3$ (at least) and some similar results have been recently obtained independently by E. REES, a student of D. B. A. EPSTEIN, by the use of K-theory. He also has some results in the non-stable case.

We shall outline very briefly some of the methods by which these results, and many others along these lines were obtained. Since the proofs of the results involve a considerable amount of detailed "calculations" the full account of these matters will have to await publication elsewhere.

The object of interest to us is

$$[S^n(r); S^k(t)] \tag{2}$$

the equivariant homotopy classes of maps $S^n(r)\to S^k(t)$. For technical reasons we consider only base point preserving maps and homotopies, where the base points are fixed under the involutions.

There is the fixed point set morphism

$$\varphi: [S^n(r); S^k(t)] \to [S^{n-r}; S^{k-t}]$$

and the forgetful morphism

$$\psi: [S^n(r); S^k(t)] \to [S^n; S^k].$$

Of course, Theorems A and C are concerned with the images of these morphisms.

This object (2) can be generalized in two ways by replacing the image or the domain by a general G-space ($G=Z_2$). We shall briefly discuss both of these generalizations.

In the case of equivariant homotopy one considers the set

$$\pi_{n,r}(X) = [S^n(r); X], \tag{3}$$

where X is some given G-space with base point in X^G. These sets were first considered by LEVINE in [8], with somewhat different notation. $\pi_{n,r}(X)$ is a group if $n-r\geqslant 1$ and is abelian when $n-r\geqslant 2$ since $S^n(r)$ is a suspension or double suspension (as a G-space) in these cases.

Let us describe some important homomorphisms associated with these groups. First, as above, there is the fixed point set functor

$$\varphi:[S^n(r);X]\to[S^{n-r};X^G];$$

that is,

$$\varphi:\pi_{n,r}(X)\to\pi_{n-r}(X^G).$$

There is also the forgetful functor

$$\psi:[S^n(r);X]\to[S^n;X]$$

(forgetting the actions); that is,

$$\psi:\pi_{n,r}(X)\to\pi_n(X).$$

Restriction to $S^{n-1}(r-1)\subset S^n(r)$ yields a homomorphism

$$\beta:\pi_{n,r}(X)\to\pi_{n-1,r-1}(X).$$

Also there is a homomorphism

$$\alpha:\pi_n(X)\to\pi_{n,r}(X)$$

defined by assigning to a map $f:S^n\to X$ the map

$$S^n(r)\to\frac{S^n(r)}{S^{n-1}(r-1)}=S^n\vee S^n\xrightarrow{f\vee f}X\vee X\xrightarrow{1\vee T}X.$$

Here $S^n\vee S^n$ has the involution interchanging factors and T denotes the involution on X.

It turns out to be quite important to consider also the groups obtained by requiring the fixed point set $S^{n-r}=S^{n-r}(0)$ of $S^n(r)$ to be sent to the base point of X by all maps and homotopies. We denote these groups by $\pi^*_{n,r}(X)$. Thus

$$\pi^*_{n,r}(X)=[S^n(r)/S^{n-r}(0);X].$$

One has the natural homomorphism

$$i:\pi^*_{n,r}(X)\to\pi_{n,r}(X).$$

As in the non-equivariant case the quotient $S^n(r)/S^{n-r}(0)$ is (equivariantly) homotopically equivalent to the mapping cone of the inclusion $S^{n-r}(0)\to S^n(r)$. Hence there is a canonical map $S^n(r)/S^{n-r}(0)\to S\,S^{n-r}(0)=S^{n-r+1}(0)$. This induces the homomorphism

$$\Delta:\pi_{n-r+1}(X^G)\to\pi^*_{n,r}(X)$$

since the equivariant maps $S^{n-r+1}(0)\to X$ are just the maps $S^{n-r+1}\to X^G$.

All the homomorphisms we have described, in fact, come from Puppe sequences associated with mapping sequences in the diagram

$$\begin{array}{ccccc}
S^0(0) & \to & S^{r-1}(r-1) & \to & S^{r-1}(r-1)/S^0(0) \\
\downarrow & & \downarrow & & \downarrow \\
S^0(0) & \to & S^r(r) & \longrightarrow & S^r(r)/S^0(0) \\
\downarrow & & \downarrow & & \downarrow \\
* & \longrightarrow & S^r \vee S^r & \longrightarrow & S^r \vee S^r
\end{array}$$

This also contains analogues of α, β and ψ for the $\pi^*_{n,r}$ groups.

It is not hard to see then that this diagram induces the following commutative "braid" diagram (Fig.) in which all four "sine curves" are exact sequences:

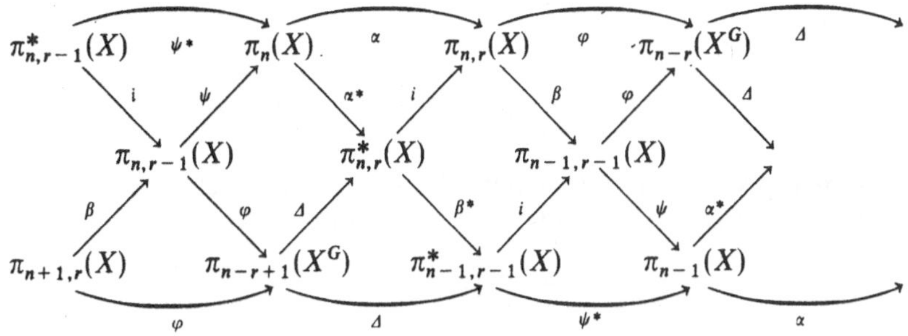

Fig.

The sequence consisting of α_*^* $\beta^*$and $\psi^*$was previously considered by Levine [8]. Note that this sequence is an exact couple and hence induces a spectral sequence (there are some sticky technical difficulties here). It is this spectral sequence, or rather one closely associated with it, that we use in our calculations.

Let us now discuss the second method of approach which was, in fact, my original point of view when first attacking these questions. In the second approach we consider the set $[X, S^m(t)]$, or rather we stabilize this situation and consider the stable equivariant cohomotopy groups

$$\{X; S^m(t)\} = \lim_{\to k} [S^k X; S^{m+k}(t)].$$

These groups form a (generalized) equivariant cohomology theory, the theory associated with the G-spectrum $\mathbf{S}(t) = \{S^n(t)\}$. That is

(4) $$\{X; S^m(t)\} = \tilde{H}^m_G(X; \mathbf{S}(t)).$$

We are interested in calculating these groups. According to [2] there is a spectral sequence

$$E_2^{p,q} = \tilde{H}^p_G(X; s^q(t)) \Rightarrow \tilde{H}^{p+q}_G(X; \mathbf{S}(t)).$$

Here $s^q(t)$ refers to the "coefficients" of the cohomology theory on the right. The E_2-term refers to the "equivariant classical comohology theory" defined in [2]. It is closely akin to the Steenrod equivariant cohomology theory but is somewhat more general. It is, however, easy to calculate this E_2-term. Moreover for $X = S^n(r)$ we can find a large number of the differentials in this spectral sequence and hence derive much information about the groups of interest to us. In general, however, the investigation of the differentials is a difficult problem. For example, the non-triviality of certain of the differentials turns out to be equivalent to the vector field problem on spheres.

We shall now turn our attention to the description of a remarkable general result concerning equivariant homotopy classes of maps of spheres. Let us use the notation

$$\begin{aligned}\pi_n(r;t) &= \lim_{\to k}\,[S^{n+k}(r); S^k(t)]\\ &= \lim_{\to k}\pi_{n+k,r}(S^k(t)) = \tilde{H}_G^{r-n}(S^r(r); \boldsymbol{S}(t)).\end{aligned}$$

Similarly we let $\pi_n^*(r,t)$ denote the analogous group where we require fixed point sets to be sent to the base point. That is

$$\begin{aligned}\pi_n^*(r;t) &= \lim_{\to k}\,[S^{n+k}(r)/S^{n+k-r}(0); S^k(t)]\\ &= \lim_{\to k}\,\pi_{n+k,r}^*(S^k(t))\\ &= \tilde{H}_G^{r-n}(S^r(r)/S^0(0); \boldsymbol{S}(t)).\end{aligned}$$

When we originally computed some of the groups $\pi_n(r,t)$ it was discovered that there was a periodicity of these groups in t but with some exceptional cases. It became clear that the exceptions resulted because of possibly non-trivial maps on the fixed point sets. This even affected the kernel of $\varphi\colon \pi_n(r;t) \to \pi_{n-r+t}$, the fixed point homomorphism, which is even closer to being actually periodic in t. This was the reason for introducing the groups $\pi_n^*(r;t)$ which exhibit an exact periodicity.

The periodicity result is that

$$\pi_n^*(r;t) \approx \pi_n^*(r; t + 2^{\Phi(r-1)}) \tag{5}$$

where Φ is the well-known function

$$\Phi(p) = \#\{k \mid 0 < k \leqslant p \quad \text{and} \quad k \equiv 0,1,2 \text{ or } 4(8)\}.$$

Let us outline the proof of this fact. First it is necessary to generalize these groups and consider

$$\begin{aligned}\pi_n(r,q;t) &= \lim_{\to k}[S^{n+k}(r)/S^{n+k-r+q}(q); S^k(t)]\\ &= \tilde{H}_G^{r-n}(S^r(r)/S^q(q); \boldsymbol{S}(t)).\end{aligned}$$

(Note that $\pi_n^*(r;t)=\pi_n(r,0;t)$.) Then one shows that the suspension with action (i.e. the involution also interchanges the vertices of the suspension) induces an isomorphism

$$\Sigma\colon \pi_n(r,q;t) \xrightarrow{\approx} \pi_n(r+1,q+1;t+1).$$

Now one constructs, by an inductive procedure, a map

$$\lambda\colon S^{p-1} \to O(k); \qquad k=2^{\Phi(p-1)}$$

such that $\lambda(-x)=-\lambda(x)$. [Proceeding one step further in the inductive construction requires that λ be inessential, forgetting equivariance. When $\pi_{p-1}(O(k))\neq 0$ this can be achieved (via a certain trick) by replacing $O(k)$ by $O(2k)$ and, according to the known stable homotopy groups of the orthogonal groups, this produces the function Φ.] By projection on the equator, λ can be extended to an equivariant map

$$\lambda\colon S^p(p) - S^0(0) \to O(k)$$

the involution on $O(k)$ being $A \to -A$.

Now suppose that X and Y are G-spaces with base points and let

$$f\colon (S^p(p) \wedge X, S^0(0) \wedge X) \to (Y,*)$$

be any equivariant map. Define

$$f^\lambda\colon (S^p(p) \wedge X \wedge S^k(0), S^0(0) \wedge X \wedge S^k(0)) \to (S^k(k) \wedge Y, *)$$

where $k=2^{\Phi(p-1)}$ by $f^\lambda(a \wedge x \wedge b)=(\lambda(a)\cdot b) \wedge f(a \wedge x)$.

We specialize to the case $Y=(S^p(p) \wedge X)/(S^0(0) \wedge X)$ with X compact and $f\colon S^p(p) \wedge X \to Y$ the canonical projection. Then it is easily seen that f^λ induces an *equivariant homeomorphism*

$$\frac{S^p(p) \wedge X \wedge S^k(0)}{S^0(0) \wedge X \wedge S^k(0)} \xrightarrow{\approx} \frac{S^k(k) \wedge S^p(p) \wedge X}{S^k(k) \wedge S^0(0) \wedge X}.$$

Now taking $X=S^r(r)$ we obtain

$$\frac{S^{p+k+r}(p+r)}{S^{k+r}(r)} \xrightarrow{\approx} \frac{S^{p+k+r}(p+k+r)}{S^{k+r}(k+r)}.$$

By composition this induces an isomorphism

$$\pi_n(p+k+r,k+r;t+k) \xrightarrow{\approx} \pi_n(p+r,r;t+k).$$

Preceding this by the isomorphism Σ^k yields

$$\pi_n(p+r,r;t) \approx \pi_n(p+r,r;t+2^{\Phi(p-1)}) \tag{6}$$

which clearly generalizes (5).

This periodicity leads to the following remarkable isomorphism:

Theorem D. *If* $t>n+r+1$ *then*

$$\pi_n^*(r;t) \approx \pi_{n-r+t}(V_{t,r})$$

where $V_{t,r}$ *is the Stiefel manifold of r-frames in euclidean t-space.*

We shall briefly indicate the proof. Let us denote $P_{t,r}=P^{t-1}/P^{t-r-1}$. Then, according to JAMES [7], we have

$$\pi_{n-r+t}(V_{t,r}) \approx \pi_{n-r+t}(P_{t,r})$$

for $t>n+r$ and the latter group is *stable* when $t>n+r+1$. According to ATIYAH [1] $P_{t,r}$ is S-dual to $SP_{r-t,r}$ (a stable "object") so that

$$\pi_{n-r+t}(V_{t,r}) \approx \{S^0; P_{t,r}\}_{n-r+t} \approx \{SP_{r-t,r}; S^0\}_{n-r+t}.$$

The latter means

$$\{P^{aj+r-t-1}/P^{aj-t-1}; S^{aj+r-t-1-n}\}$$

where $j=2^{\phi(r-1)}$, a is a certain integer whose value is immaterial, and $aj>t$. This group is just $\pi_n(aj+r-t, aj-t; 0)$ and by periodicity (6) (a times), this is $\pi_n(aj+r-t, aj-t; aj)$. By the isomorphism Σ^{t-aj}, this is the same as $\pi_n(r,0;t)=\pi_n^*(r;t)$, which proves Theorem D.

For most purposes it is sufficient to consider the groups resulting from a further stabilization. This is obtained via the suspension with action which yields a homomorphism

$$\Sigma: \pi_n(r;t) \to \pi_n(r+1;t+1)$$

which turns out to be an epimorphism if $r \geqslant n+1$ and an isomorphism for $r \geqslant n+2$. We define

$$\pi_{n,k} = \lim_{\to t} \pi_n(t+k,t) \tag{7}$$

and similarly for the "starred" groups.

The $\pi_{n,k}^*$ are periodic in k with period $2^{\Phi(n+1)}$ and moreover Theorem D implies that

$$\pi_{n,-k}^* \approx \pi_{k,r}^n \quad \text{for} \quad n<k-1 \quad \text{and} \quad n<r-1. \tag{8}$$

Here $\pi_{k,r}^n$ is the established shorthand for $\pi_{k+n}(V_{k+r,r})$.

Formula (8) together with the periodicity and HOO and MAHOWALD's calculations of $\pi_{k,r}^n$ in [6] suffice to compute $\pi_{n,-k}^*$ for $n \leqslant 13$. We shall give some further information which allows the calculation of $\pi_{n,k}$ in many cases.

Consider Fig. If we delete the arguments X and X^G, then this diagram is also valid for these doubly stable groups $\pi_{n,k}$ and yields a large amount of information.

Let us use the notation

$$\tilde{\pi}_{n,k} = \ker\{\varphi\colon \pi_{n,k} \to \pi_{n-k}\}$$

so that we have the exact sequence

$$0 \longrightarrow \tilde{\pi}_{n,k} \longrightarrow \pi_{n,k} \xrightarrow{\varphi} \pi_{n-k} \tag{9}$$

and the exact sequence

$$\pi_{n+1,k} \xrightarrow{\varphi} \pi_{n-k+1} \xrightarrow{\Delta} \pi^{*}_{n,k} \longrightarrow \tilde{\pi}_{n,k} \longrightarrow 0. \tag{10}$$

For $k \leqslant 0$ it can be seen that φ is onto and splits so that (9) yields

$$\pi_{n,k} \approx \tilde{\pi}_{n,k} \oplus \pi_{n-k} \quad \text{for} \quad k \leqslant 0. \tag{11}$$

From the argument given below Theorem B it can be seen that

$$\varphi\colon \pi_{n,k} \to \pi_{n-k} \quad \text{is onto for} \quad n \geqslant 2k \quad \text{or for} \quad k \leqslant 0. \tag{12}$$

From (10) it follows that

$$\pi^{*}_{n,k} \xrightarrow{\approx} \tilde{\pi}_{n,k} \quad \text{for} \quad k \geqslant n+2 \quad \text{or} \quad n \geqslant 2k-1. \tag{13}$$

This suffices to compute $\tilde{\pi}_{n,k}$ $(n \leqslant 13)$ except in the range

$$k-1 \leqslant n \leqslant 2k-2. \tag{14}$$

Although more detailed information is available we content ourselves here with listing these "exceptional" values of $\tilde{\pi}_{n,k}$ in the range (14) for $n \leqslant 6$. In fact the only non-zero groups in this range are

$$\tilde{\pi}_{3,4} \approx Z_{12},$$
$$\tilde{\pi}_{6,6} \approx Z_{2},$$
$$\tilde{\pi}_{6,7} \approx Z_{2}.$$

For $n \geqslant 2k$ or $k \leqslant 0$ $\varphi\colon \pi_{n,k} \to \pi_{n-k}$ is onto by (12) and in the other cases (for small n) the image of φ can be deduced from Theorems A and B. By (9) this reduces the computation of $\pi_{n,k}$ to an extension problem. In general, however, we do not know how to determine this extension. A case in which this extension does not split is given by

$$0 \longrightarrow \tilde{\pi}_{3,2} \longrightarrow \pi_{3,2} \xrightarrow{\varphi} \pi_1 \longrightarrow 0.$$

Here $\tilde{\pi}_{3,2} = \pi^{*}_{3,2} \approx Z_{12}$, but it can be shown that $\psi\colon \pi_{3,2} \to \pi_3$ is an isomorphism so that $\pi_{3,2} \approx Z_{24}$. For all other cases with $n \leqslant 6$ one can see that the extension (9) does split but this must be regarded as accidental.

References

1. ATIYAH, M. F.: Thom complexes, Proc. London Math. Soc. **11**, 291—310 (1961).
2. BREDON, G. E.: Equivariant Cohomology Theories, Lecture Notes in Math. Vol. 34, Berlin-Heidelberg-New York: Springer 1967.
3. — Equivariant stable stems, Bull. A.M.S. **73**, 269—273 (1967).
4. — Sheaf Theory, McGraw-Hill, New York (1967).
5. — Orientation in generalized manifolds and applications to transformation groups, Mich. Math. J., 35—64 (1960).
6. HOO, C. S., and M. E. MAHOWALD: Some homotopy groups of Stiefel manifolds, Bull. A.M.S. **71**, 661—667 (1965).
7. JAMES, I. M.: Spaces associated with Stiefel manifolds, Proc. London Math. Soc. **9**, 115—140 (1959).
8. LEVINE, J.: Spaces with involution and bundles over P^n. Amer. J. Math. **85**, 516—540 (1963).

Gaps in the Dimensions of Compact Transformation Groups

L. N. MANN*

Let (X, G) denote a G-space where G is a compact connected Lie group, X a connected n-dimensional manifold and the action of G on X is effective. A well-known result of MONTGOMERY and ZIPPIN [8] states that

$$\dim G \leqslant \frac{r(r+1)}{2} \leqslant \frac{n(n+1)}{2}$$

where r is the maximal dimension of the orbits of G on X. In the extreme case where $\dim G = n(n+1)/2$ it is known that G is locally isomorphic to $SO(n+1)$ and X is homeomorphic to either the n-sphere S^n or real projective n-space $P^n(R)$ [1], [3, p. 239]. Below this maximum case there is a gap of $n-2$ dimensions, at least for $n \neq 4$ and $n \geqslant 1$. In fact, we have the following result [11, p. 63], [10].

Theorem (WANG). *If* $\frac{(n-1)n}{2} + 1 < \dim G < \frac{n(n+1)}{2}$, *then* $n = 4$.

Proof. We outline a proof as follows. By (1) G acts transitively on X. Hence G acts differentiably on $X = G/H$, where H is the isotropy or stability subgroup of G at a point x in X. By BOCHNER's theorem on local linearity about x [8, p. 243],

$$H^0 \subset SO(n)$$

where H^0 denotes the identity component of H. Now

$$\dim G = \dim H^0 + n$$

and for $n \neq 4$, $H^0 = SO(n)$ or

$$\dim H^0 \leqslant \dim SO(n-1) = \frac{(n-2)(n-1)}{2}$$

(see, for example, [7]). The result follows. For $n = 4$, there is an effective action of $SU(3)/Z$ of dimension 8, Z denoting the center of $SU(3)$,

* Supported in part by NSF Grant GP-5972.

on the complex projective plane $P^2(C) = SU(3)/U(2)$ of dimension 4. As we shall see later this is the only exceptional possibility for $n=4$. It arises from the fact that

$$U(2) \subset SO(4)$$

with $4 = \dim U(2) > \dim SO(3) = 3$.

There are obvious examples of (G, X) for $\dim G = (n-1)n/2 + 1$ and $(n-1)n/2$. However, below this there is another gap in dimensions [9].

Theorem (WAKAKUWA). *If*

$$\frac{(n-2)(n-1)}{2} + 3 < \dim G < \frac{(n-1)n}{2},$$

then $n=6$.

For $n=6$, there are effective actions of the exceptional Lie group G_2 of dimension 14 on both S^6 and $P^6(R)$.

There is, in fact, a general pattern of gaps in the dimensions of G of which the ranges of the two theorems above are a part. The following theorem was established in [4].

Gap Theorem. *If*

$$\frac{(n-k)(n-k+1)}{2} + \frac{k(k+1)}{2} < \dim G < \frac{(n-k+1)(n-k+2)}{2}, \quad k=1,2,3,\ldots,$$

then there exist 3 possibilities:

1. $n=4$, $G \approx SU(3)/Z$ *and* $X \approx P^2(C)$,
2. $n=6$, $G \approx G_2$ *and* $X \approx S^6$ *or* $P^6(R)$,
3. $n=10$, $G \approx SU(6)/Z$ *and* $X \approx P^5(C)$.

Proof. The idea of the proof given in [4] is as follows. If the dimension of G is in one of the above ranges, then it is of relatively high dimension with respect to $\dim X$. It is possible to show then that G contains a simple factor subgroup S such that $\dim G - \dim S$ is small. We next investigate this dominant simple subgroup S using the classification theorem for compact simple Lie groups. If S is of type A or an exceptional Lie group, one shows that for $n \geqslant 17$,

$$\dim S < \frac{(n-k_0)(n-k_0+1)}{2},$$

where k_0 is the maximal value of k for which the inequality of the theorem is still meaningful. One arrives at this result by investigating the maximal dimensions of proper closed subgroups of compact Lie groups of type A or of exceptional type. If S is of type B, C or D, then $\dim S$ is of the form $j(j+1)/2$ for some j. Finally, suppose

$$\frac{(n-k)(n-k+1)}{2}+\frac{k(k+1)}{2}<\dim G<\frac{(n-k+1)(n-k+2)}{2}$$

for some $k, 1\leqslant k\leqslant k_0$. Then by the above remarks,

$$\dim S\leqslant\frac{(n-k)(n-k+1)}{2}$$

for $n\geqslant 17$. Now one verifies that $\dim G-\dim S$ is sufficiently small. A case by case analysis for $n<17$ turns up the three possibilities of the theorem.

The author has discovered recently that the pattern of gaps given by the theorem above is but a part of a still more general pattern of gaps.

The Gap Theorem also holds for *transitive* actions of compact connected *non-Lie* groups [5]. However, it is false for effective (non-transitive) actions.

One may apply these results to an *effective* action of a compact *non-Lie* group G on a connected *n-manifold* X. It is, of course, an outstanding question whether such an action can exist. MONTGOMERY [6] has shown that for such an action the dimension of G must be finite dimensional; in fact, his proof actually shows

$$\dim G\leqslant\frac{n(n+1)}{2}.$$

BREDON [2] has shown, on the other hand, that the maximal dimension of the orbits of such an action is at most $n-3$. Combining MONTGOMERY's, BREDON's and our results, we obtain

Theorem. *If G is a compact non-Lie group effective on a connected n-manifold X and if*

$$\dim G>\frac{(n-5)(n-4)}{2}+2,$$

then either

1. $n\geqslant 6$, $\dim G=(n-4)(n-3)/2+1$ *and G is locally isomorphic to* $A^1\oplus\mathrm{Spin}(n-3)$ *or*

2. $n=8$ *and G is locally isomorphic to* $A^1\oplus SU(3)$ *(this corresponds to* $\dim G=(n-5)(n-4)/2+3$*), where A^1 denotes a compact connected 1-dimensional abelian non-Lie group (i. e., a non-trivial inverse limit of circles).*

A detailed proof of this last theorem may be found in [5]. By the result of BREDON [2] mentioned above it is impossible for a solenoidal group A^1 to act effectively on a manifold of dimension 3 or less. However, if there exists an effective action (A^1, M^4), the product actions

1. $A^1 \oplus SO(n-3)$ on $M^4 \times S^{n-4}$,
2. $A^1 \oplus SU(3)$ on $M^4 \times P^2(C)$

correspond to the 2 possibilities of the theorem.

Further improvement of the above theorem depends exclusively on further improvement of BREDON's result [2].

References

1. BIRKHOFF, G.: Extensions of Lie groups, Math. Zeit., **53,** 226—235 (1950).
2. BREDON, G. E.: Some theorems on transformation groups, Ann. of Math. **67,** 104—118 (1958).
3. EISENHART, L. P.: Riemannian Geometry, Princeton University Press, Princeton, N.J., 1949.
4. MANN, L. N.: Gaps in the dimensions of transformation groups, Ill. J. Math. **10,** 532—546 (1966).
5. — Dimensions of compact transformation groups, Mich. Math. J., **14,** 433—444 (1967).
6. MONTGOMERY, D.: Finite dimensionally of certain transformation groups, Ill. J. Math. **1,** 28—35 (1957).
7. —, and H. SAMELSON: Transformation groups of spheres, Ann. of Math. **44,** 454—470 (1943).
8. —, and L. ZIPPIN: Topological Transformation Groups, Interscience Publishers, New York, 1955.
9. WAKAKUWA, H.: On n-dimensional Riemannian spaces admitting some groups of motions of order less than $n(n-1)/2$, Tohoku Math. J. (2) **6,** 121—134 (1954).
10. WANG, H. C.: On Finsler spaces with completely integrable equations of Killing, Journ. of the London Math. Soc. **22,** 5—9 (1947).
11. YANO, K.: The theory of Lie derivatives and its applications, Amsterdam, Bibliotheca Mathematica, 1957.

Actions of $SO(2)$ on 3-Manifolds

Peter Orlik* and Frank Raymond**

Introduction

The purpose of this report is to give a complete equivariant and topological classification of the effective actions of the circle group, $SO(2)$, on closed, connected 3-manifolds. The equivariant classification was given in [3] together with the topological classification of actions with fixed points. Actions without fixed points and without special exceptional orbits are manifolds admitting singular fiberings in the sense of Seifert [4]. Most of these manifolds were shown topologically distinct in [2]. In the last section of the present report we complete the classification of $SO(2)$ actions by considering those manifolds not treated in [2] and fixed point free actions with special exceptional orbits.

It has been known that connected, compact groups (other than the identity) acting effectively must be Lie groups. Actions with at least one orbit of dimension $\geqslant 2$ were classified by Mostert [1]***. This together with the results of the present paper give a complete classification of the effective actions of compact, connected groups on closed 3-manifolds.

We begin by listing some elementary $SO(2)$ actions. These will be some of our "building blocks". In 2 the standard actions are described (Theorem 1). In 3 we show that all actions are equivariantly equivalent to one of the standard actions, thus obtaining the equivariant classification theorem (Theorem 2). Next we identify the manifolds topologically. This is done for actions with fixed points in 4, without fixed points and without special exceptional orbits in 5 and without fixed points but in the presence of special exceptional orbits in the last part of 5.

The equivariant classification, which is guided by the work of Seifert [4], is essentially given in terms of the orbit structure. It is proved that the orbit structure invariants

$$\{b;(\varepsilon,g,\bar{h},t),(\alpha_1,\beta_1),\dots,(\alpha_n,\beta_n)\}$$

* Partially supported by NSF Grant GP-6682.

** National Science Foundation Fellow.

*** (Editor's note) See also W. D. Neumann's paper, *3-Dimensional G-manifolds with 2-Dimensional Orbits*, these Proceedings.

(whose meaning is described in the text), which take the form of a weighted 2-manifold, completely characterize the action.

Such a classification is more meaningful if one can identify in the topological sense the manifold on which the action occurs. Furthermore it is desirable to be able to list all the inequivalent actions on a given manifold.

In the presence of *fixed points* the manifold must be a *specific equivariant connected sum of handles, non-orientable handles, lens spaces and* $P^2 \times S^1 - s$. It also turns out that if M admits an action of $SO(2)$ *without fixed points, then it is impossible that M is the connected sum of two manifolds with non-trivial fundamental groups. In fact if M admits an action of* $SO(2)$ *without fixed points and is not* $S^2 \times S^1, S^3, P^2 \times S^1$, *a lens space or the non-orientable handle N, then M is either covered by the 3-sphere or it is a* $K(\pi, 1)$. *Furthermore, unless M is* $K \times S^1$ *(where K is the Klein bottle), KS (a non-trivial Klein bottle bundle over* S^1*) or a manifold admitting an action of* $SO(2)$ *with fixed points, then the given action is the only possible action on M.*

For example, $M^* \times S^1$, where M^* is a closed surface with EULER characteristic less than -1, admits only one action of $SO(2)$.

Essentially, the above statements say that M is to a great extent determined by its fundamental group (completely when the action is fixed point free and unique) which in turn is determined by the orbit structure. It follows from above that the connected sum $(K \times S^1) \# (P^2 \times S^1)$ admits no effective $SO(2)$ action. On the other hand it can be shown using [3] that

$$(S^2 \times S^1)_1 \# \cdots \# (S^2 \times S^1)_{14} \# (P^2 \times S^1)_1 \# \cdots \# (P^2 \times S^1)_5 \# L(9,2) \\ \# L(7,5) \# L(8,3) \# L(16,6)$$

admits exactly 1136 inequivalent actions with fixed points and none without fixed points. Of course some exceptional manifolds, such as $S^3, S^2 \times S^1, N$, lens spaces and $P^2 \times S^1$ admit an infinite number of distinct actions. *It will follow from our investigation that these are the only such manifolds.*

We also remark that from [4; Satz 11] it easily follows that *the 3-sphere* S^3 *is the only simply connected manifold on which* $SO(2)$ *operates.* SEIFERT has also determined the SEIFERT fiberings which are POINCARÉ spaces. It follows that, other than the 3-sphere, no POINCARÉ space admits an action of $SO(2)$ with fixed points or special exceptional orbits. In fact, [4; Satz 12], *M is an oriented homology 3-sphere, different from* S^3, *and admits an (unique) action of* $SO(2)$, *if and only if,*

$$M = \{b; (o,0,0,0); (\alpha_1, \beta_1), \ldots, (\alpha_r, \beta_r)\},$$

where $r \geqslant 3$, α_j *are all pairwise relatively prime, and*

$$b\alpha_1 \ldots\ldots \alpha_r + \beta_1 \alpha_2 \ldots\ldots \alpha_r + \ldots + \alpha_1 \alpha_2 \ldots\ldots \alpha_{r-1} \beta_r = 1.$$

Note that this equation determines b and the β_j.

The manifolds listed in Theorem 5 are a new class of $K(\pi,1)$ 3-manifolds. Their orientable double coverings are SEIFERT manifolds, but in general they do not admit a SEIFERT fibering. It seems intuitively clear that most of the manifolds of Theorems 4 and 5 are irreducible, however, we can only prove that they are $K(\pi,1)$–s. Assuming irreducibility, many can be fibered locally trivially over the circle. Another interesting fact is that if $SO(2)$ acts without fixed points on M and $\pi_1(M)$ is infinite, then its center is infinite cyclic. According to recent results of WALDHAUSEN, the latter is probably also sufficient for the existence of an $SO(2)$ action.

Summarizing some of the conclusions of this discussion we have

Theorem. *If* $SO(2)$ *acts effectively on a closed, connected* 3*-manifold* M, *then* M *is identified as*

1. $S^3, S^2 \times S^1, N, P^2 \times S^1, L(p,q)$ *admitting actions with or without fixed points,*
2. *a connected sum of the above admitting only actions with fixed points,*
3. $K \times S^1$ *or* KS *admitting only two inequivalent actions, both without fixed points,*
4. *a quotient of* $SO(3)$ *or* Sp(1) *by a finite, non-abelian, discrete subgroup, admitting a unique fixed point free action,*
5. *a* $K(\pi,1)$ *whose fundamental group has infinite cyclic center (provided it is not the* 3*-dimensional torus), admitting a unique action without fixed points and hence not homeomorphic to any other* 3*-manifold with an* $SO(2)$ *action.*

1. Examples

Parametrize the solid torus $D^2 \times S^1$ *by* $(\rho e^{i\theta}, e^{i\psi})$, *where* $0 \leqslant \rho \leqslant 1$, $0 \leqslant \theta < 2\pi$ and $0 \leqslant \psi < 2\pi$. Define an action of $SO(2)$ by

$$z \times (\rho e^{i\theta}, e^{i\psi}) \rightarrow (z^{\nu} \rho e^{i\theta}, z^{\mu} e^{i\psi}) \tag{1.1}$$

where z ranges over the complex numbers of norm 1, $0 < \nu < \mu$ and ν and μ are relatively prime integers. This is the *standard linear action* determined by (μ, ν).

Observe that the projection onto the second coordinate $D^2 \times S^1 \rightarrow S^1$ is a 2-disk bundle projection. The fiber is the disk with structural group

Z_μ. The action of the structural group Z_μ on the fiber is the linear action of the rotations generated by $z=e^{2\pi i/\mu}$, that is, for fixed ψ_0

$$(1.2) \qquad e^{2\pi i/\mu}\times(\rho e^{i\theta}, e^{i\psi_0})\to(e^{2\pi i\nu/\mu}\rho e^{i\theta}, e^{i\psi_0}).$$

An ordinary or *principal* action can be defined on $D^2\times S^1$ by

$$(1.3) \qquad z\times(\rho e^{i\theta}, e^{i\psi})\to(\rho e^{i\theta}, z e^{i\psi}).$$

An action with *fixed points* is defined by

$$(1.4) \qquad z\times(\rho e^{i\theta}, e^{i\psi})\to(z\rho e^{i\theta}, e^{i\psi}).$$

Denote the set of fixed points by F.

The subgroup $O(1)\subset SO(2)$ operates on the boundary of $D^2\times S^1$ with the principal action. If we collapse each of the principal orbits on the boundary by the action of $O(1)$ we obtain a new space N. The group $SO(2)$ still operates on N with only principal orbits in the interior of the solid torus. The orbits on the image of the boundary torus are circles doubly covered by nearby principal orbits. We shall call these *special exceptional orbits* (and their collection SE). The orbit space of this action is a disk and the action admits a cross-section. The closed 3-manifold N is just the non-orientable handle, i. e., the non-trivial S^2 bundle over S^1.

In (1.2) the center circle $(0, e^{i\psi})$ will be called an *exceptional orbit* of type (μ,ν). The principal orbits wind around it μ times and around a bounding curve $m=(e^{i\theta}, e^{i\psi_0})$, ψ_0 fixed, ν times. We may choose a curve q on the boundary torus such that q is "a cross section to the action (1.2)". It is a cross section to the principal S^1-bundle whose total space is the boundary torus. The group $SO(2)$ is naturally oriented and by choosing an orientation for the solid torus, a compatible orientation of q is determined. Orient m such that

$$(1.5) \qquad m\sim\mu q+\beta h, \quad \text{where} \quad \beta\nu\equiv 1(\mu).$$

Obviously, q may be modified by $q\sim q'+sh$ for an arbitrary integer s, so that we may assume that q was chosen such that $0<\beta<\mu$. The invariant (μ,β) coincides with the oriented SEIFERT invariant (α,β), see [4]. The collection of exceptional orbits is denoted by E.

Using the above actions as "building blocks" more complicated actions can be constructed:

a) Write the 3-sphere S^3 in terms of join coordinates $(z_1, z_2, \theta_1, \theta_2)$. Define a linear action with at most two exceptional orbits by

$$z\times(z_1, z_2, \theta_1, \theta_2)\to(z^{\alpha_1}z_1, z^{\alpha_2}z_2, \theta_1, \theta_2),$$

where α_1 and α_2 are relatively prime, positive integers. This action is the result of matching a solid torus, with standard linear action of type

(α_1, ν_1), to another solid torus, with standard linear action of type (α_2, ν_2), by an equivariant orientation reversing homeomorphism along the respective boundaries. The invariants must satisfy $\nu_1 \equiv \alpha_2 \bmod \alpha_1$ and $\nu_2 \equiv \alpha_1 \bmod \alpha_2$. If $\alpha_1 = \alpha_2 = 1$, then the action similar to above determines the Hopf fibering.

b) If the action on S^3 is given by

$$z \times (z_1, z_2, \theta_1, \theta_2) \to (z z_1, z_2, \theta_1, \theta_2),$$

then it is the standard linear action with a circle of fixed points.

c) In general, recall that a lens space can be described as the union of two solid tori, V_i, $i=1,2$, sewn together by an orientation reversing homeomorphism of their boundaries, T_i. If m_i and a_i denote meridian (bounding) and axial (center non-bounding) conjugate curves, then a homeomorphism of T_2 onto T_1, where $m_2 \sim p a_1 + q m_1$ yields the lens space $L(p,q)$. This homeomorphism can be constructed equivariantly in a number of different ways. First, suppose on V_1 we take a standard linear action of type (μ_1, ν_1) and on V_2 a principal action. If h_1 denotes a principal orbit on T_1 and q_1 denotes a cross-section to the principal $SO(2)$-fibering of T_1, then it is easily seen that $m_1 \sim \alpha_1 q_1 + \beta_1 h_1$, where $\alpha_1 = \mu_1$ and $\beta_1 \nu_1 \equiv 1 \bmod \alpha_1$. By choosing a different q_1 if necessary, but preserving the given orientation of V_1, we can normalize β_1 such that $0 < \beta_1 < \alpha_1$. On T_2 let $h_2 = a_2$ and $q_2 = m_2$. Match T_1 and T_2 by sending h_1 to h_2 and q_1 to $-q_2$. Solving the resulting homology congruences yield $m_1 \sim \alpha_1 q_1 + \beta_1 h_1 \sim -\alpha_1 q_2 + \beta_1 h_2 \sim p a_2 + q m_2$. Thus it determines the lens space $L(\beta_1, -\alpha_1 \bmod \beta_1)$. All orbits are principal except one and it is an exceptional orbit of type $(\alpha_1, \beta_1^{-1} \bmod \alpha_1)$.

Next take the standard linear action of type (μ_1, ν_1) on V_1 and the linear action with fixed points on V_2. Now m_2 is a principal orbit h_2 on T_2. Match h_2 to h_1 to obtain the relation $m_2 \sim \mu_1 a_1 + \nu_1 m_1$. It is clear that $q_2 = a_2$ is sent onto some cross section $-q_1$ in order to reverse orientation; thus we have constructed an action with exactly one exceptional orbit of type (μ_1, ν_1) and a circle of fixed points on the lens space $L(\mu_1, \nu_1)$.

d) Let M^* be one of S^2, P^2 or K and consider the standard $SO(2)$ action on it. Form $M^* \times S^1$ and let $SO(2)$ operate on the first factor by the standard action and trivially on the second factor. We obtain actions with exactly two circles of fixed points on $S^2 \times S^1$ and the orbit space is an annulus. On $P^2 \times S^1$ we have one circle of fixed points and a torus of points lying on SE-orbits. The orbit space is again an annulus. On $K \times S^1$ we have two toral components of SE-orbits and the orbit space is again an annulus.

e) Consider the standard $SO(2)$ action on S^2. Form $S^2 \times I$ and match the boundary components $S^2 \times 0$ and $S^2 \times 1$ so that the orbits which are lines of latitude are flipped across the equator of S^2. This gives the non-trivial S^2 bundle over S^1, N. The action has just one circle of fixed points and the orbit space is a Möbius band.

f) We can replace S^2 by K in the above construction. The $SO(2)$ action on K is given here by viewing K as the double of a Mobius band. The orbits on $K\times 0$ and $K\times 1$ are matched by flipping orbits around the boundary of the Möbius band. This yields a non-trivial Klein bottle bundle over S^1, denoted by KS. The action described has a torus of SE orbits and the orbit space is a Möbius band.

2. The Standard Actions

We shall consider a "weighted" 2-manifold with (possibly empty) boundary. The weights will correspond to the orbit structure of an action on a 3-manifold whose orbit space is the given 2-manifold. The objective of the next section is to show that each topological action is topologically equivalent to one of the standard actions here described.

Let M^* be a compact, connected 2-manifold with exactly $0\leqslant \bar{h}+t$ boundary components, where $\bar{h}$ and t are non-negative integers. Let the symbol ε take on the values o or $\bar{n}$ depending on whether M^* is orientable or non-orientable. If $\varepsilon=o$ then we assume that a specific orientation has been chosen for M^*. Let g denote the genus of M^*. Take n pairs of relatively prime integers $(\alpha_1,\beta_1),\ldots,(\alpha_n,\beta_n)$ with the property that if $\varepsilon=o$, then $0<\beta_j<\alpha_j$ and if $\varepsilon=\bar{n}$, then $0<\beta_j\leqslant\alpha_j/2$. If $\varepsilon=o$ and $\bar{h}+t=0$, let b be an arbitrary integer, while $b=0$ if $\bar{h}+t\neq 0$. If $\varepsilon=\bar{n}$, $\bar{h}+t=0$ and no $\alpha_j=2$, let b take on the values 0 or 1, while $b=0$ otherwise. We shall mean by

$$M^*=\{b;(\varepsilon,g,\bar{h},t);(\alpha_1,\beta_1),\ldots,(\alpha_n,\beta_n)\}^* \tag{2.1}$$

a weighted 2-manifold, where the symbols have the meaning described above.

Let $SO(2)$ operate on $M^*\times S^1$ by

$$e^{i\theta}\times(m^*\times e^{i\psi})\to(m^*\times e^{i\theta}e^{i\psi}).$$

Choose $\bar{h}$ boundary components and for each m^* in any one of these components collapse the orbit $m^*\times e^{i\psi}$ to a point by identifying $(m^*\times e^{i\psi})$ with $(m^*\times 1)$ for all ψ. For each m^* in the remaining t boundary components collapse the orbit $(m^*\times e^{i\psi})$ by identifying $(m^*\times e^{i\psi})$ with $(m^*\times e^{i(\psi+\pi)})$ for all $0\leqslant\psi<\pi$.

Clearly $SO(2)$ still operates on the identification space, which is now a closed 3-manifold with orbit space homeomorphic to M^*. The orbits corresponding to m^* in the first $\bar{h}$ boundary components are fixed points and the orbits corresponding to points in the remaining t boundary components have $O(1)$ as stability group. Note that the orientation of the identification space reverses itself near these orbits. There is a cross-section to this action which can be identified with $(M^* \times 1)$.

Choose n points $x_1^*, \dots, x_n^*$ in the interior of M^* and remove the interior of a closed disk neighborhood D_j^* around each of the points. If we remove the corresponding interiors of the solid tori over the D_j^* from the above identification space, then we obtain a compact, connected 3-manifold M_0 with n toral boundary components, T_j. Clearly $SO(2)$ still operates on M_0 since we have removed invariant tubular neighborhoods of principal orbits corresponding to the x_j^*. Moreover, we still have a cross-section $M_0^* \to M_0$ which we can still think of as $m_0^* \to m_0^* \times 1$. If we denote by q_j^* the oriented boundary curve of D_j^* and q_j its image by the cross-section, then on T_j we have an orthogonal curve system given by q_i and h, where h is any one of the principal orbits.

Now take a solid torus ${}_jV$ with the standard linear action (1.1) of type (μ_j, ν_j) and match the principal orbits on ${}_jT$ (the boundary of ${}_jV$) to the principal orbits on T_j and a cross-section to the action on ${}_jT$ to q_j (which is a cross-section to the action on T_j). If $\varepsilon = o$, we have to do this in an orientation reversing manner in order to retain local orientation. For $\varepsilon = o$, we choose (μ_j, ν_j) to be $\mu_j = \alpha_j$, $\nu_j \equiv \beta_j^{-1} \bmod \alpha_j$. For $\varepsilon = \bar{n}$, we let $\mu_j = \alpha_j$, $\nu_j \equiv \pm \beta_j^{-1} \bmod \alpha_j$ to satisfy our normalization conditions.

If $\bar{h} + t = 0$, we remove one more invariant tubular neighborhood of a principal orbit over x_0^*. On T_0, we have the cross-section curve q_0 and we equivariantly sew in a solid torus ${}_0V$ on which we have a principal action of type (1.3). The homeomorphism sending T_0 onto ${}_0T$ matches principal orbits but we send q_0 to a cross-sectional curve ${}_0q$ on ${}_0T$ satisfying the homology relation ${}_0m \sim {}_0q + bh$, where ${}_0m$ is the bouding curve in ${}_0V$, h is a principal orbit on ${}_0T$ and b is the integer described above. When $\varepsilon = o$, we send $-q_0$ to ${}_0q$ to retain orientation.

Theorem 1 [3]. *Corresponding to each compact, connected, weighted 2-manifold*

$$M^* = \{b; (\varepsilon, g, \bar{h}, t); (\alpha_1, \beta_1), \dots, (\alpha_n, \beta_n)\}^*$$

there is a closed, connected 3-manifold

$$M = \{b; (\varepsilon, g, \bar{h}, t); (\alpha_1, \beta_1), \dots, (\alpha_n, \beta_n)\}$$

and a "standard" effective action of $SO(2)$ on M such that the orbit space is M^* and the orbit structure is described by the weights.

It is clear that by choosing a differentiable structure on M^* and performing the construction differentiably the action on M is differentiable.

3. An Equivariant Classification

Theorem 2 [3]. *Let $SO(2)$ act effectively on a closed, connected 3-manifold M. Then there exists a unique 3-manifold with a standard action $M_1 = \{b;(\varepsilon, g, \bar{h}, t);(\alpha_1, \beta_1), \ldots, (\alpha_n, \beta_n)\}$ and a homeomorphism $H: M_1 \to M$ together with an automorphism $a: SO(2) \to SO(2)$, perhaps trivial, so that for all $m \in M_1$, $g \in SO(2)$*

$$H(g(m)) = a(g)H(m).$$

Furthermore, if M_1^* is oriented, the induced homeomorphism on the orbit spaces is required to be orientation preserving. If $SO(2)$ is operating differentiably on M with respect to a differentiable structure on M, then H can be chosen to be a diffeomorphism.

The proof is straightforward. First we show that the orbit structure of M is determined by the same type of invariants as the standard action, then use this information to construct the equivariant homeomorphism.

Proof. Let $G = SO(2)$ act effectively on a closed, connected 3-manifold M. Let S_x be a slice at $x \in M$. If $G(x)$, the orbit at x, is one-dimensional, then by an observation of BREDON, S_x can be chosen as a closed, topological 2-cell and the stability group Z_μ, a finite cyclic group, acts effectively on this disk. Thus an equivariant tubular neighborhood of $G(x)$ is a topological 2-disk bundle over $G(x)$ and the structure group is Z_μ imbedded in the group of homeomorphisms of the disk. (Z_μ may be the identity in which case $G(x)$ is a principal orbit.) Due to a theorem of L. E. J. BROUWER and R. KERÉKJÁRTÓ such an action on the disk is topologically conjugate to a linear action. Thus the invariant tubular neighborhood is topologically equivalent to one of the actions (1.1), (1.3) or the tubular neighborhood of SE-orbits. The latter is the non-trivial 2-disk bundle over the circle and topologically it is the product of the Möbius band with the interval. The action of $SO(2)$ is standard on the Möbius band and trivial on the second factor. All points over the center line of the Möbius band are on SE-orbits, all other points are on principal orbits. The orbits of type (1.1) are isolated; thus it is clear that the orbit space minus the fixed point set is a 2-manifold with t boundary components corresponding to the t connectedness components of the SE-orbits.

If $x \in F(G; M)$ is a fixed point, then the connected component of $F(G; M)$ containing x is a 1-manifold. Local cohomology arguments show that the action near the fixed point set is principal and that there exists a global cross-section to the action in an invariant neighborhood of each component of $F(G; M)$. Thus topologically the action near a component of the fixed point set is equivalent to (1.4). The orbit space $M^* = M/SO(2)$ is therefore a 2-manifold with $\bar{h}$ boundary components corresponding to the $\bar{h}$ connectedness components of $F(G;M)$ and t boundary components for the SE components.

If the 2-manifold M^* is orientable, choose a specific orientation on it and use this together with the orientation of $SO(2)$ to orient $M-SE$. With this orientation the exceptional orbits have oriented orbit invariants (μ_j, ν_j) as in (1.1).

If M^* is non-orientable, then the exceptional orbits have unoriented orbit invariants (μ_j, ν_j), where $0<\nu_j \leqslant \mu_j/2$. This normalization is obtained by moving the exceptional orbit with an invariant neighborhood along an orientation reversing curve on M^* missing the images of all other exceptional orbits. This isotopy reverses the orientation of the invariant tubular neighborhood. Since the orientation of the orbits cannot be reversed, the orientation of the curve q on the boundary of the invariant neighborhood has been reversed. Thus an action of type (μ, ν) of x is equivalent, via an orientation reversing homeomorphism of the invariant neighborhood, to one of type $(\mu, \mu-\nu)$.

The action on $M-(SE \cup F \cup E)$ is free and the orbit map restricted to $M-(SE \cup F \cup E)$ is a principal fibering. As long as $SE \cup F \cup E \neq \phi$, $M^*-(SE \cup F \cup E)^*$ is an open 2-manifold; hence this fibering must be trivial and we have a cross-section.

Let $x_1, \ldots, x_n$ denote n points, one from each of the n distinct exceptional orbits. Taking a closed 2-cell slice S_{x_j} at x_j as before, we observe that the image of S_{x_j} under the orbit map π is a closed 2-cell D_j^*. The boundary q_j^* of D_j^* lifts to a simple closed curve q_j on the boundary of $\pi^{-1}(D_j^*)$, and q_j is a cross-section to the principal action on the boundary torus of $\pi^{-1}(D_j^*)$. Denote the boundary of the slice S_{x_j} by m_j and a principal orbit on the boundary of $\pi^{-1}(D_j^*)$ by h. Then we have the homology relation

$$m_j \sim \alpha_j q_j + \beta_j h, \tag{3.1}$$

where $\alpha_j = \mu_j$ and $\beta_j \equiv \nu_j^{-1} \bmod \alpha_j$ for $\varepsilon = o$, respectively $\beta_j \equiv \pm \nu_j^{-1} \bmod \alpha_j$ for $\varepsilon = \bar{n}$. We assume that q_j was chosen to satisfy the normalization $0<\beta_j<\alpha_j$, respectively $0<\beta_j \leqslant \alpha_j/2$.

Provided $F \cup SE \neq \phi$ the cross-section given by $q_1 \cup q_2 \cup \cdots \cup q_n$ can be extended to all of $M' = M -$ interior of $\pi^{-1}(D_1^* \cup \cdots \cup D_n^*)$. Obviously, it can be extended to $M'-(F \cup SE)$ and since a cross-section on the boundary of an invariant neighborhood of a component of $F \cup SE$ can be extended to the whole neighborhood, it can be extended to all of M'.

If $F \cup SE = \phi$, then we choose a point x_0 on a principal orbit and perform the same construction for x_0^* as for x_j^*. Clearly, we can find a cross-section to the fibering on $M' -$ interior of $\pi^{-1}(D_0^0)$. In general, this cross-section cannot be extended to all of M'. The obstruction to so doing is the integer b, which is the "characteristic class" of the $SO(2)$ bundle over M'^* with prescribed cross-section $q_1, \ldots, q_n$. It is a cohomology class in $H^2(M'^*, q_1^* \cup \cdots \cup q_n^*; Z)$. Geometrically this means that the final solid torus is to be sewn into the torus hole T_0 according to the rule

$$m_0 \sim q_0 + bh.$$

Here b is an integer. It is reduced modulo 2 if $\varepsilon = \bar{n}$, since we can equivariantly isotop the solid torus along an orientation reversing loop and change b by an even number. In the presence of an orbit of type $(2,1)$ when $\varepsilon = \bar{n}$, further equivariant isotopies render $b = 0$.

Thus we can associate to the action of $SO(2)$ on M a set of invariants

$$\{b; (\varepsilon, g, \bar{h}, t); (\alpha_1, \beta_1), \ldots, (\alpha_n, \beta_n)\}.$$

In order that the action on M be equivalent to the standard action on some M_1 (preserving orientation where oriented), it is clearly necessary that they have the same set of invariants. By our analysis it is also sufficient because the cross-section (or partial cross-section) can be used to define part of the equivariant homeomorphism H and its extension to all of M_1 is obtained by observing that there is essentially only one way to sew in equivariantly the remaining solid tori. This completes the proof of the theorem.

In case $\varepsilon = o$, observe that reversing the orientation on $M - SE$ changes $M = \{b; (\varepsilon, g, \bar{h}, t); (\alpha_1, \beta_1), \ldots, (\alpha_n, \beta_n)\}$ to $\bar{M} = \{b'; (\varepsilon, g, \bar{h}, t), (\alpha_1, \alpha_1 - \beta_1), \ldots, (\alpha_n, \alpha_n - \beta_n)\}$ where $b' = b = 0$ if $\bar{h} + t > 0$ and $b' = -b - n$ if $\bar{h} + t = 0$.

We shall call the result of Theorem 2 an *equivariant classification* of the actions of $SO(2)$ on closed, connected 3-manifolds. It should be pointed out that a very similar result is valid for the non-closed case, see [3] for details.

Observe that the orientation of $M - SE$ (when orientable) depends on the orientation of $SO(2)$. If $SO(2)$ acts on M with opposite orientation, the orbit structure invariants cannot detect this difference. This is why we allowed an automorphism of $SO(2)$ in the theorem.

When $\bar{h} + t = 0$ the manifold M has a singular fibration in the sense of SEIFERT [4]. In fact for $\varepsilon = o$, it is the Seifert-manifold

$$(Oo, g \mid b; (\alpha_1, \beta_1), \ldots, (\alpha_n, \beta_n))$$

and for $\varepsilon = \bar{n}$ it is the Seifert-manifold

$$(N\bar{n}I, g \mid b; (\alpha_1, \beta_1), \ldots, (\alpha_n, \beta_n)).$$

The equivariant classification theorem was inspired by the fiber preserving homeomorphism classification of SEIFERT, which coincides with our theorem for $\bar{h} + t = 0$. SEIFERT's singular fiberings have four more classes but only the above two admit effective $SO(2)$ actions.

4. Equivariant Connected Sums and Topological Classification when $F \neq \phi$

In view of Theorem 2 it remains to decide when two distinct standard actions can occur on the same manifold. This is what we shall call the *topological classification.* In order to obtain a reasonable classification we must "identify" the 3-manifold M from the given data $\{b;(\varepsilon,g,\bar{h},t);(\alpha_1,\beta_1),\ldots,(\alpha_n,\beta_n)\}$. Two things will be of primary importance: equivariant connected sums when $\bar{h}>0$, and fundamental groups when $\bar{h}=0$.

According to KNESER and MILNOR, every closed, connected 3-manifold can be written uniquely up to order as the connected sum of prime 3-manifolds. A 3-manifold is *prime* if it is irreducible (every embedded 2-sphere bounds a 3-cell) or it is either $S^2\times S^1$ or N. For non-orientable M, the uniqueness requires that N can occur at most once and not at all if the decomposition contains any other non-orientable prime 3-manifold.

Theorem 3 [3]. *Let*

$$M=\{b;(\varepsilon,g,\bar{h},t);(\alpha_1,\beta_1),\ldots,(\alpha_n,\beta_n)\}$$

and assume that $\bar{h}>0$, *i.e., that* $SO(2)$ *acts on* M *with fixed points. Then* M *is homeomorphic to*

1. $S^3\#(S^2\times S^1)_1\#\cdots\#(S^2\times S^1)_{2g+\bar{h}-1}\#(P^2\times S^1)_1\#\cdots\#(P^2\times S^1)_t$ $\#L(\alpha_1,\beta_1)\#\cdots\#L(\alpha_n,\beta_n)$ *if* $(\varepsilon,g,\bar{h},t)=(o,g,\bar{h},t)$, $t\geqslant 0$;
2. $(S^2\times S^1)_1\#\cdots\#(S^2\times S^1)_{g+\bar{h}-1}\#(P^2\times S^1)_1\#\cdots\#(P^2\times S^1)_t$ $\#L(\alpha_1,\beta_1)\#\cdots\#L(\alpha_n,\beta_n)$ *if* $(\varepsilon,g,\bar{h},t)=(\bar{n},g,\bar{h},t)$, $t>0$;
3. $N\#(S^2\times S^1)_1\#\cdots\#(S^2\times S^1)_{g+\bar{h}-1}\#L(\alpha_1,\beta_1)\#\cdots\#L(\alpha_n,\beta_n)$ *if* $(\varepsilon,g,\bar{h},t)=(\bar{n},g,\bar{h},0)$.

Proof: We have already identified the following manifolds:

$\{0;(o,0,1,0)\}=S^3$; example b) of 1,

$\{0;(\bar{n},1,1,0)\}=N$; example e) of 1,

$\left.\begin{array}{l}\{0;(o,0,1,1)\}=P^2\times S^1\\ \{0;(o,0,2,0)\}=S^2\times S^1\end{array}\right\}$ example d) in 1,

$\left.\begin{array}{l}\{0;(o,0,1,0);(\alpha_1,\beta_1)\}=L(\mu_1,\nu_1)\\ \text{where } \mu_1=\alpha_1 \text{ and } \nu_1\beta_1\equiv 1 \bmod \alpha_1\end{array}\right\}$ example c) in 1.

It is known that $L(p,q)$ is homeomorphic to $L(p',q')$ if and only if $|p|=|p'|$ and $q\pm q'\equiv 0 \bmod p$ or $qq'\equiv \pm 1 \bmod p$. Hence $L(\mu_1,\nu_1)$ can also be written as $L(\alpha_1,\beta_1)$.

Let M_1 and M_2 admit effective $SO(2)$ actions with fixed points. Let M_1^* and M_2^* be the orbit spaces and let F_1^* resp. F_2^*, denote a connectivity component of the respective fixed point sets. Take an oriented 2-cell D_1^* in M_1^* which meets F_1^* in exactly one arc I_1 and such that the remaining part of D_1^* is in the interior of M_1^* meeting only images of principal orbits. Remove the interior of D_1^* and I_1. What remains of D_1^* is a boundary arc touching the former F_1^* at its endpoints. This corresponds to removing the interior of an invariant 3-cell in M_1, where the action is locally like (1.4). If we do the same for M_2^* at F_2^* and match the deleted M_1^* to the deleted M_2^* along the remaining arcs by an orientation reversing homeomorphism, we obtain the connected sum of M_1^* with M_2^*. This operation induces an *equivariant connected sum* of M_1 and M_2 along the invariant boundary 2-spheres.

Let $M_{\varepsilon,g,h,t} = \{0;(\varepsilon,g,\bar{h},t)\}$, $\bar{h}>0$. By performing the above construction n times,

$$M = \{0;(\varepsilon,g,\bar{h},t);(\alpha_1,\beta_1),\ldots,(\alpha_n,\beta_n)\}$$

can be written as

$$\begin{aligned} &M_{\varepsilon,g,\bar{h},t} \# L(\alpha_1,\beta_1) \# \cdots \# L(\alpha_n,\beta_n) \\ &= M_{\varepsilon,g,\bar{h},t} \# \{0;(o,0,1,0);(\alpha_1,\beta_1)\} \# \cdots \# \{0;(o,0,1,0);(\alpha_n,\beta_n)\}. \end{aligned}$$

To complete the proof of Theorem 3 it remains to describe $M_{\varepsilon,g,\bar{h},t}$.

Obviously,

$$M_{o,0,3,0} = M_{o,0,2,0} \# M_{o,0,2,0} = (S^2\times S^1)\#(S^2\times S^1).$$

On the other hand it is easy to see that $M_{o,1,1,0}$ is also homeomorphic to $(S^2\times S^1)\#(S^2\times S^1)$.

To find $M_{o,g,\bar{h},t}$ we proceed by first constructing

$$\begin{aligned} M_{o,0,\bar{h},0} = M_{o,0,1,0} \# (M_{o,0,2,0})_1 \# \cdots \# (M_{o,0,2,0})_{\bar{h}-1} = S^3 \# (S^2\times S^1)_1 \# \cdots \\ \# (S^2\times S^1)_{\bar{h}-1} \end{aligned}$$

and take its connected sum with

$$\begin{aligned} M_{o,g,1,0} &= (M_{o,1,1,0})_1 \# \cdots \# (M_{o,1,1,0})_g \\ &= (S^2\times S^1 \# S^2\times S^1)_1 \# \cdots \# (S^2\times S^1 \# S^2\times S^1)_g \end{aligned}$$

and finally take its connected sum with

$$M_{o,0,1,t} = (M_{o,0,1,1})_1 \# \cdots \# (M_{o,0,1,1})_t = (P^2\times S^1)_1 \# \cdots \# (P^2\times S^1)_t.$$

A similar construction gives $M_{\bar{n},g,\bar{h},t}$ as described in the statement of the theorem.

It is clear from the above that there are exactly two topologically inequivalent actions with fixed points on each oriented lens space $L(\mu,\nu)$ if $\nu^2 \not\equiv \pm 1 \bmod \mu$, while there is exactly one if $\nu^2 \equiv \pm 1 \bmod \mu$. From the list of actions it is therefore possible to classify all inequivalent actions on a given 3-manifold.

For open 3-manifolds a similar theorem can be proved, but the corresponding open $M_{\varepsilon,g,\bar{h},t}$ is a considerably more complicated 3-manifold than its closed counterpart.

5. The Topological Classification when $F=\phi$

In this section we shall identify those closed 3-manifolds which admit fixed point free effective $SO(2)$ actions. First we shall consider actions without SE-orbits. As mentioned above, these give rise to Seifert-manifolds and, apart from certain special cases, Theorem 4 below follows from a more general result of [2] concerning Seifert-manifolds. Using the proof of Theorem 4 we shall finally obtain the topological classification of fixed point free actions with SE-orbits in Theorem 5.

Theorem 4. *Let M be a closed, connected 3-manifold with an $SO(2)$ action given by $M=\{b;(\varepsilon,g,0,0);(\alpha_1,\beta_1),\ldots,(\alpha_n,\beta_n)\}$. Then*

1. $\{b;(o,0,0,0)\}=L(b,1)$,
2. $\{b;(o,0,0,0);(\alpha_1,\beta_1)\}=L(p,q)$

where $p=|b\alpha_1+\beta_1|$ *and* $q=(b/|b|)\alpha_1 \bmod p$,

3. $\{b;(o,0,0,0);(\alpha_1,\beta_1),(\alpha_2,\beta_2)\}=L(p,q)$

where $p=|b\alpha_1\alpha_2+\beta_1\alpha_2+\alpha_1\beta_2|$ *and* $q=m\alpha_2-n\beta_2$ *such that* $m(b/|b|)\alpha_1+n|b\alpha_1+\beta_1|=1$,

4. $\{0;(\bar{n},1,0,0)\}=P^2\times S^1$,
5. $\{1;(\bar{n},1,0,0)\}=N$,
6. $\{b;(\bar{n},1,0,0);(\alpha_1,\beta_1)\}=\begin{cases}P^2\times S^1 & \textit{if } |b\alpha_1+\beta_1| \textit{ is even},\\ N & \textit{if } |b\alpha_1+\beta_1| \textit{ is odd},\end{cases}$
7. $\{0;(\bar{n},2,0,0)\}=K\times S^1$,
8. $\{1;(\bar{n},2,0,0)\}=KS$,
9. *all other manifolds admit only the given $SO(2)$ action.*

Proof: First assume that for

$$\text{(5.1)}\qquad \begin{aligned} &\varepsilon=o \quad \begin{cases} g=0, & n>3,\\ g=0, & n=3, \quad \dfrac{1}{\alpha_1}+\dfrac{1}{\alpha_2}+\dfrac{1}{\alpha_3}\leqslant 1,\\ g=1, & n>0,\end{cases}\\ &\varepsilon=\bar{n} \quad \begin{cases} g=1, & n>1,\\ g=2, & n>0.\end{cases}\end{aligned}$$

Under these conditions, we can show that all the orbit structure invariants are in fact invariants of $\pi_1(M)$.

In [4], SEIFERT gave a presentation of these groups. For the situation at hand we shall use the partial cross-section obtained in 3 to find this presentation. By abuse of language, the symbols of geometric objects q and h will be employed to represent the corresponding generators of $\pi_1(M)$. In addition we introduce the symbols $a_i, b_i, i=1,\dots,g$ to generate the fundamental group of an orientable surface of genus g and the symbols v_i, $i=1,\dots,g$ to generate the fundamental group of a non-orientable surface of genus g.

Thus, for $\varepsilon = o$, we have

$$\pi_1(M) = (a_i, b_i, q_i, h | \pi_* h^{-b}, [a_i, h], [b_i, h], [q_j, h], q_j^{\alpha_j} h^{\beta_j})$$

where $i=1,\dots,g; j=1,\dots,n$ and $\pi_* = q_1 \dots q_n [a_1, b_1] \dots [a_g, b_g]$.

Similarly, for $\varepsilon = \bar{n}$,

$$\pi_1(M) = (v_i, q_j, h | \pi_* h^{-b}; [v_i, h], [q_j, h], q_j^{\alpha_j} h^{\beta_j}),$$

where $i=1,\dots,g; j=1,\dots,n$ and $\pi_* = q_1 \dots q_n v_1^2 \dots v_g^2$.

Under the conditions of (5.1), we can follow [2] to prove that

1. *the center of $\pi_1(M)$ is infinite cyclic generated by h.* Clearly (h) is in the center and $\pi_1(M)/(h) = (a_i, b_i, q_j | \pi_*, q_j^{\alpha_j})$ or $\pi_1(M)/(h) = (v_i, q_j | \pi_*, q_j^{\alpha_j})$ can be represented as an amalgamated free product where at least one of the factors is centerless. Thus, by a well-known theorem, $\pi_1(M)/(h)$ is centerless.

2. *ε and g are invariants of $\pi_1(M)/(h)$.* This follows from the theory of discontinuous groups or more directly from homology arguments and elementary ideals of ALEXANDER matrices.

3. *h is of infinite order in $\pi_1(M)$.* This observation is due to SEIFERT [4], who showed that h has finite order only if $\pi_1(M)$ is finite.

4. *the α_j are invariants of $\pi_1(M)/(h)$.* This can be proved either by a geometric argument involving the singularities of discontinuous groups or algebraically by considering the conjugacy classes of elements of finite order in $\pi_1(M)/(h)$.

Suppose now that M admits a different effective $SO(2)$ action with orbit structure invariants $\{b'; (\varepsilon', g', \bar{h}', t'); (\alpha'_1, \beta'_1), \dots, (\alpha'_{n'}, \beta'_{n'})\}$. We wish to think of this as a new manifold M' homeomorphic to M. Clearly M' is closed and connected.

1. $\bar{h}' = 0$, because otherwise $\pi_1(M')$ is either abelian or a non-trivial free product, but $\pi_1(M)$ is not abelian by the restrictions of (5.1) and not a free product by observation (1) above.

2. $t' = 0$. Suppose this is false, then $t' > 0$. For a presentation of $\pi_1(M')$ it is necessary to introduce $2t'$ new generators. Each of the t'

boundary components of M'^*, corresponding to the image of a connectedness component of SE-orbits, is represented by the generator r'_k, which is geometrically a cross-section to the action on the torus of SE-orbits in M'. For the SE-orbits we introduce h'_k on each component. For $\varepsilon' = o$, we have

$$\pi_1(M') = (a'_i, b'_i, q'_j, h', r'_k, h'_k \mid \pi'_* r'_1 \dots r'_{t'}, [a'_i, h'], \\ [b'_i, h'], [q'_j, h'], q'^{\alpha'_j}_j h'^{\beta'_j}, [r'_k, h'_k], h'^2_k h'^{-1}),$$

where $i = 1, \dots, g'$; $j = 1, \dots, n'$; $k = 1, \dots, t'$. Now either $\pi_1(M')$ is one of the groups excluded in (5.1) or its center is exactly (h').

$$\pi_1(M')/(h') = (a'_i, b'_i, q'_j, r'_k, h'_k \mid \pi'_* r'_1 \dots r'_{t'}, q'^{\alpha'_j}_j, [r'_k, h'_k], h'^2_k).$$

In this group there are elements (e. g. r'_k, h'_k) whose centralizer is not cyclic, while it is a well-known fact that the centralizer of each element of $\pi_1(M)/(h)$ is cyclic. A similar argument applies for $\varepsilon' = \bar{n}$.

It follows from 1. and 2. above that $M' = \{b'; (\varepsilon', g', 0, 0); (\alpha'_1, \beta'_1), \dots, (\alpha'_{n'}, \beta'_{n'})\}$ is again a Seifert manifold (i. e. $F = \phi$, $SE = \phi$) and from observations (2) and (4) above that in fact $M' = \{b'; (\varepsilon, g, 0, 0); (\alpha_1, \beta'_1), \dots, (\alpha_n, \beta'_n)\}$.

We shall outline the rest of the proof, details can be found in [2]. Let us denote by

$$\widehat{\pi_1(M)} = (A_i, B_i, Q_j, H \mid [A_i, H], [B_i, H], [Q_j, H]) \quad \text{or}$$
$$\widehat{\pi_1(M)} = (V_i, Q_j, H \mid [V_i, H], [Q_j, H])$$

and the obvious homomorphism $\eta: \widehat{\pi_1(M)} \to \pi_1(M)$ sending capital to lower case letters. For $\varepsilon = o$, let $\omega(x) = \omega(X) = 1$, $x \in \pi_1(M)$, $X \in \widehat{\pi_1(M)}$. For $\varepsilon = \bar{n}$ let $\omega(x) = \omega(X) = \pm 1$ corresponding to an even or odd exponent sum of the v_i in x (resp. V_i in X).

Lemma: *If* $\Phi: \pi_1(M') \to \pi_1(M)$ *is an isomorphism, then it can be lifted to an isomorphism* $\hat{\Phi}: \widehat{\pi_1(M')} \to \widehat{\pi_1(M)}$ *with the properties*

a) $\hat{\Phi}(Q'_j) = H^{\lambda_j} S_j Q^{\xi_j}_{\nu_j} S_j^{-1}$,

b) $\hat{\Phi}(\pi'_*) = H^{\lambda} S \pi^{\xi}_* S^{-1}$

where $S_j, S \in \widehat{\pi_1(M)}/(H)$; $(j \to \nu_j)$ *is a permutation such that* $\alpha_j = \alpha_{\nu_j}$; $\omega(S_j)\xi_j = \omega(S)\xi = \rho = \pm 1$ *and* $\lambda = \sum_{j=1}^{n} \lambda_j + 2\sigma$ *with* $\sigma = 0$ *if* $\varepsilon = o$.

The proof depends on a lemma of Zieschang on automorphisms of discontinuous groups and the completion of the following diagram

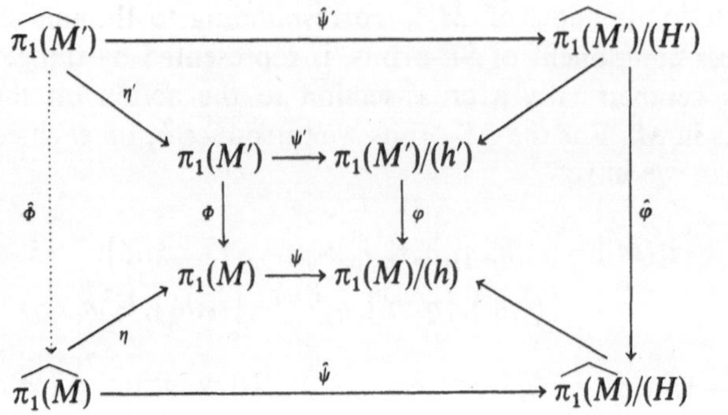

where Φ is given, φ is induced since (h) is a characteristic subgroup and all unlabeled maps are natural. ZIESCHANG's lemma provides a map $\hat{\varphi}$ with properties similar to those announced for $\hat{\Phi}$ and digram chasing gives the desired $\hat{\Phi}$.

Once the above lemma is proved, the computation is easily performed. Since $\Phi(h')=h^{\delta}$, $\delta=\pm 1$ we have

$$1=(h^{\lambda_j}s_jq_j^{\xi_j}s_j^{-1})^{\alpha_j}h^{\delta\beta'_j}=s_jq_j^{\alpha_j\xi_j}s_j^{-1}h^{\lambda_j\alpha_j+\delta\beta'_j}$$
$$=s_jh^{-\beta_j\xi_j}s_j^{-1}h^{\lambda_j\alpha_j+\delta\beta'_j}=h^{-\beta_j\xi_j+\lambda_j\alpha_j+\delta\beta'_j}.$$

By observation (3), h has infinite order, hence

$$-\beta_j\xi_j+\lambda_j\alpha_j+\delta\beta'_j=0.$$

For $\varepsilon=o$ we have $\xi_i=\rho$, hence

$$\beta_j=\rho\delta\beta'_j+\rho\lambda_j\alpha_j.$$

The conditions $0<\beta_j<\alpha_j, 0<\beta'_j<\alpha_j$ give, for $\rho\delta=1$, all $\lambda_j=0$; hence $\beta_j=\beta'_j$ for all j. For $\rho\delta=-1$, all $\rho\lambda_j=-1$; hence $\beta'_j=\alpha_j-\beta_j$ for all j corresponding to the two possible orientations of M (see (§ 3)). For $\varepsilon=\bar{n}$, the normalization requires $0<\beta_j\leqslant\alpha_j/2$ and $0<\beta'_j\leqslant\alpha_j/2$; for all j, $\beta'_j=\beta_j$ and $\lambda_j=0$.

Similarly we can compute b' by

$$1=\Phi(\pi'_* h'^{-b'})=h^{\lambda}s\pi_*^{\xi}s^{-1}h^{-\delta b'}=h^{\lambda}sh^{\xi b}s^{-1}h^{-\delta b'}=h^{\lambda+\xi b-\delta b'};$$

hence,

$$\lambda+\xi b-\delta b'=0.$$

For $\varepsilon=o$, we have $\xi=\rho$ and $\sigma=0$, thus

$$\lambda+\rho b-\delta b'=0;$$

if $\rho\delta=1$, then $\lambda=0$ and $b'=b$ if $\rho\delta=-1$, then $\delta\lambda_j=-1$ and $b'=-b-n$ corresponding to the two orientations of M.

For $\varepsilon=\bar{n}$, all $\lambda_j=0$; thus $\lambda=2\sigma$ and $b'=b \bmod 2$.

This completes the proof of the theorem under the assumptions of (5.1).

It remains to consider the special cases.

(1)
$$\{b;(o,0,0,0)\},$$
$$\{b;(o,0,0,0);(\alpha_1,\beta_1)\},$$
$$\{b;(o,0,0,0);(\alpha_1,\beta_1),(\alpha_2,\beta_2)\}$$

are $S^3, S^2\times S^1, P^3$ or $L(p,q)$. Examples of constructing them were given in § 1a) and c). The homology computation necessary to determine p and q from the orbit structure invariants was carried out in detail by von Randow [6].

(2)
$$\{b;(o,0,0,0);(\alpha_1,\beta_1),(\alpha_2,\beta_2),(\alpha_3,\beta_3)\},$$

where $1/\alpha_1+1/\alpha_2+1/\alpha_3>1$. There are four possible sets of α_j satisfying these conditions: $(2,2,\alpha_3),(2,3,3),(2,3,4),(2,3,5)$.

Since for these classes $\pi_1(M)$ is finite and non-abelian, M is not homeomorphic to any of the manifolds treated above and does not admit any $SO(2)$ action with fixed points. Since M is orientable, no SE-orbits can occur. Again h is in the center and

$$\pi_1(M)/(h)=(q_1,q_2,q_3 \mid q_1q_2q_3, q_1^{\alpha_1}, q_2^{\alpha_2}, q_3^{\alpha_3}).$$

Hence (h) is the whole center and the α_j are invariants (cf. [4, Satz 10]).

To distinguish manifolds within the same class, in the four classes, it suffices to compute $|p|=|b\alpha_1\alpha_2\alpha_3+\beta_1\alpha_2\alpha_3+\alpha_1\beta_2\alpha_3+\alpha_1\alpha_2\beta_3|$, the order of $H_1(M;Z)$. We can assume a change of orientation if necessary to obtain $p\geqslant 0$. For

$$\begin{aligned}
&(2,2,\alpha_3), && p=4\{(b+1)\alpha_3+\beta_3\};\\
&(2,3,3), && p=3\{6b+3+2(\beta_2+\beta_3)\};\\
&(2,3,4), && p=2\{12b+6+4\beta_2+3\beta_3\};\\
&(2,3,5), && p=\{30b+15+10\beta_2+6\beta_3\}.
\end{aligned}$$

Now suppose $M=\{b;(o,0,0,0);(2,1),(2,1),(\alpha_3,\beta_3)\}$ and

$$M'=\{b';(o,0,0,0);(2,1),(2,1),(\alpha_3,\beta'_3)\}$$

are homeomorphic. Then we have

$$p-p'=4\{(b-b')\alpha_3+\beta_3-\beta'_3\}=0;$$

thus $\beta'_3=\beta_3 \bmod \alpha_3$, hence $\beta'_3=\beta_3$, which in turn implies $b'=b$.

In the class (2,3,3) we could have $\beta'_2=\beta_3$ and $\beta'_3=\beta_2$ in addition to the obvious case. Otherwise the proof in the other classes follows the same outline. This analysis was first used by SEIFERT and THRELFALL [5], who also gave a geometric description of the manifolds in question.

(3) $$\{b;(o,1,0,0)\}.$$

These manifolds are principal $SO(2)$ bundles over the torus

$$\pi_1(M)=(a_1,b_1,h\,|\,[a_1,b_1]h^{-b},[a_1,h],[b_1,h]),$$
$$H_1(M;Z)=Z+Z+Z_b.$$

For $b\geqslant 0$, they are not homeomorphic to each other and $b<0$ gives the opposite orientation.

For $b=0$, we have $M=S^1\times S^1\times S^1$.

Otherwise (h) is the whole center, because every word $w\in\pi_1(M)$ can be normalized in the form $w=h^x a_1^y b_1^z$. Now suppose $w\notin(h)$, thus for example $z\neq 0$, and w is in the center. Then

$$w a_1 = a_1 w,$$
$$h^x a_1^y b_1^z a_1 = a_1 h^x a_1^y b_1^z = h^x a_1^{y+1} b_1^z.$$

Hence

$$b_1^z a_1 = a_1 b_1^z,$$
$$b_1^z a_1 = a_1 b_1 b_1^{z-1} = h^b b_1 a_1 b_1^{z-1} = \cdots = h^{zb} b_1^z a_1,$$

which implies $h^{zb}=1$ where $b\neq 0$, $z\neq 0$. This is a contradiction since $\pi_1(M)$ is infinite and therefore h has infinite order (cf. observation (3) above).

$$\pi_1(M)/(h)=(a_1,b_1\,|\,[a_1,b_1])=Z+Z$$

which is the only non-trivial abelian quotient group by the center for an orientable manifold regardless of whether the action is with or without fixed points.

(4) $$\{b;(\bar{n},1,0,0)\}$$

is clearly homeomorphic to $P^2\times S^1$ if $b=0$ and N if $b=1$.

(5) $$\{b;(\bar{n},1,0,0);(\alpha_1,\beta_1)\},$$

$$\begin{aligned}\pi_1(M)&=(v_1,q_1,h\,|\,v_1^2 q_1 h^{-b},[v_1,h],[q_1,h],q_1^{\alpha_1}h^{\beta_1})\\&=(v_1,h\,|\,[v_1,h],v_1^{-2\alpha_1}h^{b\alpha_1+\beta_1})\\&=\begin{cases}Z+Z_2 & \text{if } |b\alpha_1+\beta_1| \text{ is even},\\ Z & \text{if } |b\alpha_1+\beta_1| \text{ is odd}.\end{cases}\end{aligned}$$

The orientable double cover $\hat{M}$ of M is

$$\hat{M} = \{-1; (o,0,0,0); (\alpha_1,\beta_1), (\alpha_1,\alpha_1-\beta_1)\} = S^2 \times S^1.$$

Thus the collapsing map $\hat{M} \to M$ is just the double covering of $P^2 \times S^1$ or N.

(6) $$\{b; (\bar{n},2,0,0)\}$$

give the two principal $SO(2)$ bundles over the Klein bottle. Since both can be fibered over the circle as well, they also correspond to two Klein bottle bundles over S^1.

$$\pi_1(M) = (v_1, v_2, h \,|\, v_1^2 v_2^2 h^{-b}, [v_1,h], [v_2,h])$$
$$= \begin{cases} (v_1, v_2 \,|\, v_1^2 v_2^2) \times (h) & \text{for } b=0, \\ (v_1, v_2 \,|\, [v_1, v_2^2], [v_1^2, v_2]) & \text{for } b=1; \end{cases}$$
$$H_1(M;Z) = \begin{cases} Z+Z+Z_2 & \text{for } b=0, \\ Z+Z & \text{for } b=1. \end{cases}$$

The center of $\pi_1(M)$ is generated by $\{v_1^2, v_2^2, h\}$. Thus it properly contains the infinite cyclic group generated by h. This property, combined with the fact that $\pi_1(M)$ is non-abelian, distinguishes these manifolds from everything treated so far. It also proves that no $SO(2)$ action with fixed points is possible. We shall see below that actions with SE-orbits can occur.

This completes the proof of Theorem 4.

Theorem 5. *Let M be a closed, connected 3-manifold with an effective $SO(2)$ action given by $M = \{0; (\varepsilon, g, 0, t); (\alpha_1,\beta_1), \ldots, (\alpha_n,\beta_n)\}$, where $t > 0$. Then*

1. $\{0; (o,0,0,1)\} = N$,
2. $$\{0; (o,0,0,1); (\alpha_1,\beta_1)\} = \begin{cases} P^2 \times S^1 & \textit{if } \alpha_1 \textit{ is even}, \\ N & \textit{if } \alpha_1 \textit{ is odd}, \end{cases}$$
3. $\{0; (o,0,0,2)\} = K \times S^1$,
4. $\{0; (\bar{n},1,0,1)\} = KS$.
5. *all other manifolds admit only the given $SO(2)$ action.*

Proof. Since by assumption $t > 0$, the manifold M is non-orientable. Consider the orientable double cover $\hat{M}$. The action of $SO(2)$ lifts to the orientable double covering and commutes with the covering transformations.

For $\varepsilon = o$, we have

$$\hat{M} = \{-n; (o, 2g+t-1, 0, 0); (\alpha_1, \beta_1), \ldots, (\alpha_n, \beta_n), (\alpha_1, \alpha_1 - \beta_1), \ldots, (\alpha_n, \alpha_n - \beta_n)\},$$

and for $\varepsilon = \bar{n}$,

$$\hat{M} = \{-n; (o, g+t-1, 0, 0); (\alpha_1, \beta_1), \ldots, (\alpha_n, \beta_n), (\alpha_1, \alpha_1 - \beta_1), \ldots, (\alpha_n, \alpha_n - \beta_n)\}.$$

Since $\hat{M}$ is a Seifert-manifold, we can use the information obtained in Theorem 4. Except for the cases 1. to 4. of Theorem 5, $\hat{M}$ will satisfy the conditions of (5.1), thus the Lemma applies. Consider an M which is not one of 1. to 4.

For $\varepsilon = o$ we have as earlier

$$\pi_1(M) = (a_i, b_i, q_j, h, r_k, h_k | \pi_* r_1 \ldots r_t, [a_i, h], [b_i, h], [q_j, h], \\ q_j^{\alpha_j} h^{\beta_j}, [r_k, h_k], h_k^2 h^{-1}),$$

where $i = 1, \ldots, g$; $j = 1, \ldots, n$; $k = 1, \ldots, t$ and π_* is as before.

For $\varepsilon = \bar{n}$, we have

$$\pi_1(M) = (v_i, q_j, h, r_k, h_k | \pi_* r_1 \ldots r_t, [v_i, h], [q_j, h], q_j^{\alpha_j} h^{\beta_j}, [r_k, h_k], h_k^2 h^{-1}).$$

From the double covering we have the exact sequence

$$1 \to \pi_1(\hat{M}) \to \pi_1(M) \to Z_2 \to 1$$

and it follows from Theorem 4 that (h) is the center and it is of infinite order.

$$\pi_1(M)/(h) = (a_i, b_i, q_j, r_k, h_k | \pi_* r_1 \ldots r_t, q_j^{\alpha_j}, [r_k, h_k], h_k^2) \quad \text{or}$$
$$\pi_1(M)/(h) = (v_i, q_j, r_k, h_k | \pi_* r_1 \ldots r_t, q_j^{\alpha_j}, [r_k, h_k], h_k^2).$$

Again $\pi_1(M)/(h)$ can be represented as an amalgamated free product with at least one centerless factor. Thus $\pi_1(M)/(h)$ has no center.

Moreover, we observe that

1. the integer t is an invariant of $\pi_1(M)/(h)$, since it is the number of conjugacy classes of elements of order 2 with non-trivial centralizer;

2. if we divide $\pi_1(M)/(h)$ by the normal subgroup generated by these elements and their centralizers, then we obtain $(a_i, b_i, q_j | \pi_*, q_j^{\alpha_j})$, resp. $(v_i, q_j | \pi_*, q_j^{\alpha_j})$, thus ε, g and the α_j are invariants of $\pi_1(M)$;

3. the β_j are invariants of $\pi_1(M)$. For $\varepsilon = \bar{n}$ this is quite clear because the normalization requires $0 < \beta_j \leqslant \alpha_j/2$ and in $\hat{M}$ the collection $\{(\alpha_1, \beta_1), \ldots, (\alpha_n, \beta_n), (\alpha_1, \alpha_1 - \beta_1), \ldots, (\alpha_n, \alpha_n - \beta_n)\}$ is invariant. For each pair $(\alpha_j, \beta_j), (\alpha_j, \alpha_j - \beta_j)$, only one could come from M. For $\varepsilon = o$ the above argument does not apply since the normalization requires only $0 < \beta_j < \alpha_j$. Suppose that $M = \{0; (o, g, 0, t); (\alpha_1, \beta_1), \ldots, (\alpha_n, \beta_n)\}$ admits a

different $SO(2)$ action, say $M'=\{b';(\varepsilon',g',h',t');(\alpha'_1,\beta'_1),\ldots,(\alpha'_{n'},\beta'_{n'})\}$. By the above observations we know in fact that $M'=\{0;(o,g,0,t);(\alpha_1,\beta'_1),\ldots,(\alpha_n,\beta'_n)\}$. The homeomorphism between M' and M induces an isomorphism $\Phi:\pi_1(M')\to\pi_1(M)$. The subgroup $\pi_1(\hat{M}')$ is characteristic since it consists of the orientation preserving loops, thus Φ induces an isomorphism of the exact sequences

$$\begin{array}{ccccccc} 1\to\pi_1(\hat{M}') & \xrightarrow{i'} & \pi_1(M') & \to & Z_2 & \to & 1 \\ \downarrow{\scriptstyle\Phi_1} & & \downarrow{\scriptstyle\Phi} & & \downarrow & & \\ 1\to\pi_1(\hat{M}) & \xrightarrow{i} & \pi_1(M) & \to & Z_2 & \to & 1 \end{array}$$

Since $q'_j\in\pi_1(M')$ is an orientation preserving loop, it is the image of some $\hat{q}'_j\in\pi_1(\hat{M}')$. The injections i, i' map centers onto centers and we may assume that $i(\hat{h})=h$, $i'(\hat{h}')=h'$. Since $\hat{q}'_j$ is of order α_j in $\pi_1(\hat{M}')/(\hat{h}')$, the Lemma applies and we have

$$\Phi_1(\hat{q}'_j)=\hat{h}^{\lambda_j}\hat{s}_j\hat{q}_{v_j}^{\xi_j}\hat{s}_j^{-1}$$

where $(j\to v_j)$ is a permutation such that $\alpha_j=\alpha_{vj}$ and $\xi_1=\xi_2=\cdots=\xi_n=\rho=\pm1$. By renaming the generators of $\pi_1(\hat{M})$, choose the permutation $(j\to v_j)$ to be the identity. Let $i(\hat{s}_j)=s_j$. This gives

$$\Phi(q'_j)=h^{\lambda_j}s_jq_j^{\xi_j}s_j^{-1},$$

where $\xi_1=\cdots=\xi_n=\rho=\pm1$. The completion of 3. is a verbatim repetition of the corresponding computation in the proof of Theorem 4. This gives $\beta'_j=\beta_j$ for all j or $\beta'_j=\alpha_j-\beta_j$ for all j.

The proof of Theorem 5 is completed by noting that the special cases 1., 3. and 4. were described in 1 at the definition of SE-orbits, d) and f), respectively, and 2. is seen by lifting to the orientable double covering.

References

A more comprehensive list of references is found in [2] and [3].

1. Mostert, P. S.: On a compact Lie group acting on a manifold, Annals of Math. **65**, 447—455 (1957).
2. Orlik, P., E. Vogt, und H. Zieschang: Zur Topologie gefaserter dreidimensionaler Mannigfaltigkeiten, Topology **6**, 49—64 (1967).
3. Raymond, F.: Classification of the actions of the circle on 3-manifolds, Trans. AMS 51—78 **131** (1968).

4. Seifert, H.: Topologie dreidimensionaler gefaserter Räume, Acta math. **60,** 147—238 (1933).
5. Seifert, H., und W. Threlfall: Topologische Untersuchungen der Diskontinuitätsbereiche endlicher Bewegungsgruppen des dreidimensionalen sphärischen Raumes II, Math. Ann. **107,** 543—586 (1933).
6. von Randow, R.: Zur Topologie von dreidimensionalen Raummannigfaltigkeiten, Bonner Math. Schriften **14,** 131 (1962).

We would like to take this opportunity to correct several statements in [3].

page 51, line 2: "(15,6)" should be "(16,5)"
"5,632" should be "1136".

page 58, line 12: should read "orientation preserving homeomorphism".

page 71, line 16: "four topologically inequivalent actions" should read "two topologically inequivalent actions, in the strict sense,"

page 71, line 17: "two" should be "one".

page 72, line 5: insert "$t>0$," between "convenience" and "$\mu_i \neq \mu_j$".

The Fixed Point Set of Z_p in a $C^*(m_1, m_2, \ldots, m_{q-1}; Z_p)$-Space*

HSU-TUNG KU

1. Introduction

Let Z_p be the cyclic group of prime order p acting on the compact space X with fixed point set F. We denote by $\chi(X;Z_p)$ the Euler characteristic of the space X mod Z_p and $\hat{H}^k(Z_p;Z_p)$ the k-th Tate cohomology group of Z_p with coefficients in Z_p [2].

Definition 1.1. Let $p \neq 2$ and $q \geqslant 2$. The space X is called a *$C^*(m_1, \ldots, m_{q-1}; Z_p)$-space* if it is compact and

$$H^i(X;K_p) = Z_p \quad \text{if} \quad i = 0, m_1, \ldots, m_{q-1}; \qquad m_i > 0 \quad \text{and} \quad m_{i+1} - m_i \geqslant 2,$$
$$= 0 \quad \text{otherwise}$$

for some integers $m_1, \ldots, m_{q-1}$.

Definition 1.2. Let q be a positive integer greater than or equal to three. The compact space X is called a *$C(m_1, \ldots, m_{q-1}; K_p)$-space* if it satisfies the following conditions:

$$H^i\,(X;K_p) = \quad \text{if} \quad i = 0, m_1, \ldots, m_{q-1} \quad \text{where} \quad 0 < m_1 > m_2 < \cdots < m_{q-1}$$
$$= 0 \quad \text{otherwise,}$$

with $a_i a_j = 0$ for $i,j = 1, \ldots, q-1$, where a_i is the non-zero generator of $H^{m_i}(X;K_p)$.

The principal results of this paper are the following:

Theorem A. *Let the group Z_p act on a $C^*(m_1, \ldots, m_{q-1}; Z_p)$-space X with fixed point set F. Then the following relations hold:*

1. $\displaystyle\sum_{i \geqslant 2k}^{\infty} \dim \hat{H}^{2k-i}(Z_p; H^i(X;Z)) = \sum_{\substack{i \geqslant 2k \\ i \equiv 0(2)}} \dim H^i(X;Z_p),$

2. $\displaystyle\sum_{i \geqslant 2k+1}^{\infty} \dim \hat{H}^{2k+1-i}(Z_p; H^i(X;Z)) = \sum_{\substack{i \geqslant 2k+1 \\ i \equiv 1(2)}} \dim H^i(X;Z_p),$

---

* The results presented here are part of a doctoral dissertation written at the Tulane University under Professor P. S. MOSTERT.

3. $\sum_{\substack{i \geqslant k \\ i \equiv a(2)}} \dim H^i(F;Z_p) \leqslant \sum_{\substack{i \geqslant k \\ i \equiv a(2)}} \dim H^i(X;Z_p)$; $a=0,1$,

4. $\chi(X;Z_p)=\chi(F;Z_p)$, *and* $\chi(X,Z_p)=\chi(X/Z_p;Z_p)$ *if* $\dim_p X<\infty$.

Theorem B. *Let Z_p act on a $C(m_1,\ldots,m_{q-1};Z_p)$-space X, where $p\neq 2$. Then the fixed point set F is not empty.*

We recall in the following the structure of $\hat{H}^*(Z_p;Z_p)$ for $p\neq 2$: Additively $\hat{H}^k(Z_p;Z_p)=Z_p$ for all k. Let $t^k\in\hat{H}^k(Z_p;Z_p)$ be the generator; the unit t^0 is denoted by 1. Then the multiplication is given by

$$t^k t^{k'}=t^{k+k'} \text{ if at least one of } k,k' \text{ is even } =0 \text{ otherwise.}$$

We need the following results of SWAN and SU.

Proposition 1.3. (SWAN [5]). *Let Z_p act on a compact space X with fixed point set F. Then there are convergent spectral sequences $E_r(X)$ and $E_r(F)$ whose E_2-terms are given by*

$$E_2^{s,t}(X)=\hat{H}^s(Z_p;H^t(X;Z_p)) \quad \textit{and}$$
$$E_2^{s,t}(F)=\hat{H}^s(Z_p;H^t(F;Z_p))$$

respectively and whose E_∞-terms are associated with $J^*(X)$ and $J^*(F)$ respectively, where

$$J^*(X)=H^*(\mathrm{Hom}_{Z_p}(C_*,A^*(X)) \quad \textit{and}$$
$$J^*(F)=H^*(\mathrm{Hom}_{Z_p}(C_*,A^*(F))$$

and where C_* is a complete resolution for Z_p, $A^*(X)$, $A^*(F)$ the ALEXANDER-SPANIER-WALLACE *cochain complexes of X, F, respectively, with coefficients in Z_p. Moreover the inclusion $i:F\to X$ induces an isomorphism $i^*:J^*(X)\to J^*(F)$.*

Lemma 1.4. (SU [4]). *Let A be a Z_p-module.*

1. *If $A\otimes Z_p=\mathrm{Tor}(A;Z_p)=0$, then $\hat{H}^k(Z_p;A)=0$ for all k.*
2. *If $p\neq 2$, $A\otimes Z_p=Z_p$ and $\mathrm{Tor}(A;Z_p)=0$, then $\hat{H}^k(Z_p;A)=0$ for k odd and $\hat{H}^k(Z_p;A)=Z_p$ for k even.*
3. *If $p\neq 2$, $A\otimes Z_p=0$ and $\mathrm{Tor}(A;Z_p)=Z_p$, then $\hat{H}^k(Z_p;A)=Z_p$ for k odd and $\hat{H}^k(Z_p;A)=0$ for k even.*

2. Proof of Theorem A

Applying the universal coefficient theorem, we have

Lemma 2.1. *Let X be a $C^*(m_1,\ldots,m_{q-1};Z_p)$-space. Then*

1. $H^i(X;Z)\otimes Z_p=0$ *for* $0<i<m_1, m_j<i<m_{j+1}$,

$$j=1,2,\ldots,q-2 \quad \textit{and} \quad m_{q-1}<i,$$

2. $\operatorname{Tor}(H^i(X;Z);Z_p)=0$ *for* $0<i<m_1+1,\ m_j+1<i<m_{j+1}+1,$
$j=1,2,\ldots,q-2$ *and* $m_{q-1}+1<i.$

From now on we assume that the group Z_p acts on the space X.

Theorem 2.2. *If X is a $C^*(m_1,\ldots,m_{q-1};Z_p)$-space, then*

1. $\sum_{i=0}^{\infty} \dim \hat{H}^{-i}(Z_p;H^i(X;Z)) = \sum_{i=0(2)} \dim H^i(X;Z_p).$

2. $\sum_{i=0}^{\infty} \dim \hat{H}^{1-i}(Z_p;H^i(X;Z)) = \sum_{i=1(2)} \dim H^i(X;Z_p).$

Proof. According to Lemma 1.4, we have $\hat{H}^k(Z_p;H^i(X;Z))=0$ for $0<i<m_1,\ m_j+1<i<m_{j+1},\ i=1,\ldots,q-1$ and $m_{q-1}+1<i$ for all k.

Let $m_0=0$. For any $m_j, j=0,1,\ldots,q-1$, by the universal coefficient theorem, we see that exactly one of the following two cases can occur (except that for m_0, only the case (I) can occur):

(I) $H^{m_j}(X;Z)\otimes Z_p=Z_p$ and $\operatorname{Tor}(H^{m_j+1}(X;Z);Z_p)=0.$

(II) $H^{m_j}(X;Z)\otimes Z_p=0$ and $\operatorname{Tor}(H^{m_j+1}(X;Z);Z_p)=Z_p.$

Case (I) together with Lemma 2.1 give the following two pairs:

$H^{m_j}(X;Z)\otimes Z_p=Z_p, \operatorname{Tor}(H^{m_j}(X;Z);Z_p)=0$ and

$H^{m_j+1}(X;Z)\otimes Z_p=0,\ \operatorname{Tor}(H^{m_j+1}(X;Z);Z_p)=0.$

By Lemma 1.4 we get

3. $\hat{H}^k(Z_p;H^{m_j}(X;Z))=0$ if k is odd, $=Z_p$ if k is even
4. $\hat{H}^k(Z_p;H^{m_j+1}(X;Z))=0$ for all k.

Applying the same argument to the case (II), we have

5. $\hat{H}^k(Z_p;H^{m_j}(X;Z))=0$ for all k.
6. $\hat{H}(Z_p;H^{m_j+1}(X;Z))=Z_p$ if k is odd, $=0$ if k is even

Proof of 1.: Suppose that m_j is an odd integer and m_i is an even integer in the set $\{m_s: s=0,1,\ldots,q-1\}$. In case (I), from 3. and 4., respectively, we have

$\hat{H}^{-m_j}(Z_p;H^{m_j}(X;Z))=0,\ \hat{H}^{-m_i}(Z_p;H^{m_i}(X;Z))=Z_p$

$\hat{H}^{-(m_i+1)}(Z_p;H^{m_i+1}(X;Z))=0$ and $\hat{H}^{-(m_j+1)}(Z_p;H^{m_j+1}(X;Z))=0.$

In case (II), applying 5. and 6. respectively, we have

$\hat{H}^{-m_j}(Z_p;H^{m_j}(X;Z))=0,\ \hat{H}^{-m_i}(Z_p;H^{m_i}(X;Z))=0,$

$\hat{H}^{-(m_j+1)}(Z_p;H^{m_j+1}(X;Z))=0$

$\hat{H}^{-(m_i+1)}(Z_p;H^{m_i+1}(X;Z))=Z_p.$

Therefore in either case (I) or case (II), we obtain

$$\sum_{i=0}^{\infty} \dim \hat{H}^{-i}(Z_p;H^i(X;Z)) = \sum_{i=0(2)} \dim H^i(X;Z_p).$$

Proof of 2.: We still assume m_i even and m_j odd. In case (I), by 3., 4., we have

$\hat{H}^{1-m_i}(Z_p; H^{m_i}(X;Z))=0$ since $1-m_i$ is odd,
$\hat{H}^{1-(m_i+1)}(Z_p; H^{m_i+1}(X;Z))=0$,
$\hat{H}^{1-m_j}(Z_p; H^{m_j}(X;Z))=Z_p$ since $1-m_j$ is even and
$\hat{H}^{1-(m_j+1)}(Z_p; H^{m_j+1}(X;Z))=0$.

In the case (II), by 5. and 6., we get

$\hat{H}^{1-m_i}(Z_p; H^{m_i}(X;Z))=0$, $\hat{H}^{1-(m_i+1)}(Z_p; H^{m_i+1}(X;Z))=0$,
$\hat{H}^{1-m_j}(Z_p; H^{m_j}(X;Z))=0$ and
$\hat{H}^{1-(m_j+1)}(Z_p; H^{m_j+1}(X;Z))=Z_p$.

Thus in either case (I) or (II)

$$\sum_{i=0}^{\infty} \dim \hat{H}^{1-i}(Z_p; H^i(X;Z)) = \sum_{i=1(2)} \dim H^i(X;Z_p).$$

Proof of Theorem A. By a similar line of reasoning we can prove 1. and 2., for we still have (A), (B), 3., 4., 5. and 6. of the preceding theorem. Notice that in case (I), we have

$\hat{H}^{2k-m_j}(Z_p; H^{m_j}(X;Z))=0$, $\hat{H}^{2k-m_i}(Z_p; H^{m_i}(X;Z))=Z_p$,
$\hat{H}^{2k-(m_i+1)}(Z_p; H^{m_i+1}(X;Z))=0$ and
$\hat{H}^{2k-(m_j+1)}(Z_p; H^{m_j+1}(X;Z))=0$.

Similar modification applies to the other cases.

SWAN [5, corollary 4.2] has shown that for any integer k

$$\sum_{\substack{i \geqslant k \\ i-k\,\text{even}}} \dim H^i(F;Z_p) \leqslant \sum_{\substack{i \geqslant k \\ i-k\,\text{even}}} \dim \hat{H}^{k-i}(Z_p; H^i(X;Z)).$$

This implies 3. To prove 6., we recall that the *Heller characteristic* [5] of the space X is defined to be $\chi_H(X;Z_p)=\chi^+(X;Z_p)-\chi^-(X;Z_p)$, where

$$\chi^+(X;Z_p) = \sum_{i=0}^{\infty} \dim \hat{H}^{2n-i}(Z_p; H^i(X;Z)),$$

$$\chi^-(X;Z_p) = \sum_{i=0}^{\infty} \dim \hat{H}^{2n+1-i}(Z_p; H^i(X;Z)).$$

The values of $\chi^+(X;Z_p)$ and $\chi^-(X;Z_p)$ are independent of the choice of n. In particular we can take $n=0$. From Theorem 2.2 we have

$$\chi^+(X;Z_p) = \sum_{i=0(2)} \dim H^i(X;Z_p),$$

$$\chi^-(X;Z_p) = \sum_{i=1(2)} \dim H^i(X;Z_p)$$

so $\chi(X;Z_p)=\chi_H(X;Z_p)$. The values of $\chi^+(X;Z_p)$ and $\chi^-(X;Z_p)$ are both finite. Hence by a result of SWAN [5, Theorem 4.3], $\chi(F;Z_p) = \chi_H(X;Z_p)$. The other part follows from the following lemma:

Lemma 2.3. *Let Z_p act on the space X such that both $\dim H^*(X;Z_p)$ and $\dim_p X$ are finite. Then*

$$\chi(X;Z_p)-p\chi(X/Z_p;Z_p)+(p-1)\chi(F;Z_p)=0.$$

Proof. Since Z_p acts freely on $X-F$ and $\dim H^*(X-\mathrm{F};Z_p)$ is finite, by [1, p. 40] we have

$$\chi(X-F;Z_p)=p\chi(X/Z_p-F;Z_p).$$

From the exact cohomology sequences of the pairs (X,F) and $(X/Z_p,F)$ we see that

$$\chi(X;Z_p)=\chi(X-F;Z_p)+\chi(F;Z_p),$$

$$\chi(X/Z_p;Z_p)=\chi(X/Z_p-F;Z_p)+\chi(F;Z_p),$$

hence

$$\chi(X;Z_p)=p\chi(X/Z_p-F;Z_p)+\chi(F;Z_p)=p\chi(X/Z_p;Z_p)-(p-1)\chi(F;Z_p).$$

3. Proof of Theorem B

Let d_r be the differential of $E_r(X)$. In order to prove that F is not empty, we need the following lemma, where we use the notation of Definition 1.2:

Lemma 3.1. *Let Z_p act on a compact $C(m_1,\ldots,m_{q-1};Z_p)$-space X, where $p\neq 2$. Then $d_{m_i+1}(1\otimes a_i)=0$ for $i=1,\ldots,q-1$ and so $E_2^{k,o}(X)=E_\infty^{k,o}(X)$ for all k.*

Proof. Suppose $d_{m_1+1}(1\otimes a_1)\neq 0$. Without loss of generality, we may assume that

$$d_{m_1+1}(1\otimes a_1)=t^{m_1+1}\otimes 1$$

where

$$d_{m_1+1}:E_{m_1+1}^{o,m_i}(X)\to E_{m_1+1}^{m_1+1,m_i-m_1}.$$

For each $i=2,\ldots,q-1$, we have only the following two possibilities:

1. $d_{m_1+1}(1\otimes a_i)=0, i=2,\ldots,q-1,$ or
2. $d_{m_1+1}(1\otimes a_i)\neq 0,$ say $d_{m_1+1}(1\otimes a_i)=t^{m_1+1}\otimes a_n$.

First we show that m_1 must be odd, for if m_1 is even, then

$$0=d_{m_1+1}[(1\otimes a_1)(1\otimes a_1)]=2(t^{m_1+1}\otimes a_1)\neq 0.$$

This is a contradiction because $p \neq 2$. Applying this remark we can obtain

$$d_{m_1+1}(t^s \otimes a_1) = (-1)^s (t^{s+m_1+1} \otimes 1) \quad \text{since} \quad m_1 \quad \text{is odd.}$$

Now we show that the assumption of case 1. leads to a contradiction. If $d_{m_1+1}(1 \otimes a_i) = 0$ for some $i = 2, \ldots, q-1$, then

$$0 = d_{m_1+1}(t^s \otimes a_1 a_i) = (-1)^s (t^{s+m_1+1} \otimes a_i) \neq 0.$$

This is of course impossible.

Assume case 2. holds. Then

$$0 = d_{m_1+1}(t^s \otimes a_1 a_i) = (-1)^s (t^{s+m_1+1} \otimes a_i) \neq 0.$$

Again we get a contradiction. Hence $d_{m_1+1}(1 \otimes a_i) = 0$. Suppose now that $d_{m_i+1}(1 \otimes a_i) \neq 0$ for some $i > 1$. Let

$$d_{m_i+1}(1 \otimes a_i) = t^{m_i+1} \otimes 1.$$

Clearly $d_{m_i+1}(t^s \otimes a_j) = 0$ for $j < i$. The integer m_i must be odd by using the same argument as in the case m_1. Hence

$$0 = d_{m_i+1}(t^s \otimes a_j a_i) = (-1)^{s+m_j} (t^{s+m_i+1} \otimes a_j) \neq 0.$$

Thus the proof is complete.

Proof of Theorem B. We see immediately from Lemma 3.1 that

$$E_\infty^{k,o}(X) = E_2^{k,o}(X) = \hat{H}^k(Z_p; H^0(X; Z_p)) = \hat{H}^k(Z_p; Z_p) = Z_p.$$

Hence $\dim H^*(F; Z_p) = \dim J^k(X) \geqslant 1$.

For the circle group action, a similar argument holds using the Leray spectral sequence. Thus we can state the following:

Theorem 3.2. *Suppose that the circle group T^1 acts on a compact $C(m_1, \ldots, m_{q-1}; K_p)$-space X, where $p \neq 2$. Then the fixed point set F is not empty.*

References

1. Borel, A. et al.: Seminar on Transformation Groups, Ann. of Math. Studies **46**, 1960.
2. Cartan, H., and S. Eilenberg: Homological Algebra, Princeton University Press, Princeton, 1956.
3. Ku, Hsu-Tung: Dissertation, Tulane University, 1967.
4. Su, J. C.: Periodic transformations on the product of two spheres, Trans. Amer. Math. Soc., **112**, 369—380 (1964).
5. Swan, R. G.: A new method in fixed point theory, comment. Math. Helv. **34**, 1—16 (1960).

Group Actions and a Spectral Sequence*

I. FÁRY

1. We arrive naturally at a problem on spectral sequences when dealing with group actions [5]. Let G be a compact Lie group acting on the locally compact, Hausdorff space X. This amounts to saying, of course, that we are given a continuous map $G \times X \to X$, $(g,x) \to gx$, such that $ex = x$, for all x, if e is the neutral element of G, and that $g(hx) = (gh)x$ for all $g, h \in G$, $x \in X$. We set $Y = X/G$, the quotient set with the quotient topology; Y is locally compact and Hausdorff. We denote $f: X \to Y$, $f(x) = G(x)$, the natural map; here $G(x) = \{u \in X : u = gx\}$, which is isomorphic to G/G_x, $G_x = \{g \in G : gx = x\}$; N_x will denote the normalizer of G_x in G. f is a continuous, closed, open, and proper map.

An important and well known result of GLEASON [12, p. 222, Remark], shows that there is a sequence

$$Y = Y_0 \supset \cdots \supset Y_n \supset \cdots \qquad \Big(\bigcap_n Y_n = \phi\Big) \tag{1}$$

of closed subspaces of Y, where, in general, n runs through a set of transfinite ordinals, such that, if we set

$$X_n = f^{-1} Y_n \qquad (n \geqslant 0), \tag{2}$$

then $X_n - X_{n+1}$ is an everywhere dense, open subspace of X_n each component $U_{n\alpha}$ of which is a coordinate bundle in Steenrod's sense with base $V_{n\alpha} = f U_{n\alpha}$, fiber G/G_x, and structural group N_x/G_x. Although this would not be strictly necessary, we want to suppose that n in (1) runs through a set of positive integers. Let us consider now the Leray spectral sequence of f [1, 2, 4, 7, 10]. This has a second term E_2 which has a graded module which is the end of a spectral sequence starting with the direct sum of modules

$$H(V_{n\alpha}, H(G/G_x)). \tag{3}$$

(For notation, see below; see [7, p. 483, Proposition 1], applied to $H(Y, A)$). It is then natural to ask, if there is any single spectral sequence utilizing the above mentioned coordinate bundle structures, in particular,

---

* The work of the author has been supported in part by National Science Foundation contract GP-6974.

modules (3), going directly from a direct sum of groups (3) to the cohomology of X. The answer to this question is positive; we will formulate it more generally below.

2. Let X, Y be given locally compact, Hausdorff spaces, and

$$f: X \to Y \tag{4}$$

a given continuous map. We also suppose a sequence (1) given, where n runs through the set of positive integers. As a concrete idea motivating our research, we think about the Y_n's as successive *critical sets* of f; that is, we will consider maps

$$f|U_{n\alpha} = (f|X_n - X_{n+1})|U_{n\alpha} \quad \left(\bigcup_\alpha U_{n\alpha} = X_n - X_{n+1}\right) \tag{5}$$

having a "simpler structure" than f. As coefficient ring of cohomology we fix a commutative ring with unit R; we do not write R in our formulas. $H(S)$ denotes the *compact* cohomology ring of the locally compact space S with coefficients in R defined with a *fine couverture* of S, [1, 2, 3, 6, 7, 10]. In compact cohomology, S modulo a closed subspace T is isomorphic to the cohomology of $S-T$; hence H with two arguments stands for cohomology with coefficients in the second argument, as in (3). If S is compact, $H(S)$ is the sheaf cohomology [4]. This cohomology satisfies the Eilenberg-Steenrod axioms with compact carriers, induced homomorphisms for proper maps, continuity axiom included. Given f as above, not supposing that it is a proper map, we introduce the presheaves

$$A^q(C) = H^q(f^{-1}C), \quad A^q(C) \to A^q(D) \quad \textit{natural homomorphism} \tag{6}$$

where $C \supset D$ are compact subsets of Y, see [2]. We write A for the direct sum of the A^q's, $q \geqslant 0$. To simplify notation, we set

$$H^{p,q}(m,n) = H^p(Y_m - Y_n, A^q|Y_m - Y_n) \quad (0 \leqslant m \leqslant n \leqslant \infty); \tag{7}$$

the same notation without superscript q means direct sum over $q \geqslant 0$; again, leaving p out means summation over p.

The well known Leray spectral sequence of f has $H(0,\infty)$ as second term, and a graded algebra of $H(X)$ as E_∞ term. In our formulas we will need some of the differentials of this spectral sequence. To be explicit, we should use two more subscripts with the notation d_r below, but, *fixing the subscript* j, the following simplified notation will be workable:

$$d'_2, d'_3 \quad \textit{Leray differentials of} \quad f|X_j - X_{j+1}; \tag{8}$$

also, in formula (15) only, d'_2 stands for the Leray differential of $f|X_{j-1} - X_j$; a similar remark applies to (78):

$$d''_2 \quad \textit{denotes the Leray differential of} \quad f|X_{j-1} - X_{j+1}. \tag{9}$$

We will use furthermore, *again with j fixed*, and thereby simplified notation, the following exact triangle

$$(10)\qquad \begin{array}{ccc} H(j-1,j+1) & \xrightarrow{s} & H(j,j+1) \\ & {\scriptstyle i}\nwarrow \quad \swarrow{\scriptstyle \delta} & \\ & H(j-1,j) & \end{array} \qquad (j\ \textit{fixed})$$

and on one occasion a similar notation (see (78)), where s is the restriction homomorphism, δ the invariant coboundary, increasing the dimension by 1, and i the injection homomorphism.

3. The question raised at the end of Section 1 is answered by the following theorem from [1, 4, 7, 9]:

Theorem 1. *Given* (4) *together with* (1), *there is a spectral sequence* $\{E_r, d_r\}$, *such that*

$$(11)\qquad E_2 = \sum_{n=0}^{\infty} H(n,n+1) \qquad (\textit{direct sum}),$$

$$(12)\qquad E_\infty \subset \textit{graded module of } H(X)$$

see (6), (7); *a nonzero element of* (7) *is of filtration* $p-m$, *and of total degree* $p+q$. d_r *raises the filtration by* r *and the total degree by* 1. *If* $Y_1=\phi$, (11), (12) *is the Leray spectral sequence of* f. *If* f *is the identitiy map*, (11), (12) *is the Morse spectral sequence of* (1). *The spectral sequence is a topological invariant of the data* (4), (1). *Also, topologically non-invariant terms* E_0, E_1 *are defined.*

Theorem 2. *We use the notation of Theorem* 1. *Given a cohomology class*

$$(13)\qquad h \in H^q(j,j+1),$$

the second differential d_2 *is given by*

$$(14)\qquad d_2 h = \delta h + d_2' h;$$

furthermore,

$$(15)\qquad d_2'\delta + \delta d_2' = 0,$$

holds true (*see* (8), (10) *and remarks after* (8)).

In [4] an important generalization of these theorems is proved, but we cannot go into that.

4. We apply now Theorem 1 to the special case of the map $X \to X/G$ of Section 1. Let us prove that (7) is the direct sum of modules (3). By GLEASON's results, $A|V_{n\alpha}$ is locally isomorphic to a locally constant presheaf [7, 2], and by [11], the fundamental group $\pi_1(V_{n\alpha}, y)$ acts trivi-

ally on this presheaf. This proves our statemant in virtue of the general theory [2, 10]. Let us mention an easy corollary of this. If the coefficient ring R is a field, (3) is isomorphic to

(16) $$H(V_{n\alpha})\oplus_R H(G/G_x) \quad (R\text{: field}).$$

(See [2,11].)

In this particular case, (11) is thus explicitly given: it is composed of clearly $d_2' h$, which we can see by writing down (36) for the Leray spectral scription of the group action.

Let us call attention to the following. No homotopy invariants (not even fundamental groups) enter in forming (11), and, somewhat loosely, we may also say that "all homology invariants" appear there. We found, as a "heuristic principle" the following remark always useful: the general theory of spectral sequences does not solve problems, but raises "the right questions". In our present case the right question concerns the rings (3).

The second term of the Leray spectral sequence of f cannot be described with this precision. This was originally [5] one of the motivations for introducing the spectral sequences (11), (12).

5. In the present paper we determine the d_3 of the spectral sequence of Theorem 1. One would like to have, of course, formulas similar to (20) for all differentials, and it may seem unnecessary to publish such an incomplete result. One justification for publishing this paper is that in the spectral sequence of [8] the "last differential" is precisely d_3, see [8, p. 31], hence (20) is meaningful for the Lefschetz theory. Another justification could conceivably be that the determination of d_r will be an induction requiring special discussion of d_3.

In connection with Theorem 2, Professor LERAY told me (half in jest, I suppose): in any spectral sequence, the differentials are the natural homomorphisms we can find which are compatible with the degree requirements. In case of (14), for example, $d_2 h$ will be of filtration degree $p+2$, where $p=q-j$. Hence we must find an element in $H^{q+1}(j-1,j)$ for which δh is a good candidate, as well as an element in $H^{q+2}(j,j+1)$ for which $d_2' h$ is the natural candidate; these two elements are also possible when total degree is considered. "Cohomology operations" (not in the technical sense) generally increase the dimension; hence we expect no other component for $d_2 h$. In a similar vein we can guess $d_3 \iota h$, where

(17) $\iota: Z(E_2)\to E_3$ *is the natural homomorphism.*

In order that ιh be defined we must have

(18) $$\delta h = 0,$$

(19) $$d_2' h = 0,$$

simultaneously, as E_2 is a direct sum and (14) holds true. By exactness of (10), $s^{-1}h$ is then a non-empty set, and $\delta k \in \delta s^{-1}h$ will be of the "right" filtration $p+3$ for any choice $k \in s^{-1}h$. Again, $d_2'' k$, see (9), is of the "right" total degree, but it is not in the "right" group. However, $s d_2'' k = d_2' h = 0$ by (19), thus $i^{-1} d_2'' s^{-1} h$ is a non-empty set having the "right" filtration and total degree. Finally (19) also shows that $d_3' h$ is defined; it has the "right" total degree and filtration degree. For the same reasons as in the case of d_2, we do not expect components in the other terms of (11). We arrive thus, using purely the degree properties as guides, to the formula

$$(20) \qquad d_3 \imath h = \imath(\delta s^{-1} h + i^{-1} d_2'' s^{-1} h + d_3' h),$$

with a somewhat ambiguous notation. This formula, if properly interpreted, turns out to be true.

Theorem 3. *We consider the spectral sequence* (11), (12), *and an* h *in* (13) *for which* $d_2 h = 0$ *in* (14). *Then* $d_3 \imath h$, *see* (17), *is defined and* (20) *holds true using notations* (8), (9), (10), (17) *and the following conventions: we choose first an arbitrary element* $k \in s^{-1} h$, *then an arbitrary element* $k_1 \in i^{-1} d_2'' k$ *following this by a suitable choice of* $k_2 \in d_3' h$.

Remark. Our excuse for writing the ambiguous formula (20) is simple: we arrive at (20) naturally, the formula is relatively simple and suggestive, and it is a trivial matter to write down the precise formula implied, once the conventions concerning the notations and choices of elements are stated. Specifically, (20) is to be interpreted as follows:

$$(20') \quad d_3 \imath h = \imath(\delta k + k_1 + k_2), \quad k \in s^{-1} h, \quad k_1 \in i^{-1} d_2'' k, \quad k_2 \in d_3' h$$

where $k_2 \equiv k_2(k, k_1)$ modulo $d_2' a$, $a \in H(j, j+1)$, $\delta a = 0$. Let us also notice that $d_3' h$ is not the usual notation: we should write $k_2 \in \imath_1^{-1} d_3' \imath_1 h$ using a natural homomorphism similar to (17). The correct value of the last term in (20) can thus be found by adding an appropriate d_2'-coboundary to any given element compatible with the notation $d_3' h$. In the proof of Theorem 3 we justify (20) for particular choices of the terms on the right, and deduce (20') using (14) only. We postpone this proof to the end, although it could be inserted here.

I started work on this paper under the influence of discussions with G. E. Bredon, W. C. Hsiang, and W. Y. Hsiang during the conference. The Hsiang brothers pointed out to me, that, once the general theory of this spectral sequence is sufficiently advanced, the following problem may become manageable: compute the spectral sequence fully in case G acts by representation on R^n. If this could be done, a natural next step would be to consider how much of this structure is preserved under more general conditions, if, for example, G acts topologically or differentiably on R^n. I also talked with K. JÄNICH about similar problems connected with his results.

6. In our proofs we use the same method as in [7], in particular, we use the technique of couvertures. See [2, 3, 6] as introduction to this method, as well as [10] for the general theory. Our method of proof of Theorem 3 necessitates the repetition of the main construction used in [7] in the proof of Theorem 2, but certain details must be done differently to prepare the proof of Theorem 3. For this reason it is simpler to give a complete proof of Theorem 2, than it would be to give references. We give thus complete proofs of both theorems. We also must recall certain facts concerning E_0 and E_1, but it will not be necessary to prove Theorem 1 anew.

Let us recall that a couverture on X is a cochain algebra K with supports $S(k)$ compact in X; operations on k and $S(k)$ are compatible. We denote K_U, for $U \subset X$, the subalgebra $\{k \in K: S(k) \subset U\}$, and for $F = X - U$, $FK = K/K_U$. A very simple and suggestive notation is to use F to denote the natural homomorphism $K \to K/K_U$, which we will call *restriction*. Then we have formulas

$$(21) \qquad F(k+l) = Fk + Fl, \quad F(kl) = Fk \cdot Fl, \quad Fdk = dFk,$$

where d is the appropriate differential. We will use (21) mainly for the case $F = x$, point of X, see (23), (26), (27), etc. below.

Let us introduce an appropriate couverture of X, and its filtration. Let K be a fine R-couverture of X, and $\bar{L}$ a fine Z-couverture of Y, where Z is the ring of integers. We set $L = f^{-1}\bar{L}$; this is a couverture of X; it is not fine, in general. By definition, $L \circ K$ is a quotient of the Z-tensor product $L \otimes K$, and is a fine couverture of X. The filtration introduced by the ideals

$$(22) \qquad \sum_{m \geq p} L^m \circ K,$$

defines the Leray spectral sequence of f, with $H(Y, A)$ as second term and a graded algebra of $H(X)$ as E_∞ term see (6), [1, 2, 4, 7, 10, 11]. Also, $E_0 = \mathrm{Gr}(L \circ K)$, graded algebra of $L \circ K$, d_0 being the differential "with respect to K". $E_1 = L \circ A$, and d_1 is the differential "with respect to L".

7. Given (1), we "modify" the filtration (22) as follows. For $a \in L \circ K$, we have

$$(23) \qquad xa \in xL \otimes xK = xL \circ xK.$$

Let the filtration $\varphi_x(xa)$ be defined by:

$$(24) \qquad \textit{for} \quad x \in X_n, \ \varphi_x(xa) \geq p \quad \textit{iff} \quad xa \in \sum_{u \geq p+n} xL^u \circ xK.$$

The filtration of $L \circ K$ is defined then by setting

(25) $$\varphi(a) = \inf \varphi_x(xa); \qquad (\varphi(0) = +\infty)$$

similarly $\varphi(Fa)$ can be defined by (25) taking inf over $x \in F$. Then the restriction homomorphism increases the filtration, and this fact, of course, motivates the definition (25), as the general theory of spectral sequences shows. When x varies in $X_n - X_{n+1}$, filtrations (22), (25) differ by the constant n only, thus somewhat vaguely stated, "over this part" the corresponding spectral sequences "essentially agree". We will constantly use below a precise form of this important fact.

An important "continuity property" of filtration (25) can be worded as follows: If $\varphi(X_n a) \geqslant p$, then X_n has a neighborhood V such that $\varphi_x(xa) \geqslant p+1$ for $x \in V - X_n$. Proof: For $x = x_0 \in X_n$,

(26) $$xa \in \sum_{u \geqslant p+n} xL^u \circ xK.$$

Then x_0 has a compact neighborhood U, such that

(27) $$Ua \in \sum_{u \geqslant p+n} UL^u \circ UK,$$

and then (26) follows for every $x \in U$, which means $\varphi_x(xa) \geqslant p+1$ for $x \in U - X_n$. The V of our statement is the union of the U's appearing in (27).

The filtration (25) defines the spectral sequence (11), (12). This filtration, contrary to (22), is not compatible, in the usual sense, with cup products.

We will use the following notation, definitions, and facts concerning spectral sequences:

(28) $$d: L \circ K \to L \circ K \quad \textit{is the total differential,}$$

(29) $$C_r^p = \{a \in L \circ K : \varphi(a) \geqslant p,\ \varphi(da) \geqslant p + r\} \qquad (p, r \textit{ integers}),$$

(30) $$E_r^p = C_r^p / (C_{r-1}^{p+1} + d\,C_{r-1}^{p-r+1}),$$

(31) $$E_r = \sum_p E_r^p,$$

(32) $$d_r(a + C_{r-1}^{p+1} + d\,C_{r-1}^{p-r+1}) = da + C_{r-1}^{p+r+1} + d\,C_{r-1}^{p+1},$$

(33) $$\iota: \ker d_r \to E_{r+1},$$

where p is *not* restricted to be positive as usual, as it is, in particular, in the Leray spectral sequence. Let us write down explicitly some special cases for future reference:

(34) $$E_0^p = C_0^p / (C_0^{p+1} + d\,C_0^{p+1}) \qquad (C_{-1}^p = C_0^p),$$

(35) $$E_1^p = C_1^p / (C_0^{p+1} + d\,C_0^p),$$

(36) $$E_2^p = C_2^p/(C_1^{p+1} + dC_1^{p-1}),$$

(37) $$E_3^p = C_3^p/(C_2^{p+1} + dC_2^{p-2}),$$

(38) $$E_3^{p+3} = C_3^{p+3}/(C_2^{p+4} + dC_2^{p+1}).$$

8. We come now to the proof on Theorem 2. Let be given the cohomology class h in (13). By Theorem 1, this is an element of E_2; hence by (36) we can choose

(39) $$a \in C_2^p \quad \textit{class of} \quad a \quad \textit{in} \quad E_2 \quad \text{is} \quad h.$$

We will prove now that a can be so choosen that

(40) $$S(a) \cap X_{j+1} = \phi,$$

holds true. Proof: If we restrict f to $X - X_{j+1}$, and replace the sequence (1) by $Y_t - Y_{i+1}, t = 0, \ldots, j+1$, then we can form the corresponding spectral sequence of Theorem 1. The h in (13) is an element of the second term of this spectral sequence, and is represented by an $a' \in L_U \circ K_U$ where $U = X - X_{j+1}$. Such an element satisfies both (39) and (40).

By (22), (24), (25), (26), (27) and by the definition of the spectral sequence (compare (39) with (29)):

(41) $$xa \in \sum_{m \geqslant p+j} xL^m \circ xK, \qquad \text{if} \quad x \in X_j - X_{j+1},$$

(42) $$xda \in \sum_{m \geqslant p+j+2} xL^m \circ xK, \quad \text{if} \quad x \in X_j - X_{j+1}.$$

For each x, there is a compact neighborhood U_x having "similar properties", i.e. such that (41), (42) hold true, if we replace x by U_x. Let U be the union of these U_x's. Let us choose a local unit $u \in L^0$, i.e. an element, such that: (α) $S(u) \subset U$; (β) the restriction of u to $S(a) \cap X_j$ is the unit of the restriction of L to this set. Now L operates from the left on $L \circ K$ by cup product on the first factor, hence ua is defined. From properties (α), (β) of u follows then

(43) $$xua \in \sum_{m \geqslant p+j} xL^m \circ xK \qquad (x \in X),$$

(44) $$xuda \in \sum_{m \geqslant p+j+2} xL^m \circ xK \qquad (x \in X),$$

for all $x \in X$. In fact, if $x \in U$, then (41), (42) apply by definition of U, if $x \notin U$, $xu = 0$, hence (43), (44) are trivially true. Let us show now that

(45) $$ua \in C_2^p.$$

By (43) $\varphi_x(xa \geqslant p$ for $x \in X_j$, and by the construction of U, $\varphi_x(xa) \geqslant p+1$ for $x \notin X_j$. Hence $\varphi(ua) \geqslant p$. As to the coboundary

(46) $$d(ua) = du \cdot a + uda$$

of this element, we write

(47) $$xd(ua) = xdu \cdot a + xuda.$$

Now $xdu \in xL^1$, and (41). If $x \in S(a) \cap X_j$, $xdu = 0$, $xuda = xda$, and (41) shows that this element is of filtration $\geqslant p+1$. Hence the product $xdu \cdot xda$ is of filtration $\geqslant p+2$. We have thus proved that the first term on the right of (46) is of filtration $\geqslant p+2$. As to the second term, (42) shows this inequality. Summing up, we have

(48) $$\varphi(du \cdot a) \geqslant p+2,$$

(49) $$\varphi(uda) \geqslant p+2.$$

We have thus proved (45). As a corollary we have:

(50) $$a - ua$$

belongs to C_2^p. Let us show that the element (50) determines the zero class in E_2. Now for $x \in X_j$, xu is the unit thus $xa = xu \cdot xa$, hence

(51) $$S(a-ua) \cap X_j = \phi.$$

Examining the expression of $a-ua$, we find, that its filtration is $\geqslant p+1$, and the filtration of its coboundary is $\geqslant p+2$. Thus (50) belongs to C_1^{p+1}, its class in E_2 is thus null. Hence $d_2 h$ is represented by the class of (46) in E_2^{p+2}. We want to prove that both terms on the right in (50) belong to C_1^{p+2}. In view of (48), (49), it is enough to prove

(52) $$\varphi_x(xdu \cdot da) \geqslant p+4.$$

Proof of (52): If $x \in X_j$, $xdu = 0$, and (52) holds trivially. If $x \notin X_j$, $\varphi_x(xda) \geqslant p+3$ (see (26), (27) and definition of u), and $xdu \in xL^1$. This completes the proof of (52). Let us give now an interpretation of the classes of the elements

(53) $$du \cdot a \in C_2^{p+2},$$

(54) $$uda \in C_2^{p+2}.$$

We consider first (53). It is known that the E_1 term of the spectral sequence has the following form:

(55) $$E_1 = \Sigma (X_m - X_{m+1}) L_{(X - X_{m+1})} \circ (A | X_m - X_{m+1}).$$

The differential d_1 is the standard one on the first factor of each term. Let us consider now

(56) $$ua + C_0^{p+1} + dC_0^p \in E_1$$

$(X_j - X_{j+1})ua = (X_j - X_{j+1})a$ is a cocycle of $X_j - X_{j+1}$ representing h, hence $du \cdot a$ is a cocycle of $X - X_{j+1}$ and its restriction to $X_{j-1} - X_j$ represents δh. This accounts for the δh in (14).

We consider now (54). $(X_j - X_{j+1})uda = (X_j - X_{j+1})da$ represents clearly $d_2' h$, which we can see by writing down (36) for the Leray spectral sequence of $f|X_j - X_{j+1}$.

The component of (46) in $H(m, m+1)$ is clearly zero for $m \geqslant j+1$ in view of (40). For $m \leqslant j-2$, $x \in X_m - X_{m+1}$, the filtration of (47) is $\geqslant p+3$, hence the corresponding component in E_1 is zero. This is carried over to E_2. Thus the proof of (14) is complete. Equation (15) follows easily from $d_2 h = 0$, and from the fact that E_2 is a direct sum.

9. In order to prove Theorem 3, we suppose h in (13) given, such that (18), (19) hold true. Hence $\iota h \in E_3$. We choose now

(57) $$a \in C_3^p, \quad \textit{class of} \quad a \quad \textit{in} \quad E_3 \quad \textit{is} \quad \iota h.$$

We can, furthermore, take an a, for which

(58) $$S(a) \cap X_{j+1} = \phi.$$

The reason for this is the same as it was for (40): h belongs to the spectral sequence considered after (40), and (18), (19) hold true in this spectral sequence, hence the conclusion.

For the element a in (57) we write relations analogous to (41), (42). Specifically, we have (41), and

(59) $$x\,da \in \sum_{m \geqslant p+3} xL^m \circ xK;$$

the only difference between (42) and (59) being the lower limit of the summation. As in the previous section, we can choose, for every $x \in X_j - X_{j+1}$ a compact neighborhood U_x such that both (41) and (59) hold true, if we replace x by U_x. We denote U the union of the U_x's: U is a neighborhood of $X_j - X_{j+1}$, and (41), (59) hold true for every $x \in U$. We choose an element u of L^0, such that Cu is the unit of CL, $C = S(a) \cap X_j$, and $S(u) \subset U$.

Now $C_3^p \subset C_2^p$; hence the results of the previous section apply to ua: $ua \in C_2^p$, a and ua have the same cohomology class in E_2, $du \cdot a$ represents the component $\delta h = 0$, and uda represents the component $d_2' h = 0$. (We do not claim $ua \in C_3^p$, which is not true in general.) Now (18) and (36) show

(60) $$du \cdot a = c_1^{p+3} + dc_1^{p+1} \qquad (c_v^u \in C_v^u).$$

We notice, furthermore, that this equation also holds true in the spectral sequence of $f|X - X_{j+1}$, $Y_m - Y_{j+1}$, $m \leqslant j+1$ used after (40). As in the case of (40), we can infer that

(61) $$S(c_1^{p+3}) \cap X_{j+1} = \phi, \quad S(c_1^{p+1}) \cap X_{j+1} = \phi$$

hold true for appropriate choices of the elements in (60); we choose and fix these elements satisfying (61). Taking the coboundary of (60) and taking restriction to $x \in X$, we get

$$x\,du \cdot da = x\,dc_1^{p+3}.$$

If $x \in X_j$, $x\,du = 0$, hence the filtration is $+\infty$, if $x \notin X_j$, $x \in S(u)$, $\varphi_x(x\,da) \geqslant p+4$, hence the filtration on the left is $\geqslant p+5$. This shows $c_1^{p+3} \in C_2^{p+3}$, hence we will denote c_1^{p+3} by c_2^{p+3}:

(62) $$du \cdot a = c_2^{p+3} + dc_1^{p+1} \qquad (S(c_v^u) \cap X_j = \phi).$$

Let us introduce now a local unit v as follows. Let

$$C = (S(a) \cup S(c_1^{p+1}) \cup S(c_2^{p+3})) \cap X_{j-1}.$$

We take a neighborhood V of C such that the remark in connection with (26), (27) applies to all elements a, c_1^{p+1}, c_2^{p+3} involved. Then, by definition, v is such that: (α) Cv is the unit; (β) $S(v) \subset V$. Let us prove then

(63) $$v(ua - c_1^{p+1}) \in C_3^p;$$

this is the relation analogous to (45) in case of d_3. Now direct computation gives:

(64) $$d(v(ua - c_1^{p+1})) = dv \cdot (ua - c_1^{p+1}) + vc_2^{p+3} + vu\,da;$$

we also need the restriction by x:

(65) $$x\,d(v(ua - c_1^{p+1})) = x\,dv \cdot (ua - c_1^{p+1}) + xvc_2^{p+3} + xvu\,da.$$

If $x \in X_{j-1}$, $x\,dv = 0$, hence the first term on the right in (65) is zero. If $x \notin X_{j-1}$, $\varphi_x(xua) \geqslant p+2$, hence the first term on the right is of filtration $\geqslant p+3$; for the other two terms this is clear by the definition of a, c_2^{p+3}, see (57), (60), (63). This completes the proof of (63).

We will show now that the element

(66) $$v(ua - c_1^{p+1})$$

can replace a in computing $d_3 h$. In view of (64), and (37), it will be enough to prove that

(67) $$v(ua - c_1^{p+1}) - a \in C_2^{p+1}.$$

As both a and (66) belong to C_3^p, the coboundary of their difference is of filtration $\geqslant p+3$. Therefore, in order to prove (67), it is enough to show that

(68) $$xv(ua - c_1^{p+1}) - xa$$

has filtration $\geqslant p+1$ for all $x\in X$. If $x\in X_j$, $xu=xv$ is the unit, and $xc_1^{p+1}=0$ in view of (61), thus (68) is zero in this case. If $x\in X_j$, $\varphi_x(xa)\geqslant p+1$, unless $xu=0$, and $\varphi_x(xc_1^{p+1})\geqslant p+1$ by definition. This completes the proof of (67), which shows

$$d_3\iota h = d(v(ua-c_1^{p+1}))+C_2^{p+4}+dC_2^{p+1}. \tag{69}$$

We consider next the coboundary of (66) given in (64), and prove the following membership relations:

$$dv\cdot(ua-c_1^{p+1})\in C_2^{p+3}, \tag{70}$$

$$vc_2^{p+3}\in C_2^{p+3}, \tag{71}$$

$$vuda\in C_2^{p+3}. \tag{72}$$

We will need the differential of (65):

$$\begin{cases} 0=-xdv(c_2^{p+3}+uda)+xdv\cdot c_2^{p+3}+xdc_2^{p+3} \\ \quad +xdv\cdot u\cdot da+xvdu\cdot da. \end{cases} \tag{73}$$

The first term on the right in (73) is the coboundary of the element in (70). If $x\in X_{j-1}$, $xdv=0$, hence this term is of filtration $+\infty$, if $x\notin X_{j-1}$, $\varphi_x(xc_2^{p+3})\geqslant p+4$, $\varphi_x(xuda)\geqslant p+5$, hence the filtration of the first term on the right is also $\geqslant p+5$. This proves (70). The sum of the second and third terms in (73) is the coboundary of the element in (71). The third term is clearly of filtration $\geqslant p+5$; if $x\in X_{j-1}$, $xdv=0$, and if $x\notin X_{j-1}$, $\varphi_x(xc_2^{p+3})\geqslant p+4$, thus the second term is also of filtration $\geqslant p+5$. As the first three terms on the right are of filtration $\geqslant p+5$, the same holds true for the sum of the last two terms. Thus the proof of (71, 72) is complete.

We proved (69), and that the coboundary (64) of the element (66) is the sum of the three elements in (70), (71), (72). The next step is to give invariant interpretation of the elements (70), (71), (72).

We will deal first with the element in (71). We set $b=ua-c_1^{p+1}$, and we consider the

(74) *Leray spectral sequence of* $f|X_{j-1}-X_{j+1}$.

We write the non-invariant terms E_0, E_1 of (74) in terms of the couvertures $X_{j-1}L$, $X_{j-1}K$. By (18), $s^{-1}h$ is not empty and a choice k in this coset belongs to the second term of (74). We claim: if k is *appropriately* chosen, it is represented by the element

$$(X_{j-1}-X_{j+1})b=(X_{j-1}-X_{j+1})vb \tag{75}$$

of the couverture defining (74). Now (75) has Leray filtration $\geqslant p+j$, and its coboundary has Leray filtration $\geqslant p+j+2$; this is contained in the proofs above and will not be repeated here. (75) represents thus some cohomology class in the second term of (74). Taking the restriction of (75) to $X_j - X_{j+1}$, we get $(X_j - X_{j+1})a$ in view of (61). The class represented by (75) has thus a restriction represented by $(X_j - X_{j+1})a$. This completes the proof: if $k \in s^{-1} h$ is appropriately chosen, it is represented by (75). Then the coboundary of (75)

$$(76) \qquad (X_{j-1} - X_{j+1})(du \cdot a + u\,da - dc_1^{p+1})$$

represents $d_2'' k$. At the same time in the second term of (74) the element (76) determines the same class as

$$(77) \qquad (X_{j-1} - X_{j+1})(du \cdot a - dc_1^{p+1}) = (X_{j-1} - X_{j+1}) v c_2^{p+3}$$

as the Leray filtration of $xu \cdot da$, $x \in X_{j-1} - X_{j+1}$, is $\geqslant p+j+3$, with coboundary of filtration $\geqslant p+j+4$, thus can be neglected. Now the support of (77) does not intersect X_j; hence (77) represents a class in $i^{-1} d_2'' k$. This gives an interpretation of the second term on the right of (64), and shows that it has the form indicated in (20).

In the proof of Theorem 2, we already remarked that $dv \cdot b$ represents δk if b represents the cohomology class k. Together with the preceding paragraph, this gives the interpretation of (20) for the first term on the right of (64).

As to the last term in (20), it is clear that $X_j da$ represents $d_3' \iota h$ in the Leray spectral sequence of $f | X_j - X_{j+1}$. We have thus completed the proof of formula (20) in the following sense: there are cohomology classes $k \in s^{-1} h$, $k_1 \in i^{-1} d_2'' k$, and $k_2 \in d_3' h$ so that (20′) holds true.

The following direct computation, based on (14) and on the validity of (20′) for particular choices of elements will complete the proof of Theorem 3. We take arbitrary $l \in H\,(j-1, j)$, $m \in H(j, j+1)$, and a $c \in H(j, j+1)$ such that $\delta c = 0$. Then

$$(78) \qquad \begin{aligned} d_2(l+m+c) &= \delta l + (d_2' l + \delta m) + (d_2' m + d_2' c) \\ &= \delta i l + (i^{-1} d_2'' i l + \delta m) + (d_2' m + d_2' c). \end{aligned}$$

We can change k arbitrarily in (20′) by a suitable choice of l added to the right hand side, then we can change by an added δm the choice in i^{-1}, finally the last term can be changed using a c, $\delta c = 0$. By (78) above the argument of ι has been changed by $d_2(l+m+c)$, hence the right hand side of (20′) has not been changed. Let us add that the individual terms on the right of (20) or (20′) do not necessarily belong to E_3.

References

1. BOREL, A.: Seminar on Transformation Groups, Princeton University Press, Princeton, N.J., 1960.
2. — Cohomologie des espaces localement compacts d'après J. Leray, 3ième edition, Berlin-Göttingen-Heidelberg-New York: Springer 1964.
3. BOURGIN, D. G.: Modern Algebraic Topology, The MacMillan Co., New York, N. Y., 1963.
4. BREDON, G. E.: Sheaf Theory, McGraw-Hill, New York, N.Y., 1967.
5. FÁRY, I.: Sur les anneaux spectraux de certaines classes d'applications. V. Application de X sur X/G. Sur le term E_2, C. R. Acad. Sci. Paris **236,** 1224—1226 (1953).
6. — Notion axiomatic de l'algèbre de cochaînes dans la théorie de J. Leray, Bull. Soc. Math. France **82,** 97—135 (1954).
7. — Valeurs critiques et algébres spectrales d'une application, Ann. of Math., **63,** 437—490 (1956).
8. — Cohomologie des variétés algébriques, Ann. of Math. **65,** 21—73 (1957).
9. — Spectral Sequences of Certain Maps, Symposium Internacional de Topologia Algebraica, Mexico 1958.
10. LERAY, J.: L'anneau filtré d'homologie d'un espace localement compact et d'une application continue, J. Math. Pures Appl. **29,** 1—139 (1950).
11. — L'homologie d'un espace fibré dont la fibre est connexe, J. Math. Pures Appl. **29,** 169—213 (1950).
12. MONTGOMERY, D., and L. ZIPPIN: Topological Transformation Groups, Interscience Publishers, New York, N.Y., 1955.

Exotic PL Actions which are Topologically Linear

FRANK RAYMOND*

Theorem. *Let L be an n-dimensional lens space whose fundamental group is isomorphic to* $\mathbb{Z}_p$ *and where* $n \geqslant 5, p \geqslant 5, p \neq 6$. *Then, for each L there exists an infinite number of PL actions of* $\mathbb{Z}_p$ *on* S^{n+1}, *free except for 2 fixed points. The actions are all mutually PL inequivalent but all are topologically equivalent to the linear action determined by L. Furthermore, by deleting one of the fixed points the actions are smooth and all are smoothly equivalent to the linear action determining L.*

Proof. In [1], MILNOR described an infinite number of mutually inequivalent smooth free actions of the finite cyclic group $\mathbb{Z}_p$ on the n-sphere S^n where $p \geqslant 5, p \neq 6$ and $n \geqslant 5$. MILNOR took a lens space L with fundamental group $\mathbb{Z}_p$ and an h-cobordism $W = (W; L, L')$ where the torsion of the inclusion of L in W corresponded to a unit in the group ring $Z(\mathbb{Z}_p)$. An infinite number of such smooth manifolds L' can be constructed so they are not PL homeomorphic to any lens space and all are mutually non-PL homeomorphic. The universal covering space $\tilde{W}$ of W is diffeomorphic to $S^n \times I$ and the covering transformations restricted to the boundary components gave the desired examples.

Since the covering transformations are smooth on $\tilde{W}$, we can, by choosing a smooth triangulation of W, attach cones to both boundary components of W. Call this new complex W'. Corresponding to attaching the cones on L and L' we may equivariantly attach cones on $S^n \times 0$ and $S^n \times 1$, the boundary components of $\tilde{W}$. The resulting manifold $\tilde{W}'$ is PL-homeomorphic to S^{n+1}. Let $\tilde{x}$ and $\tilde{y}$ denote the respective vertices of the cones attached to $S^n \times 0$ and $S^n \times 1$, respectively. The smooth action of $\mathbb{Z}_p$ on $\tilde{W}$ extends to an action on $\tilde{W}'$ and is free except for fixed points $\tilde{x}$ and $\tilde{y}$. The action on the $(n+1)$-cell attached to $S^n \times 0$ (corresponding to $\tilde{L}$) is smoothly equivalent after smoothing near $\tilde{x}$, to the linear action on the $(n+1)$-cell which determines the cone over L. Thus, on $\tilde{W}' - \tilde{y}$, the action is smoothly equivalent to a linear action on $\mathbb{R}^{n+1}$ no matter what L' may be (The orbit space $W' - (x \cup y)$ is diffeomorphic to $L \times \mathbb{R}^1$.) The action on $\tilde{W}'$ is the topological extension of

* Partially supported by NSF GP 08105.

this linear action from $\tilde{W}' - \tilde{y}$ and consequently is topologically equivalent to a fixed linear action on S^{n+1}.

The PL-action on $\tilde{W}'$ has orbit space W'. Since, in any subdivision of W', the boundary of the star of the vertex y is PL homeomorphic to L', the action on S^{n+1} can never be PL equivalent to a linear action. Furthermore, by varying L', as MILNOR did, the resulting actions on S^{n+1}, while topologically equivalent, must all be mutually PL inequivalent.

References

1. MILNOR, J.: Some free actions of cyclic groups on spheres, Proc. Bombay Colloquium in Differential Analysis, 37—42 (1964).

The Lefschetz Fixed Point Theorem for Involutions*

HSU-TUNG KU and MEI-CHIN KU

The purpose of this note is to show that the Lefschetz fixed point theorem holds for involutions on locally compact spaces. The Alexander-Spanier-Wallace cohomology with compact supports will be used. Let X be a locally compact space. The Lefschetz number Λ_f of a map $f: X \to X$ is defined by

$$\Lambda_f = \sum_i (-1)^i \operatorname{Trace}(f^* | H^i(X; R)),$$

where $f^*: H^i(X; R) \to H^i(X; R)$ and R is the field of real numbers.

Lefschetz Fixed Point Theorem. *Suppose that a finite solvable group G acts freely on a locally compact space X with $\dim_Z X < \infty$ and $H^*(X; Z)$ finitely generated. Then*

$$\sum_{\substack{g \in G \\ g \neq e}} \Lambda_g = 0$$

where e is the identity element of G. In particular, if $G = Z_2$, then $\Lambda_g = 0$.

Proof. Define

$$\chi^G(X; R) = \sum_i (-1)^i \dim H^i(X; R)^G.$$

By [1, p. 38] we have

$$H^*(X/G; R) = H^*(X; R)^G.$$

Hence

$$\chi^G(X; R) = \chi(X/G; R).$$

Let order $(G) = q$. First we show that for each $i \geqslant 0$

$$\dim H^i(X; R)^G = \left\{ \sum_{g \in G} \operatorname{Trace} g^* | H^i(X; R) \right\} / q.$$

---

* This work was supported by the National Science Foundation.

To prove this, let

$$\varphi(v)=\left\{\sum_{g\in G} g^*(v)\right\}/q \quad \text{for} \quad v\in H^i(X,R).$$

Then $\varphi\cdot g^*=g^*\cdot\varphi=\varphi$ for every $g\in G$, and $\varphi^2=\varphi$. Moreover $v\in H^i(X;R)^G$ if and only if $\varphi(v)=v$. Thus

$$\dim H^i(X;R)^G=\operatorname{Trace}(\varphi)=\left\{\sum_{g\in G}\operatorname{Trace} g^*|H^i(X;R)\right\}/q.$$

By using this result, we have

$$\chi^G(X;R)=\left\{\sum_{i,g}(-1)^i\operatorname{Trace} g^*|H^i(X;R)\right\}/q=\left\{\chi(X;R)+\sum_{\substack{g\in G\\ g\neq e}}\Lambda_g\right\}/q.$$

Thus

$$\chi(X/G;R)=\left\{\chi(X;R)+\sum_{\substack{g\in G\\ g\neq e}}\Lambda_g\right\}/q$$

or

$$\sum_{\substack{g\in G\\ g\neq e}}\Lambda_g=q\chi(X/G;R)-\chi(X;R).$$

But $q\chi(X/G;R)-\chi(X,R)=0$ by [1, p. 46]. Hence the result follows.

As an easy application we have the following (suggested by Professor D. Montgomery):

Corollary 1. (Smith [3]). *Let G be a cyclic group of prime period acting on a euclidean n-space X. Then G has a fixed point.*

Proof. Since

$$\sum_{\substack{g\in G\\ g\neq e}}\Lambda_g\neq 0.$$

More generally we have

Corollary 2. *No finite abelian group can act freely on a euclidean space.*

Conjecture. Under the hypotheses of the theorem, if G is finite cyclic, then $\Lambda_g=0$ for every $g\neq e$.

References

1. Borel, A. et al.: Seminar on Transformation Groups, Ann. of Math. Studies. **46**, Princeton Univ. Press, Princeton, 1960.
2. Greenberg, M.: Lectures on Algebraic Topology, Benjamin, New York, 1967.
3. Montgomery, D., and L. Zippin: Topological Transformation Groups, Interscience, New York, 1955.

Involutions of the n-Cell

KYUNG WHAN KWUN*

In [2], it is shown that there exists an involution of the n-cell, for each $n \geqslant 4$, such that the orbit space and the fixed point set are an n-cell and an $(n-1)$-cell respectively, but the action is inequivalent to the standard one. We outline below how this may be modified to produce uncountably many inequivalent involutions having the same property.

Recall that if $a \in A \subset X$, we say $X - A$ is *locally simply connected at* a if for any neighborhood U of a, there is a smaller neighborhood V of a such that every loop in $V - A$ is null-homotopic in $U - A$.

1. Case $n=4$

Let A be a closed subset of the 2-sphere S^2 such that $\overline{\operatorname{int} A} = A$. Combining techniques of [1, p. 81–84] and [2] with some care, it is possible to produce an involution T_4 of a 4-cell I^4 such that:

1. the orbit space of T_4 is a 4-cell;
2. the fixed point set F of T_4 is a 3-cell;

and

3. the closure of the set of points of F at which $\operatorname{Bd} I^4 - F$ is locally simply connected is homeomorphic to A.

2. Case $n \geq 5$

As in [2], let T_n be the $(n-4)$-fold suspension of T_4. Then

1. the orbit space of T_n is an n-cell;
2. the fixed point set F of T_n is an $(n-1)$-cell;

and

3. the closure of the set of points of F at which $\operatorname{Bd} I^n - F$ is locally simply connected is homeomorphic to the $(n-4)$-fold suspension of A.

To see 3., it suffices to observe that the suspension of a space X is, except for the suspension points, homeomorphic to $X \times (0,1)$ and that

---

* Supported in part by NSF Grant GP-5868.

$X-A$ is locally simply connected at a if and only if $(X-A)\times(0,1)$ is at $(a,\frac{1}{2})$.

3. Choices of A

Let B, B' be a 2-cell and an annulus. Let Λ be the set of sequences whose terms are B or B'. For each $\lambda\in\Lambda$, consider a sequence of sets $C_1, C_2, \ldots \subset S^2$ converging to $p_\lambda \in S^2$ such that C_i is B or B' according as the i-th term of λ is B or B', $C_i \cap C_j \neq \phi$ only if $|i-j|\leqslant 1$ and $C_i \cap C_{i+1}$ consists of two points. Let $A_\lambda = C_1 \cup C_2 \cup \cdots \cup p_\lambda$. A_λ has the property that if two points separate A_λ then they must be $C_i \cap C_{i+1}$. Likewise if an n-sphere in the n-fold suspension $\Sigma^n A_\lambda$ separates $\Sigma^n A_\lambda$, then the n-sphere must be the one corresponding to some $\Sigma^n(C_i \cap C_{i+1})$. Hence for a fixed $n\geqslant 4$, $\Sigma^n A_\lambda$ are all topologically distinct.

References

1. Bing, R. H.: Inequivalent families of periodic homeomorphisms of E^3, Ann. of Math. **80**, 78—93 (1964).

2. Kwun, K. W.: An involution of the n-cell, Duke Math. Jour. **30**, 443—446 (1963).

Topological Actions do not Necessarily have Reasonable Slices

FRANK RAYMOND

Let $\mathbb{Z}_6$ act differentiably on $\mathbb{R}^n$ without fixed points. Such actions exist according to the constructions of E. E. FLOYD and J. KISTER. Extend this action to S^n by adding a point at infinity. If we could choose an open invariant acyclic set S, then the complement C, which would also be acyclic, could not have the fixed point property. In particular, there exists no closed slice which is a generalized n-cell. (A generalized n-cell is an orientable, compact acyclic n-cm with boundary.) To see the last remark, it is only necessary to observe that the closure of the complement of the generalized cell would be a generalized n-cell. But a generalized cell does have the fixed point property.

A Simple Proof of the Maximal Tori Theorem of E. CARTAN*

HSU-TUNG KU and MEI-CHIN KU

The purpose of this note is to give a new proof of the maximal tori theorem without using Lie algebra theory. Our method is based on the Euler characteristic formula for total group actions. More general results in this direction can be found in [3, 4]. Throughout this paper, Alexander-Spanier-Wallace cohomology with compact supports will be used.

Lemma 1. [3, 4]. *Let the torus group T^k act on a locally compact Hausdorff space X with fixed point set F such that both* $\dim_0 X$ *and* $\dim H^*(X; K_0)$ *are finite (where K_0 denotes the field of characteristic zero). Then*

$$\chi(X; K_0) = \chi(F; K_0).$$

Lemma 2. *Let G be a compact connected Lie group, and T any maximal toral subgroup of G. Then* $\chi(G/T; K_0) = \operatorname{ord} W(G)$, *where* $W(G)$ *denotes the Weyl group of G.*

Proof. Let T act on G/T via left translation $(t, gT) \to tgT$. Then the fixed point set is $N(T)/T$. That is, $F(T, G/T) = W(G)$. Lemma 1 implies

$$\chi(G/T; K_0) = \chi(F(T, G/T), K_0) = \operatorname{ord} W(G).$$

Theorem 1. *Let G be a compact connected Lie group. Then all maximal tori are mutually conjugate.*

Proof. Let T and T' be any two maximal tori of G, and let T' act on G/T by left translation as in the proof of Lemma 2. Then

$$\chi(F(T', G/T); K_0) = \chi(G/T; K_0) = \operatorname{ord} W(G) \neq 0$$

by Lemmas 1 and 2, and so $F(T', G/T)$ is not empty. Hence T' is conjugate to T.

---

* This work was supported by the National Science Foundation.

Appendix. Remarks on the Euler Characteristic Formula

We shall use the notation of [5]. In [6], SWAN has shown that if Z_p, the cyclic group of prime order p, acts on a compact Hausdorff space X such that Z_p acts trivially on $H^*(X;Z)$ and $\dim H^*(X;Z_p)<\infty$, then $\chi(X;Z_p)=\chi(F(Z_p;X);Z_p)$. We can generalize this result as follows:

Theorem 2. *Let the elementary commutative p-group Z_p^k (or cyclic group Z_{p^k}) act on a compact Hausdorff space X. Suppose Z_p^k (or Z_{p^k}) acts trivially on $H^*(X;Z)$ and* $\dim H^*(X;Z_p)$ *is finite. Then* $\chi(X;Z_p)$ $=\chi(F;Z_p)$.

Proof. We only prove the result for Z_p^k. Assume $k>1$. Let Z_p be a subgroup of Z_p^k. Under the hypotheses, Z_p acts trivially on $H^*(X;Z)$, hence by [6], we have $\chi(X;Z_p)=\chi_H(X;Z_p)$. But for Z_p-actions, the following relation holds [6]:

$$\chi_H(X;Z_p)=\chi_H(F(Z_p;X);Z_p)=\chi(F(Z_p;X);Z_p),$$

so we may consider the subgroup $Z_p^2=Z_p\times Z_p'$ of Z_p^k, and the action of Z_p' on $F(Z_p;X)$. Thus we have

$$\begin{aligned}\chi_H(F(Z_p;X);Z_p)&=\chi_H(F(Z_p';F(Z_p;X);Z_p)=\chi_H(F(Z_p^2;X);Z_p)\\&=\chi(F(Z_p^2;X);Z_p).\end{aligned}$$

Hence $\chi(X;Z_p)=\chi(F(Z_p^2;X);Z_p)$. By repeating the same argument, we have

$$\chi(X;Z_p)=\chi(F(Z_p^k;X);Z_p).$$

A simple application is the following:

Proposition (BOREL [1, p. 170]). *Let G be a compact connected Lie group without p-torsion. Then every elementary commutative p-group of G is contained in a torus of G.*

Proof. Let Z_p^k be any elementary commutative p-group of G, and T a maximal torus of G. Let Z_p^k act on G/T by left translation. Since Z_p^k acts trivially on $H^*(X;Z)$, and G/T is torsion free it follows that

$$\chi(F(Z_p^k;G/T);Z_p)=\chi(G/T;Z_p)=\operatorname{ord}W(G)\neq 0.$$

This completes the proof.

Question. The proof of W. Y. HSIANG in [2, p. 132] implies the following result: *Let G be a compact connected Lie group acting differentiably* *on a compact connected smooth manifold X. Suppose that*

$\chi(G/G_x; Z_p) = 0 \bmod p$ for $x \notin F$. *Then* $\chi(X; Z_p) = \chi(F; Z_p) \bmod p$.

We conjecture that the above result holds for topological actions.

References

1. BOREL, A. et al.: Seminar on Transformation Groups, Ann. of Math. Studies **46**, Princeton Univ. Press, Princeton 1960.
2. HSIANG, W. Y.: On the principal orbit type and P. A. Smith theory of $SU(p)$ actions, Topology **6**, 125—135 (1967).
3. KU, HSU-TUNG: An Euler characteristic formula for compact group actions, Hung-Ching Chow sixty-fifth Anniversary Volume, Mathematics Research Center, Taiwan University, Taipei, 81—88 (1967).
4. — A generalization of the Conner inequalities, these Proceedings.
5. — The fixed point set of Z_p in a $C^*(m_1, m_2, \ldots, m_{q-1}; Z_p)$-space, these Proceedings.
6. SWAN, R. G.: A new method in fixed point theory, Comment. Math. Helv. **34**, 1—16 (1960).

A Relation between Group Actions and Index

JAMES A. SCHAFER

In this note we wish to state and prove a result* relating the action of a group acting freely on a $4k$-manifold to the index of the manifold given by the bilinear form determined by cup product. All the manifolds will be differentiable and orientable and the group actions referred to will all be differentiable and preserve orientation. Q will denote the group of rational numbers.

Lemma. *Let π be a finite group acting freely on M^{4k}. Then if π acts trivially on $H^{2k}(M^{4k};Q)$, the index $\sigma(M^{4k})$ is zero.*

Proof. It is well known [1, 3], that if $p\colon M^{4k}\to M^{4k}/\pi$ is the covering map, then p^*: maps $H^*(M^{4k}/\pi;Q)$ isomorphically onto $H^*(M^{4k},Q)^\pi$, the invariant elements of $H^*(M^{4k},Q)$. *Since π acts trivially on $H^{2k}(M^{4k},Q)$*, p^*_{2k} is an isomorphism of $H^{2k}(M^{4k}/\pi;Q)$ onto $H^{2k}(M^{4k};Q)$. Let $[M^{4k}]$, $[M^{4k}/\pi]$ denote the fundamental classes of M^{4k} and M^{4k}/π respectively in $H_{4k}(-,Z)$. It is clear that $p_*[M^{4k}]=|\pi|[M^{4k}/\pi]$, where $|\pi|$ denotes the order of the group π. Let ϕ, ψ denote the bilinear forms given by cup product of the manifolds M^{4k} and M^{4k}/π respectively.

Let u, $v\in H^{2k}(M^{4k},Z)$, then we have

$$\begin{aligned}|\pi|\,\psi(u,v)&=|\pi|(u\cup v)[M^{4k}/\pi]=(u\cup v)\,p_*[M^{4k}]\\&=(p^*u\cup p^*v)[M^{4k}]=\phi(p^*u,p^*v).\end{aligned}$$

Therefore if ψ' denotes the bilinear form $|\pi|\cdot\psi$ defined on $H^{2k}(M^{4k}/\pi,Z)$, we have that $\phi(p^*u,p^*v)=\phi'(u,v)$ for all $u,v\in H^{2k}(M^{4k}/\pi,Z)$. If we denote the induced forms on $H^{2k}(M^{4k};Q)$ and $H^{2k}(M^{4k}/\pi,Q)$ by ϕ_Q and ψ'_Q respectively, then since $p^*\colon H^{2k}(M^{4k}/\pi;Q)\to H^{2k}(M^{4k};Q)$ is an isomorphism onto, we see that the forms ϕ_Q and ψ_Q are equivalent. But clearly ψ'_Q and ψ_Q are equivalent forms and therefore ψ_Q and ϕ_Q have the same index; i.e. $\sigma(M^{4k}/\pi)=\sigma(M^{4k})$.

However it follows easily from the Hirzebruch index theorem that $\sigma(M^{4k})=|\pi|\cdot\sigma(M^{4k}/\pi)$ [2]. If π is not the trivial group this is only possible if $\sigma(M^{4k})=\sigma(M^{4k}/\pi)=0$.

---

* Since submitting this note, I have been informed that a similar result has recently been obtained by H.-T. and M.-C. KU.

Theorem. *Let π be a connected group which is not torsion free; then π cannot act freely on any manifold of index different from zero.*

Proof. If π acted freely on a manifold M^{4k}, then any subgroup would act freely and since π is not torsion free, a finite cyclic group π' would act freely on M^{4k}. But π' would also act trivially on $H^{2k}(M^{4k}; Q)$ since $\pi' \leqslant \pi$ which is connected. By the Lemma $\sigma(M^{4k}) = 0$.

Corollary. *No manifold of index different from zero admits a free circle action.*

References

1. BOREL, A., et al.: Seminar on Transformation Groups, Annals of Math. Studies **46**, 1960.
2. CHERN, S. S., F. HIRZEBRUCH, and J. P. SERRE: On the index of a fibered manifold, Proc. A.M.S. **8**, 587—596 (1957).
3. ECKMAN, B.: Coverings and Betti Numbers. Bull. A.M.S. **55**, 95—101 (1949).
4. HIRZEBRUCH, F.: Topological Methods in Algebraic Geometry, Berlin-Heidelberg-New York: Springer 1966.

An Example

J. C. Su

In [1], Professor Bredon asked the following question: "Is there an involution on some $S^n \times S^m$ with fixed point set a complex projective 3-space CP^3? Can this happen when n and m are even and the action on $H^*(S^n \times S^m; Z)$ is trivial?" The following construction seems adequate to provide such examples. Let $S^3 \to S^7 \to S^4$ be the Hopf fibering. Peassing to the (complex) projective space bundle, we obtain the fibering $S^2 \longrightarrow CP^3 \xrightarrow{\pi} S^4$. Let ξ^3 be the 3-plane bundle over S^4 associated with π, and let η^{m-2} be any $(m-2)$-plane bundle over S^4 such that $\xi^3 \oplus \eta^{m-2}$ is trivial. Let T be the involution on $\xi^3 \oplus \eta^{m-2}$ which is the identity on ξ^3 and is (-1) on η^{m-2}. The restriction of T on the unit sphere bundle of $\xi^3 \oplus \eta^{m-2}$ gives an involution on $S^4 \times S^m$ with CP^3 as fixed point set. Notice that the diagram

$$\begin{array}{ccc} CP^3 & \xrightarrow{i} & S^4 \times S^m \\ & {\scriptstyle \pi}\searrow \quad \swarrow & \\ & S^4 & \end{array}$$

is commutative. Since $\pi^*: H^4(S^4; Z) \to H^4(CP^3; Z)$ is an isomorphism, $i^*: H^4(S^4 \times S^m; Z) \to H^4(CP^3; Z)$ is also an isomorphism. This implies that T^* is the identity on $H^4(S^4 \times S^m; Z)$. If we take m even, then dimension parity implies that T is orientation preservation so that T^* is the identity on $H^{m+4}(S^4 \times S^m; Z)$. It follows that T^* is also the identity on $H^m(S^4 \times S^m; Z)$.

Reference

1. Bredon, G. E.: Cohomological aspects of transformation groups, (these Proceedings).

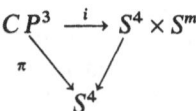

Problems

1. What should one mean by a *piecewise linear* (PL) action? If the action is free, it means that the orbit space is triangulable. What should it be in general? – W. BROWDER.

2. Do cobordant knots K, K' give rise to h-cobordant manifolds $\kappa_n(K)$, $\kappa_n(K')$? – F. HIRZEBRUCH.

3. Formulate an abstract equivarient homology (and its allied cohomology) theory of which the equivariant bordism theory of CONNER-FLOYD and the equivariant K-theory of ATIYAH-SEGAL are examples. Can one relate an abstract equivariant theory with its associated non-equivariant theory? Partial answers are given in recent work by G. BREDON. – C. N. LEE.

4. Let X be a compact space with $\dim_p X < \infty$. If the cohomology ring $H^*(X:Z_p)$ has n generators, does $H^*(F:Z_p)$ have less than or equal to n generators, where F is a component of the fixed point set of Z_p acting on X? – G. BREDON.

5. Under the same hypotheses as the previous problem, if $H^*(X:Z_p)$ is generated by one element over the Steenrod Algebra $\mathscr{A}_p$, what can one say about $H^*(F:Z_p)$? – G. BREDON.

6. Under the same hypotheses as the previous problem, If $H^*(X:Z_p)$ is generated by some element a and its Bockstein $\beta(a)$, what can one say about $H^*(F:Z_p)$? (Examples show that, in general, it does not have to have this same form.) – G. BREDON.

7. Is there some cohomological hypothesis, related to a manifold being Kählerian, which is inherited by fixed point sets? – G. BREDON.

8. Let X be a compact Hausdorff space whose cohomology ring is isomorphic to that of a lens space. Let the circle group S^1 (or T^n) act effectively on X. Is it true that the fixed point set $F(S^1, X)$ is a cohomology lens space if it is non-empty? (With more conditions on X, the answer is yes.) – J. PAK.

9. It is known that if Z_p, $p \neq 2$, acts (topologically) on QP^n, then each component of the fixed point set F is a homology QP^r or a homology CP^r over Z_p for some $r \geqslant 0$. If $n \geqslant p-2$, then it is known that at most one component of F is a homology QP^r, $r \geqslant 1$, the rest being homology CP^r's. Is this true without the restriction $n \geqslant p-2$? G. BREDON.

10. Let G be a compact Lie group acting on E^n with at least one fixed point. Is it true that the complement of the set of points on principal

orbits is connected? (This is true if G is abelian.) – F. RAYMOND.

11. Let S^1 act on E^n and let p be a fixed point. Let A be the set of orbits of one particular type. Is $p \in \bar{A}$? – F. RAYMOND.

12. Let the torus T^k act on E^n and let $G \subset T^k$ be a subgroup. Is $F(G, E^n)$ connected? Is it connected when $k=1$? (there are examples of T^2 acting on S^n in such a way that a subgroup $Z_2 \times Z_3$ has fixed point set two disjoint circles. But in these examples T^2 has no fixed point, so they do not permit the making of counter examples.) If the answer to this question is positive, it will give a positive answer to HSIANGS' Problem 16. – P. S. MOSTERT.

PART III

Compact (Non-Lie) Transformation Groups

Cohomological and Dimension Theoretical Properties of Orbit Spaces of p-Adic Actions

FRANK RAYMOND*

1. Introduction

The following statement, sometimes known as the generalized Hilbert-Smith conjecture, remains unsolved.

Conjecture. *If G is a compact group and acts effectively on a connected n-manifold M, then G is a Lie group.*

The conjecture is true if any one of the following conditions holds:

a) G acts differentiably on M and M is a differentiable manifold, MONTGOMERY and ZIPPIN [16; p. 208];

b) every element of G has finite order, NEWMAN [18], P. A. SMITH [23], MONTGOMERY [14];

c) $n \leqslant 2$, MONTGOMERY and ZIPPIN [16, p. 233],
$n \leqslant 3$ and the connected component of the identity is not the identity, MONTGOMERY and ZIPPIN [16, p. 254];

d) all orbits are locally connected, MONTGOMERY and ZIPPIN [16, p. 244];

e) some orbit is of dimension $\geqslant n-2$, BREDON [5];

f) the dimension of the orbit space $M/G \leqslant n-k$, where k is the maximum dimension of any orbit, RAYMOND [19].

Perhaps the simplest case for which the answer is unknown is the possibility of a free action of the dyadic group on Euclidean 3-space. In fact, the conjecture is true if one can show that the p-adic group A_p, for each prime p, can not operate effectively on M. This reduction of the problem to the p-adic group stems from the structure theorems of locally compact finite dimensional groups together with results of NEWMAN, SMITH, MONTGOMERY and others on transformation groups. Essentially, if G is a compact non-Lie group acting effectively on an n-manifold then G contains (in the center of G) a p-adic group A_p, for some prime p. See [16] Chapter VI] for many of the details.

* National Science Foundation Fellow.

One way of studying this conjecture is to assume the existence of a p-adic action on a manifold and investigate the consequences. Of course, one hopes for a contradiction. On the other hand, if such an action exists, this investigation sheds light on the nature of the example and may even lead one to its construction. The orbit space of such an action would have to be a reasonably nice space and yet exhibit many bizarre properties. These properties are mostly cohomological in nature and, unfortunately, seem to be theoretically possible. Furthermore, the existence of such (orbit) spaces appears to be intertwined with other difficult problems in dimension theory, cohomology theory and cohomology manifolds.

Recall that A_p is defined as the inverse limit of the sequence

$$Z_p \leftarrow Z_{p^2} \leftarrow Z_{p^3} \leftarrow \cdots$$

where Z_{p^i} is the multiplicative cyclic subgroup of the complex numbers of norm 1 consisting of the p^i-th roots of unity. The homomorphism $Z_{p^i} \leftarrow Z_{p^{i+1}}$ is generated by sending the first primitive p^{i+1}-th root of unity onto the first p^ith primitive root of unity. The elements of the p-adic group are often called the p-adic integers. The topology of A_p is the topology induced from the inverse limit.

The sequence is imbedded in an inverse sequence of fiberings:

$$\begin{array}{ccccccccc}
Z_p & \xleftarrow{p} & Z_{p^2} & \longleftarrow \cdots \longleftarrow & Z_{p^i} & \longleftarrow \cdots \longleftarrow & A_p \\
\downarrow & & \downarrow & & \downarrow & & \downarrow \\
S_1 & \xleftarrow{p} & S_2 & \longleftarrow \cdots \longleftarrow & S_i & \longleftarrow \cdots \longleftarrow & \Sigma_p \\
\downarrow & & \downarrow & & \downarrow & & \downarrow \\
S_1/Z_p & \longleftarrow & S_2/Z_{p^2} & \longleftarrow \cdots \longleftarrow & S_i/Z_{p^i} & \longleftarrow \cdots \longleftarrow & \Sigma_p/A_p = S
\end{array}$$

The middle horizontal maps are homomorphisms of the circle group onto the circle group of degree p. We are thinking of the circle group as the multiplicative group of complex numbers of norm 1. The bottom horizontal maps are isomorphisms of degree 1. This limit fibering is a principal A_p-fibering. Notice that any open and closed subgroup of A_p is again a p-adic group. In fact the proper open subgroups are the proper closed subgroups and are of the form $p^j A_p = A_p^j$ where $A_p/A_p^j \approx Z_{p^j}$. *The Cech cohomology of Σ_p is easily computed:*

a) $H^1(\Sigma_p; Z) = p$-adic rationals, Q_p, (i.e. fractions m/p^n, where m contains no factor divisible by p).

b) $H^1(\Sigma_p; Z_p) = 0$, since the maps in the sequence are all of degree p.

c) $H^1(\Sigma_p; L) = L$, where L is a field whose characteristic $\neq p$.

As it turns out, if A_p acts on a manifold, the dimension of the orbit space must be greater than the dimension of the manifold. The following theorem was first proved by C. T. Yang [30] by using an extension of Smith theory. Other proofs using the classifying space B_{A_p} for A_p were given by Bredon, Raymond and Williams [8] and C. N. Lee [13].

Theorem 1. *Let $G=A_p$ act effectively on a locally compact Hausdorff space X such that $\dim_Z X \leqslant n$. Then $\dim_Z X/A_p \leqslant n+3$. If $X=M^n$, an n-manifold, then $\dim_Z M^n/A_p = n+2$.*

We obtain this result later for a free action on a manifold and we shall refer the reader to the papers just mentioned for the general cases.

In general, let A_p act on a locally compact space X and form $X'=X\times_{A_p}\Sigma_p$ as usual. (If $x\in X$, $y\in\Sigma_p$, then $(x,y)\sim(x',y')$ if and only if there exists $g\in A_p\subset\Sigma_p$ such that $gx=x'$ and $gy=y'$. Denote the equivalence class of (x,y), the elements of X', by $[x,y]$). Define an action $\Sigma_p\times X'\to X'$ by $g[x,y]=[x,gy]$.

This action is well defined. We have the following commutative diagram with the obvious meanings, cf; [8; Prop. 2.1]:

$$\begin{array}{ccccc}
X & \xleftarrow{\pi_1} & X\times\Sigma_p & \xrightarrow{\pi_2} & \Sigma_p \\
\downarrow & & \downarrow & & \downarrow \\
X'/\Sigma_p = X/A_p & \xleftarrow{\pi_1'} & X\times_{A_p}\Sigma_p & \xrightarrow{\pi_2'} & S^1 \\
 & \nwarrow^{\pi_1} & \swarrow & S^1=\Sigma_p/A_p & \\
 & & X'/A_p = X/A_p\times S^1 & &
\end{array} \tag{1.1}$$

Here π_i and π_i' are projections and maps induced by projections. The map π_2' is a locally trivial fibering with fiber X and structure group A_p. The inverse image of $x^*\in X/A_p$ is $\Sigma/(A_p)_x$, where x lies on the orbit x^* and $(A_p)_x$ denotes the stability group at x.

2. The Orbit Space as a Cohomology Manifold

We now state a proposition which, when coupled with Theorem 1, is instrumental in validating case f) of the conjecture. The proposition, which is easily deduced from standard techniques of finite transformation groups, states that the orbit map does not destroy local cohomological properties provided that the coefficient group is sufficiently unrelated to the order of the transformation group.

Proposition 1 [19, Theorem 1]. *Let G be a separable totally disconnected compact group which acts effectively on a connected orientable n-cm M over a field L. Suppose that G acts trivially on $H_c^n(M,L)$.*

1. *If G is finite and the characteristic of L is 0 or prime to the order of G, or* 2. *if $G=A_p$, and the characteristic of $L\neq p$,*

or 3. *if G is an arbitrary totally disconnected group and the characteristic of L is* 0,

then the orbit space M/G is an orientable n-cm *over L.*

Proposition 1 also implies that the dimension raising of Theorem 1, if it is going to occur, must be determined by the p-torsion of the orbit space. For, suppose that $X=M^n$ and A_p acts effectively on M. If M is not orientable, lift the action of A_p to the oriented double covering $\tilde{M}$ of M. The action of the covering transformation commutes with the lifted action. The lifted action, of course, preserves the orientation and M/A_p is homeomorphic to $\tilde{M}/A_p/Z_2$, where $/Z_2$ denotes the orbit space of the covering transformations, see [6] and [8; § 4]. In particular, $\dim_L M/A_p = \dim_L \tilde{M}/A_p$, for any coefficient domain. If L is a field whose characteristic is $\neq p$, then Proposition 1, 2. implies that $\tilde{M}/A_p$ is an n-cm over L and hence $\dim_L M/A_p = n$.

If $\mathscr{L}^*$ denotes the family of fields Z_q, where q runs through all primes and 0 (Z_0 denoting the rational integers), then $\dim_{\mathscr{L}^*} X$, called the *field dimension of* X, is defined to be $\max_{L\in\mathscr{L}^*}\{\dim_L X\}$. The following inequality is easily proved.

$$(2.1) \qquad \dim_{\mathscr{L}^*} X \leqslant \dim_Z X \leqslant \dim_{\mathscr{L}^*} X + 1,$$

see, for example, [20; 3.1]. Thus, if $\dim_Z M^n/A_p = n+2$, according to Theorem 1, then $n+1 \leqslant \dim_{Z_p} M^n/A_p \leqslant n+2$. Thus dimension raising of the orbit map must be intimately connected with the prime p.

Theorem 2 [19; Theorem 2]. *Let A_p act freely on the n-manifold M^n. Then M^n/A_p is an n*-cm *over a field whose characteristic $\neq p$, and an* $(n+1)$-cm *over a field whose characteristic is p.*

Proof. First, assume that M^n is orientable and that the action of A_p preserves the orientation. Then we have already seen that Proposition 1 2. implies the first part. Form $M' = M^n \times_{A_p} \Sigma_p$ as in 1.1. M' is an orientable $(n+1)$-manifold fibered over S^1. Σ_p acts on M' and the orbit map $M' \to M'/\Sigma_p = M/A_p$ is a Vietoris map with respect to coefficients in a field of characteristic p. Wilder's monotone mapping theorem [26], which is a consequence of the Vietoris mapping theorems and Poincare duality implies that M'/Σ_p is an orientable $(n+1)$-cm over a field whose characteristic is p. If now M^n is orientable but A_p reverses orientation, then $p=2$. However, $A_2^{\frac{1}{2}}$ preserves the orientation and consequently $M^n/A_2^{\frac{1}{2}}$ is an n-cm over L, characteristic $L\neq 2$, and an $(n+1)$-cm over Z_2. The map $M^n/A_2^{\frac{1}{2}} \to M^n/A_p$ is a 2 to 1 covering map and consequently M^n/A_2 is a cm. If M^n is not orientable, lift the action of A_p to the orientable double covering $\tilde{M}^n$, [6]. The lifted action is free and orientation preserving. Furthermore, the cyclic group of order 2 of deck transformations acts freely on $\tilde{M}^n/A_p$ and yields M^n/A_p as orbit space. Thus the theorem is proved.

No example of spaces X which are cm's over each field but for which the dimension is not constant are known.

3. The Orbit Map as a Fiber Map

It was pointed out by C. N. LEE that if *A_p acts freely, then the orbit map $X \to X/A_p$ is a fiber map in the sense of* HUREWICZ. *This* can easily be seen by considering the sequence of p-fold covering maps:

$$X/A_p \leftarrow X/A_p^1 \leftarrow X/A_p^2 \leftarrow \cdots$$

The finite composition of maps in the sequence is a covering map and the total composition $X/A_p \leftarrow X/\bigcap_j A_p^j = X$ has the path lifting property.

Theorem 3 [21]. *If A_p acts freely on an n-manifold M^n, then no closed subset C of M^n/A_p with $\dim_Z C > n-2$ can be deformed, in M^n/A_p, to a point.*

Proof. Suppose that C is a closed subspace of M^n/A_p such that C can be deformed to a point in M^n/A_p. Then, by the covering homotopy property, we may lift the homotopy and obtain a cross section $\chi: C \to M^n$. Thus C can be imbedded as a closed subspace of both M^n and M^n/A_p.

Recall that M^n/A_p is an n-cm over any field whose characteristic is not p and an $(n+1)$-cm over a field whose characteristic is p. Furthermore, for any cm over a principal ideal domain L, closed subspaces of (cohomological) codimension 0, (respectively, codimension 1), are identical with those subspaces having non-empty interior, (respectively, those subspaces with empty interior) and that locally separate the cm. Obviously, these last mentioned properties are topological properties. Hence, $\dim_Z C \leqslant n-2$. This completes the proof.

Since $\dim_Z M^n/A_p$ is actually $n+2$, this means that every closed subspace deformable to a point in M^n/A_p must have (cohomological) codimension at least 4—a rather unpleasant property. In particular, M^n/A_p contains no polyhedron of codimension <4.

4. Dimension of Closed Subspaces of the Orbit Space

Theorem 4 [21]. *Let A_p act freely on M^n and let C be a closed subspace of M^n/A_p. Then,*

1. $\dim_Z C = n+2 \Rightarrow \dim_{Z_p} C = n+1, \dim_L C = n,$
2. $\dim_Z C = n+1 \Rightarrow \dim_{Z_p} C = n, \dim_L C = n-1,$
3. $\dim_Z C = n \Rightarrow \begin{cases} \dim_{Z_p} C = n, \dim_L C = n-1, \text{ or} \\ \dim_{Z_p} C = n-1, \dim_L C < n-1, \end{cases}$
4. $\dim_Z C = n-1 = \dim_{Z_p} C \leqslant n-1, \dim_L C < n-1,$

where L is a field of characteristic $\neq p$.

Proof. Recall that in § 2 we defined the field dimension of a space. Therefore, if $\dim_Z C = n+2$, which is the maximum possible because of Theorem 2 and formula 2.1, then C has interior in M^n/A_p. Consequently 1. holds. If $\dim_Z C = n+1$, then $\dim_{\mathscr{L}^*} C = n+1$ or n. If $\dim_{\mathscr{L}^*} C = n+1$, then once again $\dim_{Z_p} C = n+1$ and $\dim_L C = n$. If $\dim_{\mathscr{L}^*} C = n$, then $\dim_{Z_p} C = n$. and $\dim_L C = n-1$. For, if not, then there exists some field L whose characteristic $\neq p$ and $\dim_L C = n$. But this means that C as a subset of M^n/A_p, an n-cm over L, has interior. Consequently, $\dim_{Z_p} C = n+1$ since M^n/A_p is an $(n+1)$-cm over Z_p which contradicts $\dim_{\mathscr{L}^*} C = n$. Thus $\dim_{Z_p} C = n$. But, then $\dim_L C = n-1$, since C locally separates M^n/A_p.

Suppose $\dim_{\mathscr{L}^*} C = n-1$. If $\dim_L C = n-1$, for some L, then C must locally separate M^n/A_p. Consequently $\dim_{Z_p} C = n$, which contradicts $\dim_{\mathscr{L}^*} C = n-1$. Therefore, $\dim_{Z_p} C = n-1$, and $\dim_L C < n-1$ for each L whose characteristic differs from p.

If $\dim_{\mathscr{L}^*} C < n-1$, then $\dim_L C < n-1$ and $\dim_{Z_p} C < n-1$ and C must be void of interior and fail to locally separate M^n/A_p.

From these observations 3. and 4. follow immediately. Also 2. follows provided that we show that $\dim_Z C = n+1$ implies that $\dim_{\mathscr{L}^*} C = n$. This is a consequence of the fact that $\dim_Z C = n+2$, if and only if, C has interior in M^n/A_p which, in turn, is a consequence of the proof that $\dim_Z M^n/A_p = n+2$. The "only if" part has already been demonstrated.

R. F. Williams has proved a related result in the non-free case, see [29]. Our result is stronger in the free case.

5. Calculation of $H^*(B_{A_p}; Z)$

To complete the proof of 2. in Theorem 4 and to indicate the proof of Theorem 1, at least in the free case, we compute the integral cohomology of the universal classifying space B_{A_p} for the p-adic group A_p. Actually, what is really computed is the integral chohmology of the N-universal classifying space of A_p where N is very large. The N-fold join of A_p and its orbit space are not "nice spaces" but they do have the virtue of being compact and so suffice for the cohomological conclusions of Theorem 1.

It is not known whether or not M^n/A_p has finite covering dimension. If an action of A_p does exist and M^n/A_p has infinite covering dimension, then the analysis of M^n/A_p using an N-fold of A_p would probably be inadequate. If we took the infinite join of A_p and computed the cohomology of the resulting universal classifying space B_{A_p}, then we unfortunately would still have difficulties in applying this information since neither the spaces B_{A_p} nor M^n/A_p would be "nice enough" to permit traditional usage of Künneth theorems, Vietoris mapping theorems, or the Leray spectral sequence of a map.

We give here the computation of C. N. LEE [13]. (A slightly different method is used in [8]. This can be simplified following a suggestion of BREDON). By $B_{\Sigma p}$ and B_{A_p} we mean the N-universal (compact) classifying spaces for Σ_p and A_p, respectively. From the diagram

$$\begin{array}{ccc} E & \xrightarrow{A_p} & B_{A_p} \\ \downarrow{\scriptstyle \Sigma_p} & & \downarrow{\scriptstyle S^1=\Sigma_p/A_p} \\ B_{\Sigma_p} & = & B_{\Sigma_p} \end{array}$$

where E denotes the N-fold join of Σ_p, and maps are principal bundle maps we may consider the Gysin sequences

$$\begin{array}{ccccccccc} \cdots \to & H^i(B_{\Sigma_p}) & \to & H^i(B_{A_p}) & \to & H^{i-1}(B_{\Sigma_p})\otimes Z & \to & H^{i+1}(B_{\Sigma_p}) & \to \cdots \\ & \downarrow & & \downarrow & & \downarrow{\scriptstyle \phi_i} & & \downarrow{\scriptstyle \approx} & \\ \cdots \to & H^i(B_{\Sigma_p}) & \to & H^i(E) & \to & H^{i-1}(B_{\Sigma_p})\otimes Q_p & \to & H^{i+1}(B_{\Sigma_p}) & \to \cdots \end{array}$$

where the coefficient groups are all Z, and $Q_p=H^1(\Sigma_p;Z)$. Since the homomorphism $H^1(S^1)\to H^1(\Sigma_p)$ takes 1 to 1 (or -1), ϕ_1 is injective. Hence, $H^1(B_{A_p})=0$. Note from the lower sequence that $H^{i-1}(B_{\Sigma_p})\otimes Q_p \approx H^{i+1}(B_{\Sigma_p})$, all $i\geqslant 1$, and $H^1(B_{\Sigma_p})=0$. Hence, $H^{2i+1}(B_{\Sigma_p})=0$ for all i. Tensoring the exact sequence, $0\to Z\to Q_p\to Q_p/Z\to 0$, with Q_p it follows immediately that $Q_p\otimes Q_p\approx Q_p$ and $Q_p\otimes Q_p/Z=0$. Hence, $H^{2i}(B_{\Sigma_p})\approx Q_p$, for $i\geqslant 1$.

Since ϕ_i is just a coefficient homomorphism and $H^{2i}(B_{\Sigma_p})\approx Q_p$, ϕ_i is bijective for all $i\geqslant 2$. The diagram then immediately yields $H^{2i+1}(B_{Ap})=0$ for all i. Therefore $H^{i-1}(B_{\Sigma_p})\otimes Z\to H^{i+1}(B_{\Sigma_p})$ is an isomorphism, for all $i\geqslant 2$. This implies $H^i(B_{A_p})=0$, for all $i>2$. Examination of the diagram when $i=1$, yields $H^2(B_{A_p})=Q_p/Z$, the p-adic rationals mod 1. We have obtained for the N-universal classifying spaces the Proposition 2 [8, 13].

$$H^i(B_{A_p};Z) = \begin{cases} Z, & i=0 \\ Q_p/Z, & i=2 \\ 0, & \textit{otherwise.} \end{cases}$$

$$H^i(B_{\Sigma_p};Z) = \begin{cases} Z, & i=0 \\ Q_p, & i \text{ even} \\ 0, & i \text{ odd.} \end{cases}$$

To show that $\dim_Z M^n/A_p = n+2$ when A_p acts freely we consider the diagram 1.1 but with Σ_p replaced by the N-universal space E:

$$\begin{array}{ccccc} M^{n'} & \xleftarrow{\pi_1} & M^{n'} \times E & \xrightarrow{\pi_2} & E \\ \downarrow & & \downarrow & & \downarrow \\ M^n/A_p \times S^1 = M^{n'}/A_p & \xleftarrow{\pi_1'} & M^{n'} \times_{A_p} E & \xrightarrow{\pi_2'} & E/A_p = B_{A_p}. \end{array}$$

The map π_2' is a locally trivial fibering with fiber $M^{n'}$ and where the structure group A_p acts trivially on the cohomology of $M^{n'}$. (We can assume again that M^n is orientable and A_p does not reverse orientation and hence M^n is orientable). The spectral sequence of π_2' implies that $H_c^{n+3}(M^{n'} \times_{A_p} E; Z) \approx Q_p/Z$. However the map π_1' is essestially a Vietoris map over Z, since A_p is acting freely. Thus, $H_c^{n+3}(M^{n'}/A_p; Z) \approx Q_p/Z$. Hence, by the Künneth theorem, $H_c^{n+2}(M^n/A_p; Z) \approx Q_p/Z$. Consequently, $\dim_Z M^n/A_p = n+2$. (We know already that $\dim_Z(M^n/A_p) \leqslant n+2$). This yields Theorem 1 in the free case. Note that the argument also shows that each non-empty open subspace of M^n/A_p also has dimension $n+2$ over Z. The proof 2. of Theorem 4 is thereby completed since a closed subset C of M^n/A_p has interior in M_n/A_p only if $\dim_Z C = n+2$.

6. Strange Product Behavior

Pontryagin [17] described a sequence of spaces Φ_p, one for each prime, to disprove the product law $\dim X \times Y = \dim X + \dim Y$ for compact spaces. Each space was 2 dimensional with the property that $\dim_{Z_p} \Phi_p = 2$ and $\dim_{Z_q} \Phi_p = 1$, $q \neq p$. Furthermore, $\dim_Z \Phi_p \times \Phi_q = 3$, $q \neq p$. Note also that $\dim_{\mathscr{L}^*} \Phi_p \times \Phi_q = 3$ Montgomery observed that Kolmogoroff's adaption [12] of Pontryagin's construction of Φ_2 was the orbit space of an action of A_2 on a 1-dimensional space, the universal plane curve, see [27; III] for a description. In [22] Raymond and Williams modified and generalized Kolmogoroff's construction to obtain an action of A_p (not free) on a compact n-dimensional space X, for each $n \geqslant 2$, such that the covering dimension of $X/A_p = n+2$. In fact, $\dim_{Z_p} X/A_p = n+2$ and $\dim_{Z_q} X/A_p = n$, $p \neq q$. No examples of an action of A_p are known which raise the cohomology dimension by 3 although this is the theoretical limit, [30] and [8].

The spaces X^n described in [22] are dimensionally deficient in the sense of PONTRYAGIN. It is shown in [22], by the methods of BOCKSTEIN [2], and DYER [9], that if X^n and X^m are n and m-dimensional spaces on which A_p and A_q act as described in [22], then $\dim(X^n/A_p \times X^m/A_q) = n+m+2$ if $p \neq q$, and $n+m+4$ if $p=q$. On the other hand, if X^n and X^m are n and m-manifolds with effective, not necessarily free, actions of A_p and A_q on X^n and X^m respectively then

$$\dim_Z(X^n/A_p \times X^m/A_q) = \begin{cases} n+m+2 \text{ if } p \neq q, \text{ and} \\ n+m+3 \text{ if } p=q. \end{cases}$$

Again this is obtained from the formulae of BOCKSTEIN.

We wish to generalize these last mentioned results in the free case. BOLTYANSKI [3] was the first to describe a compact 2 dimensional space X such that $\dim X \times X < 2 \dim X$. A slight generalization of a result of BOCKSTEIN and FARY [10] states

Proposition 3 [20; § 3. Corollary]. *If C is a locally compact space, then*

$$\dim_Z C^k = \begin{cases} k \dim_Z C, \text{ if } \dim_{\mathscr{S}^*} C = \dim_Z C, \\ k(\dim_Z C - 1) + 1, \text{ if } \dim_{\mathscr{S}^*} C < \dim_Z C. \end{cases}$$

Here C^k means the product of C with itself k-times.

A locally compact Hausdorff space C is said to be *dimensionally deficient in the* sense of BOLTYANSKI if $\dim_{\mathscr{S}^*} < \dim_Z C$. Hence $\dim_Z C \times C = 2 \dim_Z C - 1$. C_p is said to be *dimensionally deficient in the sense of* PONTRYAGIN if there exists exactly one prime $p \neq 0$ so that $\dim_{Z_p} C_p = \dim_Z C_p$ and $\dim_{Z_q} C_p < \dim_Z C_p$, where $q \neq p$. Hence, $\dim_Z C_p \times C_p = 2 \dim_Z C_p$ but $\dim_Z C_p \times C_q < \dim_Z C_p + \dim_Z C_q$, if $q \neq p$. The latter inequality can be obtained from [9; 2.2 (c)].

If A_p acts freely on an n-manifold M^n, we combine the results of Proposition 3 and Theorem 4 to obtain

Theorem 5. *Every closed subspace C of M^n/A_p with $\dim_Z C > n-2$ is dimensionally deficient in the sense of* PONTRYAGIN *or* BOLTYANSKI. *If $\dim_Z C > n$, then C must be dimensionally deficient in the sense of* BOLTYANSKI.

In [22], the examples are dimensionally deficient in the sense of PONTRYAGIN although the orbit map raised dimension by 2. No examples of A_p actions on locally compact spaces which raise dimension by 2 and for which the orbit space is dimensionally deficient in the sense of Boltyanski seem to be known. In fact, R. F. WILLIAMS has recently shown [28] that it is impossible to construct from our techniques in [22] such an example. Nevertheless, he does succed to construct, in this paper, an example of a free action of A_p on an n-dimensional space X such that $\dim X/A_p = 1 + \dim X$ and $H_c^{n+1}(U;Z) \approx Q_p/Z$ for each connected open subset of X/A_p. We claim that $\dim_{\mathscr{L}*} X/A_p < \dim_Z X/A_p$, for $H_c^{n+1}(U;Z_p) = H_c^{n+1}(U;Z) \otimes Z_p = Q_p/Z \otimes Z_p = 0$. It follows, by Proposition 3, that WILLIAMS' examples are dimensionally deficient in the sense of BOLTYANSKI.

7. Sufficiency of a Polyhedral Example

In preparation of the examples exhibited in [22] we attempted to construct an action on a space X with an orbit space having the properties that an action on a manifold would necessitate. At the same time we tried to make X as nearly like a cohomology manifold as possible. (We may substitute cohomology manifolds in the statement of the conjecture without getting any closer to the answer. The results of this report are also valid for cohomology manifolds). However, for a direct attempt to produce an example on a manifold we shall now show that it is sufficient to produce one on a finite complex.

Let K be a finite complex of dimension $m \leqslant n$; G be a compact group acting effectively on K. It is known that G must be finite dimensional and metric. Write K as $M^m(K) \cup (K - M^m(K))$, where $M^m(K)$ is the (open) set of points of K which have neighborhoods homeomorphic to Euclidean m-space. Set $K^{m-1} = K - M^m(K)$. It is a subcomplex of dimension $< m$. Define inductively, $M^i(K)$ as the set of points of K^i which have neighborhoods homeomorphic to Euclidean i-space and K^{i-1} as $K^i - M^i(K)$. The sets K^i and $M^i(K)$ are all G-invariant sets and $K = \bigcup_{i=0}^{m} M^i(K)$. Because each manifold $M^i(K)$ consists of a finite number of components, G can not have arbitrarily small finite subgroups and still act effectively. Thus, if G is not a Lie group it must contain, in its center, a p-adic group A_p for some prime p. Select a sufficiently small p-adic subgroup $A_p^j \subset A_p$ such that each connected component of each $M^i(K)$ is left invariant under the action of A_p^j for each $i = 0, 1, \ldots, m$. Now assume whenever a p-adic group acts on a connected manifold of dimension $\leqslant n$ that the action is not effective. Then A_p^j must act effectively as a finite group on each component. Hence there exists some subgroup $A_p^k \subset A_p^j$ such that A_p^k leaves everything fixed. Thus, G could not have been acting effectively on K. Note that the same argument coupled with c) of the introduction shows that G must be a Lie group if it acts effectively on a finite 2-dimensional complex or if G is connected and acts effectively on a finite 3-dimensional complex.

References

16. contains an extensive bibliography on the entire range of the subject of transformation groups. We have also included in this list of references a few papers we have not used.

1. ANDERSON, R. D.: Zero dimensional compact groups of homeomorphisms, Pac. Journ. Math., **7**, 799 (1957).
2. BOCKSTEIN, M.: A complete system of fields of coefficients for the Δ-homological dimension, Doklady Akad. Nauk. SSSR (N.S.), **38**, 187—189 (1943).
3. BOLTYANSKI, V.: An example of a two-dimensional compactum whose topological square is three-dimensional, Amer. Math. Soc. Transl. **48**, 3—6 (1951); [Dokl. Akad. Nauk. SSSR **67**, 597—599 (1944)].
4. BOREL, A. et al.: Seminar in Transformation groups, Annals of Mathematics Study, **46**, Princeton, 1960.
5. BREDON, G. E.: Some theorems on transformation groups, Ann. of Math., **67**, 104—118 (1958).
6. — Orientation in generalized manifolds and applications to the theory of transformation groups, Mich. Math. J., **7**, 35—64 (1960).
7. — Cohomology fiber spaces, The Smith-Gysin sequence and orientation in generalized manifolds, Mich. Math. J., **10**, 321—333 (1963).
8. —, F. RAYMOND, and R. F. WILLIAMS: p-adic transformation groups, Trans. Amer. Math. Soc., **99**, 488—498 (1961).
9. DYER, E.: On the dimension of products, Fund. Math., **47**, 141—160 (1959).
10. FÁRY, I.: Dimension of the square of a space, Bull. Amer. Math. Soc., **67**, 135—137 (1961).
11. HAHN, F. J.: On the action of a locally compact group on E_n, Pac. J. Math., **11**, 221—223 (1961).
12. KOLMOGOROFF, A.: Über offene Abbildungen, Ann. of Math. **38**, 36—38 (1937).
13. LEE, C. N.: Compact 0-dimensional transformation groups, dissertation, University of Virginia (1959).
14. MONTGOMERY, D.: Pointwise periodic homeomorphisms, Amer. J. Math. **59**, 118—120 (1937).
15. — Finite dimensionality of certain transformation groups, Ill. Journ. of Math., **1**, 28—35 (1957).
16. —, and L. ZIPPIN: Topological Transformation groups. Interscience, 1955.
17. PONTRYAGIN, L.: Sur une hypothese fondamentale de la theorie de la dimension, C. R. Acad. Sci. Paris, **190**, 1105—1107 (1930).
18. NEWMAN, M. H. A.: A theorem on periodic transformations of spaces, Quart. J. Math., **2**, 400—416 (1931).
19. RAYMOND, F.: The orbit space of totally disconnected groups of transformations on manifolds, Proc. Amer. Math. Soc., **12**, 1—7 (1961).
20. — Some remarks on the coefficients used in the theory of homology manifolds, Pac. J. of Math., **15**, 1365—1376 (1965).
21. — Two problems in the theory of generalized manifolds, Mich. Math. J., **14**, 353—356 (1967).

22. —, and R. F. WILLIAMS: Examples of p-adic transformation groups, Ann. of Math., **78**, 92—106 (1963).
23. SMITH, P. A.: Transformations of finite period, III, Newman's theorem, Ann. of Math., **42**, 446—456 (1942).
24. — Periodic and nearly periodic transformations, Lectures in Topology, The Univ. of Mich. Conference of 1940, Edited by R. L. Wilder and W. L. Ayres, Univ. of Mich. Press, 1941,
25. WILDER, R. L.: Topology of Manifolds, Amer. Math. Soc. Colloquium Publications 32, Amer. Math. Soc., Providence, R. I. 1949.
26. — Monotone mappings of manifolds, II, Mich. Math. J., **5**, 19—23 (1958).
27. WILLIAMS, R. F.: A useful functor and three famous examples in topology, Trans. Amer. Math. Soc., **106**, 319—329 (1963).
28. — The construction of certain 0-dimensional transformation groups, to appear in the Trans. Math. Soc.
29. — Compact non-Lie groups, these Proceedings.
30. YANG, C. T.: p-adic transformation groups, Mich. Math. J., **7**, 201—218 (1960).

Compact Non-Lie Groups

R. F. WILLIAMS*

The principal unsolved problem in this area is the famous *Hilbert-Smith Conjecture: If a compact group G acts freely** on a manifold, then G is a Lie group.* To prove this conjecture, it would suffice, [4, 8 or 1], to show that no p-adic group can act freely on a manifold. Thus, in the past, researchers have looked for whatever surprising consequences they could find, from the assumption that a p-adic group A_p acts freely on an n-manifold X.

In 1940, P. A. SMITH [4] found,

A_0) $\dim X/A_p \neq \dim X$.

Later, C. T. YANG [7] found

A_2) $\dim X_n/A_p = m = \dim X + 2$ (or ∞); and

B) $H_c^m(U) = Z_{p^\infty}$, U any connected open subset of X/A_p.

These two properties were also proved in [1] as easy (at least in the free case) consequences of the computation

$$(*) \qquad H^q(B_{A_p}) \doteq \begin{cases} Z, & \text{if} \quad q=0 \\ Z_{p^\infty}, & \text{if} \quad q=2 \\ 0, & \text{otherwise}, \end{cases}$$

where B_{A_p} is the classifying space of A_p. The proof is as follows: using (*), a well-known spectral sequence has Z_{p^∞} as a corner term, and out pops A_2 and B. Thus, these two properties are the two most salient consequences of the assumption that A_p acts on a manifold.

In [3], FRANK RAYMOND and the author gave an example in which A_2 holds. The space of course was not a manifold and B was far from true.

In [6], we give an example of a 1-dimensional space X (1 for simplicity; similar constructions work for any n) and a free action of the p-adic group A_p on X such that

* Supported in part by NSF Grant GP 5591.

** Here and below we consider only the free case for simplicity. Similar remarks apply to *effective* transformation groups.

A_1) $\dim X/A_p = m = \dim X + 1$.

B) $H_c^m(U) = Z_{p^\infty}$, for any connected open subset U of X/A_p.

This example represents the author's best attempts to construct an example of a p-adic transformation group having properties A_2 and B.

Indeed, it is conjectured that A_2 and B cannot occur together for any space X, at least if $\dim X/A_p < \infty$. A weak version of this conjecture is proved in Part II, to wit: *the current technique of construction cannot yield an example satisfying A_2 and B.*

Though very weak, this result is of interest for three reasons: 1. the existence of the example satisfying A_1 and B; 2. the "obstruction" is a rather delicate one; and 3. the success of this technique of construction is *very nearly* necessary for the existence of such an example. That is, this line of reasoning ("if there is an example, then it can be constructed in such-and-such a way") could well lead to a proof of the Hilbert-Smith conjecture.

In [6], we give another reason for a renewed attack on this old problem in the form of a compact, 2-dimensional "classifying space" B'_{A_p}, due to E. E. FLOYD The point is, ordinary cohomology does not distinguish between a "real" classifying space and B'_{A_p}, though all the results on this conjecture can be obtained via $H^*(B_{A_p})$ or its isomorph, $H^*(B'_{A_p})$. However, complex K-theory *does* distinguish between B_{A_p} and B'_{A_p}. Perhaps K-theory and some version of the spectral sequence alluded to above lead to new results on this problem.

In a paper written some time ago, and only recently submitted for publication [7] a different approach to this conjecture is begun. The rough idea is as follows. First, if A_p acts on a manifold M^n, then the p-adic solenoid acts on the "mapping torus" of $f: M^n \to M^n$, where f "generates" the group A_p, see [1]. As this is an $(n+1)$-manifold, it suffices to consider actions of solenoids, say Σ acting on M. Then $\dim M/\Sigma = \dim M + 1$. *If* one could find a fairly nice, subspace of M/Σ of dimension $\dim M - 1$, then a contradiction would result. More precisely,

Definition. *A space X is (m, Z)-flat iff X is a locally compact, locally connected Hausdorff space such that for any sufficiently small connected open set U, (H_c^* denotes compact Čech cohomology):*

a) $H_c^m(U; Z) \approx Z$;
b) *The map* $H_c^m(U; Z) \to H_c^m(X; Z)$ *induced by inclusion is an isomorphism; and*
c) $H_c^{m+1}(U; Z) = 0$.

Generalized orientable m-dimensional cohomology manifolds over Z are (m, Z)-flat, but *not* conversely; thus (m, Z)-flat spaces are closed under certain limiting procedures which makes it more likely that they can be constructed in certain well behaved spaces.

Example. If a 2-manifold, say a 2-sphere is triangulated, and a small handle added to each triangle, then triangulated again and a handle added to each new triangle, etc., then the sequence of manifolds so defined converges to an $(2, Z)$-flat space.

For the next theorem, we assume that a (p-adic) solenoid Σ acts on a space X and that X/Σ is (m, Z)-flat; $\pi: X \to X/\Sigma$ denotes the orbit map.

Computational Theorem. *Suppose $x_0 \in X$ and that V is a connected open neighborhood of $y_0 = \pi(x_0)$. Then $H_c^m(\pi^{-1}(V); Z)$ contains a subgroup $G' \approx H'(\Sigma x_0; Z)$ which maps monomorphically under the map*

$$H_c^{m+1}(\pi^{-1}(V); Z) \to H_c^{m+1}(X; Z)$$

induced by inclusion.

The proof proceeds by showing this holds (in a stronger form) in the special case of a "near free action" (see below), then using a factorization theorem to show the special case implies the more general.

Next assume a p-adic group A acts on a space X, $x_0 \in X$ and $\pi: X \to X/A$ is the orbit map. Let $A_i \subset A$ be the kernel of the natural map $A \to Z_{p^i}$.

Definition. *The action of A is said to be near free at x_0 iff there exist connected open sets $X/A = U_0 \supset U_1 \supset U_2 \supset \ldots$, all containing $\pi(x_0)$, such that the stability group*

$$A_x = \begin{cases} A_i, & \text{for} \quad x \in \pi^{-1}(U_{i-1} - U_i) \\ 1, & \text{for} \quad x \in \bigcap_i \pi^{-1}(U_i). \end{cases}$$

Near free factorization lemma. If $A_{x_0} = 1$ and X/A is locally connected, then there is an action of $A' \approx A$ on X and an action of A'' on X/A' such that $(X/A')/A'' = X/A$ and A'' is near-free at the image of x_0 under $\pi': X \to X/A'$.

In conclision, note that this does prove the theorem of MONTGOMERY-ZIPPIN [2] that a p-adic group cannot act effectively on a 2-manifold. For then a solenoid Σ would act on a 3-manifold M. Now M/Σ being locally connected and connected contains an arc B, which is, of course $(1, Z)$-flat. By the computational theorem, one has a hold on the cohomology of $\pi^{-1}(B) = X$. It follows that X cannot be imbedded in a 3-manifold, by an argument due to FRANK RAYMOND.

References

1. BREDON, G. E., FRANK RAYMOND, and R. F. WILLIAMS: p-adic transformation groups, Trans. Amer. Math. Soc. **99**, 488—498 (1961).
2. MONTGOMERY, D., and L. ZIPPIN: Topological Transformation Groups, Interscience Publishers, N.Y. (1955).
3. RAYMOND, F., and R. F. WILLIAMS: Examples of p-adic transformation groups, Annals of Math. **78**, 92—106 (1963).
4. SMITH, P. A.: Periodic and nearly periodic transformations, Lectures in Topology, Univ. of Michigan Press, 1941.
5. WILLIAMS, R. F.: A useful functor and three famous examples in topology. Trans. Amer. Math. Soc. **106**, 319—329 (1963).
6. — Concerning the construction of certain 0-dimensional transformation groups, to appear in the Trans. Amer. Math. Soc. **129**, 140—156 (1967).
7. — Solenoidal transformation groups, to appear.
8. YANG, C. T.: p-adic transformation groups, Mich. Math. J. **7**, 201—218 (1960).

Applications of Transformation Groups to Problems in Topological Semigroups

KARL H. HOFMANN* and PAUL S. MOSTERT

Frequently mathematicians think of the theory of topological semigroups (if they think of it at all) as a generalization of the theory of topological groups. From such a vantage point there appears to be a discouraging scarcity of rich structure, such as, for instance, is encountered in the theory of Lie groups or of compact connected groups, and a hopeless profusion of examples exposing a vast variety of pathological phenomena. Contrary to this initial impression however, a slight shift in the observer's stand from the viewpoint of topological groups to that of topological transformation groups reveals more structure than expected and brings some order into the chaos.

Let us take, for example, a topological semigroup with identity 1. The set $H(1)=H$ of all elements with two sided inverse relative to 1 is a group, called the *group of units*. In many interesting cases, this group is topological. This is true when S is compact or when $H(1)$ is locally compact [2]. (However, there is an example where this is not the case, even when S is locally compact [15]. Of course $H(1)$ is not locally compact in that example.) The group $H \times H$ acts on the right of S under what we have called the *double action*

(D) $$(s,(g,h)) \mapsto g^{-1} s h : S \times (H \times H) \to S.$$

The group $H \times H$ is the semidirect product of the normal subgroup $1 \times H$ and the diagonal subgroup $\Delta = \{(h,h) : h \in H\}$, both of which are isomorphic to H. By restriction to these subgroups we obtain the more familiar action of H by right translation

(R) $$(s,h) \mapsto s h : S \times H \to S,$$

and by inner automorphisms

(I) $$(s,h) \mapsto h^{-1} s h : S \times H \to S.$$

* The first named author is an ALFRED P. SLOAN Fellow and the second is a National Science Foundation Senior Postdoctoral Fellow. This work was also partially supported by NSF Grant GP 6219.

All of these actions play a crucial role in the structure theory of S. Is it then surprising to find that the additional constraints given by the multiplication should yield for appropriately restricted classes of topological semigroups quite adequate structural descriptions, some of which show striking similarities to the constructional descriptions emerging recently in the theory of transformation groups?*

Let us lead into the topic by considering a few natural questions which arise in the context of the group of units and its actions. The class of connected locally compact topological semigroups with identity is certainly a natural one to consider. Within this class, a conspicuous subcategory is furnished by semigroups living on a closed manifold. What can be said about their structure? The answer is quite simple. They are all groups, compact connected Lie groups of course, and so, from the standpoint of semigroups, cease to be of further interest. There are, in fact, some more general results which outline the realm of what might be expected in this direction:

Theorem 1 (MOSTERT and SHIELDS [18]). *Let S be a semigroup with identity on a manifold. Then the maximal subgroup is open in S and is a topological group.*

One can generalize this result quite a bit by replacing the manifold by a topological space which only has 'intrinsic points'. To explain what this means is a real difficulty and has little to do with transformation groups, but quite a bit with algebraic topology. We discussed the question in [6] as the question of *peripherality*.

Theorem 2 (HUDSON and MOSTERT [9]). *Let S be a compact connected finite dimensional semigroup with identity on which the group of homeomorphisms is transitive. Then S is a group.*

The proof of this fact used transformation groups in order to produce in any compact connected semigroup with identity a compact connected subset which meets $H(1)$ exactly in 1. This question suggests the for-

* It should also be remarked that the application of transformation groups to compact semigroups with separately continuous multiplication (a class that occurs frequently in analysis) is in the same way possible by virtue of Ellis' theorem [2], which states that the group H is a topological group if it is locally compact. (Contrary to the nice situation for compact topological semigroups, the group H in fact is not always compact, or even locally compact.) For a description of the degree to which this theory has been developed, the reader is referred to BERGLUND and HOFMANN, *Compact Semitopological Semigroups and Weakly Almost Periodic Functions*, Lecture Notes in Mathematics 42, Springer-Verlag, 1967.

mulation of the following problem, the complete state of which is unknown:

Problem 1. Let G be a compact group acting on a continuum X. For $x \in X$, does there exist a non-degenerate continuum $C \subset X$ such that $G(x) \cap C = x$ (providing, of course, that $X \neq G(x)$)?

For a Lie group the answer follows from the slice theorem. If G is metric and X is semi-locally one connected, a positive answer was given by MOSTERT [14] and, under the additional assumption that $G(x)$ is a principal orbit, by ANDERSON and HUNTER [1]. For semigroups this issue is no longer of importance because a much superior semigroup theoretic result is available as we shall see presently.

Theorem 2 is false without the hypothesis of finite dimensionality as is exposed by a countable product of ordinar unit intervals under multiplication; the homeomorphism group on the Hilbert cube is indeed transitive. However, the following result still remains true:

Theorem 3 (MADISON [10]). *Let S be a compact connected semigroup with identity on a homeogeneous space of a compact group. Then S is a group.*

This result for group spaces was first announced by SELDEN (AMS Notices, 1965); a proof was later published by SELDEN and MADISON [11] while in the meantime we had an independent proof [6; p. 171] under these more special circumstances. The following question remains open, even though a solution does not promise to be much harder than the proof of the preceding theorem:

Problem 2. If S is a compact connected semigroup with identity on the homogeneous space of a locally compact group, is it necessarily a group?

These results, however, do not exhaust the topic of semigroups on manifolds by any means. The following fact, for instance, uses a result about fixed points of one parameter group actions in the plane [13]:

Proposition 4 (MOSTERT and SHIELDS [18]). *Let S be a semigroup whose underlying space is $\mathbb{R}^2$ and let H_0 denote the identity component of $H(1)$. Then there is an idempotent in the boundary of H_0 (unless, of course, S is itself a group).*

Proposition 5. (HORNE [7, 8]). *Let S be a semigroup with identity on three dimensional closed half space whose interior is $H(1)$. Then the boundary contains an idempotent.*

One is thus led to the following problem which was posed in 1957 by MOSTERT and SHIELDS [17] and is still unanswered except in the above cases:

Problem 3. Let S be a semigroup with identity on a connected manifold. Suppose that S has no idempotents other than the identity. Is then S a group?

We have by no means exhausted the questions than can be asked about semigroups on manifolds*. We shall return to the subject later. However, now let us become a little more systematic. We observe that in a semigroup, groups are even more prevalent than our concentration on the group of units may indicate. Following Green, in each semigroup one can introduce an equivalence relation $\mathcal{H}$ as follows: $s\mathcal{H}t$ if and only if $s\cup Ss=t\cup St$ and $s\cup sS=t\cup tS$. The coset of s modulo $\mathcal{H}$ is denoted by $H(s)$. If e is an idempotent, then $H(e)$ is exactly the group of units of the subsemigroup eSe, and is also the maximal group containing e. If S is commutative, and in some more general instances, $\mathcal{H}$ is a congruence relation and $S/\mathcal{H}$ is a semigroup, but in general this is not the case and $S/\mathcal{H}$ is just a set with a partial order deduced from a quasiorder on S whose definition is given by $s\leqslant_{\mathcal{H}}t$ if and only if $s\in t\cup(St\cap tS)$. If S is a compact topological semigroup, then $\mathcal{H}$ has a closed graph and $S/\mathcal{H}$ is a compact Hausdorff space. For an arbitrary element $s\in S$, the set $\{t\in S: H(s)t\subset H(s)\}$ is a subsemigroup T_s of S on which the relation $R_s=\{(t,t'): ht=ht'\}$ for all $h\in H(s)$ is a congruence relation. As SCHÜTZENBERGER has observed [19], the semigroup $\Gamma_s=T_s/R_s$ is in fact a group and acts in the obvious fashion on $H(s)$. The action is simply transitive, and if e is an idempotent, then the function $h\mapsto R_e(h): H(e)\to\Gamma_e$ is an isomorphism of groups. If S is a compact topological semigroup, then Γ_s has a natural compact group topology and the action on $H(s)$ is continuous.

In some of the deepest investigations concerning compact semigroups, two concepts reduce to one and the same fundamental tool: The fixed point set of a transformation group and the centralizer of a subgroup of a semigroup. The centralizer of $H(1)$, for instance, is the set of all $s\in S$ such that $sh=hs$ for all $h\in H(1)$. But this is clearly the fixed point set for the action (I) of $H(1)$ on S under inner automorphisms. It is one of the most important outstanding problems in the theory to gain more insight into the nature of the centralizer of the maximal group.

---

* Applications of transformation groups to other areas involving topological-algebraic objects on manifolds have also been made. For example, S. N. HUDXLN has made significant use of these methods in the study of certain types of *loop* structures on manifolds (Lie loops with invariant uniformities, I and II, Trans. Amer. Math. Soc. **115**, 417—432 (1965), and **118**, 526—533 (1965)) in his study of analytic loops. (A loop is a continuous, but not necessarily associative, multiplication on a space such that the equations $ax=b$ and $ya=b$ have unique solutions continuous in a and b. Such objects arise naturally in geometry.)

Notably, a positive solution to the following problem would revolutionalize substantial parts of the entire theory:

Problem 4. If S is a compact connected semigroup with identity 1 and zero 0, is the centralizer of $H(1)$ connected?

Using the fact, that such a semigroup is acyclic over the rationals and a result of BOREL and CONNER about the fixed points of a circle group acting on a rationally acyclic space we have obtained the following

Theorem 6 [6; p. 62]. *If S is a compact connected semigroup with identity* 1 *and* 0 *and if A is a compact connected abelian subgroup of $H(1)$, then the centralizer of A is connected.*

Since 0 and 1 belong to this centralizer, the centralizer is thus in a very definite sense nondegenerate. In the wide range between this theorem and a positive answer to Problem 4 we obtained the following theorem with the aid of substantially transformation group theoretic methods, a discussion of which would lead us too far into technicalities:

Theorem 7 [6; p. 177]. *If S is a compact connected semigroup with* 1 *and* 0 *and no other idempotents, and if the quotient space $S/(H \times H)$ of S modulo the double action (D) is a totally ordered space, then the centralizer of $H(1)$ is connected.*

The importance of this theorem is not likely to be evident to anyone who has never contemplated compact semigroups at some length. Yet some of the applications which we shall explicate in a moment perhaps give a fair measurement of its significance. However let us first mention the most important application of Theorem 6, for which researchers in the field have labored quite a while.

Theorem 8 [6]. *Let S be a compact connected semigroup with* 1 *and* 0. *Then there is a compact connected abelian subsemigroup T containing* 0 *and* 1 *such that $T \cap H(1) = 1$ and that $T/\mathcal{H}_T$ is totally ordered.*

Some remarks are in place. Firstly, the hypothesis about the presence of a zero is no restriction whatsoever in this as well as in the preceding results. Every compact semigroup has a unique minimal ideal*

* The structure of the minimal ideal is fully known for compact semigroups, but since there is little contact with transformation group aspects of the theory of compact semigroups in its description, we have not chosen to discuss it here. The interested reader is referred to the First Fundamental Theorem [6; p. 56].

which may be collapsed to a point. One thus obtains a semigroup with zero, and the results which we have just described allow an easy retranslation into the general situation. Secondly we observe, that T is exactly a continuum of the kind which was postulated in Problem 1 when we consider the action (R) of $H(1)$ under right translation on S. Thirdly, by a standard minimality argument, it is elementary to see that S contains minimal compact connected subsemigroups containing 0 and 1. Theorem 8 says that all of these have to be abelian, a fact for which there is certainly no obvious and superficial reason. Lastly, the fact that $T/\mathscr{H}_\tau$ is a totally ordered semigroup (recall that $\mathscr{H}_T$ is a congruence for abelian T) gives a first indication that the structure of T may be amenable. This is in fact the case to a surprising degree, and practically all details of the structure of T when chosen minimal can be considered as known. The importance of this information is perhaps only comparable with the knowledge about maximal tori in compact Lie groups or of Sylow groups in finite groups. Yet this assertion is more readily stated than the details are explained.

In order to give at least a vague idea of the structure of T we have to indicate the structure theory of a special class of compact semigroups which plays the decisive role in this context. A semigroup S is built up from simpler building blocks which are joined together in a fashion not unsimilar to some of the adjunction procedures which are now frequently used in the theory of transformation groups. Let us first give a rough description of the building blocks. Each such has two basic constituents, the first one being a compact group G, the second one being a semigroup Σ of a particular nature: Let $\mathbb{H}$ denote the semigroup of all non-negative real numbers under addition and $\mathbb{H}^*$ the compact semigroup obtained by adding infinity acting as a zero: $\infty + x = x + \infty = \infty$ for all $x \in \mathbb{H}$. Then $x \to e^{-x}$: $\mathbb{H}^* \to [0,1]$ with $e^{-\infty} = 0$ is an isomorphism of compact semigroups. Let U be the universal compact solenoidal group (i. e. the character group $(\mathbb{R}_d)^\wedge$ of the discrete reals) and $f: \mathbb{R} \to U$ the universal compact one-parameter group (i. e. the adjoint function to the identity map $\mathbb{R}_d \to \mathbb{R}$). The designation of f as *universal* stems from the fact that any one parameter group $\mathbb{R} \to C$ with compact [8]odomain factors uniquely through f. In the compact semigroup $\mathbb{H}^* \times U$, let Σ be the compact abelian subsemigroup

$$\{(h, f(h)) : h \in \mathbb{H}\} \cup \infty \times U,$$

and define $F: \mathbb{H} \to \Sigma$ by $F(h) = (h, f(h))$. Then F is a universal compact one parameter semigroup in the sense that each one-parameter semigroup with compact codomain factors uniquely through F. Geometrically the semigroup Σ may be visualized as the union of a closed half line whose free end spirals down onto the group U from its outside, following the path of the one parameter subsemigroup $f(\mathbb{H})$ in S more and more closely. If, in Σ, one collapses the ideal $\infty \times U$ to a point, one obtains a semigroup which again is isomorphic to $\mathbb{H}^*$.

Now let us form $G \times \Sigma$. The quotient semigroups of this semigroup are very well understood, although there are some non-trivial delicate points. For a general insight, may it suffice to say that they are not too different from $G \times \Sigma$ itself and are essentially obtained by collapsing in the orbits $G \times x$ of G the orbits of a subgroup $G(x)$ of G which increases with h in an admissible fashion, where $x = (h, f(h))$ and reaches its maximum for $x = (\infty, s)$. There are other complications which we neglect for the time (whoever is really interested in a detailed discussion is referred to [6], Chapter B, Section 2, and to [16]). All the semigroups B so obtained share the following properties: 1. There is a homomorphism η onto $\mathbb{H}^*$ or the semigroup $\mathbb{H}_r$, which is the interval $[0,r]$, $r \in \mathbb{R}$, with the semigroup operation $(x,y) \to \min(x+x,r)$. 2. The inverse image $\eta^{-1}(0)$ of 0 is G (up to natural isomorphism) and is the group of units $H_B(1)$. 3. The inverse image $\eta^{-1}(\infty)$ (resp. $\eta^{-1}(r)$) is a group $M(B)$.

Now imagine that we have two such semigroups B and B' with varying ingredients, but with the property, that there is an injective morphism of semigroups from $M(B)$ into $H_{B'}(1)$. Then one may glue B and B' together using this morphism and arrive at a new semigroup $B \# B'$ containing B and B' such that $B \# B'$ now allows a homomorphism onto a semigroup obtained by joining the two intervals $\eta_B(B)$ and $\eta_{B'}(B)$ together at one of their endpoints. One may regard this construction as a process by which we start from the two element semigroup $X = \{0,1\}$ (which is a very special totally ordered compact semilattice) and insert at each element of X a semigroup of the type B making special provisions for pasting pieces together. One of the crucial facts of the theory is, that starting from any totally ordered compact semilattice X whatsoever one can arrive at a semigroup with identity by replacing each element of X with one of a compatible family of semigroups of type B and gluing the pieces together to obtain a semigroup S. The semilattice X will actually be contained in S upon identification through a suitable map $X \to S$. Such a semigroup will then allow a morphism $\eta: S \to I$ onto a semigroup I which is a totally ordered compact connected space whose one endpoint is a zero, whose other endpoint is an identity, and which contains an isomorphic copy of X. Such semigroups are very well known; they are called I-semigroups (for interval semigroups). It turns out that two elements are mapped onto the same point under η if and only if they are $\mathscr{H}$-equivalent in the sense which we discussed before. Thus η is a congruence and $S/\eta \cong I$. We have called a semigroup S which is built up in such a fashion, a *hormos* alluding to a Greek word meaning something like an ornamental chain. What is interesting about this observation concerning $S/\mathscr{H}$ is the fact that it characterizes a hormos. Indeed, we have the following.

Theorem 9 [6]. *Let S be a compact semigroup with identity. Then the following two statements are equivalent:*

a) *S is a hormos.*
b) *η is a congruence on S and $S/\mathscr{H}$ is an I-semigroup.*
c) *S is totally quasi-ordered relative to $\mathscr{H}$ and $S/\mathscr{H}$ is connected.*

Notably, any semigroup of the kind that we mentioned in Theorem 8 is an abelian hormos. As the construction indicates, it generally teems with groups and group actions.

If a hormos has the additional property, that it does not contain a proper compact connected subsemigroup containing the identity and meeting its minimal ideal $\eta^{-1}(0)$, then more precise statements can be made. Let us call such a hormons *irreducible*. (By Theorems 8 and 9, every irreducible semigroup is an irreducible hormos.) Starting from a totally ordered semilattice X as before, it is possible to construct an irreducible hormos $\operatorname{Irr} X$, an I-semigroup $I(X)$ a homomorphism $\Phi\colon \operatorname{Irr} X \to I(X)$ such that $X \to \operatorname{Irr} X \to I(X)$ with a suitable first function given by the construction is an injection preserving identity and zero, and finally that every irreducible semigroup whose subsemigroup of idempotents is isomorphic to X is obtainable as a homomorphic image of $\operatorname{Irr} X$. The groups which go into the construction of $\operatorname{Irr} X$ can be amazingly big even in an immediate vicinity of the identity. It is the more surprising, that in $\operatorname{Irr} X$ the group of units is degenerate, a fact which is part of Theorem 8. At this point it may be a good thing to recall that all these remarks served to implement Theorem 8 and to give an idea, that an extensive structure theory has been developed to back up the power of this theorem, which started from a purely transformation group theoretical theorem.

Now let us give some comment on Theorem 7, which, together with the results that we have stated above, yields some quite surprising facts.

Suppose that a semigroup satisfies the hypotheses of Theorem 7. (There are in fact quite a few assumptions that are natural from the viewpoint of semigroup theory and which imply these hypotheses.) One can show that there is a one parameter semigroup $\phi\colon \mathbb{H}^* \to S$ such that $\phi(0) = 1$, $\phi(\infty) = 0$, and that $\phi\mathbb{H}$ is in the centralizer of $H(1)$. The compact semigroup $H(1) \times \mathbb{H}^*$ is then mapped homomorphically into S under the map $(g,h) \mapsto gf(h)$; and the image fills out a whole neighborhood of 1. Thus the local structure of S is well known and is in fact of the type of the semigroups B which we discussed in building the hormos.

Perhaps the most interesting applications of these facts arise in the following situation: Suppose that S is a compact topological semigroup with identity, and suppose that there is a compact group $G \subseteq H(1)$ of units such that the space S/G of orbits under the action by left translation is a totally ordered connected space. Then the orbits of G are the cosets of a congruence relation and the space S/G is a compact semigroup. Then S contains a totally ordered compact subsemigroup T in the centralizer of G. The function $(g,t) \mapsto gt: G \times T \to S$ is a surjective monomorphism of compact semigroups, and the composition of maps $T \to G \times T \to S \to S/G$ is an isomorphism of semigroups. Since the structure of totally ordered compact connected semigroups with identity is quite well understood, we have here a practically complete structure theory for S. The following sample of applications is representative for the power of this result:

Theorem 10 [6]. *Let S be a connected locally compact semigroup with identity* 1. *Suppose that the following two conditions are satisfied:*

a) *S is topologically embeddable in an n-dimensional manifold.*

b) *S contains an* $(n-1)$*-dimensional compact group of units.*

Then one of the following cases occurs:

a) *S is a Lie group.*

b) *S is the direct product of a connected compact Lie group L and a semigroup T with identity on a one-dimensional totally ordered separable or non-separable manifold with no, one, or two endpoints.*

c) *S is the quotient semigroup of the direct product of a compact connected Lie group L with a semigroup T with identity on a separable one-manifold with zero as one endpoint and with or without a further endpoint, modulo a congruence relation, which in* $L \times T$ *collapses all sets* $gN \times 0$ *to single points, where N is a normal subgroup of L which is isomorphic to a* 0, 1, *or* 3*-sphere.*

The hypotheses of this theorem are satisfied if S is a connected locally compact semigroup with identity 1 on an n-manifold with a regular boundary (i.e. a boundary on which every point has a neighborhood homeomorphic to Euclidean half space in n dimensions) and if 1 is a boundary element such that the component of 1 in the boundary is a compact subsemigroup. In the special case where this boundary is connected, this result was obtained by MOSTERT and SHIELDS [17] in a paper which was the forerunner to all such questions and indeed to the application of transformation groups to semigroups.

For the formulation of the next theorem along this line, let $\mathbb{R}^*$ be the semigroup of all reals with the point ∞ adjoined to the positive end as a semigroup zero. Let $f: \mathbb{R} \to U$ be again the universal one parameter group with compact codomain.

Theorem 11 [4]. *Let S be a locally compact semigroup with identity such that the group of units H is a dense open subgroup without compact open subgroup. Assume that $S-H$ is non-empty and compact. Then there is a compact group G and a closed normal subgroup C of $U \times G$ such that S is isomorphic to the quotient semigroup of the semigroup $\{(r, f(r), g): r \in \mathbb{R}, g \in G\} \cup \infty \times U \times G \subseteq \mathbb{R}^* \times U \times G$ modulo the congruence whose cosets are the sets $\infty \times (k, g)C \subset \infty \times U \times G$ and are singleton otherwise*

The following is a corollary of the preceding results:

If S is a locally compact semigroup with identity 1 and zero 0 on a space with a transitive homeomorphism group such that $S-0$ is a group, then S is isomorphic to the multiplicative semigroup of the real, the complex, or the quaternion field.

There are many other techniques that have not been mentioned here. For example, the so called "peripherality" theorems and problems have alluded to once but nowhere discussed. These results have, however, played an important role in many of those theorems that have been described here. A brief survey of some of the "peripherality" theory can be found in [3] and [16]. A thorough (but still unsatisfactory) study is contained in [6]. But the emphasis here has been on the applications of transformation groups, and as a final remark on this way, it is perhaps worth mentioning that the paper [12] and its follow-up [5] were written for the purpose of supplying the information needed to study certain classes of semigroups on subspaces of manifolds.

References

1. Anderson, L. W., and R. P. Hunter: Small continua at certain orbits, Archiv der Math., **14**, 350—353 (1963).
2. Ellis, R.: Locally compact transformation groups, Duke Math. J. **24**, 119—125 (1957).
3. Hofmann, K. H.: The structural components of a compact connected semigroup with identity, Seminaire Dubreil-Pisot (Algébre et Theorie des nombres) 19e annee, 1965/66, n° 20.
4. — Locally compact semigroups in which a subgroup with compact complement is dense, Trans. Amer. Math. Soc., **106**, 19—51 (1963).
5. —, and P. S. Mostert: Compact groups acting with (n—1)-dimensional orbits on subspaces of n-manifolds, Math. Ann. **167**, 224—239 (1966).
6. —, — Elements of Compact Semigroups, Chas. E. Merrill Books, Inc., Columbus, 1966.
7. Horne, J. G.: Flows that fibre and some semigroup questions, to appear.
8. — A locally compact connected group acting on the plane has a closed orbit, Illinois J. Math., **9**, 644—650 (1965).

9. HUDSON, A. L., and P. S. MOSTERT: A finite dimensional homogeneous clan is a group, Annals of Math., **78**, 41—46 (1963).
10. MADISON, B.: Semigroups on coset spaces, Duke Math. J., to appear.
11. —, and J. SELDEN: Clans on group supporting spaces, Proc. Amer. Math. Soc., **18**, 540—545 (1967).
12. MOSTERT, P. S.: On a compact Lie group acting on a manifold, Annals of Math., **65**, 447—455, and 589 (1957).
13. — One-parameter transformation groups in the plane, Proc. Amer. Math. Soc., **9**, 462—463 (1958).
14. — Continua meeting an orbit at a point (with corrections), Fundamenta Math., **52**, 319—321 (1963); **56**, 221—222 (1964).
15. — Untergruppen von Halbgruppen, Math. Zeitschr., **82**, 29—36 (1963).
16. — The structure of topological semigroups-revisited, Bull. Amer. Math. Soc., **72**, 601—618 (1966).
17. —, and A. L. SHIELDS: On the structure of semigroups on a compact manifold with boundary, Annals of Math., **65**, 117—143 (1957).
18. —, — Semigroups with identity on a manifold, Trans. Amer. Math. Soc., **96**, 380—389 (1959).
19. SCHÜTZENBERGER, M. P.: $\mathscr{D}$ représentation des demi-groupes, C. R. Acad. Sci. Paris, **244**, 1994—1996 (1957).

On the Action of Compact Groups

MEI-CHIN KU

1. Introduction

In this paper, certain cohomological properties of compact transformation groups are studied. Particular emphasis is given to results without finite cohomology dimension assumptions on the space.

Let G be a compact group acting on a locally compact Hausdorff space X with fixed point set F and orbit space X/G. The following condition will simply be denoted by (I_0).

(I_0). *The identity component* $(G_x)_0$ *of the isotropy group* G_x *is contained in the center* $C(G)$ *of* G *for any* $x \notin F$.

In the study of compact group actions, the actions of compact totally disconnected groups play an important role, so we discuss this in Section 2. We investigate compact group actions in Section 3. For example, let G be a compact group acting on a compact space X. Suppose that $H^i(X; K_0) = 0$ for $i \geqslant N$, N a positive integer, and that condition (I_0) is satisfied. Then $H^i(F; K_0) = 0$ for $i \geqslant N$ if G is connected; $H^i(X/G; K_0) = 0$ for $i \geqslant N$ if G is a finite dimensional abelian group. In Section 4, we give a necessary and sufficient condition for the existence of a nonempty fixed point set. We study cohomologically locally connected spaces in Section 5. Some properties of finite group actions are proved in Section 6.

Section 7 is concerned with the cohomology structure of orbit spaces. We consider the case in which X is a cohomology sphere, and G is a compact connected Lie group or a cyclic group of order p. The cohomology group structure is completely determined. Notice that we give a new proof of a theorem of LIAO and also of CONNER-FLOYD (under slightly different hypotheses [1]).

We shall use the notation of [1]. Throughout this paper we let p be either a prime or zero, and let K_p be a field of characteristic p, Z the ring of integers, Z_p the prime field, and Q the field of rational numbers; $\dim_p X$ will denote the cohomology dimension over K_p. By a space we

---

* The results presented here are part of a doctoral dissertation written at the Tulane University under Professor P. S. MOSTERT.

shall mean a locally compact Hausdorff space. Alexander-Spanier-Wallace cohomology with compact supports is used throughout this paper.

2. The Action of Compact Totally Disconnected Groups

We shall generalize some results in Chapter III of [1] from finite group actions to compact totally disconnected group actions.

Theorem 2.1. *Let G be a compact totally disconnected group acting on a space X and $\pi: X \to X/G$ be the orbit map. Then $\pi^*: H^*(X/G; K_0) \approx [H^*(X; K_0)]^G$, where*

$$[H^*(X;K_0)]^G = \{h \in H^*(X;K_0) | gh = h \quad \text{for all} \quad g \in G\}.$$

Proof. It is well known that $(G,X) = \varprojlim (G/H, X/H)$ where each G/H is a finite group. By [1, III (2.3)], we have

$$H^*(X/G; K_0) \approx [H^*(X/H; K_0)]^{G/H}.$$

By passage to the direct limit, we obtain

$$H^*(X/G; K_0) \approx [H^*(X; K_0)]^G.$$

We get the following three corollaries immediately.

Corollary 2.2. *Let G be a compact totally disconnected group acting on a space X such that $H^i(X;K_0)=0$ for $i \geqslant N$. Then $H^i(X/G;K_0)=0$ for $i \geqslant N$. In particular, if X is acyclic over K_0, then so is the orbit space X/G.*

Corollary 2.3. *Let G be a compact totally disconnected group acting on a space X. If $H^*(X;K_0)$ is of finite type, then so is $H^*(X/G;K_0)$.*

Corollary 2.4. *Let G be a compact totally disconnected group acting on a space X. If $\dim_0 X$ is finite, then so is $\dim_0 X/G$.*

Proof. Let $\dim_0 X = n < \infty$. Let U' be any open subset of X/G and $\pi: X \to X/G$ be the orbit map. Let $U = \pi^{-1}(U')$. Then U is an open subset of X and is invariant under G. Hence $H^i(U;K_0)=0$ for $i \geqslant n+1$. Thus $H^i(U';K_0)=0$ for $i \geqslant n+1$ by Theorem 2.1. This implies that $\dim_0 X/G$ is finite.

Theorem 2.5. *Let G be a compact totally disconnected group acting on a space X. Then $\dim_0 X/G = \dim_0 X$.*

Proof. From the proof of Corollary 2.4, we have $\dim_0 X/G \leqslant \dim_0 X$. That $\dim_0 X/G \geqslant \dim_0 X$ follows from the fact that a proper light map cannot lower dimension.

3. The Action of Compact Groups

Lemma 3.1. 1. *Let G be a locally compact group acting on a space X and let H be a normal subgroup of G. In a natural way the group G/H acts on X/H. Then $F(G,X)$ is homeomorphic to a subspace of $F(G/H, X/H)$.*

2. *Let G be a compact connected group acting on a space X and H be any compact totally disconnected normal subgroup of G. In an obvious way G/H acts on X/H. Then $F(G,X) \approx F(G/H, X/H)$.*

Theorem 3.2. *Suppose $H^i(X; K_0) = 0$ for $i \geqslant N$ and $\dim_0 X < \infty$. Let G be a finite dimensional compact connected group acting on X such that condition (I_0) is satisfied. Then $H^i(F; K_0) = 0$ and $H^i(X/G; K_0) = 0$ for $i \geqslant N$. In particular, if X is acyclic over K_0, then so is X/G.*

Proof. There is a compact totally disconnected normal subgroup H of G such that G/H is a Lie group. By Corollary 2.2 and Corollary 2.4, we have $H^i(X/H; K_0) = 0$ for $i \geqslant N$ and $\dim_0 X/H < \infty$. Consider G/H acting on X/H:

Case 1. Suppose G/H is abelian: We prove the assertion by induction. If $\dim G/H = 1$, that is, $G/H = T^1$, then we apply [1, p. 60] and Lemma 3.1 to get $H^i(F; K_0) = 0$, $H^i(X/G; K_0) = 0$ for $i \geqslant N$. Suppose now the assertion is true for $\dim G/H = n-1$, that is $G/H = T^{n-1}$. Let $\dim G/H = n$ and K be a circle subgroup of G/H. Then $H^i(F(K, X/H); K_0) = 0$ for $i \geqslant N$, $\dim_0 F(K, X/H) < \infty$ and $F(G/H, X/H) = F(G/HK, F(K, X/H))$. By the inductive hypothesis, we have $H^i(F; K_0) = 0$ for $i \geqslant N$. Also $(G/H)/K$ acts on $(X/H)/K$, so $H^i(X/G; K_0) = 0$ for $i \geqslant N$.

Case 2. Suppose G/H is any arbitrary compact connected Lie group: Since $(G_x)_0 \subset C(G)$ for $x \notin F$, it follows that $(G/H)_{H(x)} = G_x H/H$ and $((G/H)_{Hx})_0 = (G_x)_0 H/H \subset C(G)H/H \subset C(G/H)$. Thus

$$(*) \qquad ((G/H)_{H(x)})_0 \subset C(G/H) \quad \text{for} \quad H(x) \notin F(G/H, X/H).$$

Let K be the identity component of $C(G/H)$. We have $\dim_0 F(K, X/H) < \infty$. Since K is a compact connected abelian Lie group, by case 1, we have $H^i(F(K, X/H); K_0) = 0$ for $i \geqslant N$. Consider $(G/H)/K$ acting on $F(K, X/H)$. By (*), we have that

$$((G/H)/K)_{KH(x)} = ((G/H)/K)_{H(x)} \quad \text{is finite for} \quad H(x) \in F(K, X/H)$$

and $H(x) \notin F((G/H)/K, F(K, X/H))$. From the identity $F(G/H, X/H) = F((G/H/K, F(K, X/H))$ and [1, p. 60], we obtain $H^i(F; K_0) = 0$ for $i \geqslant N$.

For the orbit space X/G, let us consider $(G/H)/K$ acting on $(X/H)/K$. Since $\dim_0 (X/H)/K < \infty$, $H^i((X/H)/K; K_0) = 0$ for $i \geqslant N$ and by Lemma 3.1 1., $F(G/H, X/H) \subset F((G/H)/K, (X/H)/K)$, hence $(G/HK)_{K(Hx)}$ is finite for $KH(x) \notin F((G/H)/K, (X/H)/K)$. Thus [1, p. 60] implies that $H^i(X/G; K_0) = 0$ for $i \geqslant N$.

Lemma 3.3. *Let G be a finite dimensional compact non-totally disconnected group acting on a space X such that the following conditions are satisfied:*

1. (I_0),
2. *G/G_0 is finite, where G_0 is the identity component of G,*
3. *There is a compact totally disconnected normal subgroup H of G such that G_0/H is a non-abelian Lie group.*

Then $F(G_0, X) = F(G, X)$.

Proof. By Lemma 3.1, we have $F(G_0, X) = F(G_0/H, X/H)$. We want to show that $F(G/H, X/H) = F(G_0/H, X/H)$. Conditions (2) and (3) imply that G/H is a Lie group. Let K be the identity component of $C(G/H)$. Note that G_0/H is the identity component of G/H, hence $K \subset G_0/H$. This implies that K is a normal subgroup of G_0/H. It is seen that $((G/H)/K)_{Hx}$ is finite for $Hx \notin F((G/H)/K, F(K, X/H))$. Hence $F((G_0/H)/K, F(K, X/H)) = F((G/H)/K, F(K, X/H))$. For otherwise, if $Hx \in F((G_0/H)/K, F(K, X/H)) - F((G/H)/K, F(K, X/H))$, then $((G/H)/K)_{K(Hx)} \supset (G_0/H)/K$. But $(G_0/H)/K$ is connected and $(G_0/H)/K \neq$ identity, hence $((G/H)/K)_{K(Hx)}$ is not finite, a contradiction. Therefore, $F(G_0/H, X/H) = F(G/H, X/H) \approx F(G_0, X)$. Now, let $x \in F(G_0, X)$, then $G_0 x = x$, so $Hx = x$. Hence $Hx \in F(G/H, X/H)$, and for any $g \in G$, $(gH)(Hx) = H(gx) = Hx = x$. Thus we have shown that $H(gx) = x$ for all $g \in G$, and so $Gx = x$, that is, $x \in F(G, X)$. Therefore $F(G_0, X) = F(G, X)$.

Theorem 3.4. *Let G be a finite dimensional compact non-totally disconnected group acting on X such that $\dim_0 X < \infty$ and $H^i(X; K_0) = 0$ for $i \geqslant N$. Moreover, assume that the following conditions are satisfied:*

1. (I_0),
2. *G/G_0 finite,*
3. *There is a compact totally disconnected normal subgroup H of G such that G_0/H is a non-abelian Lie group.*

Then $H^i(F; K_0) = 0$ for $i \geqslant N$.

Proof. By Lemma 3.3, $F(G_0, X) = F(G, X)$. Clearly $((G_0)_x)_0 \subset C(G) \cap G_0 \subset C(G_0)$ if $x \notin F(G_0, X)$. Hence by Theorem 3.2, the result follows.

Remark. The condition "X is a space and $\dim_0 X < \infty$" can be replaced by "X is a compact space" as we shall see in Theorem 3.17.

In passing to the actions of general compact connected group, we need the following lemma [9].

Lemma 3.5. (HOFMANN-MOSTERT). *If G is a compact connected group and A the set of all finite dimensional compact connected normal subgroups, then A is a semilattice under the product of subgroups, and the union A is dense in G.*

Theorem 3.6. *Suppose $\dim_0 X < \infty$ and $H^i(X;K_0)=0$ for $i \geqslant N$. Let G be a compact connected group acting on the space X such that condition (I_0) is satisfied. Then $H^i(F;K_0)=0$ for $i \geqslant N$.*

Proof. For arbitrary compact subgroups H, H' of G, we have $F(HH')=F(H)\cap F(H')$. By Lemma 3.5, the group G is generated by all finite dimensional compact connected normal subgroups H, and their fixed point sets $F(H)$ satisfy $H^i(F(H);K_0)=0$ for $i \geqslant N$. Hence the result follows from the continuity of the colomology.

Lemma 3.7. *Let G be a compact group acting on a space X such that condition (I_0) is satisfied. If G_0 is not contained in $C(G)$, then $F(G_0,X) = F(G,X)$.*

Proof. Suppose $F(G_0,X) \supsetneqq F(G,X)$. Then for any $x \in F(G_0,X) - F(G,X)$, we have $G_0 \subset G_x$. Hence $G_0 \subset (G_x)_0 \subset C(G)$. This is a contradiction. Thus $F(G_0,X)=F(G,X)$.

Theorem 3.8. *Let G be a compact group acting on a space X such that $H^i(X;K_0)=0$ for $i \geqslant N$, $\dim_0 X < \infty$ and condition (I_0) is satisfied. If G_0 is not contained in $C(G)$, then $H^i(F;K_0)=0$ for $i \geqslant N$.*

Proof. By Lemma 3.7, $F(G_0,X)=F(G,X)$. We see that $((G_0)_x)_0 \subset C(G_0)$, hence by Theorem 3.6, we get $H^i(F;K_0)=0$ for $i \geqslant N$.

Proposition 3.9. *Let G be a finite dimensional compact group acting on a space X such that $H^i(X;K_0)=0$ for $i \geqslant N$, $\dim_0 X < \infty$, and condition (I_0) is satisfied. Then $H^i(X/G;K_0)=0$ for $i \geqslant N$.*

Proof. Let G_0 be the identity component of G. Then G/G_0 is a compact totally disconnected group. By Theorem 3.2, $H^i(X/G_0;K_0)=0$ for $i \geqslant N$. Now G/G_0 acts on X/G_0, hence the assertion follows from Corollary 2.2.

The remainder of this section, we do not assume finite cohomology dimensionality of the space X. One of our main results is that the assumption of [1, IV 3.6)] is fulfilled for the classical compact Lie groups A_n, B_n, C_n, D_n and also for G_2.

Let the group $G=T^1$ (resp. S^3) act on a space X such that $H^*(B_{G_x};K_p)$ is trivial for any $x\in X$. Consider the spectral sequence of the orbit map $\pi:X\to X/G$. The E_2-term is given by $E_2=H^*(X/G;\underline{H^*(G;K_p)})$ and the Leray sheaf $\underline{H^*(G;K_p)}$ is constant [1, p. 61]. Thus

$$E_2=H^*(X/G;K_p)\otimes H^*(G;K_p).$$

Hence we can apply the well-known argument [2, p. IX-8] to get the following lemma:

Lemma 3.10. *Let the group $G=T^1$ (resp. S^3) act on a space X such that $H^*(B_{G_x};K_p)$ is trivial for any $x\in X$. Then there is a Gysin sequence of the form*

$$\cdots\longrightarrow H^i(X/G)\xrightarrow{\delta} H^{i+k+1}(X/G)\longrightarrow H^{i+k+1}(X)\longrightarrow H^{i+1}(X/G)\longrightarrow\cdots$$

with the homomorphism δ defined by $\delta(e^i)=c\cup e^i$, where $k=1$ (res. 3) and $c\in H^{k+1}(X/G)$.

P. E. CONNER [7] has proved the following theorem in the case $G=T^1$ and X a Peano Continum. Note that the assumption of X being metrizable is not used in the proof, so we can use the same argument to get:

Theorem 3.11. *Let the group $G=T^k$ (resp. $(S^3)^k$), where $k\geqslant 1$, act on a compact space X. If $H^*(B_{Gx};K_p)$ is trivial for $x\notin F$ and if $H^i(X;K_p)=0$ for $i\geqslant N$, then $H^i(F;K_p)=0$ and $H^i(X/G;K_p)=0$ for $i\geqslant N$.*

Proof. By an inductive argument, it is sufficient to prove that the assertion is true for the case $k=1$.

For $i\geqslant N$, we have the following commutative diagram

$$\begin{array}{ccccccccc} & & 0 & \to & H^i(F) & \to & H^{i+1}(X-F) & \longrightarrow & 0 \\ & & \uparrow & & \uparrow\approx & & \uparrow\pi^* & & \uparrow \\ \cdots & \to & H^i(X/G) & \to & H^i(F) & \to & H^{i+1}(X/G-F) & \to & H^{i+1}(X/G)\to\cdots \end{array}$$

where the coefficient field K_p is omitted.
It is easily seen that

$$\pi^*:H^i(X/G-F;K_p)\to H^i(X-F;K_p)\ \text{is onto for}\ i\geqslant N+1.$$

Under the hypotheses we have the Gysin sequence by Lemma 3.10,

$$\cdots\longrightarrow H^i(X/G-F)\xrightarrow{\delta} H^{i+s+1}(X/G-F)\longrightarrow H^{i+s+1}(X-F)$$
$$\longrightarrow H^{i+1}(X/G-F)\longrightarrow\cdots$$

where the homomorphism δ is given by $\delta(e^i)=c^{s+1}\cup e^i$ for $e^i\in H^i(X/G-F)$ and c^{s+1} is a non-zero element of $H^{s+1}(X/G-F;K_p)$, $s=1$ (resp. 3).

Hence δ is one to one for $i \geqslant N+1-s$. Thus if $H^{N+1-s}(X/G-F) \neq 0$, then $(c^{s+1})^j \cup e^{N+1-s} = \delta^j(e^{N+1-s}) \neq 0$ for all non-zero elements e^{N+1-s} in $H^{N+1-s}(X/G-F)$ and all $j \geqslant 1$. This contradicts [7, Lemma 2.3], [9]. Hence $H^i(X/G-F;K_p)=0$ for $i \geqslant N+1-s$. Thus $H^i(X-F;K_p)=0$ for $i \geqslant N+1$. By the cohomology e xact sequence of the pair (X,F) and $(X/G,F)$ respectively, we get $H^i(F;K_p)=0$ and $H^i(X/G;K_p)=0$ for $i \geqslant N$.

The following Lemma 3.12 1. is contained in [10].

Lemma 3.12. 1. *Let G be a compact non-totally disconnected group acting on a compact space X such that* $H^*(B_{G_x};K_0)=0$ *for* $x \notin F$. *If* $H^i(X;K_0)=0$ *for* $i \geqslant N$, *then* $H^i(F;K_0)=0$ *for* $i \geqslant N$.

2. *Let G be an infinite compact Lie group acting on a compact space X such that* $H^*(B_{G_x};K_p)=0$ *for* $x \notin F$. *If* $H^i(X;K_p)=0$ *for* $i \geqslant N$, *then* $H^i(F;K_p)=0$ *for* $i \geqslant N$.

Proof. Let T^1 be any circle subgroup of G. Then $F(G,X)=F(T^1,X)$. Applying Theorem 3.11, we get the result. Notice that our proofs in Theorem 3.11 and Lemma 3.12 are also valid for $H^i(X;K_p)=0$ for $i \geqslant N$. Thus if X is acyclic over K_p, then F is acyclic over K_p.

Lemma 3.13. *Let G be a compact connected Lie group acting on a space X such that* G_x *is finite for* $x \notin F$. *Let H be a proper connected (non-trivial) subgroup of G. Then the Leray sheaf of the map* $\pi:(X-F)/H \to (X-F)/G$ *is constant.*

Proof. Let G act on $(X-F)/H$ by $g(Hx)=H(gx)$. For $y=\pi(x)$, $\pi^{-1}(y)=G/G_xH$, and the natural map $G/H \to G/G_xH$ yields an isomorphism $\alpha_y: H^*(\pi^{-1}(y)) \approx H^*(G/H)$ which is independent of the choice of x in $\pi^{-1}(y)$. Thus we can apply [1, XII (2.2)] to obtain the result.

Theorem 3.14. *Let the group* $G=SO(n)$, $n \geqslant 2$, *(resp.* $O(n)$, $n \geqslant 2$; $SU(n)$, $U(n)$, $Sp(n)$, Spin(n), G_2) *act on a compact space X such that* $H^*(B_{G_x};K_p)$ *is trivial for* $x \notin F$. *If* $H^i(X;K_p)=0$ *for* $i \geqslant N$, *then* $H^i(X/G;K_p)=0$ *for* $i \geqslant N$. *Here we assume that* $p \neq 2$, *if* $G=O(n)$ *or* Spin(n).

Proof. We omit the coefficient field K_p in the proof. Let $G=SO(n)$: Consider the following sequence of maps:

$$\pi_1: X-F \to (X-F)/SO(2)$$
$$\pi_2: (X-F)/SO(2) \to (X-F)/SO(3)$$
$$\cdots\cdots\cdots$$
$$\pi_{n-1}: (X-F)/SO(n-1) \to (X-F)/SO(n).$$

Then we have the following commutative diagram:

$$\begin{array}{ccccccccc}
\cdots\to & H^i(X) & \to & H^i(F) & \to & H^{i+1}(X-F) & \to & H^{i+1}(X) & \to\cdots \\
 & \uparrow & & \uparrow & & \uparrow{\scriptstyle \pi_1^*} & & \uparrow & \\
\cdots\to & H^i(X/SO(2)) & \to & H^i(F) & \to & H^{i+1}(X/SO(2)-F) & \to & H^{i+1}(X/SO(2)) & \to\cdots \\
 & \uparrow & & \uparrow & & \uparrow{\scriptstyle \pi_2^*} & & \uparrow & \\
\cdots\to & H^i(X/SO(3)) & \to & H^i(F) & \to & H^{i+1}(X/SO(3)-F) & \to & H^{i+1}(X/SO(3)) & \to\cdots \\
 & \uparrow & & \uparrow & & \uparrow{\scriptstyle \pi_3^*} & & \uparrow & \\
 & \vdots & & \vdots & & \vdots & & \vdots & \\
 & \uparrow & & \uparrow & & \uparrow{\scriptstyle \pi_{n-1}^*} & & \uparrow & \\
\cdots\to & H^i(X/SO(n)) & \to & H^i(F) & \to & H^{i+1}(X/SO(n)-F) & \to & H^{i+1}(X/SO(n)) & \to\cdots
\end{array}$$

For $i \geqslant N$: By Theorem 3.11, we have $H^i(X/SO(2); K_p)=0$ for $i \geqslant N$. Since $H^*(\pi_2^{-1}(\bar{x})) \approx H^*(SO(3)/SO(2)) \approx H^*(S^2)$ for $\bar{x} \in (X-F)/SO(3)$. Hence by Lemma 3.13, and Lemma 3.10, we have the Gysin exact sequence

$$\begin{aligned}
\cdots \longrightarrow H^i(X/SO(3)-F) &\xrightarrow{\delta} H^{i+2+1}(X/SO(3)-F) \\
&\xrightarrow{\pi_2^*} H^{i+2+1}(X/SO(2)-F) \longrightarrow H^{i+1}(X/SO(3)-F) \longrightarrow \cdots
\end{aligned}$$

Therefore δ is one to one for $i \geqslant N+1-2$ since we see that π_2^* is onto for $i \geqslant N+1$. Applying the same argument as in the proof of Theorem 3.11, and we get $H^i(X/SO(3); K_p)=0$ for $i \geqslant N$ just as in Theorem 3.11. Continuing the same process, finally π_{n-1}^* is onto for $i \geqslant N+1$. Since $H^*(\pi_{n-1}^{-1}(\bar{x})) \approx H^*(SO(n)/SO(n-1)) \approx H^*(S^{n-1})$, we still have the Gysin sequence. By the same argument as above, we obtain $H^i(X/SO(n); K_p)=0$ for $i \geqslant N$.

For the groups $G=SU(n)$, $U(n)$, $Sp(n)$ and G_2, we apply the same argument as in the case of $SO(n)$. This is possible because we have the following relations [3]:

$$\frac{SU(n)}{SU(n-1)} = \frac{U(n)}{U(n-1)} = S^{2n-1} \quad \text{for} \quad n \geqslant 2,$$

$$\frac{Sp(n)}{Sp(n-1)} = S^{4n-1}, \quad \text{and} \quad \frac{G_2}{SU(3)} = S^6.$$

If $G=0(n)$ and $\mathrm{Spin}(n)$, then the result follows from Corollary 2.2, since $O(n)/SO(n) \approx Z_2$ and $\mathrm{Spin}(n)/Z_2 \approx SO(n)$.

Proposition 3.15. *Let F_4 act almost freely on a compact connected space X; that is, G_x is finite for any $x \in X$. If $H^i(X; K_0)=0$ for $i \geqslant N$, then $H^i(X/F_4; K_0)=0$ for $i \geqslant N$.*

Proof. It is known that $F_4/\mathrm{Spin}(9)$ is the Caley projective plane, so that the Leray sheaf of the map $\pi: X/\mathrm{Spin}(9) \to X/F_4$ is the constant sheaf $X/F_4 \times H^*(F_4/\mathrm{Spin}(9);K_0)$. The E_2-term of the Leray spectral sequence of the map π is given by

$$E_2^{i,j} = H^i(X/F_4;K_0) \otimes H^j(F_4/\mathrm{Spin}(9);K_0).$$

Since $H^*(F_4/\mathrm{Spin}(9);K_0) = K_0[a]/(a^3)$, $\deg a = 8$, and since $H^i(X/\mathrm{Spin}(9);K_0) = 0$ for $i \geqslant N$, if we can show that $E_2 \approx E_\infty$. then $E_\infty \approx H^*(X/\mathrm{Spin}(9);K_0)$ additively, and the result will follow. So it remains to show that the spectral sequence is trivial.

Suppose now that $E_2 \neq E_\infty$; then $d_9 \neq 0$ or $d_{17} \neq 0$. Thus we can let $0 \neq b \in H^i(X/F_4;K_0)$ such that

$$d_9(b \otimes a) \neq 0 \quad \text{or} \quad d_9(b \otimes a^2) \neq 0.$$

If $d_9(b \otimes a) \neq 0$, then we can assume that $d_9(1 \otimes a) = b' \otimes 1$, so that

$$\begin{aligned} d_9(b \otimes a) &= d_9((b \otimes 1)(1 \otimes a)) \\ &= (-1)^{\deg b}(b \otimes 1)\, d_9(1 \otimes a) \\ &= (-1)^{\deg b}(b \otimes 1)(b' \otimes 1) \\ &= (-1)^{\deg b}(b b' \otimes 1) \neq 0. \end{aligned}$$

Thus, $bb' \neq 0$. Since b is arbitrary, we can see that $(b')^n \neq 0$ for all $n > 0$. This is of course impossible.

If $d_9(b \otimes a^2) \neq 0$, then we still let $d_9(1 \otimes a) = b' \otimes 1$, and so

$$\begin{aligned} d_9(b \otimes a^2) &= d_9((b \otimes a)(1 \otimes a)) \\ &= d_9(b \otimes a)(1 \otimes a) + (b \otimes a)\, d_9(1 \otimes a) \\ &= (-1)^{\deg b}(b b' \otimes 1)(1 \otimes a) + (b \otimes a)(b' \otimes 1) \\ &= (-1)^{\deg b}(b b' \otimes a) + (b b' \otimes a). \end{aligned}$$

Now, if $\deg b$ is odd, then $d_9(b \otimes a^2) = 0$, a contradiction. If $\deg b$ is even, then $d_9(b \otimes a^2) = 2(bb' \otimes a) \neq 0$. Thus $bb' \neq 0$. Again we get $(b')^n \neq 0$ for any integer $n > 0$. Hence $d_9 = 0$.

Now, if there is $0 \neq b \in H^i(X/F_4;K_0)$ such that $d_{17}(b \otimes a^2) \neq 0$, then

$$\begin{aligned} d_{17}(b \otimes a^2) &= d_{17}((b \otimes a)(1 \otimes a)) \\ &= d_{17}(b \otimes a)(1 \otimes a) + (b \otimes a)\, d_{17}(1 \otimes a) = 0 \end{aligned}$$

since $d_{17}(b \otimes a) = 0$ and $d_{17}(1 \otimes a) = 0$. This completes the proof of the theorem.

Theorem 3.16. *Let G be a finite dimensional compact connected group acting on a compact space X such that condition (I_0) is satisfied. If $H^i(X;K_0) = 0$ for $i \geqslant N$, then $H^i(F;K_0) = 0$ for $i \geqslant N$.*

Proof. Apply the same argument as in Theorem 3.2 by using Lemma 3.12 instead of [1, p. 60].

From Theorem 3.16 and Lemma 3.5, we obtain the following theorem:

Theorem 3.17. *Let G be a compact connected group acting on a compact space X such that condition (I_0) is satisfied. If $H^i(X;K_0)=0$ for $i \geqslant N$, then $H^i(F;K_0)=0$ for $i \geqslant N$.*

Proof. The proof is similar to that of Theorem 3.6.

Remark 3.18. The statement of Theorem 3.4 and Theorem 3.8 remain true if we replace "X is a space and $\dim_0 X < \infty$" by "X is a compact space".

Theorem 3.19. *Let G be a finite dimensional compact abelian group acting on a compact space X such that $H^i(X;K_0)=0$ for $i \geqslant N$. Then $H^i(X/G;K_0)=0$ for $i \geqslant N$.*

Proof. Case 1, if G is connected, then there is a compact totally disconnected normal subgroup H of G such that G/H is a torus group. Consider G/H acting on X/H, and obtain the assertion by Theorem 3.11.

Case 2, if G is not connected, then G/G_0 is a totally disconnected group which acts on X/G_0. By case 1, $H^i(X/G_0;K_0)=0$ for $i \geqslant N$. Hence the result follows from Corollary 2.2.

K. H. HOFMANN and P. S. MOSTERT [9, p. 332] have shown that if G is a compact group acting on a compact space X which is acyclic over Q such that the condition (I_0) is satisfied, then the fixed point set is acyclic over Q. Now we can prove the orbit space is also acyclic over Q under the same hypothesie, by using the fact [9] that if G is compact connected Lie group acting on a acyclic space X over Q and if G_x is finite for $x \notin F$, then X/G is acyclic over Q.

Proposition 3.20. *Let G be a finite dimensional compact group acting on a compact space X which is acyclic over K_0. If the condition (I_0) is satisfied, then X/G is acyclic over K_0.*

Proof. The same method as in Theorem 3.2 and Proposition 3.9 is applied here.

Theorem 3.21. *Let G be a compact connected Lie group acting on a compact space X such that $H^*(B_{G_x};K_p)$ is trivial for $x \notin F$. Suppose $H^i(X;K_p)=0$ for $i \geqslant N$ and $H^*(X;K_p)$ is of finite type, then $H^*(F;K_p)$ and $H^*(X/G;K_p)$ are of finite type*

Proof. We know that G contains a circle subgroup T^1 and $F(T^1,X) = F(G,X)$. By Theorem 3.11 and [1, p. 54], we have

$$H^i(X_{T_1};K_p) \approx H^i(F_{T^1};K_p) \quad \text{for} \quad i \geqslant N.$$

Thus $H^i(F_{T_1};K_p)$ is finitely generated for $i \geqslant N$ by [1, p. 60]. By [1, IV (2.4)] and the Künneth rule, each $H^s(F;K_p)$ is embedded in some $H^i(F_{T^1};K_p)$ for suitably large i, and therefore $H^*(F;K_p)$ is of finite type.

[1, p. 60] implies that $H^*(F_G;K_p)$ and $H^*(X_G;K_p)$ are of finite type. Hence, by the cohomology sequence, $H^*((X-F)_G;K_p)$ is of finite type. Using the isomorphism $H^*((X-F)_G;K_p)\approx H^*(X/G-F;K_p)$, it follows that $H^*(X/G-F;K_p)$ is of finite type. Finally, the cohomology sequence of $X/G \bmod F$ shows that $H^*(X/G;K_p)$ is also of finite type.

Theorem 3.22. *Let G be a finite dimensional compact connected group acting on a compact space X such that condition* (I_0) *is satisfied. Suppose* $H^i(X;K_0)=0$ *for* $i\geqslant N$ *and* $H^*(X;K_0)$ *is of finite type. Then* $H^*(F;K_0)$ *and* $H^*(X/G;K_0)$ *are of finite type.*

Proof. There is a compact totally disconnected normal subgroup H of G such that G/H is a Lie group. By Corollary 2.2 and Corollary 2.3, we have $H^i(X/H;K_0)=0$ for $i\geqslant N$ and $H^*(X/H;K_0)$ is of finite type. Consider G/H acting on X/H. By Lemma 3.1, $F(G/H,X/H)=F(G,X)$.

Case 1, if G/H is abelian, then we prove the assertion by induction on its dimension.

Case 2, if G/H is an arbitrary compact Lie group, then let K be the identity component of $C(G/H)$. We have the identity $F(G/H,X/H) = F((G/H)/K,\ F(K,X/H))=F'$. Since $(G_x)_0\subset C(G)$, $((G/H)/K)_{K(Hx)}$ is finite for $K(Hx)\notin F'$. Hence $H^*(F;K_0)$ is of finite type by case 1 and Theorem 3.21. Similarly $H^*(X/G;K_p)$ is of finite type if we consider the group $(G/H)/K$ acting on $(X/H)/K$.

Corollary 3.23. *Let G be a finite dimensional compact group acting on a compact space X such that condition* (I_0) *is satisfied. Suppose* $H^i(X;K_0)=0$ *for* $i\geqslant N$ *and* $H^*(X;K_0)$ *is of finite type. Then* $H^*(X/G;K_0)$ *is of finite type.*

4. Characterization of Non-Empty Fixed Point Sets

The work of this section was inspired by the result of Conner and Floyd [8]. We recall that an endomorphism

$$f^*:\tilde{H}^i(X;K_0)\to\tilde{H}^i(X;K_0)$$

is nilpotent if for each $h\in\tilde{H}^i(X;K_0)$ we have $f^{*n}(h)=0$ for n sufficient large, where $\tilde{H}^i(X;K_0)$ is the reduced cohomology group.

Lemma 4.1. (Ku [10]). *Let G be a compact non-totally disconnected group acting on a compact space X which is acyclic over* K_0. *If* G_x *is finite for* $x\notin F$, *then* $F\neq\phi$ *and is acyclic over* K_0.

Theorem 4.2. *Let G be a compact non-totally disconnected group acting on a compact space X. If* G_x *is finite for* $x\notin F$, *then* $F\neq\phi$ *if and* *only if there exists an equivariant map* $f:X\to X$ *such that* $f^*:H^i(X;K_0)\to\tilde{H}^i(X;K_0)$ *is nilpotent for* $i=0,1,2,\ldots$

Proof. If $F\neq\phi$, then we can define $f:X\to X$ by $f(x)=x_0$ for any fixed $x_0\in F$ and all $x\in X$. It is seen that $fg(x)=x_0$, $gf(x)=gx_0$ for all $x\in X$, $g\in G$. Therefore $gf=fg$ for all $g\in G$. Since $f:X\to X$ is the composition of the maps $X\to x_0$ and inclusion $i:x_0\to X$, $f^*:\tilde{H}^i(X;K_0)\to\tilde{H}^i(X;K_0)$ is the composition of the maps $\tilde{H}^i(X;K_0)\to\tilde{H}^i(x_0;K_0)$ and $\tilde{H}^i(x_0;K_0)\to\tilde{H}^i(X;K_0)$ which is a zero map. Therefore f^* is nilpotent for $i=0,1,2,\ldots$

To prove the converse, we consider the inverse system

$$X\xleftarrow{f}X\xleftarrow{f}X\xleftarrow{f}\cdots$$

Let Y be the inverse limit space whose points are the sequences (x_i), where $x_i\in X$ with $f(x_{i+1})=x_i$. The inverse system induces a direct system of cohomology groups as follows:

$$\tilde{H}^i(X;K_0)\xrightarrow{f^*}\tilde{H}^i(X;K_0)\xrightarrow{f^*}\tilde{H}^i(X;K_0)\xrightarrow{f^*}\cdots$$

Since f^* is nilpotent, $H^i(Y;K_0)=\lim\{\tilde{H}^i(X;K_0),f^*\}=0$ for $i=0,1,2,\ldots$

Define G to act on Y by $g(x_i)=(gx_i)$. Let $F'=F(G,Y)$. If $x=(x_i)\notin F'$, $x\in Y$, then there exists at least one index i_0 such that $x_{i_0}\notin F$ because $F'=\lim(F,f|F)$. Thus, if $g\in G_x$, then $g(x_i)=(gx_i)=(x_i)$, i.e., $gx_i=x_i$ all i. This implies $g\in\cap_i G_{x_i}\subset G_{x_{i0}}$, and $G_x\subset G_{x_{i0}}$ for $x_{i_0}\notin F$. By hypothesis, G_{x_i} is finite for $x_{i_0}\notin F$, and hence G_x is finite. Therefore we may apply Lemma 4.1 to get $F'\neq\phi$ and so $F\neq\phi$.

Corollary 4.3. *Let G be a compact connected Lie group acting on a compact space $X\neq\phi$. If G_x is finite for $x\notin F$ and if there exists an equivariant map $f:X\to X$ such that $f^*:\tilde{H}^i(X;K_0)\to\tilde{H}^i(X;K_0)$ is nilpotent for $i=0,1,2,\ldots$, then $F\neq\phi$ and $(f|F)^*:\tilde{H}^i(F;K_0)\to\tilde{H}^i(F;K_0)$ is also nilpotent for $i=0,1,2,\ldots$*

Proof. As in the proof of Theorem 4.2, we have $F'=\lim(F,f|F)$, and hence $H^i(F':K_0)=\lim\{\tilde{H}^i(F;K_0),(f|F)^*\}$. But F' is acyclic, and thus $(f|F)^*$ is nilpotent for $i=0,1,2,\ldots$

Theorem 4.4. *Let G be a finite dimensional compact connected group which acts on a compact space $X\neq\phi$ and suppose condition (I_0) is satisfied. Then $F\neq\phi$ if and only if there is an equivariant map $f:X\to X$ such that $f^*:\tilde{H}^i(X;K_0)\to\tilde{H}^i(X;K_0)$ is nilpotent for $i=0,1,2,\ldots$*

Proof. It is enough to show the if part. There is a compact totally disconnected normal subgroup of G such that G/H is a compact connected Lie group. Now we consider the group G/H acting on X/H. The map f induces the map $\bar{f}$ which is given by $\bar{f}(Hx)=H(fx)$ such that the following diagram commutes:

$$\begin{array}{ccc} X & \xrightarrow{f} & X \\ {\scriptstyle\pi}\downarrow & & \downarrow{\scriptstyle\pi} \\ X/H & \xrightarrow{\bar{f}} & X/H \end{array}$$

The map $\bar{f}$ is G/H-equivariant, for $\bar{f}[(gH)(Hx)]=\bar{f}[H(gx)]=H(fgx)$ $=H(gfx)=(gH)(Hfx)=(gH)\bar{f}(Hx)$. By Theorem 3.3 and the fact that G acts trivially on $\tilde{H}^*(X;K_0)$ [6], $\tilde{H}^*(X/H;K_0)\approx[\tilde{H}^*(X;K_0)]^H$ $=\tilde{H}^*(X;K_0)$. Since f^* is nilpotent, hence $\bar{f}^*$ is also nilpotent.

Case 1. If G/H is abelian, then we prove the theorem by induction on its dimension.

Case 2. If G/H is any arbitrary compact connected Lie group then let K be the identity component of $C(G/H)$. We consider $(G/H)/K$ acting on $F(K,X/H)$, which is not empty by case 1. We see that $F(G/H,X/H)$ $=F((G/H)/K,F(K,X/H))$. Let $f_2=\bar{f}|F(K,X/H)$. Then f_2 is equivariant with respect to $(G/H)/K$ and f_2^* is nilpotent by Corollary 4.3. As we have shown before, under the hypotheses $(G_x)_0\subset C(G)$, we have $((G/H)/K)_{HKx}$ finite for $HK(x)\notin F((G/H)/K,\ F(K,X/H))$. Hence $F((G/H)/K,\ F(K,X/H))\neq\phi$ and so $F\neq\phi$.

Theorem 4.5. *Let G be a compact connected group acting on a compact space $X\neq\phi$ such that condition (I_0) is satisfied. Then $F\neq\phi$ if and only if there is an equivariant map $f:X\to X$ such that $f^*:\tilde{H}^i(X;K_0)$ $\to\tilde{H}^i(X;K_0)$ is nilpotent for $i=0,1,\ldots$*

Proof. The group G is generated by all finite dimensional compact connected subgroups H whose fixed point set $F(H)\neq\phi$ by Theorem 4.4. Hence $F(G,X)=\cap_H F(H,X)\neq\phi$.

Corollary 4.6. *Let G be a compact group acting on a compact space $X\neq\phi$ such that condition (I_0) is satisfied and G_0 is not contained in $C(G)$. Then $F\neq\phi$ if and only if there exists a map $f:X\to X$ which is equivariant with respect to G_0 such that $f^*:\tilde{H}^i(X;K_0)\to\tilde{H}^i(X;K_0)$ is nilpotent for $i=0,1,2,\ldots$*

Proof. By Lemma 3.7, $F(G,X)=F(G_0,X)$. Also $((G_0)_x)_0\subset C(G_0)$. Hence the result follows from Theorem 4.5.

To obtain the final main results of this section, we need two lemmas of [10]:

Lemma 4.7 (Ku). *Let G be a compact connected Lie group acting on an acyclic compact space X over K_0 and* $\operatorname{rank} G_x<\operatorname{rank} G$ *for any $x\notin F$. Then the fixed point set F is acyclic over K_0.*

Lemma 4.8 (Ku). *Let G be a compact connected Lie group acting on an acyclic compact space X over Z_p. Suppose that p-rank $G_x<p$-rank G for any $x\notin F$. Then F is acyclic over Z_p.*

We state the following two theorems. The proofs are similar to the proof of Theorem 4.2.

Theorem 4.9. *Let G be a compact Lie group acting on a compact space X such that $\operatorname{rank} G_x < \operatorname{rank} G$ for any $x \notin F$. Then $F \neq \phi$ if and only if there exists an equivariant map $f: X \to X$ such that $f^*: \tilde{H}^i(X; K_0) \to \tilde{H}^i(X; K_0)$ is nilpotent for $i = 0, 1, 2, \ldots$*

Theorem 4.10. *Let G be a compact connected Lie group acting on a compact space X such that p-rank $G_x <$ p-rank G for any $x \notin F$. Then $F \neq \phi$ if and only if there exists an equivariant map $f: X \to X$ such that $f^*: \tilde{H}^i(X; Z_p) \to \tilde{H}^i(X; Z_p)$ is nilpotent for $i = 0, 1, 2, \ldots$*

5. On Cohomologically Locally Connected Spaces

Theorem 5.1. *Let G be a compact totally disconnected group acting on a space X which is* clc *over K_0 and $\dim_0 X < \infty$. Then the orbit space X/G is* clc *over K_0.*

Proof. First, we show that for any pair $U^* \supset V^*$ of open connected subspaces of X/G with $\bar{V}^*$ compact $\subset U^*$, $\operatorname{Im} j_{U^*V^*}$ is finitely generated. Let $\pi: X \to X/G$ be the orbit map. Let U, V be a component of $\pi^{-1}(U^*)$, $\pi^{-1}(V^*)$ respectively such that $V \subset U$. Then $V \subset U$, then $V \subset \bar{V} \subset U$ and $\bar{V}$ compact. Since X is clc over K_0, by [1, chap. I (2.2)], $\operatorname{Im} j^*_{UV}$ is finitely generated.

Let G_1, G_2 be the maximal open and closed subgroups of G which leave U, V resp. invariant and such that U/G_1, V/G_2 map homeomorphic onto U^*, V^*, resp. This is possible by [13, p. 2]. Clearly, $G_2 \subset G_1$. The commutative diagram

$$\begin{array}{ccc} U/G_2 & \xrightarrow{\pi'} & U/G_1 \\ {\scriptstyle \pi_2}\nwarrow & & \nearrow{\scriptstyle \pi_1} \\ & U & \end{array}$$

induces the following commutative diagram

$$\begin{array}{ccc} H^*(U/G_2) & \xleftarrow{\pi'^*} & H^*(U/G_1) \\ {\scriptstyle \pi_2^*}\searrow & & \swarrow{\scriptstyle \pi_1^*} \\ & H^*(U) & \end{array}$$

where the coefficient field K_0 is omitted. By Theorem 2.1, π_1^*, is one to one, hence π'^* is one to one.

Consider the following commutative diagram:

$$\begin{array}{ccccccc}
H^*(V^*) & & \xrightarrow{\quad\approx\quad} & & H^*(V/G_2) & \xrightarrow{1:1} & H^*(V) \\
\downarrow{\scriptstyle j^*_{U^*V^*}} & & & & \downarrow & & \downarrow{\scriptstyle j^*_{UV}} \\
H^*(U^*) & \xrightarrow{\approx} & H^*(U/G_1) & \xrightarrow{\pi'^*} & H^*(U/G_2) & \xrightarrow{\pi_2^*} & H^*(U)
\end{array}$$

Since $\operatorname{Im} j^*_{UV}$ is finitely generated, $\operatorname{Im} j^*_{U^*V^*}$ is finitely generated.

Next, let U^* be any given open neighborhood of $x^* = \pi(x)$, $x \in X$. Then we can choose an open connected subset U^*_α of U^* such that $x^* \in U^*_\alpha$. We have shown that for each U^*_α, there exists a V^*_α with $x^* \in V^*_\alpha$, $\bar{V}^*_\alpha$ compact $\subset U^*_\alpha$ and $\operatorname{Im} j^*_{U^*_\alpha V^*_\alpha}$ finitely generated. From the following commutative diagram

$$\begin{array}{ccc}
H^*(V^*_\alpha) & \xrightarrow{\quad j^*_{U^*V^*_\alpha}\quad} & H^*(U^*) \\
{\scriptstyle j^*_{U^*_\alpha V^*_\alpha}}\searrow & & \nearrow \\
& H^*(U^*_\alpha) &
\end{array}$$

we see that $\operatorname{Im} j^*_{U^*_\alpha V^*_\alpha}$ is finitely generated. Thus by [1, Chapter I (2.2)], we get the orbit space X/G is clc over K_0.

Theorem 5.2. *Suppose X is* clc *over K_0 and $\dim_0 X < \infty$. Let G a finite dimensional compact connected group acting on X such that condition (I_0) is satisfied. Then F and X/G are* clc *over K_0.*

Proof. One uses Theorem 5.1, [1, p. 68] and applies the same argument as in the proof of Theorem 3.2.

Corollary 5.3. *Suppose X is* clc *over K_0 and $\dim_0 X < \infty$. Let G be a finite dimensional compact group acting on X such that condition (I_0) is satisfied. Then the orbit space X/G is* clc *over K_0.*

6. Remarks on Finite Group Actions

Swan [14] has showed that if the group Z_p acts on an acyclic compact space X over Z_p, then the fixed point set F is acyclic over Z_p. In this section we shall prove that the orbit space is also acyclic over Z_p. We also make some other remarks. Note that we do not assume the finite cohomology dimensionality of the space X.

Lemma 6.1. *If the cyclic group Z_p acts on a space X, then*

$$\dim H^n(X/Z_p - F; Z_p) \leqslant \sum_{j=0}^{n} \dim H^j(X - F; Z_p).$$

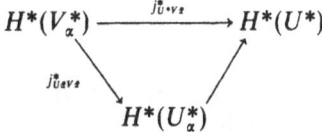

Proof. The E_2-term of the spectral sequence of the covering map $\pi: X-F \to (X-F)/Z_p$ is given by

$$E_2^{i,j} = H^i(Z_p; H^j(X-F; Z_p))$$

and $E_\infty = \operatorname{Gr} H^*((X-F)/Z_p; Z_p)$. Hence

$$\dim H^n((X-F)/Z_p; Z_p) = \sum_{i+j=n} \dim E_\infty^{i,j} \leqslant \sum_{i+j=n} \dim E_2^{i,j} \leqslant \sum_{j=0}^{n} \dim H^j(X-F; Z_p).$$

Theorem 6.2. *If a finite group G acts on a compact space X which is acyclic over Z_p, then the orbit space X/G is also acyclic over Z_p.*

Proof. We prove the theorem first in the case $G=Z_p$. Since the fixed point set F is acyclic over Z_p [14], from the cohomology exact sequence of the pair (X,F), we see that $H^i(X-F;Z_p)=0$ for $i\geqslant 0$. Therefore by Lemma 6.1, we have $H^i(X/Z_p-F;Z_p)=0$ for $i\geqslant 0$. Thus the cohomology exact sequence of the pair $(X/G,F)$ gives that X/Z_p is acyclic over Z_p. By induction, we can prove the result for $G=(Z_p)^k$, where $k\geqslant 1$.

Finally, if G is any finite group, then the proof is exactly the same as the proof of [1, p. 45].

Theorem 6.3. *Let G be a finite group acting on a compact space such that $H^i(X;Z_p)=0$ for $i\geqslant N$. If $H^*(X;Z_p)$ is of finite type, then $H^*(X/G;Z_p)$ is also of finite type.*

Proof. We prove the theorem first in the case $G=Z_p$. By [14, Corollary 4.1], we have

$$\sum_{j\geqslant k} \dim H^j(F;Z_p) \leqslant \sum_{j\geqslant k} \dim H^j(X;Z_p) \quad \text{for all } k.$$

Hence $H^*(F;Z_p)$ is of finite type. Thus [1, p. 60] implies that $H^*(F_{Z_p};Z_p)$ and $H^*(X_{Z_p};Z_p)$ are of finite type. By the cohomology sequence of the pair (X_{Z_p},F_{Z_p}) and the Vietoris-Begle mapping theorem, we obtain $H^*(X/Z_p-F;Z_p)$ is of finite type. Finally, the cohomology sequence of the pair $(X/G,F)$ gives that $H^*(X/G;Z_p)$ is of finite type.

By induction, the assertion is true for $G=(Z_p)^k$, $k\geqslant 1$. If G is any finite group, we apply the same argument as in the proof of [1, p. 45].

Corollary 6.4. *Let G be a compact Lie group acting on a compact space X and $H^i(X;Z_p)=0$ for $i\geqslant N$. If $H^*(B_{G_x};Z_p)$ is trivial for $x\notin F$ and $H^*(X;Z_p)$ is of finite type, then $H^*(X/G;Z_p)$ is of finite type.*

Proof. We know that G/G_0 is a totally disconnected group, where G_0 is the identity component of X/G_0. By Theorem 3.21, $H^*(X/G_0;Z_p)$

is of finite type. Consider G/G_0 acting on X/G_0. Then the result follows from Theorem 6.3, since G/G_0 is finite group.

7. The Orbit Space of a Cohomology Sphere

Lemma 7.1. *Let G be a compact Lie group (assume p divides the order of G if G is a finite group) acting on a space X such that* $\dim_p X$ *is finite. Suppose that* $H^*(B_{G_x}; K_p)$ *is trivial for* $x \notin F$ *and that X is totally non-homologous to zero in* X_G *rel. to* K_p. *Then the following sequence is exact for all integers* $i \geqslant 0$.

$$0 \to H^i(X_G; K_p) \to H^i(F_G; K_p) \to H^{i+1}(X/G - F; K_p) \to 0.$$

In particular, if $G = T^1$ *or* S^3, *we can replace* $\dim_p X < \infty$ *by* $H^i(X; K_p) = 0$ *for* $i \geqslant N$, *and X is compact.*

Proof. Consider the map $\pi_1: (X-F)_G \to X/G - F$. By the Vietoris-Begle mapping theorem, we have $\pi_1^*: H^*(X/G - F; K_p) \approx H^*((X-F)_G; K_p)$. Now let $j: F_G \to X_G$ be the map induced by the inclusion. It is enough to show that the homomorphism $j^*: H^*(X_G; K_p) \to H^*(F_G; K_p)$ is injective. We know from [4, p. 484], we may let $\{\alpha_j\}$ be a set of homogeneous elements of $H^*(X_G; K_p)$ such that $H^*(X_G; K_p)$ is the free $H^*(B_G; K_p)$-module generated by $\{\alpha_j\}$. It suffices to show that $j^*(\alpha_j) \neq 0$ for any $\alpha_j \in \{\alpha_j\}$. Suppose there exists some α_j such that $j^*(\alpha_j) = 0$. Choose a non-zero element $t^n \in H^n(B_G; K_p)$ for $n > \dim_p X$. This is possible by [1, IV (2.4)]. Clearly $j^*(t^n \alpha_j) = 0$. But $t^n \alpha_j$ is a non-zero element of $H^{n+s}(X_G; K_p)$ where $s = \deg \alpha_j$, and j* is an isomorphism in degree $> \dim_p X$. Thus we obtain $t^n \alpha_j = 0$, a contradiction. Hence j^* is injective. From the cohomology exact sequence of the pair (X_G, F_G) and since j^* is injective, π_1^* is an isomorphism, hence the sequence

$$0 \to H^i(X_G; K_p) \to H^i(F_G; K_p) \to H^{i+1}(X/G - F; K_p) \to 0$$

is exact for all $i \geqslant 0$.

In case $G = T^1$ or S^3, if $H^i(X; K_p) = 0$ for $i \geqslant N$, then $H^i(X/G - F; K_p) = 0$ for $i \geqslant N$ by Theorem 3.11 and Theorem 3.14, hence [1, p. 54] holds. By the same reason, in the following lemma if $G = T^1$ or S^3, we don't need the condition that $\dim_p X < \infty$ when the space X is compact.

Lemma 7.2 (Ku [10]). *Let G be a compact connected Lie group acting on a cohomology n-sphere X over* K_p. *Suppose that* $\dim_p X < \infty$ *and that* $H^*(B_{G_x}; K_p)$ *is trivial for* $x \notin F$, *then F is a cohomology r-sphere over* K_p *and* $n - r$ *is even.*

Theorem 7.3. *Let G be a compact connected Lie group acting on a cohomology n-sphere X over K_p. Suppose that $\dim_p X < \infty$ and $H^*(B_{G_x};K_p)$ is trivial for $x \notin F \neq \phi$. Then*

$$\begin{aligned} H^i(X/G;K_p) &= H^{i-r-1}(B_G;K_p) \quad \textit{for} \quad r+3 \leqslant i \leqslant n \\ &= K_p \quad \textit{for} \quad i=0 \\ &= 0 \quad \textit{otherwise.} \end{aligned}$$

Proof. We omit the coefficient field K_p in the proof. By Lemma 7.2, the fixed point set F is a cohomology r-sphere and $n-r$ is even. If $F \neq \phi$, then $\pi_2^*: H^*(B_G;K_p) \to H^*(X_G;K_p)$ is injective [1, p. 53]. Consider the fibre bundle $\pi_2: X_G \to B_G$. There is an exact sequence of the form

$$\cdots \longrightarrow H^n(B_G) \xrightarrow{\pi_2^*} H^n(X_G) \xrightarrow{j^*} H^0(B_G) \otimes H^n(X) \longrightarrow H^{n+1}(B_G) \otimes H^0(X) \longrightarrow \cdots$$
$$H^n(X_G) \xrightarrow{i^*} H^n(X) \cong H^0(B_G) \otimes H^n(X)$$

Since π_2^* is one to one [1, p. 53], i^* is onto. That is, X is totally non-homologous to zero in X_G rel. to K_p. Therefore we can apply Lemma 7.1. Assume that $0 \leqslant i < n$.

a) $0 \leqslant i \leqslant r: \dim H^i(X/G-F) = \dim H^{i-1}(B_G) \dim H^0(F) - \dim H^{i-1}(B_G) \dim H^0(X) = 0,$

b) $r+1 \leqslant i \leqslant n: \dim H^i(X/G-F) = \dim H^{i-1}(B_G) \dim H^0(F) + \dim H^{i-1-r}(B_G) \dim H^r(F) - \dim H^{i-1}(B_G) \dim H^0(X) = \dim H^{i-1-r}(B_G),$

c) $i \geqslant n+1: \dim H^i(X/G-F) = 0.$

By the cohomology exact sequence, we get

$$H^i(X/G;K_p) \approx H^{i-1-r}(X/G-F;K_p) \quad \text{for} \quad 1 \leqslant i \leqslant r-1, i \geqslant r+2.$$

It remains to show that $H^r(X/G) = H^{r+1}(X/G) = 0$, which, however, follows from the commutative diagram:

$$\begin{array}{ccccccccc} & & 0 & \longrightarrow & H^r(F) & \longrightarrow & H^{r+1}(X-F) & \longrightarrow & 0 \\ & & & & \uparrow \approx & & \uparrow \pi_1^* & & \\ 0 & \longrightarrow & H^r(X/G) & \longrightarrow & H^r(F) & \xrightarrow{\bar{\delta}^*} & H^{r+1}(X/G-F) & \longrightarrow & H^{r+1}(X/G) \longrightarrow 0. \end{array}$$

If $n=r, \dim H^i(X/G-F)=0$ for all i, then $H^i(X/G) \approx H^i(F)$ for all i.

Corollary 7.4. *Let $G = T^1$ act on a compact cohomology n-sphere X over K_p. Let the fixed point set F be a cohomology r-sphere over K_p, and suppose $n-r$ is even. If $H^*(B_{G_x};K_p)$ is trivial for $x \notin F$, then*

$$\begin{aligned} H^i(X/G;K_p) &= K_p \quad \textit{for} \quad i = 0, r+3, r+5, \ldots, n-1. \\ &= 0 \quad \textit{otherwise.} \end{aligned}$$

Proof. Since $H^*(B_G;K_p)=K_p[t]$, $\deg t=2$, the result follows from Theorem 7.3.

Corollary 7.5. *Let $G=S^3$ act on a compact cohomology n-sphere X over K_p. Let the fixed point set F be a cohomology r-sphere over K_p and suppose $n-r$ is even. If $H^*(B_{G_x};K_p)$ is trivial for $x\notin F$, then*

$$\begin{aligned} H^i(X/G;K_p) &= K_p \quad \textit{for} \quad i=0, r+5, r+9, \ldots, n-3 \\ &= 0 \quad \textit{otherwise.} \end{aligned}$$

Proof. It is known that $H^*(B_G;K_p)=K_p[t]$, $\deg t=4$ [3]. Since we assume that $F\neq\phi$, the spectral sequence of $\pi_2: X_G\to B_G$ is trivial, that is, $E_2=E_\infty$. Also for $m>n$, we have $H^m(X_G;K_p)\approx H^m(F_G;K_p)$. Hence

$$\dim H^m(F_G;K_p)=\sum_{i+j=m}\dim E_2^{i,j}=\sum_{i+j=m}\dim H^i(B_G;K_p)\dim H^j(F;K_p),$$
$$\dim H^m(X_G;K_p)=\sum_{i+j=m}\dim E_\infty^{i,j}=\sum_{i+j=m}\dim H^i(B_G;K_p)\dim H^j(X;K_p).$$

By taking $i=0$ (4), we see that

$$\sum_{j=a(4)}\dim H^j(F;K_p)=\sum_{j=a(4)}\dim H^j(X;K_p) \quad \text{for each} \quad a=0,1,2,3.$$

But X is a cohomology n-sphere and F is a cohomology r-sphere, and thus if $n=a$ (4), then $r=a$ (4). Hence $n=r$ (4). Thus the result follows from Theorem 7.3 and above observation.

By the same argument we can prove:

Theorem 7.6 (Liao). *Let $G=Z_p$ act on a cohomology n-sphere X mod K_p and suppose that $\dim_p X$ is finite. Let the fixed point set F be a cohomology r-sphere mod K_p. Then*

$$\begin{aligned} H^i(X/G;K_p) &= H^{i-r-1}(B_G;K_p) \quad \textit{for} \quad r+2\leqslant i\leqslant n \\ &= 0 \quad \textit{otherwise if} \quad i\neq 0. \end{aligned}$$

References

1. Borel, A. et al.: Seminar on Transformation Groups, Ann. of Math. Studies, **46,** 1960
2. — Cohomologie des éspace localement compacts d'apres J. Leray, Lecture Notes in Mathematics 2, Berlin-Göttingen-Heidelberg-New York: Springer 1964.
3. — Sur la cohomologie des éspace fibrés principaux et des éspaces homogènes des groupes de Lie compacts, Ann. of Math. **57,** 115—207 (1953).
4. —, and F. Hirzebruch: Characteristic classes and homogeneous spaces I, Amer. J. Math. **80,** 458—538 (1958).

5. BREDON, G. E.: The cohomology ring structure of a fixed point set, Ann. of Math. **80,** 524—537 (1964).
6. —, F. RAYMOND, and R. F. WILLIAMS: *p*-adic groups of transformations, Trans. of Amer. Math. Soc. **99,** 488—498 (1961).
7. CONNER, P. E.: The action of the circle group, Michigan J. Math. **4,** 241—247. (1957).
8. —, and E. E. FLOYD: On the construction of periodic maps without fixed points, Proc. of Amer. Math. Soc. **10,** 354—360 (1959).
9. HOFMANN, K. H., and P. S. MOSTERT: Elements of Compact Semigroups, Appendix II, Charles E. Merrill Books, Columbus, 1966.
10. KU, HSU-TUNG: Dissertation, Tulane University, 1967.
11. KU, MEI-CHIN: Dissertation, Tulane University, 1967.
12. MONTGOMERY, D., and L. ZIPPIN: Topological Transformation Groups, Interscience publishers, New York, 1955.
13. RAYMOND, F.: The orbit spaces of totally disconnected groups of transformations on manifolds, Proc. of Amer. Math. Soc. **12,** 1—7 (1961).
14. SWAN, R. G.: A new Method in fixed point theory, Comment Math. Helv. **34,** 1—16 (1960).

A Generalization of the Conner Inequalities*

HSU-TUNG KU

1. Introduction

Let G be a compact group which acts on a locally compact Hausdorff space X. The main object of this paper is to study the relationship between the structure of the space and its fixed point set. The rank of compact connected Lie groups acting freely or effectively on cohomology manifolds is also investigated.

In the study of the cohomology structure of the fixed point set, we begin with a generalization of the Conner inequalities. For example, we will show that if G is a compact non-totally disconnected group acting on a compact space X with fixed point set F such that $\dim_0 X < \infty$ and G_x is finite for $x \notin F$, then for any integer $k \geqslant 0$

$$\sum_{\substack{j \geqslant k \\ j = a(2)}} \dim H^j(F; K_0) \leqslant \sum_{\substack{j \geqslant k \\ j = a(2)}} \dim H^j(X; K_0), \quad a = 0, 1.$$

We examine the cohomology ring structure of the fixed point set of the circle group acting on a cohomology product of two spheres over K_0 in Section 3.

In Section 4, we go on to discuss the rank of a compact connected Lie group acting on a cohomology manifold. There are many known results concerning elementary p-group $(Z_p)^k$ actions. We prove corresponding results for compact connected Lie group actions.

All spaces considered will be locally compact Hausdorff unless otherwise stated. Throughout this paper, K_p denotes a field of characteristic p (p prime or zero), Z_p the prime field of characteristic p, and $\dim_p X$ the cohomology dimension of the space X over K_p [1]. The Alexander-Spanier-Wallace cohomology with compact supports will be used. The notation and terminology that we use will be found in [1].

---

* The results presented here are part of a doctoral dissertation written at the Tulane University under Professor P. S. MOSTERT.

2. A Generalization of the Conner Inequalities

Theorem 2.1. *Let G be a compact Lie group acting on a compact space X with fixed point set F such that* $\dim_p X = n < \infty$ *and* $H^*(B_{G_x}; K_p)$ *is trivial for* $x \notin F$. *Then for any integer* $m > \dim_p X$ *and any integer* $k \geqslant 0$

$$\sum_{\substack{j\geqslant k\\ i+j=m}} \dim E_2^{ij}(\pi_2') \leqslant \sum_{\substack{j\geqslant k\\ i+j=m}} \dim E_2^{ij}(\pi_2),$$

where $E_2^{ij}(\pi_2')$ *and* $E_2^{ij}(\pi_2)$ *are the Leray spectral sequences of the maps* $\pi_2: X_G \to B_G$ *and* $\pi_2': F_G \to B_G$ *respectively.*

Proof. The inclusion map $i: F_G \to X_G$ induces $i^*: E_r^{ij}(\pi_2) \to E_2^{ij}(\pi_2')$, and the spectral sequence $E_r(\pi_2')$ is always trivial. If $m > \dim_p X$, then $i^*: H^m(X_G; K_p) \approx H^m(F_G; K_p)$ and i^* maps $j^{m-k+1,k-1}(X_G)$ into $j^{m-k+1,k-1}(F_G)$, so i^* induces an epimorphism

$$\frac{H^m(X_G:K_p)}{J^{m-k+1,k-1}(X_G)} \longrightarrow \frac{H^m(F_G:K_p)}{J^{m-k+1,k-1}(F_G)}$$

where $H^n(X_G:K_p) = J^{o,n}(X_G) \supset J^{1,n-1}(X_G) \supset \cdots \supset J^{n,o}(X_G) \supset 0$ is the filtration with $E_\infty^{i,j}(\pi_2) = J^{i,j}(X_G)/J^{i+1,j-1}(X_G)$ [2, 3]. Thus we have

$$\begin{aligned}
\sum_{\substack{j\geqslant k\\ i+j=m}} \dim E_2^{i,j}(\pi_2') &= \sum_{\substack{j\geqslant k\\ i+j=m}} \dim E_\infty^{i,j}(\pi_2')\\
&= \dim H^m(F_G; K_p)/J^{m-k+1,k-1}(F_G)\\
&\leqslant \dim H^m(X_G; K_p)/J^{m-k+1,k-1}(X_G)\\
&= \sum_{\substack{j\geqslant k\\ i+j=m}} \dim E_\infty^{i,j}(\pi_2) \leqslant \sum_{\substack{j\geqslant k\\ i+j=m}} \dim E_2^{i,j}(\pi_2)
\end{aligned}$$

As a corollary, we get the following generalization of CONNER's results. He proved the case when $k = p = 0$ [1, Chap. IV].

Corollary 2.2. *Let* $G = T^1$, $\dim_p X$ *be finite and suppose that* $H^*(B_{G_x}; K_p)$ *is trivial for* $x \notin F$. *Then for any integer* $k \geqslant 0$, *we have*

$$\sum_{\substack{j\geqslant k\\ j=a(2)}} \dim H^j(F; K_p) \leqslant \sum_{\substack{j\geqslant k\\ j=a(2)}} \dim H^j(X; K_p), \quad a = 0, 1.$$

Proof. By assumption, the Leray sheaf $\underline{H^*(X;K_p)}$ is constant [1, p. 61]. Hence

$$\sum_{\substack{j\geqslant k\\ i+j=m}} \dim H^i(B_G;K_p)\dim H^j(F;K_p)\leqslant \sum_{\substack{j\geqslant k\\ i+j=m}} \dim H^i(B_G;H^j(X;K_p)).$$

Applying [3, Theorem 19.1], $H^*(B_G;K_p)=K_p[t]$, $\deg t=2$. Hence the result follows.

Corollary 2.3. *Let G be a compact non-totally disconnected group acting on a compact space X with fixed point set F such that* $\dim_0 X<\infty$ *and* G_x *is finite for* $x\notin F$. *Then*

$$\sum_{\substack{j\geqslant k\\ j=a(2)}} \dim H^j(F;K_0)\leqslant \sum_{\substack{j\geqslant k\\ j=a(2)}} \dim H^j(X;K_0),\quad a=0,1$$

for any integer $k\geqslant 0$.

Proof. One uses the same argument as in [8], and Corollary 2.2. In fact, we can replace the condition $\dim_0 X<\infty$ by $H^i(X;K_0)=0$ for $i\geqslant N$, N a positive integer. (See [8] [1, Chap. IV. 3.6].)

Corollary 2.4. *Let* $G=T^s$ *and let* $\dim_0 X$ *be finite. Then for any integer* $k\geqslant 0$,

$$\sum_{\substack{j\geqslant k\\ j=a(2)}} \dim H^j(F;K_0)\leqslant \sum_{\substack{j\geqslant k\\ j=a(2)}} \dim H^j(X;K_0),\quad a=0,1.$$

Proof. This follows from corollary 2.2 by induction on s. A simple application is to prove the following [1. Chap. IV. 5.2].

Corollary 2.5. *Let* $\dim_p X$ *be finite, where G is an infinite compact connected Lie group. Assume that* $H^*(B_{G_x};K_p)$ *is trivial for any* $x\notin F$ *and* $H^i(X;K_p)=0$ *for* $i\geqslant N$. *Then the same is true for* $H^*(F;K_p)$.

Proof. Suppose there exists an integer $m\geqslant N$ such that $H^m(F;K_p)\neq 0$. Choose the integer n sufficiently large so that $H^n(B_G;K_p)\neq 0$. Then

$$\begin{aligned} 0&<\dim H^n(B_G;K_p)\dim H^m(F;K_p)\\ &\leqslant \sum_{j\geqslant m} \dim H^{n+m-j}(B_G;K_p)\dim H^j(F;K_p)\\ &\leqslant \sum_{j\geqslant m} \dim H^{n+m-j}(B_G;H^j(X;K_p))=0. \end{aligned}$$

This is of course a contradiction.

Theorem 2.6. *Let G be a compact connected Lie group acting on a space X such that $\dim H^*(X;K_0)<\infty$ and $\operatorname{rank} G_x<\operatorname{rank} G$ for any $x\notin F$. Then $\chi(X;K_0)=\chi(F;K_0)$.*

Proof. For any maximal torus subgroup T, we have $F(T;X)=F(G;X)$. The result then follows from [8].

Theorem 2.7. *Let G be a compact connected Lie group acting on a compact space X. Suppose that $\dim_0 X<\infty$ and $\operatorname{rank} G_x<\operatorname{rank} G$ for any $x\notin F$. Then for any integer $k\geqslant 0$*

$$\sum_{\substack{j\geqslant k\\ j=a(2)}} \dim H^j(F;K_0)\leqslant \sum_{\substack{j\geqslant k\\ j=a(2)}} \dim H^j(X;K_0),\quad a=0,1.$$

Proof. This follows from corollary 2.3.

Remark. Theorem 2.6 and Theorem 2.7 remain true if we replace "G is a compact connected Lie group and $\operatorname{rank} G_x<\operatorname{rank} G$ for $x\notin F$" by "G is a finite dimensional compact connected group and $\chi(G/G_x;K_0)=0$ for $x\notin F$" using Proposition 2.10.

From Theorem 2.6 and Theorem 2.7, we get easily the following:

Corollary 2.8. *Let G be a compact connected Lie group acting on a compact acyclic space (resp. cohomology n-sphere) over K_0. Suppose that $\dim_0 X<\infty$, $\operatorname{rank} G_x<\operatorname{rank} G$ for $x\notin F$. Then F is acyclic over K_0 (resp. a cohomology r-sphere over K_0 and $n-r$ is even).*

Remark. Corollary 2.8 remains true without the hypotheses $\dim_0 X<\infty$ by using [10].

Theorem 2.9 *Let G be a compact connected Lie group acting on a space X with $\dim_p X<\infty$, $\dim H^*(X;Z_p)<\infty$. Suppose that p-rank $G_x<p$-rank G for $x\notin F$. Then*

1. $\chi(F;Z_p)=\chi(X;Z_p) \bmod p$.
2. $\sum_{j\geqslant k} \dim H^j(F;Z_p)\leqslant \sum_{j\geqslant k} \dim H^j(X;K_p)$ *for any integer $k\geqslant 0$.*

Proof. This follows using an easy induction of [1, Chap. III, 4.4] and [14, Corollary 4.1] respectively.

Proposition 2.10 (HOPF-SAMELSON). *Let H be a closed subgroup of a compact connected Lie group G. Then $\chi(G/H;K_0)\neq 0$ if and only if* $\operatorname{rank} G=\operatorname{rank} H$.

Proof. BOREL [1. p. 169] has shown that $\operatorname{rank} G = \operatorname{rank} H$ if and only if the characteristic homomorphism

$$H^*(B_H; K_0) \to H^*(G/H; K_0)$$

is an epimorphism. Consider the orbit map $\pi: B_{H_0} = E_H/H_0 \to (E_H/H_0)/(H/H_0) = B_H$. Since H/H_0 is finite, by Conner's lemma [1, Chap. III], $\pi^*: H^*(B_H; K_0) \approx H^*(B_{H_0}; K_0)^{H/H_0}$. Thus $H^*(B_H; K_0)$ contains only elements of even degree. Hence $\chi(G/H; K_0) \neq 0$.

Conversely, if $\chi(G/H; K_0) \neq 0$, let T^k be a maximal torus subgroup of G acting on G/H by left translation. Then $\chi(F(T^k; G/H); K_0) = = \chi(G/H; K_0) \neq 0$. Thus, T^k is conjugate to a subgroup of H. Hence $\operatorname{rank} G = \operatorname{rank} H$.

Proposition 2.11. *Suppose that G is a compact connected Lie group and that H is a closed subgroup. If $\chi(G/H; Z_p) \neq 0 \bmod p$, then H is of maximal* p-rank. *Moreover* $\chi(G/H_0; Z_p) = 0 \bmod p$ *if and only if* $\operatorname{ord}(H/H_0) = 0 \bmod p$.

Proof. Let $(Z_p)^k$ be a maximal elementary p-group. Let $(Z_p)^k$ act on G/H via $(g', gH) \to g'gH$. Then $\chi(F((Z_p)^k; G/H); Z_p) = \chi(G/H; Z_p) \neq 0 \bmod p$. Hence F is not empty, and some conjugate of $(Z_p)^k$ is contained in H. Thus H is of maximal p-rank. To prove the second statement, we consider the principal fibre map $G/H_0 \to G/H$ with fibre H/H_0, and get $\chi(G/H_0; Z_p) = \chi(G/H; Z_p) \operatorname{ord}(H/H_0)$. Hence the result follows.

Remark 2.12. For $p=2$, $\chi(G/H; Z_2)$ is a Stiefel-Whitney number of G/H. By a result of CONNER and FLOYD [17, 43.9], we also know that H is of maximal 2-rank. Now let G have no torsion. It is known that if H_0 is the maximal torus subgroup of G, then $\chi(G/H_0; Z_p) = \operatorname{ord}(W(G))$, and so $\chi(G/H; Z_p) = \operatorname{ord}(W(G))/\operatorname{ord}(H/H_0)$, where $W(G)$ means the Weyl group of G. Therefore $\chi(G/H; Z_p) \neq 0 \bmod p$ if and only if $\operatorname{ord}(W(G))/\operatorname{ord}(H/H_0) \neq 0 \bmod p$. In particular we state the following example: If $G = SU(n)$, then $\chi(G/H; Z_p) \neq 0 \bmod p$ if and only if $n!/\operatorname{ord}(H/H_0) \neq 0 \bmod p$ for $\operatorname{ord}(W(SU(n))) = n!$.

3. The Fixed Point Set of the Circle Group in a Cohomology Product of two Spheres

Let the circle group $G = T^1$ act on a compact cohomology product of two spheres over K_0; that is, $H^*(X; K_0) = H^*(S^{m_1} \times S^{m_2}; K_0)$, and $\dim_0 X$ is finite. We shall study the cohomology ring structure of the fixed point set F in some particular cases. These results are needed in Section 4. For the action of the group Z_p, the fixed point sets of such actions were completely determined by J. C. SU [13].

Our discussion relies on the Leray spectral sequence of the map $\pi_2: X_G \to B_G$. We treat the spectral sequence according to the case that it is non-trivial or trivial, respectively. For convenience, we assume that $H^*(B_G; K_0) = K_0[t^2]$, where $\deg t^2 = 2$ and $t^n = 0$ if n is odd [3].

Theorem 3.1. *Under the above conditions, if the Leray spectral sequence of the map π_2 is non-trivial, then the fixed point set F is a cohomology sphere over K_0.*

Proof. Clearly $\dim H^*(F; K_0) < 4$, for if $\dim H^*(F; K_0) = 4$, then the spectral sequence of π_2 is trivial [1, Chap. IV, 5.4]. Note that at least one of the m_1, m_2 is odd, for otherwise $\chi(F; K_0) = \chi(X; K_0) = 4$ by Theorem 2.6. Hence $\chi(F; K_0) = 0$ and $\dim H^*(F; K_0) = 0$ or 2.

Suppose now that the Leray spectral sequence of the map π_2 is trivial and that at least one of the m_1, m_2 is odd. The space X is totally non-homologous to zero in X_G rel. to K_0. Hence the inclusion map $i: X \to X_G$ induces an epimorphism $i^*: H^*(X_G; K_0) \to H^*(X; K_0)$. Let $\alpha_k \in H^*(X_G; K_0)$ be such that $i^*(\alpha_k) = a_k$ for $k = 1, 2, 3$ where a_k is the homogeneous generator of $H^*(X; K_0)$ with $\deg a_k = m_k$, $m_3 = m_1 + m_2$. We may choose $\alpha_3 = \alpha_1 \alpha_2$. Then $H^*(X_G; K_0)$ is the free $H^*(B_G; K_0)$-module generated by $\{1, \alpha_1, \alpha_2, \alpha_3\}$, [6], [4, p. 484]. Let $b_0 = 1, b_1, b_2, b_3$ be the homogeneous generators of $H^*(F; K_0)$ such that $\deg b_k = n_k$ for $k = 1, 2, 3$. By Corollary 2.2, we may assume that $n_i \leqslant m_i$, $i = 1, 2, 3$ and both m_i, n_i are even or odd. If $N > \dim_0 X$, then $j^*: H^N(X_G; K_0) \approx H^N(F_G; K_0)$. We will use the notation $t^s b$ instead of $t^s \otimes b$ as an element of $H^s(B_G; K_0) \otimes H^q(F; K_0)$.

Theorem 3.2. *If the Leray spectral sequence of the map π_2 is trivial and F is connected, then F is a cohomology product of two spheres, say, $H^*(F; K_0) = H^*(S^{n_1} \times S^{n_2}; K_0)$. More precisely, if m_1, m_2 are both odd, then n_1, n_2 are also odd, and if exactly one of the m_1, m_2 is odd, then so is precisely one of n_1 and n_2.*

Proof. Let N be even and $N > \dim_0 X$. It is sufficient to consider the following two cases:

1. m_1 and m_2 are both odd,
2. exactly one of m_1, m_2 is odd.

Assume case (i). The anti-commutativity of the cup product implies

$$\alpha_1^2 = \alpha_2^2 = \alpha_3^2 = \alpha_1 \alpha_3 = \alpha_2 \alpha_3 = \alpha_1 \alpha_2 + \alpha_2 \alpha_1 = 0.$$

Let

$$t^{N+1-n_1}b_1=j^*(At^{N+1-m_1}\alpha_1+Bt^{N+1-m_2}\alpha_2),$$
$$t^{N+1-n_2}b_2=j^*(Ct^{N+1-m_1}\alpha_1+Dt^{N+1-m_2}\alpha_2),$$
$$t^{N-n_3}b_3=j^*(Et^{N-m_3}\alpha_3)$$

where A, B, C, D, E belong to K_0. We see that $t^{2N+2-2n_1}b_1^2=0$, and this implies that $b_1^2=0$. Similarly $b_2^2=b_3^2=b_1b_3=b_2b_3=0$. Since $\alpha_1\alpha_2\neq 0$, we must have $b_1b_2\neq 0$, so we may choose $b_3=b_1b_2$. Consider case (ii). We may suppose m_1 even, m_2 odd and n_1 even. Then

$$t^{N-n_1}b_1=j^*(At^{N-m_1}\alpha_1).$$
$$t^{N+1-n_2}b_2=j^*(Bt^{N+1-m_2}\alpha_2+CT^{N+1-m_3}\alpha_3).$$
$$t^{N+1-n_3}b_3=j^*(Dt^{N+1-m_2}\alpha_2+Et^{N+1-m_3}\alpha_3).$$

As m_2, m_3 are odd, $\alpha_2^2=\alpha_3^2=\alpha_2\alpha_3=\alpha_1\alpha_3=0$, whence $b_2^2=b_3^2=b_2b_3=0$. Thus

$$j^*(\alpha_1)=A^{-1}j^*(t^{m_1-n_1}b_1),$$
$$j^*(\alpha_2)=Lj^*(t^{m_2-n_2}b_2)+Mj^*(t^{m_2-n_2}b_3).$$

Since $\alpha_1\alpha_2\neq 0$, it follows that $b_1b_2\neq 0$ or $b_1b_3\neq 0$. It is impossible to have both of them, for otherwise $\dim H^*(F;K_0)$ is greater than four. If $b_1b_2\neq 0$, then $b_1b_3=0$. Choose $b_3=b_1b_2$ (which is possible by dimension count). The element b_1^2 is clearly zero because there is exactly one homogeneous base element of even degree which is greater than zero. Therefore F is a cohomology product of two spheres over K_0 in either case.

We consider now the case when the fixed point set F is disconnected.

Theorem 3.3. *If the Leray spectral sequence of the map π_2 is trivial and the fixed point set F is disconnected, then F is the disjoint union of two cohomology spheres over K_0.*

Proof. It is enough to eliminate the following case (for $H^*(F;K_0) = H^*(x\cup P;K_0)$, where x is a point and P is a cohomology projective 2-space of even degree, is impossible):

$$(*)\qquad H^*(F;K_0)=H^*(x\cup(S^{n_1}\vee S^{n_2});K_0),\ n_1,n_2 \text{ odd}$$

where $\cup$ means the disjoint union. If m_1, m_2 are both odd, then

$$j^*(\alpha_1)=At^{m_1-n_1}b_1+Bt^{m_1-n_2}b_2,$$
$$j^*(\alpha_2)=Ct^{m_2-n_1}b_1+Dt^{m_2-n_2}b_2.$$

But $j^*(\alpha_1\alpha_2)\neq 0$, this contradicts (*). If m_1 is even and m_2 odd, let

$$j^*(\alpha_1)=At^{m_1}b_0+Bt^{m_1}b_0',$$
$$j^*(\alpha_2)=Ct^{m_2-n_1}b_1+Dt^{m_2-n_2}b_2$$

where $b_0,b_1,b_2\in H^*(S^{n_1}\vee S^{n_2};K_0)$ with $\deg b_0=0$ and $b_0'\in H^0(x;K_0)$. Then

$$j^*(\alpha_1\alpha_2)=ACt^{m_1+m_2-n_1}b_1+ADt^{m_1+m_2-n_2}b_2=j^*(At^{m_1}\alpha_2).$$

Thus $\alpha_1\alpha_2=At^{m_1}\alpha_2$ because the homomorphism j^* is injective. Applying the homomorphism i^*, we obtain

$$a_1a_2=i^*(\alpha_1\alpha_2)=i^*(At^{m_1}\alpha_2)=0.$$

This is a contradiction. A similar proof holds for m_1 odd and m_2 even.

4. The Actions of Compact Connected Lie Groups on Cohomology Manifolds

We shall apply the arguments of MANN and SU [11] to prove Theorems 4.1 and 4.4.

Theorem 4.1. *Let G be a compact connected Lie group of rank k acting on a compact connected space X such that $H^*(B_{G_x};K_p)$ is trivial for all $x\in X$, $\dim H^*(X;K_p)<\infty$, $\dim_p X=n<\infty$. Let $B=\sum_{i=1(2)}\dim H^i(X;K_p)$, then*

1. $k\leqslant(2+(n+1)(B-1))/2$, *if n is odd;*
2. $k\leqslant(2+(n+2)(B-1))/2$, *if n is even.*

In particular, if the space X is a cohomology n-sphere over K_p, then n is odd and $k\leqslant 1$.

Proof. Let T^k be a maximal torus subgroup of G. Without loss of generality, we may assume that $G=T^k$ and $k\geqslant 1$. Consider the Leray spectral sequence of the map $\pi_2: X_G\to B_G$. The E_2-term is given by

$$E_2^{i,j}=H^i(B_G;K_p)\otimes H^j(X;K_p)$$

and

$$E_{r+1}^{i+1,o}=E_r^{i+1,o}/\mathrm{Im}(E_r^{i+1-r,r-1}\xrightarrow{d_r}E_r^{i+1,o}),\quad r\geqslant 2.$$

To prove (i), notice that

$$\dim E_\infty^{n+1,o}\leqslant\dim H^{n+1}(X_G;K_p)=\dim H^{n+1}(X/G;K_p)=0$$

because

$$\pi_1^*: H^*(X/G; K_p) \approx H^*(X_G; K_p)$$

by the Vietoris-Begle mapping theorem and $H^{n+1}(X/G; K_p) = 0$ by [1, Chap. IV 5.2]. We see that

$$\begin{aligned}\dim E_{r+1}^{n+1,o} &\geqslant \dim E_r^{n+1,o} - \dim E_r^{n+1-r,r-1} \\ &\geqslant \dim E_r^{n+1,o} - \dim E_2^{n+1-r,r-1},\end{aligned}$$

so $\dim E_r^{n+1,o} \leqslant \dim E_{r+1}^{n+1,o} + \dim E_2^{n+1-r,r-1}$.

By taking $r = 2, 3, \ldots, n+1$, we obtain

(1) $$\begin{aligned}\dim E_2^{n+1,o} &\leqslant \dim E_2^{n-1,1} + \dim E_2^{n-3,3} + \cdots + \dim E_2^{0,n} \\ &\leqslant \dim H^{n-1}(B_G; K_p) B.\end{aligned}$$

Thus we have

$$\binom{\frac{n+1}{2} + k - 1}{k-1} \leqslant \binom{\frac{n-1}{2} + k - 1}{k-1} B.$$

Therefore, $k \leqslant (2 + (n+1)(B-1))/2$.

Proof of (2.). By repeating the same argument as in the proof of 1., we have

(2) $$\dim E_2^{n+2,o} \leqslant \dim E_2^{n,1} + \dim E_2^{n-2,3} + \cdots + \dim E_2^{2,n-1}.$$

We can simplify as before to get the result.

Theorem 4.2. *Let G be a compact connected Lie group of rank k acting on a compact connected space X with* $\dim_p X = n + m < \infty$. *If* $H^*(X; K_p) = H^*(S^n \times S^m; K_p)$ *and* $H^*(B_{G_x}; K_p)$ *is trivial for all* $x \in X$, *then*

1. $k = 0$ *if m and n are even,*
2. $k \leqslant 2$ *if m and n are odd, or m even and* $n = 1$,
3. $k \leqslant 1$ *otherwise.*

Proof. As in Theorem 4.1, we may assume that $G = T^k$. 1. is the obvious consequence of the identity $\chi(X; K_p) = \chi(F; K_p)$. If m, n are odd, applying (2) we obtain

$$\dim E_2^{n+m+2,o} \leqslant \dim E_2^{m+1,n} + \dim E_2^{n+1,m};$$

that is

$$\binom{\frac{m+n+2}{2} + k - 1}{k-1} \leqslant \binom{\frac{m+1}{2} + k - 1}{k-1} + \binom{\frac{n+1}{2} + k - 1}{k-1}.$$

Letting $k=3$ and simplifying, we get $(m+1)(n+1)\leqslant 4$. Thus both m and n must be equal to 1. If $m=n=1$, then $\dim T^3(x)>\dim_p X=2$. This is a contradiction. Now if m is even and n odd, by (1) we have

$$\dim E_2^{n+m+1,o}\leqslant \dim E_2^{m,n}+\dim E_2^{0,m+n},$$

so

$$\binom{\frac{m+n+1}{2}+k-1}{k-1}\leqslant\binom{\frac{m}{2}+k-1}{k-1}+1.$$

Thus, if $k=2$, then $n\leqslant 1$, and if $k=3$, then

$$(n+1)^2/4+m(n+1)+(3/2)n\leqslant 1/2.$$

This is impossible.

By a similar argument we can show:

Theorem 4.3. *Let G be a compact connected Lie group of rank k acting on a compact space X, $\dim_p X=3n$ and $H^*(X;K_p)=H^*((S^n)^3;K_p)$. Suppose moreover, $H^*(B_{G_x};K_p)$ is trivial for all $x\in X$. Then $k\leqslant 3$ if $n\geqslant 7$.*

Proof. Assume $G=T^k$. In this case n must be odd and from (1) we get

$$\dim E_2^{3n+1,o}\leqslant \dim E_2^{2n,n}+\dim E_2^{0,3n}.$$

If $k=4$, then

$$\left(\frac{3n+1}{2}+3\right)\left(\frac{3n+1}{2}+2\right)\left(\frac{3n+1}{2}+1\right)\leqslant 3(n+3)(n+2)(n+1)+6.$$

Obviously, if $n\geqslant 7$, this inequality can not hold.

Theorem 4.4. *Let G be a compact connected Lie group of rank k acting effectively on a first countable connected cohomology n-manifold X over K_0. Suppose $\dim_0 F(G,X)=r\geqslant 0$. Then $k\leqslant (n-r)/2$.*

Proof. Let $G=T^k$. Since G acts effectively on X and $r\geqslant 0$, $F(G,X)$ is a closed proper subset of X and $r<n$ [1, Chap. I 4.6]. By a theorem of BOREL [1, XIII, 4.3], there must exist a closed connected subgroup H_1 of G of codimension 1 (that is $H_1\approx T^{k-1}$) such that $r_1=n(H_1)>n(G)=r$. By [1, V. 3.2], $n-r_1$ and $n-r$ are even. Hence r_1-r is even and $r_1\geqslant r+2$. Now we consider the effective action of H_1 on X. Repeat the argument as before; there exists a subgroup $H_2 \approx T^{k-2}$ of H_1 such that $r_2 = n(H_2) > n(H_1) = r_1$. Again $r_2 \geqslant r_1 + 2$. Continuing this process, finally we have a closed connected subgroup H_{k-1} of codimension $k-1$ such that $r_{k-1} = n(H_{k-1}) \geqslant n(H_{k-2}) + 2 = r_{k-2} + 2$. Thus

$$n \geqslant r_{k-1} + 2 \geqslant (r + 2(k-1)) + 2 = r + 2k.$$

Hence $k \leqslant (n-r)/2$.

Theorem 4.5. *Let G be a compact connected Lie group of rank k acting effectively on a first countable connected compact generalized cohomology n-sphere over K_0. Then $k \leqslant (n+1)/2$.*

Proof. Let T^k be a maximal torus subgroup of G. From now on, we consider the action of T^k on X. If the fixed point set F is not empty, the theorem follows from Theorem 4.4. If the fixed point set F is empty, we consider the following two cases:

1. The isotropy group is finite for any point of the space X. In this case, by Theorem 4.1, it holds.

2. There exists an isotropy group containing the torus subgroup $T^{k'}$, with k' maximal such that $k' < k$ and $k' \neq 0$. The group $T^k/T^{k'}$ acts on $F(T^{k'}, X)$ with all isotropy subgroups finite by assumption. The space $F(T^{k'}, X)$ is a cohomology sphere, hence by Theorem 4.1, $k - k' \leqslant 1$. Note that

$$F(T^k, X) = F(T^k/T^{k'}, F(T^{k'}, X)) = \phi,$$

so that $\dim_0 F(T^{k'}, X) - (-1)$ is even, whence $\dim F(T^{k'}, X) \geqslant 1$. Now, applying Theorem 4.4, we obtain

$$k' \leqslant (n - \dim_0 F(T^{k'}, X))/2 \leqslant (n-1)/2.$$

Therefore

$$k \leqslant k' + 1 \leqslant (n-1)/2 + 1 = (n+1)/2.$$

Corollary 4.6. *Let $G = T^k$ act effectively on a compact first countable connected generalized cohomology n-sphere over K_0. Then*

1. *$k \leqslant (n-1)/2$ if $F \neq \phi$,*

2. *$k \leqslant (n+1)/2$ if $F = \phi$ and moreover, if $k = (n+1)/2$, there exists a soubgroup T^{k-1} of T^k with $F(T^{k-1}, X) \neq \phi$. $F(T^{k-1}, X)$ is a generalized cohomology 1-sphere over K_0 in this case.*

Proof. It remains to consider the case $k = (n+1)/2$. The existence of of a subgroup T^{k-1} of T^k with $F(T^{k-1}, X) \neq \phi$ is clear from Theorem 4.1.

If $\dim_0 F(T^{k-1},X)>1$, then

$$k-1 \leqslant \frac{n-\dim_0 F(T^{k-1},X)}{2} \leqslant (n-2)/2.$$

Thus $k \leqslant n/2$, a contradiction.

Theorem 4.7. *Let X be a compact connected first countable $(n+m)-\mathrm{cm}$ over K_0 and $H^*(X;K_0)=H^*(S^m\times S^n;K_0)$. If G is a compact connected Lie group of rank k acting effectively on the space X, then*

$$k \leqslant \left[\frac{m+1}{2}\right]+\left[\frac{n+1}{2}\right],$$

where $[s]$ denotes the largest integer not exceeding s.

Proof. It is enough to show the case $G=T^k$. If the fixed point set F is not empty, then by Theorem 4.4, we have

$$k \leqslant \frac{m+n}{2} \leqslant \left[\frac{m+1}{2}\right]+\left[\frac{n+1}{2}\right].$$

If all the isotropy groups are finite, then $k \leqslant 2$ by Theorem 4.2. So it remains to consider the case that T^k acts effectively but not all the isotropy groups are finite and $F=\phi$. Let $T^{k'}$ be the torus subgroup of T^k of maximal rank $k'<k$, $k'\neq 0$ such that $F(T^{k'},X)\neq\phi$. Let $r=\dim_0 F(T^{k'},X)$, then by Theorem 4.4, we have

$$k' \leqslant (m+n-r)/2.$$

We may assume that at least one of m, n is odd, for otherwise F is not empty. The results of Section 3 still hold for the torus group action by induction. Therefore $F(T^{k'},X)$ must be one of the following.

1. A connected cohomology product of two spheres over K_0,
2. A disjoint union of two cohomology spheres over K_0,
3. A cohomology sphere over K_0.

Proof in case 1. The group $T^k/T^{k'}\approx T^{k-k'}$ acts on $F(T^{k'},X)$ with all the isotropy groups finite by assumption, so Theorem 4.2 gives

$$k-k' \leqslant 1, \text{ if one of } m, n \text{ is odd (say } n \text{ and } n>1),$$
$$k-k' \leqslant 2 \text{ otherwise.}$$

Since $F(T^{k'},X)$ is a connected cohomology product of two spheres over K_0, $r\geqslant 2$, we have

$$k\leqslant k'+1\leqslant (m+n-r)/2+1\leqslant (m+n)/2 \leqslant \left[\frac{m+1}{2}\right]+\left[\frac{n+1}{2}\right]$$

in the first case. Otherwise,

$k \leqslant k'+2 \leqslant (m+n-r)/2+2 \leqslant (m+n+2)/2$ if both m and n are odd and $k \leqslant (m+n+1)/2$ if m is even and $n=1$, since $r \geqslant 3$ in this case.

Hence

$$k \leqslant \left[\frac{m+1}{2}\right] + \left[\frac{n+1}{2}\right]$$

still holds in these cases.

Proof in case 2., 3. Since $T^k/T^{k'}$ acts on $F(T^{k'}, X)$, by dimension parity $r \geqslant 1$. Using Theorem 4.1, we obtain $k-k' \leqslant 1$. Hence

$$k \leqslant k'+1 \leqslant (m+n-r)/2+1 \leqslant (m+n+1)/2 \leqslant \left[\frac{m+1}{2}\right] + \left[\frac{n+1}{2}\right].$$

This completes the proof of the theorem.

Theorem 4.8. *Let G be a compact connected Lie group of rank k acting effectively on a first countable connected compact generalized cohomology complex or quaternion projective n-space X over K_0. Then $k \leqslant n/2$.*

Proof. Apply Theorem 4.4 to the maximal torus subgroup T^k of G.

References

1. Borel, A. et al.: Seminar on Transformation Groups, Ann. of Math. Studies **46**, 1960.
2. — Cohomologie des espaces localement compacts d'apres J. Leray, Lecture Notes in Mathematics **2**, Berlin-Göttingen-Heidelberg-New York: Springer 1964.
3. — Sur la cohomologie des fibrés principaux et des espaces homogènes de groupes de Lie compacts, Ann. of Math. **57**, 115—207 (1953).
4. —, and F. Hirzebruch: Characteristic classes and homogeneous spaces I, Amer. J. Math. **80**, 458—538 (1958).
5. Bredon, G. E.: Cohomology ring structure of a fixed point set, Ann. of Math. **80**, 524—537 (1964).
6. Cartan, H., and S. Eilenberg: Homological Algebra, Princeton University Press, Princeton, 1956.
7. Conner, P. E., and E. E. Floyd: Differential Periodic Maps, Berlin-Göttingen-Heidelberg-New York: Springer 1964.
8. Ku, Hsu-Tung: An Euler characteristic formula for compact group actions, Hung-Ching Chow Sixty-Fifth Anniversary Volume, Math. Res. Center, Taiwan Univ., 81—88 (1967).
9. — Dissertation, Tulane University, 1967.
10. Ku, Mei-Chin: Dissertation, Tulane University, 1967.

11. MANN, L. N., and J. C. SU: Actions of elementary p-groups on manifolds, Trans. Amer. Math. Soc. **106**, 115—126 (1963).
12. MONTGOMERY, D., and L. ZIPPIN: Transformation Groups. Interscience, New York, 1955.
13. SU, J. C.: Periodic transformations on the product of two spheres, Trans. Amer. Math. Soc. **112**, 369—380 (1964).
14. SWAN, R. G.: A new method in fixed point theory, Comment. Math. Helv. **34**, 1—16 (1960).

SHORT NOTES

Some Simple Observations about A_p Actions

PAUL S. MOSTERT*

The homotopy theorem for Čech cohomology is stated in the following way: Let $\varphi: X \times A \to Y$ be a map of topological spaces. Denote $\varphi_a = \varphi(a, -)$. If A is an arc, then

$$H^*(\varphi_a) = H^*(\varphi_b): H^*(Y; L) \to H^*(X; L)$$

for any a, $b \in A$ and any group L of coefficients. Under appropriate conditions on X and Y, the "generalized homotopy theorem" states that A can be a (compact) connected space. However, a more accurate statement, and the one that is actually proved, is the statement given below. Here sheaf cohomology is used, but of course when X and Y are paracompact (or when paracompactifying supports are used), ČECH and sheaf cohomology coincide.

Homotopy Theorem. *Let X and Y be spaces with respective support families Φ and Ψ. Let A be a space and χ a support family on A. Suppose $\varphi: X \times A \to Y$ is a map which is compatible with Φ and Ψ. (That is, each φ_a defines naturally a map $H^*_\Psi(Y; L) \to H^*_\Phi(X; L)$.) Let $\pi: X \times A \to A$ be the natural projection and let $\mathscr{H}^*(\pi)$ be the Leray sheaf of this map. Suppose finally that for each $a \in A$, the map*

$$(*) \qquad \mathscr{H}^*(\pi)_a \to H^*_\Phi(X \times \{a\}; L)$$

is an isomorphism. Then the function

$$A \times H^*_\Psi(Y; L) \to H^*_\Phi(X; L)$$

given by

$$(a, h) \mapsto H^*(\varphi_a)(h)$$

is continuous when all cohomology groups are given the discrete topology.

Proof. It is enough to show that the map

$$A \times H^*_{\Phi \times \Psi}(X \times A; L) \to H^*_\Phi(X; L)$$

* NSF Senior Postdoctoral Fellow. This work was also partially supported by NSF grant GP 6219.

given by

$$(a,h) \mapsto H^*_{\Phi \times \Psi}(i_a)(h),$$

where $i_a: X \to X \times A$ is defined by $i_a(x) = (x,a)$, is continuous. Let $i_V: X \times V \to X \times A$, $i_{aV}: X \to X \times V$ be in the first case the injection and in the second case the map $x \mapsto (x,a)$, $a \in V \subset X$. Let $\bar{h} \in H^*_{\Phi \times \Psi}(X \times A; L)$. Since $\pi i_a = 1$, there is an element $h_a \in \operatorname{Ker} H^*(i_a)$ such that $\bar{h} = H^*(\pi)(h) + h_a$, where $h = H^*(i_a)(\bar{h})$. By hypothesis, there is a neighborhood V of a in A such that $H^*(i_V)(h_a) = 0$. If now $b \in V$, $H^*(i_b)(\bar{h}) = h + H^*(i_{bV}) H^*(i_V)(h_a) = h = H^*(i_a)(\bar{h})$, and this proves the theorem.

The hypothesis (*) on the Leray sheaf is satisfied under either of the following hypotheses:

1. X is locally compact $\operatorname{clc}_L^\infty$, A, χ and Φ are arbitrary.
2. X and A are locally compact, Φ and Ψ are compact supports.

In particular, if X is a manifold, (*) is satisfied. For the proofs of these statements, the reader is referred to [1; pp. 145–146 and p. 138].

Now we show how the homotopy theorem in this form applies to the Hilbert-Smith conjecture.

Observation 1. *If G is a topological group acting on a locally compact* clc_L^n *space X with* $H^*_\Phi(X;L)$ *finitely generated (*Φ *any support family), then there is an open subgroup of G which acts trivially on* $H^*_\Phi(X;L)$.

This is an immediate consequence of the Homotopy Theorem as stated above. Its connection to the Hilbert-Smith conjecture is as follows. First, we know the Hilbert-Smith conjecture can be reduced to the consideration of A_p actions. Now any open subgroup H of A_p is isomorphic to A_p, and if H cannot act effectively on X then neither can A_p. Thus, the Hilbert-Smith conjecture is equivalent to the following:

Conjecture: *Let* A_p *act on a manifold M in such a way that* A_p *acts trivially on* $H^*_\Phi(M;\mathbb{Z})$, *where* Φ *is either compact supports or closed supports. Then the action is not effective.*

Various results in the literature have been given which require the hypothesis that A_p acts trivially on the cohomology (see, e. g., [2], [3] and [4]). These results can thus be reinterpreted in light of the above observation.

Much attention has been given to free A_p actions. Clearly, the observation that the homotopy theorem is in reality a theorem about continuity of maps has its counterpart in the observation that the Lefschetz number $\lambda(g) = \Sigma(-1)^n \operatorname{Tr}(H^n(g))$ is continuous in g and is locally constant. Hence, we have from the Lefschetz fixed point theorem.

Observation 2. *Let G be a topological group acting freely on a compact manifold with non-vanishing Euler characteristic. Then G is discrete.*

Remark. There is a form of the generalized homotopy theorem which is an immediate consequence of the Vietoris mapping theorem and hence uses the connectivity (as well as the compactness and acyclicity) of A. (See [1; p. 56].) It is useful in that it is valid for any pair of spaces X and Y.

References

1. Bredon, G. E.: Sheaf Theory, McGraw-Hill, New York, 1967.
2. —, F. Raymond, and R. F. Williams: p-adic transformation groups, Trans. Amer. Math. Soc. **99**, 488—498 (1961).
3. Raymond, F.: The orbit space of totally disconnected groups of transformations on manifolds, Pac. J. of Math. **15**, 1365—1376 (1965).
4. — Cohomological and dimension theoretical properties of orbit spaces of p-adic actions, these Proceedings.

Problems

1. Much of the study of A_p-actions on manifolds is concerned with free actions. There is the general feeling, but no proof, that if one could solve the Hilbert-Smith conjecture for free actions, then it would not be difficult to extend the proof to arbitrary actions. Is there a proof that the validity of the conjecture for free actions implies that for arbitrary actions? – P. S. MOSTERT.

2. Is there a compact metric space X of dimension 1 and an action of a pro-p group G on X such that $\dim X/G = n$, $n = 4,5,\ldots$? (Examples are known for $n = 2,3$.) (A pro-p group is a projective limit of p-groups.) – R. F. WILLIAMS.

3. Let G be a compact (connected) group acting on a compact (necessarily acyclic) Hopf space X with identity and zero in such a way as to preserve the multiplication (i. e. $g(xy) = g(x)g(y)$). Is the fixed point set connected? Such a result would have very important applications to the structure theory of compact semigroups. (When G is connected and abelian, the result is known even without the assumption that X is a Hopf space, proveding it is acyclic over the rationals.) – K. H. HOFMANN and P. S. MOSTERT.

4. Let G be a compact group of dimension n embedded in $\mathbb{R}^{n+1}$. It is known that G must be a Lie group. What groups admit such an embedding? If G is a compact group and H is a closed subgroup with $\dim G/H = n$ and G/H is embedded in $\mathbb{R}^{n+1}$, is then G/H locally connected? Assuming the answer is yes, can one describe these coset spaces? Answers to the first and third of these questions would largely complete the picture of groups acting on subsets of R^n with n-dimensional orbits. (See HOFMANN and MOSTERT, *Compact groups acting with* $(n-1)$*-dimensional orbits on subspaces of n-manifolds*, Math. Annalen 167, 224–239 (1966).) Some information about the first question has recently been obtained by L. N. MANN and J. SICKS (To appear.) – K. H. HOFMABB and P. S. MOSTERT.

5. A compact space X is called a *limit manifold* if it is a projective limit of a system $\{S_i, \pi_{ij}, A\}$ of compact manifolds such that the natural maps $\pi_i: X \to X_i$ induce non-zero homomorphismus $\pi_i^n: H^n(X_i:\mathbb{Z}) \to H^n(X:\mathbb{Z})$, where $n = \dim X_i$. A compact group is a limit manifold. Is the coset space of a compact group a limit manifold? – K. H. HOFMANN and P. S. MOSTERT

PART IV

Non-Compact Lie Transformation Groups

On Hilbert's Theory of Closed Group Actions*

I. FÁRY

1. Introduction. Let us call two triangles A, B, C and P, Q, R "congruent up to ε", if there are congruent triangles A', B', C' and P', Q', R', such that the pairs of vertices A and $A', \ldots, R$ and R' are at a distance $<\varepsilon$. A strong "continuity", or rather "closedness", property of geometry is this: "congruence up to ε", for all $\varepsilon > 0$, implies congruence. This property, valid in absolute geometry, i. e. independently of the parallel axiom, can be formulated using the topology of the plane only, without referring to its metric. In the usual axiomatic presentation of geometry, continuity comes at the end, nearly as an afterthought. HILBERT made a systematic and creative study of this classical approach [11]. Then, inspired by Klein's Erlangen Program and Lie's theory, he opened up a new approach to plane geometry, based on the topology of the plane, and on the "closedness" property of the group of isometries referred to above, see [12]. He presented his result as a first step toward a solution of the famous "Hilbert's V-th".

It is hardly necessary to recall the importance of Hilbert's problem for the development of various parts of mathematics, and the fruitfulness of its solution (see [35] and references there). Before the complete solution of "Hilbert's V-th", an extensive survey [51] shows the bearing of [12] on later development. We feel it would be useful to have a new appraisal of the older results in this field, taking into account the general structure theory of locally compact groups. The last sections of [35] are, of course, devoted to this question, but I am sure that, because of lack of space, the authors were not able to report as fully as they could. The present paper deals with a "two-dimensional aspect" of Hilbert's original theory, which is not covered by the theory of locally compact groups. We formulate and discuss results from [10]. Although our discussion here is completely elementary, we would like to suggest that new methods of topology should have an interesting application here (see Section 4). We hope to contribute in this way to an awakening of

* The work of the author has been supported in part by National Science Foundation contract GP-6974.

interest in this field. This aim should also excuse the extensive Bibliography.

2. Definitions. Let us fix our terminology, and recall some facts from general topology. If $g: X \to X$ is a continuous map, we denote $\Gamma(g)$ its graph, closed subspace of $X \times X$. Given a family G of maps g, we set

$$\Gamma(G) = \bigcup \{\Gamma(g): g \in G\}, \quad (\Gamma(G) \subset X \times X). \tag{1}$$

If (1) is *closed* in $X \times X$, G itself will be called 1-*closed.* In case the topology of X can be described by sequential convergence, 1-closedness amounts to this:

$$\begin{cases} \textit{if} \quad y_\alpha = g_\alpha x_\alpha, \quad g_\alpha \in G, \quad \textit{and} \quad x = \lim_{\alpha\to\infty} x_\alpha, \quad y = \lim_{\alpha\to\infty} y_\alpha, \\ \textit{then there exists a} \quad g \in G, \quad \textit{such that} \quad y = gx. \end{cases} \tag{2}$$

We write

$$G(x) = \{y \in X: y = gx \quad \textit{for some} \quad g \in G\}, \quad (G(x) \subset X); \tag{3}$$

$$G_x = \{g \in G: gx = x\}, \quad (G_x \subset G), \tag{4}$$

as usual. If G is 1-closed, $G(x)$ is closed, and furthermore the decomposition space $Y = \{G(x): x \in X\}$ inherits some properties of X. It is not required in (2) that the sequence $\{g_\alpha\}$ converges, but if it is pointwise convergent and $\lim g_\alpha$ belongs to G, $g = \lim g_\alpha$ satisfies the condition in (2).

Let us define [12, 51] k-closedness,* $k \geqslant 1$. We set $X^{(k)} = X \times \cdots \times X$, k factors, $g^{(k)}(x_1, \ldots, x_k) = (gx_1, \ldots, gx_k)$, and $G^{(k)} = \{g^{(k)}: g \in G\}$. Then G is called *k-closed*, if and only if

$$\Gamma(G^{(k)}) = \bigcup \{\Gamma(g^{(k)}): g \in G\}, \quad (\Gamma(G^{(k)}) \subset X^{(k)} \times X^{(k)}) \tag{5}$$

is *closed* in $X^{(k)} \times X^{(k)}$. In other words: *G is k-closed, iff* $G^{(k)}$ *is* 1-*closed.* If G is k-closed, and $l \leqslant k$, then it is l-closed, in particular it is always 1-closed, hence trajectories (3) are closed in X. Proof: intersect (5) with the closed subspace $x_l = \cdots = x_k$. We will see later nontrivial cases where 2-closedness implies k-closedness for all $k \geqslant 1$. It is also instructive to consider, as examples, the affine groups acting on R^n; as for nontrivial results concerning affine groups, see [36]. In terms of sequential convergence, if this applies, k-closedness is the following property:

---

* Correction of the terminology of [10]: "k-rigid" should be replaced by "k-closed" throughout. (I did not have [51] on hand when I wrote [10], and misused this term. Of course, "k-closed" is the natural and the traditional term.)

$$(6)\quad \begin{cases} \text{if } y_{\alpha i}=g_\alpha x_{\alpha i},\ i=1,\dots,k, \text{ and } x_i=\lim\limits_{\alpha\to\infty} x_{\alpha i},\ y_i=\lim\limits_{\alpha\to\infty} y_{\alpha i},\ i=1,\dots,k, \\ \text{then there exists a } g\in G, \text{ such that } y_i=g x_i,\ i=1,\dots,k \text{ holds true.} \end{cases}$$

For metric spaces, k-closedness can be formulated in terms of G-congruence up to ε: the conclusion in (6) amounts to requiring $\{x_1,\dots,x_k\}$ G-congruent to $\{y_1,\dots,y_k\}$, the condition amounts to requiring G-congruence up to ε, for all $\varepsilon>0$, see [12, 51].

If d is a metric for X, we say that G is *d-equicontinuous at* $x_0\in X$, if there is a function $\delta(x,\varepsilon)>0$ defined for $x=x_0$, and $\varepsilon>0$, such that $d(x_0,x_1)<\delta(x_0,\varepsilon)$ implies $d(g x_0, g x_1)<\varepsilon$, for all $g\in G$. If the equicontinuity modulus $\delta(x,\varepsilon)$ is defined for all $x\in X$, we say that G is d-equicontinuous; our condition is sometimes called "pointwise-equicontinuity". Following [19] we say that a homeomorphism g is *regular* at x if $G=\{g^n: n\in Z\}$ is equicontinuous at x. If d_i, $i=1,2$, induce the same topology, but different uniformities, d_i-equicontinuity may be different for $i=1,2$. Furthermore, if x is a regular point of G, hgh^{-1} may fail to be regular at hx, where $h: X\to X$ is also a homeomorphism. If X is locally compact, we avoid this circumstance by choosing an "extreme uniformity" defined by a metric extending to a metric of the one point compactification of X.

If G is a group, $G\times X\to X$, $(g.x)\to gx$ is a map such that $ex=x$ for all x and the neutral element $e\in G$, $g(hx)=(gh)x$ for all g, h, x, and the maps $x\to gx$ are continuous, we say that G acts on X; if G has a topology, we require $G\times X\to X$ to be continuous. G is effective iff it is a group of homeomorphisms; we will *always* suppose this. We say that

$$(7)\quad \begin{cases} \text{the action of } G \text{ in } X \text{ is topologically rigid, if given different} \\ \text{points } x, y, z \text{ there are neighborhoods } U_x, U_y, U_z \text{ of } x, y, z \text{ re-} \\ \text{spectively, such that for every } g\in G \text{ at least one of the sets} \\ g U_z\cap U_x,\ g U_z\cap U_y \text{ is empty.} \end{cases}$$

We refer to [51], p. 211–218, for a discussion of this definition, in particular with regard to quasi rotations ($\theta=\theta(\xi^2+\eta^2)$, $\alpha=\beta=0$ in (8)). For compact, metric space X, the relationship between k-closedness, equicontinuity, rigidity, and compactness of G is quite clear, in the more general case one has to be careful in distinguishing these concepts; a number of non-trivial questions arise. Let us recall only one result of [5] following the wording in [51], p. 213:

Theorem of van Dantzig and van der Waerden. If X is a locally compact, locally connected, and connected space satisfying the second countability axiom, and the action of G in X is topologically rigid, then the set $\bar{G}$ of homeomorphisms $X\to X$ which are sequential limits of elements of G is a locally compact transformation group of X satisfying the second countability axiom; $\bar{G}$ is topologically rigid, and k-closed for every $k \geqslant 1$; $\bar{G}$ is closed in the group H of all homeomorphisms $X \to X$ taken with the compact-open topology. If $\{g_n x_0\}$, $g_n \in \bar{G}$, has a cluster point in X, $\{g_n\}$ has a cluster point in $\bar{G}$. If $G(x_0)$ is compact for some $x_0 \in X$, then $\bar{G}$ is compact.

A locally compact transformation group need not be topologically rigid, of course (see [36] for affine maps); hence regidity, k-closedness, or equivalent concepts are necessary in topological foundations of geometry. In view of the extensive theory of locally compact groups, the result above is clearly important.

For given X, let us denote by H the group of all ontohomeomorphisms $X \to X$ with the compact-open topology. The most restrictive equivalence relation for transformation groups $G \subset H$ is conjugacy by an element of H: G_1, G_2 are equivalent, if $G_2 = h^{-1} G_1 h$ for some $h \in H$. For example, let $X = R^2$, G_E the group of sense preserving isometries

$$\text{(8)} \qquad \begin{cases} \xi' = \xi \cos\theta + \eta \sin\theta + \alpha \\ \eta' = -\xi \sin\theta + \eta \cos\theta + \beta . \end{cases}$$

If $f \in H$ is conjugate to $g \in G_E$ given by (8) in Cartesian coordinates, then f is expressed by the same functions (8) in appropriate coordinates. In the general case, conjugacy in H conserves all properties of maps and sets of maps which can be termed topological; the most precise results concerning transformation groups are thus formulated in terms of conjugacy classes. To prove that two transformation groups are conjugate is more difficult than to prove that two spaces are homeomorphic, in some sense. For example, the homeomorphism problem for surfaces has been solved long ago [17], but the homeomorphism problem for surface transformations is not solved yet. The main results of some papers in the Bibliography are solutions of special cases of this problem, see [1, 2], ([5, 6, 7]), [8, 9, 10, 12, 15, 17, 18, 19, 20, 21], ([26]), [32], ([36, 38]), [40, 41, 44, 45], ([46]), [47, 48], ([50]), [51]. (general theories in parentheses).

3. Main Results. As usual we use *surface* to mean a metric C^0, 2-manifold without boundary. We proved in [10]:

Theorem A. *If G acts equicontinuously on the surface M, then M carries a Riemann metric of constant curvature in which G is a group of isometries.*

The emphasis in this theorem is, of course, on *constant curvature* (compare with [10, 24]). This theorem solves the homeomorphism problem for groups acting equicontinuously on surfaces, because we may consider known *all subgroups* of the group of isometries of *geometries with constant curvature* (see Theorem B below). We do not want to elaborate more on this aspect of Theorem A, because it would be somewhat long-winded.

Let H be the group of all surjective homeomorphisms $R^2 \to R^2$ with the compact-open topology, and H_+ the closed, open, invariant subgroup of orientation preserving h's. Let G_E be the group of all sense preserving isometries (8) of the usual Euclidean metric. If $f: R^2 \to B$ is a homeomorphism of R^2 onto the plane B of the Bolyai geometry, or geometry of constant curvature $\kappa < 0$, and L is the group of all orientation preserving isometries of B, then $G_B = f^{-1} L f$ is a subgroup of H_+ determined up to conjugacy in H. The groups G_E and G_B are three dimensional, the space of each is homeomorphic to $S^1 \times R^2$, and are distinguished by a purely group theoretical property: G_E contains an invariant abelian subgroup R^2, the group of translations ($\theta = 0$ in (8)); G_B, however, has no such subgroup. The stereographic projection of R^2 onto $S^2 - \omega$ defines a metric d on R^2; *this is the only metric we will consider when dealing with equicontinuity*. All subgroups of G_E, G_B are d-equicontinuous, hence, if G acts on R^2, and either

(9) $$h^{-1} G h \subset G_E \quad \text{or} \quad k^{-1} G k \subset G_B \qquad (h, k \in H)$$

holds true, then G is d-equicontinuous. Excluding the cases $M = S^2, P^2$, a rather trivial application of the theory of covering spaces shows that Theorem A is equivalent to the following result:

Theorem B. *If G acts d-equicontinuously on R^2, then it is conjugate to a subgroup of G_E or to a subgroup of G_B; that is to say* (9) *holds true.*

In other words, the only way to get d-equicontinuous groups acting on R^2 is to conjugate a subgroup of G_E or G_B. For some subgroups, for example for subgroups generated by an element, both inclusions in (9) are possible. This is not the case when

(10) $$h^{-1} G h = G_E \quad \text{or, else} \quad k^{-1} G k = G_B \qquad (h, k \in H)$$

holds true.

We formulate now some statements equivalent to Theorem B. To simplify the discussion, we will always suppose

(11) $$G \subset H_+$$

and we make one of the following additional hypotheses:

(12) *G acts d-equicontinuously on R^2,*

(13) *G is topologically rigid,*

(14) *G is 2-closed.*

A group (11) which satisfies one of the conditions (12), (13), (14) *is contained in a group maximal with respect to this property.* A maximal group of this sort is conjugate to G_E or to G_B. Similar statement: a group (11) satisfying one of the conditions (12), (13), (14) is contained in a group of the same sort of dimension 3; if $\dim G=3$, (10) holds true.

Let us formulate now *special cases* of Theorem B. It is easy to see that G_E, G_B have no proper closed subgroups for which $G_x(y)$ is infinite for every pair of points x, y $(y \neq x)$. In fact there is no proper closed subgroup such that

(15) $\quad G_x(y)$ *is infinite for* $(x,y)=(x_0,y_0), (x_1,y_1), \quad x_0 \neq x_1.$

Supposing thus (11), (14), (15), Theorem B gives (9), hence (10). We see thus that Theorem B gives a known sharpening of

Hilbert's Theorem. *If the action* $G \subset H_+$ *is 3-closed, and* $G_x(y)$ *is infinite for all pairs of points* x, y $(y \neq x)$, *then G is the group of all orientation preserving isometries of a Euclidean or Bolyai geometry.*

Remark. As a point of fact, Hilbert formulated his theorem for a G acting on an open connected subspace of R^2. This case is, of course, covered by Theorem A. See also [37].

Although Hilbert's Theorem is a special case of Theorem B, it is an important tool in the proofs of [10], and it is not proven anew there. Among other classically known special cases of Theorem A, we mainly use Hilbert's Lemma, and Kerékjártó's Theorem formulated below.

Hilbert's Lemma. *Let* $G=G_a$ $(a \in R^2)$ *be a* 2*-closed group* (11) *such that* $G_a(x)$ *is infinite for some* x. *Then* $h^{-1}Gh$ *is the full rotation group* (8) $(\alpha=\beta=0)$ *for appropriate* $h \in H$.

Kerékjártó's Theorem. *If* $g \in H_+$ *is regular, i.e.* $\{g^n : n \in Z\}$ *satisfies* (12), *there is an* $h \in H_+$ *such that* $h^{-1}gh$ *is an isometry* (8).

Again, Hilbert's Lemma is formulated in § 20, p. 198 in [12] supposing 3-closedness and that (15) holds true for $x=a$ and for $y \neq a$. From Kerékjártó's theorem follows easily, however, that the original formulation and the Lemma above are equivalent. We need thus only § 1–§ 7 from Hilbert's paper, and Kerékjártó's theorem to prove completely the Lemma above. Discussion in [35] of three-dimensional analogous problems shows that we cannot hope for a much simpler conjoint proof of these results. See also [10]. We will formulate more special cases of Theorem B in Section 5, where we outline its proof.

4. On the proof of Hilbert's Theorem. Hilbert's original proof [12] of his theorem is not outdated because it is direct, elementary, and potentially an axiomatic discussion of the subject. He used only the Jordan curve theorem, Cantor's theorem on the order type of R, and similar basic results without proof. The proof of his Lemma contains a method of proving Schoenfliess' theorem on the converse of the Jordan curve theorem, see [41]. On the other hand, his proof is very long and arduous, hence it may be of some interest to give a proof using modern tools.

In what follows, we relax the conditions in Hilbert's Theorem, supposing only (11), (14), (15). We define the map

(16) $$f: G \to R^2 \times R^2, \ f(g) = (g x_0, g x_1) \qquad \text{(see (15))}.$$

Then $f(g) = f(h)$ implies that x_0, x_1 are fixed points of $h^{-1} g$, hence this map is the identity by Kerékjártó's Theorem: f is an injection. By (14) fG is a closed subspace of R^4; also, fG is a topological transformation group if it is taken with the induced topology. (We do *not* use here the van Dantzig, van der Waerden theorem, but the topologies obtained are identical, of course.) In the sequel we take G with the topology defined by f^{-1}. Let us introduce now the map

(17) $$p: G \to R^2, p(g) = g x_0, \qquad (g \in G).$$

By Hilbert's Lemma, we may suppose that G_{x_0} is the rotation group (8), with $\alpha = \beta = 0$. Let us denote G^e the component of e in G. $G^e_{x_1}(x_0)$ is a Jordan curve passing through x_0, hence its images under the maps G_{x_0} cover a neighborhood of x_0; every point of this neighborhood belongs to pG^e. Hence pG^e is an open set. Also, by (14), pG^e is closed, hence $pG^e = R^2$. Given $x \in R^2$, $x = g x_0$, and $G_x = g G_{x_0} g^{-1}$, hence G_x is $\cong S^1$; in other notation, $p^{-1} x$ is homeomorphic to S^1. At the same time $(p|G^e)^{-1}(x)$ is homeomorphic to S^1, hence $p^{-1} x = (p|G^e)^{-1} x$ for all x, thus $G = G^e$, and G is connected. G contains 1-dimensional compact subgroups; let K be a maximal compact subgroup of G; K is connected by Iwasawa's theorem [14, 35], p. 188, and the map

(18) $$\varphi: K \times R^l \to G, \ \varphi(k, h_1, \ldots, h\ ) = h_1, \ldots, h_l\, k$$

is a homeomorphism; $\dim K = m \geqslant 1$. By [39], p. 216, Example 59, there is an m-dimensional Lie group L and a monomorphism of abstract groups $\psi: L \to K$ such that: ψ is continuous, but ψ^{-1} is not necessarily continuous (hence ψ is *not* open); ψL is everywhere dense in K. Let $M \subset L$ be a one parameter subgroup, $a \in M$. Then $\psi a \in K$, hence ψa is conjugate to a rotation of R^2 by Kerékjártó's theorem. $\psi a^{1/2}$ must also be a rotation with the same center as ψa; taking rational powers of a, we conclude $\psi M = G_x$, for some x. Similarly, for another one parameter subgroup M' of L, $\psi M' = G_y$ for some y. Supposing $y \neq x$, the argument after (16) shows that $p(K)$ is an open and closed set, which is a contradiction. Hence $L = M \cong S^1$, thus $K \cong S^1$. (18) shows now by [35], p. 184, Theorem, that G is a Lie group. We also know that G is homeomorphic to $S^1 \times R^2$. We can thus introduce, in the usual manner, a Riemann metric of R^2 invariant under G; this metric has, of course, a constant curvature. Also, once we have shown that G is a Lie group, there are many other ways to complete the proof. Let us only add, that once the theory of locally compact groups has been applied, the end of the proof of Hilbert's theorem is, essentially, Lie's original work.

One word about this proof. There is a more general form of Kerékjártó's theorem, which concerns maps regular up to a finite number of points [20]. We feel that the result of [20] gives the best information in two dimensions, and should not be avoided. This theorem, however, is not true in higher dimensions (see [35], p. 229), hence, in order to use general principles in the proof above, it could be replaced by the van Dantzig, van der Warden theorem. After these changes, it is even clearer that the proof hinges on Hilbert's Lemma and on the general structure theory of locally compact groups.

5. On the proof of Theorem A. It is in the nature of Theorem A that the theory of locally compact groups cannot be of real help throughout the proof. In [10] we discuss separately, and with different methods, the cases corresponding to $\dim G$ being 0, 1, 2, and $\geqslant 3$.

Case $\dim G = 0$. In this case we use Brouwer's theorem on finite groups acting on S^2, see [17], as well as some results of Kerékjártó, in order to describe the quotient space R^2/G. We prove that the set of fixed points of $g \in G$ is discrete in R^2 and its image in R^2/G. Then we show that the quotient space is a surface. The construction of the invariant metric of constant curvature is then quite simple; in fact, we show that there is an invariant complex structure for orientable M and orientation preserving g's.

Case $\dim G = 1, 2$. Classical results of Brouwer, in particular of his solution of "Hilbert's V-th" for two-dimensional groups, come into play. It is worthwhile to repeat this: Theorem A contains the solution of "Hilbert's V-th" for groups of dimension $\leqslant 2$, when applied to the maps $x \to gx$ of a group $G (x, g \in G)$. Also, we have to deal with groups, which are less usual, for example, groups (8) with $\theta \equiv k(2\pi/6) \pmod{2\pi}$; we have groups containing both translations and rotations in Bolyai geometry too.

Finally, in case $\dim G \geqslant 3$, some results from dimension theory, Hilbert's Lemma and Hilbert's theorem give the necessary information.

References

1. Brouwer, L. E. J.: Die Theorie der endlichen kontinuierlichen Gruppen, unabhängig von den Axiomen von Lie, Math. Ann. **67**, 246—267 (1909).

2. — Beweis des ebenen Translationssatzes, Mathematische Annalen, **72,** 37—54 (1912).
3. CAMERON: Almost Periodic Transformations, Trans. Amer. Math. Soc., **36,** 276—291 (1934).
4. CARTAN, É.: La topologie des groupes de Lie, Actualités scientifiques et industrielles, Vol. 8.
5. VAN DANTZIG, D., and B. L. VAN DER WAERDEN: Über metrische homogene Räume, Abhandlungen Hamburg, **6,** 367—376 (1928).
6. DIEUDONNÉ, J.: On topological groups of homeomorphisms, Amer. Journ. of Math., **70,** 659—680 (1948).
7. EISENHART, L. P.: Continuous Groups of Transformations, Princeton, Univ. Press, 1933.
8. FÁRY, I.: Sur les groupes d'homéomorphismes également continus du plan, Comptes rendus de l'Académie, Paris, **228,** 534—536 (1949).
9. — Sur la dimension des groupes d'homéomorphismes également continus du plan, Comptes rendue de l'Académie, Paris, **228,** 801—803 (1949).
10. — On Topological Characterization of Plane Geometries, Cahiers du Seminaire dirigé par Ch. Ehresmann, **7,** 1—33 (1964).
11. HILBERT, D.: Grundlagen der Geometric, 4th ed., Teubner, Berlin 1913.
12. — Anhang IV, Über die Grundlagen der Geometrie, in Grundlagen der Geometrie, 4th ed., Teubner, Berlin, 1913.
13. HUREWICZ, W., and H. WALLMAN: Dimension Theory, Princeton, University Press, 1948.
14. IWASAWA, K.: On some types of topological groups, Ann. of Math. **50,** 507—557 (1949).
15. KAPLAN, S.: Regular curve families filling the plane, I, Duke Math. Journ., **7,** 154—185 (1940); Part II, ibid., **8,** 11—46 (1941).
16. KELLY, J. L.: General Topology, New York, van Nostrand, 1955.
17. KERÉKJÁRTÓ, B. von: Vorlesungen über Topologie, I, Springer, Berlin 1923.
18. — On a geometrical theory of continuous groups, II, Ann. of Math., **29,** 169—179 (1928).
19. — Über die fixpunktfreien Abbildungen der Ebene, Acta Szeged **6,** 226—234 (1934).
20. — Topologische Characterisierung der linearen Abbildungen, Acta Szeged, **6,** 235—262 (1934).
21. — Sur la structure des transformations topologiques des surfaces en elles mêmes, L'enseignement mathématique, **35,** 297—316 (1936).
22. KOLMOGOROFF, A.: Zur topologisch-gruppentheoretischen Begründung der Geometrie, Nach. Akad. Wiss., Göttingen Math. Phys. Klasse, 1930, 208—210.
23. — Über offene Abbildungen, Ann. of Math., **38,** 36—39 (1937).
24. MAYER, S. B., and N. E. STEENROD: The group of isometrics of a riemannian manifold, Ann. of Math., **40,** 400—416 (1939).
25. MONTGOMERY, D.: Continuity in topological groups, Bull. Amer. Math. Soc., **42,** 879—882 (1936).
26. — Pointwise periodic homeomorphisms, Amer. Journ. Math. **59,** 118—120 (1937).

27. — Almost periodic transformation groups, Trans. Amer. Math. Soc., **42,** 322—332 (1937).
28. —, and L. ZIPPIN: Periodic one-parameter groups in three-space, Trans. Amer. Math. Soc., **40,** 24—36 (1936).
29. — Translation groups in three-space, Amer. Journ. Math., **49,** 121—128 (1937).
30. — Compact abelian transformation groups, Duke Math. Journ., **4,** 363—373 (1938).
31. — Non-abelian compact, connected groups of three-space, Amer. Journ. Math., **61,** 375—387 (1939).
32. — A theorem on the rotation group of the two-sphere, Bull. Amer. Math. Soc., **46** (1940).
33. — Topological transformation groups I, Ann. of Math., **41,** 778—791 (1940).
34. — Topological group foundations of rigid space geometry, Trans. Amer. Math. Soc., **48,** 21—49 (1940).
35. — Topological transformation groups, Interscience Publ. New York, N. Y., 1955.
36. MOORE, C.: Distal affine transformation groups, To appear later in Amer. Journ. of Math.
37. MOORE, R. L.: On the Lie-Riemann-Helmholtz-Hilbert problem, Amer. Journ. Math., **41,** 299—319 (1919).
38. NEWMAN, M. H. A.: A theorem on periodic transformations of spaces, Quart. Journ. of Math., Oxford Series, **2,** 1—8 (1931).
39. PONTRYAGIN, L.: Topological groups, Princeton University Press, 1939.
40. RADÓ, T.: Über den Begriff der Riemannschen Fläche, Acta Szeged, **2** (1925).
41. RIESZ, F.: Über einen Satz der Analysis Situs, Math. Annalen, **59,** 409—415 (1904).
42. SAMELSON, H.: Topology of Lie groups, Bull. Amer. Math. Soc., **58,** 2—37 (1952).
43. SEIFERT, H.: Topologie dreidimensionaler gefaserter Räume, Acta Math., **60,** 147—238 (1932).
44. STOÏLOW, S.: Leçous sur les principes topologiques de la théorie des fonctions analytiques, Gauthier-Villards Ed., Paris, 1938.
45. TERASAKA, H.: Ein Beweis des Brouwerschen ebenen Translationssatzes, Japanese Journ. of Math. **7,** 61—69 (1930).
46. WEYL, H.: Die Idee der Riemannschen Fläche, Berlin, 1913.
47. WHITNEY, H.: Regular families of corves, Ann. of Math. **34,** 270 (1933).
48. — Cross sections of curves in three-space, Duke Math. Journ., **4,** 222—226 (1938).
49. WHYBURN, G. T.: Interior surface transformations, Duke Math. Journ. **4,** 626—634 (1938).
50. WILDER, R. L.: Topology of manifolds, Amer. Math. Soc. Publ., New York, 1949.
51. ZIPPIN, L.: Transformation groups, Wilder Editor, Lectures in Topology,

On the Variation of Isotropy Subalgebras

ROGER RICHARDSON

Introduction

If the compact Lie group G acts continuously on a manifold M, then there are a number of theorems which give detailed information on the behavior of isotropy subgroups. In particular, there are the conjugacy theorem of MONTGOMERY-ZIPPIN [13, p 251], the theorem on the existence of a slice of MONTGOMERY-YANG [11] and MOSTOW [14], and the principal orbit theorem of MONTGOMERY-SAMELSON-YANG [9, 12]. If, in addition, M is compact, there is the theorem of FLOYD [4] and MOSTOW [15] on the finiteness of the set of orbit types. However, simple examples involving linear representations show that none of these theorems is valid for the case of a non-compact Lie group G acting differentiably on a differentiable manifold M (although Palais [17] has proved the existence of a slice under the severe restriction that G acts properly on M). For example, there exist a rational representation of a semi-simple real algebraic group G on a vector space V and a locally closed algebraic submanifold M of V which is stable under G such that all orbits of G on M have the same dimension and that there is an uncountable family of pairwise non-isomorphic (and hence non-conjugate) isotropy subgroups for the action of G on M.

Indeed, in the case of non-compact G, virtually nothing seems to be known about the behavior of isotropy subgroups. In this paper we prove some beginning results in this direction. Unfortunately, our results are given for the isotropy subalgebras rather than the isotropy subgroups, since the behavior of the isotropy subalgebras is more tractable.

Let the Lie group G act differentiably on the differentiable manifold M and let $\mathfrak{g}$ be the Lie algebra of G. Let M_n be the set of $x \in M$ such that the orbit $G(x)$ is n-dimensional and, for each $x \in M$, let $\mathfrak{g}_x$ be the isotropy subalgebra at x. Then we show that the isotropy subalgebras $\mathfrak{g}_x$ depend continuously (with respect to the appropriate topology) on x. Furthermore, the isotropy subalgebras vary differentiably on each M_n. One interesting consequence of these results is

Proposition 3.2. *Let $x \in M$ and let $E(x)$ be the set of all $z \in M$ such that $\mathfrak{g}_z$ is conjugate to $\mathfrak{g}_x$. Let y belong to the closure of $E(x)$. Then $\mathfrak{g}_y$*

contains a subalgebra which is a contraction of $\mathfrak{g}_x$ *in* $\mathfrak{g}$. (See 3.2 for the definition of a contraction.)

All of the results above are of an elementary nature. However, when combined with the results of [18, 19] on deformations of subalgebras, they yield a number of interesting consequences. As a typical example we have.

Let the semi-simple Lie group G act differentiably on M and let $x \in M_n$ *be such that* $\mathfrak{g}_x$ *contains a Cartan subalgebra of* $\mathfrak{g}$. *Then there exists a neighborhood U of x in* M_n *such that, if* $y \in U$, *then* $\mathfrak{g}_y$ *is conjugate to* $\mathfrak{g}_x$.

I would like the thank BOB HERMANN for this many contributions to this paper. He stimulated my interest in these questions, he conjectured Proposition 3.2 and he corrected a number of mistaken ideas of my own on this subject.

1. Examples

In this section we give some simple examples of non-compact transformation groups. Although the behavior of the isotropy subgroups may seem pathological to one used to working exclusively with compact transformation groups, we would like to emphasize that these examples represent typical behavior in the non-compact case.

1.1 Let $G = SL(2, R)$ act on the Riemann sphere S by fractional linear transformations in the usual manner. Then there are three orbits, the upper and lower hemispheres and the equator. If x lies in either the upper or lower hemisphere then the isotropy subgroup G_x is a circle subgroup of G and hence is compact. If x is on the equator, then G_x is a simply connected two-dimensional solvable subgroup of G and hence G_x does not contain any compact subgroups. Consequently, the conjugacy theorem for isotropy subgroups of MONTGOMERY-ZIPPIN does not hold here. Note also that a lower dimensional orbit, the equator, separates S, which contradicts one of the conclusions of [10] for compact transformation groups.

1.2 Let Q denote the quadratic form on R^4 given by

$$Q(x_1, \ldots, x_4) = x_1^2 + x_2^2 + x_3^2 - x_4^2$$

and let G be the group of automorphisms of Q with determinant 1; G is the Lorentz group $SO(3, 1)$. We let G act on R^4 by the identity representation. For each $t \in R$, let $X_t = \{x \in R^4 | Q(x) = t\}$ and let $X'_0 = X_0 - \{0\}$. Then the orbits of G on R^4 are $\{0\}$, X'_0 and the X_t's for $t \neq 0$. If $x \in X_t$ and $t < 0$, then G_x is conjugate to the rotation group $SO(3)$ (embedded in the obvious way in $SO(3, 1)$), which is a compact connected semi-simple Lie group. If $x \in X_t$ with $t>0$, then G_x is conjugate to $SO(2,1)$ which is non-compact, semi-simple and has two components. If $x \in X'_0$, then G_x is a three-dimensional solvable connected Lie group, which is isomorphic to the group of rigid proper motions of the plane. We note that all orbits of G on $R^4 - \{0\}$ are three-dimensional.

Let $t \to x_t$ be a differentiable curve in $\boldsymbol{R}^4 - \{0\}$ such that $x_t \in X_t$ for every t. For $t>0$ (or $t<0$), all isotropy subgroups G_{x_t} are conjugate, but the isotropy subgroups "jump" at $t=0$. If x_t approaches x_0 through negative values of t, then the isotropy subgroups G_{x_t}, which are compact, approach as a "limit" G_{x_0} which is non-compact. If x_t approaches x_0 through positive t, then the second component of the isotropy subgroup G_{x_t} "disappears" at x_0. (This cannot happen in the case of compact Lie transformation groups, since, if all orbits are of the same dimension, the number of components of G_x is an upper semi-continuous function of x.)

1.3 An easy argument shows that if a compact Lie group acts linearly on Euclidian space $\boldsymbol{R}^n$, then there is only a finite number of orbit types. However, this is no longer true for non-compact Lie groups, even for the case of algebraic subgroups of $GL(n,\boldsymbol{R})$ acting on $\boldsymbol{R}^n$. In this case the isotropy subgroups can vary "continuously" through an infinite family of non-conjugate subgroups. The following example was pointed out by MOSTOW [14]. Let G be the abelian linear algebraic subgroup of $GL(3,\boldsymbol{R})$ consisting of all matrices of the form

$$M(a,b) = \begin{pmatrix} 1 & a & b \\ 0 & 1 & 0 \\ 0 & 0 & 1 \end{pmatrix}.$$

For $t \in \boldsymbol{R}$, let v_t denote the vector $(0,1,t)$. Then the isotropy subgroup G_{v_t} consists of all $M(a,b)$ such that $a+bt=0$. Thus all isotropy subgroups G_{v_t} are distinct and, since G is abelian, non-conjugate.

In the above example all isotropy groups G_{v_t} are conjugate under the group of outer automorphisms of G. However, by a simple modification of this example, we can obtain a linear action of a Lie group with an infinite family of pairwise non-isomorphic isotropy subgroups. Let $\mathfrak{s}$ be the subalgebra of the Lie algebra $\mathfrak{gl}(3,\boldsymbol{R})$ of all 3×3 real matrices spanned by the matrix units E_{11}, E_{33}, E_{12} and E_{13} and let S be the corresponding analytic subgroup of $GL(3,\boldsymbol{R})$; S is simply connected. We define a representation $d\rho: \mathfrak{s} \to \mathfrak{gl}(3,\boldsymbol{R})$ by $d\rho(E_{12}) = d\rho(E_{13}) = 0$, $d\rho(E_{11}) = E_{12}$ and $d\rho(E_{33} = E_{13}$. Let $\rho: S \to GL(3,\boldsymbol{R})$ be the corresponding Lie group representation and let v_t be as above. Then for the action of S on $\boldsymbol{R}^3$ defined by ρ, the isotropy subalgebra $\mathfrak{s}_{v_t}$ is spanned by E_{12}, E_{13} and $tE_{11}-E_{33}$. It follows from the classification of three-dimensional Lie algebras [7, p. 12] that $\mathfrak{s}_{v_t}$ and $\mathfrak{s}_{v_s}$ are isomorphic if and only if $st=1$. Thus we have an infinite family of pairwise non-isomorphic (and hence non-conjugate) isotropy subgroups.

The above action of S on $\boldsymbol{R}^3$ is not effective. However, if we let τ denote the direct sum of ρ and the identity representation of S on $\boldsymbol{R}^3$, then the action of S on $\boldsymbol{R}^6$ determined by τ is effective. Furthermore, if $w_t=(v_t,0)\in\boldsymbol{R}^3\times\boldsymbol{R}^3$, then $S_{w_t}=S_{v_t}$ and consequently there are an infinite number of pairwise non-isomorphic isotropy subgroups.

Similar examples exist for linear actions of semi-simple groups, although it seems to be difficult to write them out explicitly. We shall outline an "existence proof" for such examples. It is known (see, e. g., [1, p. 122, Ex. 18]) that the set of isomorphism classes of nilpotent Lie algebras is uncountable. By Ado's theorem, every nilpotent Lie algebra $\mathfrak{n}$ admits a faithful representation $\rho:\mathfrak{n}\to\mathfrak{gl}(n,\boldsymbol{R})$ such that $\rho(x)$ is a nilpotent matrix for every $x\in\mathfrak{n}$; hence $\mathfrak{n}'=\rho(\mathfrak{n})\subset\mathfrak{sl}(n,\boldsymbol{R})$. It follows from [3, p. 123, Prop. 14] that $\mathfrak{n}'$ is an algebraic (Lie) subalgebra of $\mathfrak{gl}(n,\boldsymbol{R})$. Furthermore, it is a consequence of [2, p. 84, Thm. 1] that $\mathfrak{n}'$ occurs as an isotropy subalgebra for a linear representation of $SL(n,\boldsymbol{R})$. It is not difficult to show that this implies that, for sufficiently large n, there exists a (rational) linear representation $\tau: SL(n,\boldsymbol{R})\to GL(m,\boldsymbol{R})$ such that an uncountable number of pairwise non-isomorphic nilpotent subgroups of $SL(n,\boldsymbol{R})$ occur as isotropy subgroups for the corresponding linear action. Moreover, it can be shown by elementary algebraic geometry that there exists a locally closed (in the Zariski topology) algebraic submanifold M of $\boldsymbol{R}^m$ which is stable under the action of $SL(n,\boldsymbol{R})$ and is such that all orbits of $SL(n,\boldsymbol{R})$ on M are of the same dimension and that there is an uncountable number of pairwise non-isomorphic isotropy subgroups for the action of $SL(n,\boldsymbol{R})$ on M.

2. The Variation of $\mathfrak{g}_x$

2.1 *Notation.* All differentiable manifolds and differentiable maps will be assumed to be of class C^∞. If M is a differentiable manifold and $x\in M$, then $T(M,x)$ denotes the tangent space of M at x. If V is a finite-dimensional real vector space, then $GL(V)$ is the Lie group of automorphisms of V and $\mathfrak{gl}(V)$ is the Lie algebra of endomorphisms of V.

Let G be a Lie group which acts differentiably on a differentiably manifold M and let $\mathfrak{g}$ be the Lie algebra of G. If $x\in M$, then $G(x)$ denotes the orbit of x under G, G_x denotes the isotropy subgroup of G at x and $\mathfrak{g}_x$ denotes the Lie algebra of G_x; $\mathfrak{g}_x$ is the *isotropy subalgebra* of $\mathfrak{g}$ at x. For each non-negative integer n, let M_n denote the set of points of M

which lie on n-dimensional orbits of G (here $\dim G(x)$ is the dimension of $G(x)$ as a submanifold of M, i. e., $\dim G(x)=\dim G-\dim G_x$).

2.2 *Variation of* $\mathfrak{g}_x$. For each $x\in M$, let $\varphi_x\colon G\to M$ be defined by $\varphi_x(g)=g\cdot x$ and, to simplify notation, let $T_x\colon \mathfrak{g}\to T(M,x)$ denote the differential of φ_x at the identity element $e\in G$. Then it is well known (and easy to prove) that $\mathfrak{g}_x$ is the kernel of T_x. Now let $\alpha\colon U\to V$ be a diffeomorphism of an open subset U of M onto an open subset V of m-dimensional Euclidian space $\boldsymbol{R}^m$. For each $x\in U$, let $S_x\colon \mathfrak{g}\to T(V,\alpha(x))$ be the differential of the composite map $\alpha\circ\varphi_x$. If, for every $v\in V$, we identify $T(V,v)$ with $\boldsymbol{R}^m$ in the usual way, then S_x is a linear map of $\mathfrak{g}$ into $\boldsymbol{R}^m$. It follows easily from the fact that G acts differentiably on M that $x\to S_x$ is a differentiable map of U into the vector space $L(\mathfrak{g},\boldsymbol{R}^m)$ of all linear maps of $\mathfrak{g}$ into $\boldsymbol{R}^m$. Moreover, since the differential $d\alpha_x$ is an isomorphism for every $x\in U$, the kernel of S_x is just the isotropy subalgebra $\mathfrak{g}_x$. Thus we see that, roughly speaking, $\mathfrak{g}_x$ varies with x in the same way that the kernel of a linear transformation T varies with T. As an immediate consequence of these remarks, we have

Proposition 2.1. *For each* n, $\bigcup_{p\geqslant n} M_p$ *is an open subset of* M.

The proof follows from the fact that the set of $T\in L(\mathfrak{g},\boldsymbol{R}^m)$ such that dim kernel $(T)\leqslant n$ is an open subset of $L(\mathfrak{g},\boldsymbol{R}^m)$ and from the continuity of the map $x\to S_x$.

2.3 *Variation of the kernel of a linear map.* In this section we shall show that the kernel of a linear map T of fixed rank depends differentiably on T. Although this result is well-known, it is hard to give a satisfactory reference. Since our proof is phrased in terms of transformation groups, it seems appropriate to include it here.

Let V and W be finite-dimensional vector spaces over $\boldsymbol{R}$, let $L(V,W)$ be the vector space of all linear maps of V into W and let $q=\dim V$. We denote by $L(V,W)_n$ the subset of $L(V,W)$ consisting of all maps of rank n, i. e., $T\in L(V,W)_n$ if image (T) is of dimension n. There is a natural action of the group $GL(W)\times GL(V)$ on $L(V,W)$ defined as follows: if $(g,h)\in GL(W)\times GL(V)$ and $T\in L(V,W)$, then $(g,h)\cdot T=g\circ T\circ h^{-1}$. Furthermore, the orbits of $GL(W)\times GL(V)$ on $L(V,W)$ are precisely the subsets $L(V,W)_n$. Since, for every n, the union of the $L(V,W)_p$ with $p\geqq n$ is an open subset of $L(V,W)$, it follows easily that each $L(V,W)_n$ is a locally closed submanifold of $L(V,W)$.

We denote by $\Gamma_n(V)$ the Grassmann manifold of n-dimensional subspaces of V. There is a natural transitive action of $GL(V)$ on $\Gamma_n(V)$ defined by $h\cdot Z=h(Z)$ for $h\in GL(V)$ and $Z\in\Gamma_n(V)$. We extend this to a transitive action of $GL(W)\times GL(V)$ on $\Gamma_n(V)$ by setting $(g,h)\cdot Z=h(Z)$

for $(g,h)\in GL(W)\times GL(V)$ and $Z\in\Gamma_n(V)$. Now let $T_0\in L(V,W)_n$ and let $W_0=\mathrm{kernel}(T_0)$. We define maps $\alpha_0: GL(W)\times GL(V)\to L(V,W)_n$ and $\beta_0: GL(W)\times GL(V)\to\Gamma_{q-n}(V)$ by $\alpha_0(g,h)=(g,h)\cdot T_0$ and $\beta_0(g,h)=(g,h)\cdot W_0$. If $k_n: L(V,W)_n\to\Gamma_{q-n}(V)$ denotes the map $T\to\mathrm{kernel}(T)$, then it is easy to check that $k_n\circ\alpha_0=\beta_0$. Let K denote the isotropy subgroup of $GL(W)\times GL(V)$ at T_0 for the action of $GL(W)\times GL(V)$ on $L(V,W)$. If H denotes the isotropy subgroup of $GL(V)$ at W_0, then the isotropy subgroup of $GL(W)\times GL(V)$ at W_0 is $GL(W)\times H$. Thus α_0 and β_0 determine diffeomorphisms

$$\alpha: (GL(W)\times GL(V))/K\to L(V,W)_n \quad \text{and}$$
$$\beta: (GL(W)\times GL(V))/(GL(W)\times H)\to\Gamma_{q-n}(V)$$

It is easy to verify that $K\subset GL(W)\times H$. Hence we have an induced map

$$\pi: (GL(W)\times GL(V))/K\to(GL(W)\times GL(V))/(GL(W)\times H).$$

Since $k_n\circ\alpha_0=\beta_0$, the following diagram is commutative.

$$\begin{array}{ccc}(GL(W)\times GL(V))/K & \xrightarrow{\pi} & (GL(W)\times GL(V))/(GL(W)\times H)\\ {\scriptstyle\alpha}\downarrow & & \downarrow{\scriptstyle\beta}\\ L(V,W)_n & \xrightarrow{k_n} & \Gamma_{q-n}(V)\end{array}$$

Since α and β are diffeomorphisms and π is differentiable, it follows that k_n is differentiable. Since $L(V,W)_n$ is a locally closed submanifold of $L(V,W)$, it follows (e. g., by taking a tubular neighborhood) that k_n can be extended to a differentiable map of an open neighborhood of $L(V,W)_n$ in $L(V,W)$ into $\Gamma_{q-n}(V)$.

Let S be a subset of M. A map ψ of S into a differentiable manifold N is differentiable if there exists an open neighborhood U of S in M and a differentiable map $\varphi: U\to N$ such that $\varphi(s)=\psi(s)$ for $s\in S$. The proposition below follows immediately from the results of this section and the previous one.

Proposition 2.2. *The map $x\to\mathfrak{g}_x$ is a differentiable map of M_n into $\Gamma_{d-n}(\mathfrak{g})$, where $d=\dim\mathfrak{g}$.*

Thus we see that the isotropy subalgebras vary differentiably on each M_n.

2.4 *The topology of $\Gamma(V)$.* Let V be a finite-dimensional real vector space and let $\Gamma(V)$ denote the union of the Grassmann manifolds $\Gamma_n(V)$,

$0 \leqslant n \leqslant \dim V$. We wish to define a topology on $\Gamma(V)$. Intuitively, a subspace X of V lies in a neighborhood of a subspace Y if the intersection of X with the unit sphere S of V (with respect to an inner product on V) lies in a neighborhood of $S \cap Y$ in S or, equivalently, if $P(X)$, the projective space corresponding to X, lies in a neigborhood of $P(Y)$ in $P(V)$. It is easy to make this intuitive definition precise. Let x_0 denote the subspace $\{0\}$ of V. For each open set U of $P(V)$, let

$$U^* = \{x_0\} \bigcup \{X \in \Gamma(V) | P(X) \subset U\}.$$

We let the sets of the form U^*, where U ranges over the open sets of $P(V)$, be a basis for the open sets in the topology of $\Gamma(V)$. If $X \in \Gamma(V)$, then all sets of the form U^*, where U is an open neighborhood of $P(X)$ in $P(V)$, form a basis for the neighborhood of X. We observe that the topology of $\Gamma(V)$ is non-Hausdorff. The following properties are easy to verify:

(2.4.a) *The topology of $\Gamma(V)$ induces the usual topology on each $\Gamma_n(V)$.*

(2.4.b) *If $X \in \Gamma(V)$ then the closure of $\{X\}$ consists of all $Y \in \Gamma(V)$ such that $X \subset Y$.*

We shall also need the following result:

(2.4.c) *Let (X_j) be a sequence in $\Gamma_n(V)$ such that $X_j \to X$ and $X_j \to Y$ with $X \in \Gamma_n(V)$. (Here the arrows denote convergence in $\Gamma(V)$.) Then $X \subset Y$.*

Proof: Assume that $q = \dim(X \cap Y) < n$. Then an easy argument shows that there exist open subsets U_1 and U_2 of $P(V)$ with $P(X) \subset U_1$ and $P(Y) \subset U_2$ such that $U_1^* \cap U_2^* = (U_1 \cap U_2)^*$ does not meet $\Gamma_r(V)$ if $r > q$. But for sufficiently large j, we have $X_j \in U_1^* \cap U_2^*$, which gives a contradiction. Hence $\dim X \cap Y = n = \dim X$. But this implies $X \subset Y$.

2.5 *Continuity of isotropy subalgebras.* Let V, W and $L(V, W)$ be as in 2.3. Let $k: L(V, W) \to \Gamma(V)$ denote the map $T \to \text{kernel}(T)$.

Proposition 2.3. *k is a continuous map.*

The proof is trivial and will be omitted.

As an immediate consequence of this proposition and the results of 2.2 we have

Proposition 2.4. *The map $x \to \mathfrak{g}_x$ is a continuous map of M into $\Gamma(V)$.*

Thus we have shown that the isotopy subalgebras $\mathfrak{g}_x$ depend continuously on x.

3. Deformations of Subalgebras

3.1 *Rigid subalgebras.* (See [18].) Let G and $\mathfrak{g}$ be as above and let $\mathscr{A}_n$ be the subset of $\Gamma_n(\mathfrak{g})$ consisting of all n-dimensional subalgebras of $\mathfrak{g}$. Then $\mathscr{A}_n$ is a Zariski closed subset of $\Gamma_n(\mathfrak{g})$. The adjoint representation $Ad: G \to GL(\mathfrak{g})$ defines an action of G on $\Gamma_n(\mathfrak{g})$ by $g \cdot W = (Ad\,g)(W)$ for $g \in G$ and $W \in \Gamma_n(\mathfrak{g})$, and $\mathscr{A}_n$ is stable under the action of G. Let $\mathfrak{a}$ be an n-dimensional subalgebra of $\mathfrak{g}$ and, to avoid confusing notation, let a denote the corresponding point of $\mathscr{A}_n$. Then $\mathfrak{a}$ is a *rigid* subalgebra of $\mathfrak{g}$ if the orbit $G(a)$ is an open subset of $\mathscr{A}_n$. (It follows in this case that $G(a)$ is a union of components of a Zariski open subset of $\mathscr{A}_n$ [18, P. 102, Prop. 11.2]. In particular, there is only a finite number of conjugacy classes of rigid subalgebra of $\mathfrak{g}$.) That is, $\mathfrak{a}$ is a rigid subalgebra if every subalgebra of $\mathfrak{g}$ near $\mathfrak{a}$ is conjugate to $\mathfrak{a}$. The following result (Corollary 11.5 of [18]) gives a sufficient condition for $\mathfrak{a}$ to be a rigid subalgebra in terms of Lie algebra cohomology.

If $H^1(\mathfrak{a}, \mathfrak{g}/\mathfrak{a}) = 0$, *then* $\mathfrak{a}$ *is a rigid subalgebra of* $\mathfrak{g}$.

(Here the $\mathfrak{a}$-module structure of $\mathfrak{g}/\mathfrak{a}$ is determined by the adjoint representation of $\mathfrak{a}$ on $\mathfrak{g}$.)

The result below is an immediate consequence of the continuity of the isotropy subalgebras.

Proposition 3.1. *Let* $x \in M_n$ *be such that the isotropy subalgebra* $\mathfrak{g}_x$ *is a rigid subalgebra of* $\mathfrak{g}$. *Then there exists a neighborhood* U *of* x *in* M_n *such that if* $y \in U$, *then* $\mathfrak{g}_y$ *is conjugate to* $\mathfrak{g}_x$.

We list below some examples from [18] of rigid subalgebras $\mathfrak{a}$ of a Lie algebra $\mathfrak{g}$.

a) $\mathfrak{a}$ is a semi-simple subalgebra of $\mathfrak{g}$.

b) $\mathfrak{a}$ is a Cartan subalgebra of $\mathfrak{g}$.

c) $\mathfrak{g}$ is a semi-simple Lie algebra and $\mathfrak{a}$ is a subalgebra of $\mathfrak{g}$ of maximal rank (i. e., $\mathfrak{a}$ contains a Cartan subalgebra of $\mathfrak{g}$).

d) $\mathfrak{a}$ is a reductive subalgebra of $\mathfrak{g}$ which is equal to its own normalizer in $\mathfrak{g}$.

We remark that the Proposition quoted at the end of the Introduction follows from c) and Proposition 3.1.

3.2 *Contractions of subalgebras.* (See [5, Chap. 11]). Let $L = (V, \mu)$ be a Lie algebra over $\boldsymbol{R}$ with underlying (finite-dimensional) vector space V and multiplication μ and let $\mathscr{M}$ denote the set of all Lie algebra multiplications on V; $\mathscr{M}$ is an algebraic set in the vector space $A^2(V, V)$ of all alternating bilinear maps of $V \times V$ into V. The group $G = GL(V)$ has a natural representation on $A^2(V, V)$ and $\mathscr{M}$ is stable under the corresponding action of G on $A^2(V, V)$ (see [16]). The orbits of G on $\mathscr{M}$

are just isomorphism classes of Lie algebras with underlying vector space V. A Lie algebra $L'=(V,\mu')$ is a *contraction* of $L=(V,\mu)$ if μ' lies on the closure of the orbit $G(\mu)$. Intuitively L' is a contraction of L if L' is a limit of Lie algebras isomorphic to L. More generally, if $L=(V,\mu)$ and $L_1=(W,\mu_1)$ are Lie algebras whose underlying vector spaces are isomorphic, then L_1 is a contraction of L if L_1 is isomorphic to a Lie algebra $L'=(V,\mu')$ which is a contraction of L. If L' is a contraction of L and if L is abelian (resp. nilpotent, solvable), then L' is abelian (resp. nilpotent, solvable).

In dealing with isotropy subalgebras, it is convenient to specialize the notion of contraction of Lie algebras. Let G be a connected Lie group with Lie algebra $\mathfrak{g}$ and let $\mathfrak{a}$ and $\mathfrak{b}$ be n-dimensional subalgebras of $\mathfrak{g}$. To avoid confusing notation, we denote by a and b the points of $\Gamma_n(\mathfrak{g})$ corresponding respectively to $\mathfrak{a}$ and $\mathfrak{b}$. G acts in the usual way on $\Gamma_n(\mathfrak{g})$. We say that $\mathfrak{b}$ is a *contraction of* $\mathfrak{a}$ *in* $\mathfrak{g}$ if b lies in the closure of the orbit $G(a)$ in $\Gamma_n(\mathfrak{g})$. That is, $\mathfrak{b}$ is a contraction of $\mathfrak{a}$ in $\mathfrak{g}$ if $\mathfrak{b}$ is a limit of subalgebras of $\mathfrak{g}$ which are conjugate to $\mathfrak{a}$. If $\mathfrak{b}$ is a contraction of $\mathfrak{a}$ in $\mathfrak{g}$, then $\mathfrak{b}$ is a contraction of $\mathfrak{a}$.

Proposition 3.2. *Let the Lie group G act differentiably on the differentiable manifold M. Let $x\in M$ and let $E(x)$ be the set of $z\in M$ such that $\mathfrak{g}_z$ is conjugate to $\mathfrak{g}_x$. If $y\in M$ is in the closure of $E(x)$, then $\mathfrak{g}_y$ contains a subalgebra $\mathfrak{a}$ which is a contraction of $\mathfrak{g}_x$ in $\mathfrak{g}$.*

Proof: There exists a sequence (z_j) of points of $E(x)$ such that $z_j\to y$. Let $m=\dim\mathfrak{g}_x$. Since $\Gamma_m(\mathfrak{g})$ is compact we may, by choosing a subsequence, assume that $\mathfrak{g}_{z_j}$ converges to a limit $\mathfrak{a}$ in $\Gamma_m(\mathfrak{g})$. Since the set $\mathscr{A}_m$ of m-dimensional subalgebras of $\mathfrak{g}$ is closed in $\Gamma_m(\mathfrak{g})$, $\mathfrak{a}$ is a subalgebra of $\mathfrak{g}$. But, since $z_j\in E(x)$, we have $\mathfrak{g}_{z_j}=(\operatorname{Ad} g_j)(\mathfrak{g}_x)$ for some $g_j\in G$. Hence $(\operatorname{Ad} g_j)(\mathfrak{g}_x)\to\mathfrak{a}$ and consequently $\mathfrak{a}$ is a contraction of $\mathfrak{g}_x$. Since $\mathfrak{g}_{z_j}\to\mathfrak{a}$ and, by the continuity of isotropy subalgebras, $\mathfrak{g}_{z_j}\to\mathfrak{g}_y$ (where, in this case, the arrow denotes convergence in $\Gamma(g)$), it follows from 2.4.c that $\mathfrak{a}\subset\mathfrak{g}_y$. This completes the proof.

Remark. We have formulated our results for the case of Lie groups acting differentiably on differentiable manifolds. However, with the obvious trivial modifications, all of our proofs and results hold for the case of complex Lie groups acting holomorphically on complex analytic manifolds and for the case of algebraic groups acting algebraically on non-singular algebraic varieties (over an algebraically closed field of characteristic 0).

The following result of Kostant [7, p. 1003] is an easy consequence of Proposition 3.2.

Corollary 3.3. *Let* $\mathfrak{g}$ *be a complex semi-simple Lie algebra of rank* r *and let* $y \in \mathfrak{g}$. *Then the centralizer* $\mathfrak{g}_y$ *of* y *in* $\mathfrak{g}$ *contains an* r*-dimensional abelian subalgebra.*

Proof. Let G be the adjoint group of $\mathfrak{g}$ and let G act on $\mathfrak{g}$ in the obvious manner. If $x \in \mathfrak{g}$, then the isotropy subalgebra of $\mathfrak{g}$ at x for the action of G on $\mathfrak{g}$ is just the centralizer $\mathfrak{g}_x$ of x in $\mathfrak{g}$. Let $\mathscr{R}$ be the set of regular elements of $\mathfrak{g}$. Then $\mathscr{R}$ is a non-empty Zariski-open, and hence dense, subset of $\mathfrak{g}$. If $x \in \mathscr{R}$, then $\mathfrak{g}_x$ is a Cartan subalgebra of $\mathfrak{g}$, thus $\mathfrak{g}_x$ is in particular an r-dimensional abelian subalgebra of $\mathfrak{g}$. But all Cartan subalgebras of $\mathfrak{g}$ are conjugate. Therefore, if $x \in \mathscr{R}$ and $y \in \mathfrak{g}$, y is in the closure of $E(x)$. Consequently, by Proposition 3.2, $\mathfrak{g}_y$ contains a subalgebra $\mathfrak{a}$ which is a contraction of $\mathfrak{g}_x$ in $\mathfrak{g}$. But the set of r-dimensional abelian subalgebras of $\mathfrak{g}$ is a closed subset of $\Gamma_r(\mathfrak{g})$. Hence $\mathfrak{a}$ is an abelian subalgebra of $\mathfrak{g}$.

We remark that Kostant's proof of this result is essentially the same, for the particular case at hand, as our proof of Proposition 3.2.

The key point of the above proof is that every contraction of an abelian subalgebra is abelian. In general, very little is known about the problem of finding all contractions of a subalgebra in a Lie algebra. However, HERMANN has obtained some interesting results [5, Chapter 11; 6, Thm. 4.1] concerning contractions of a maximal compactly embedded, subalgebra $\mathscr{R}$ in a semi-simple Lie algebra $\mathfrak{g}$.

Relatively stable subalgebras 3.4. (See [19].) Let $\mathfrak{g}$ be a Lie algebra over R, let $\mathfrak{b}$ be an n-dimensional subalgebra of $\mathfrak{g}$ and let $\mathfrak{a}$ be a subalgebra of $\mathfrak{b}$. Then $\mathfrak{a}$ is a *stable subalgebra of* $\mathfrak{g}$ *relative to* $\mathfrak{b}$ if there exists a neighborhood U of $\mathfrak{b}$ in the set $\mathscr{A}_n$ of n-dimensional subalgebras of $\mathfrak{g}$ such that if $\mathfrak{c} \in U$, then $\mathfrak{c}$ contains a subalgebra $\mathfrak{a}'$ of $\mathfrak{g}$ which is conjugate to $\mathfrak{a}$ in $\mathfrak{g}$. That is, every subalgebra of $\mathfrak{g}$ near $\mathfrak{b}$ contains a subalgebra conjugate to $\mathfrak{a}$. It is shown in [19] that if the Lie algebra cohomology space $H^1(\mathfrak{a}, \mathfrak{g}/\mathfrak{b})$ vanishes (here $\mathfrak{g}/\mathfrak{b}$ is considered as an $\mathfrak{a}$-module by means of the adjoint representation of $\mathfrak{a}$ on $\mathfrak{g}$), then $\mathfrak{a}$ is a stable subalgebra of $\mathfrak{g}$ relative to $\mathfrak{b}$. The result below is an easy consequence of this theorem and the continuity of the isotropy subalgebras.

Proposition 3.5. *Let the Lie group* G *act differentiably on the differentiable manifold* M *and let* $x \in M_n$. *Let* $\mathfrak{a}$ *be a subalgebra of* $\mathfrak{g}_x$ *such that* $H^1(\mathfrak{a}, \mathfrak{g}/\mathfrak{g}_x) = 0$. *Then there exists an open neighborhood* U *of* x *in* M_n *such that if* $y \in U$, *then* $\mathfrak{g}_y$ *contains a subalgebra of* $\mathfrak{g}$ *which is conjugate to* $\mathfrak{a}$.

We list below several examples of relatively stable subalgebras:

a) Let $\mathfrak{b}$ be a subalgebra of $\mathfrak{g}$ and let $\mathfrak{a}$ be a semi-simple subalgebra of $\mathfrak{b}$. Then $\mathfrak{a}$ is a stable subalgebra of $\mathfrak{g}$ relative to $\mathfrak{b}$.

b) Let $\mathfrak{a}$ be a Cartan subalgebra of $\mathfrak{g}$ and let $\mathfrak{b}$ be a subalgebra of $\mathfrak{g}$ which contains $\mathfrak{a}$. Then $\mathfrak{a}$ is stable relative to $\mathfrak{b}$.

Thus, for example, if G acts differentiably on M and if $x \in M_n$ is such that $\mathfrak{g}_x$ contains a Cartan subalgebra $\mathfrak{a}$, then there is a neighborhood U of x in M_n such that $\mathfrak{g}_y$ contains a conjugate of $\mathfrak{a}$ for every $y \in U$.

References

1. Bourbaki, N.: Groupes et algèbres de Lie, Chap. 1, Algèbres de Lie, Paris, Hermann, 1960.
2. Chevalley, C.: Théorie des groupes de Lie, II, Groupes Algebriques, Paris, Hermann, 1951.
3. — Theorie des groupes de Lie, III, Theoremes generaux sur les algebres de Lie, Paris, Hermann, 1955.
4. Floyd, E. E.: Orbits of torus groups operating on manifolds, Ann. of Math., **65**, 505—512 (1957).
5. Hermann, R.: Lie groups for Physicists, New York, Benjamin, 1966.
6. — Compactification of homogeneous spaces, I, J. Math. and Mech., **14**, 655—678 (1965).
7. Jacobson, N.: Lie Algebras, New York, Interscience, 1962.
8. Kostant, B.: The principal three-dimensional subgroups and the Betti numbers of a complex simple Lie group, Amer. J. Math., **81**, 973—1032 (1959).
9. Montgomery, D., H. Samelson, and C. T. Yang: Exceptional orbits of highest dimension, Ann. of Math., **64**, 131—141 (1956).
10. — —, and L. Zippin: Singular points of a compact transformation group, Ann. of Math., **63**, 1—9 (1956).
11. —, and C. T. Yang: The existence of a slice, Ann. of Math., **65**, 108—116 (1957).
12. Orbits of highest dimension, Trans. Amer. Math. Soc., **87**, 284—293 (1958).
13. —, and L. Zippin: Topological transformation groups, New York, Interscience (1955).
14. Mostow, G. D.: Equivariant embeddings in euclidean space, Ann. of Math., **65**, 432—446 (1957).
15. — On a conjecture of Montgomery, Ann. of Math., **65**, 513—516 (1957).
16. Nijenhuis, A., and R. Richardson: Deformations of Lie Algebra structures, J. Math. and Mech., **17**, 89—106 (1967).
17. Palais, R.: On the existence of slices for actions of non-compact Lie groups, Ann. of Math., **73**, 295—323 (1961).

18. RICHARDSON, R.: A rigidity theorem for subalgebras of Lie and associative algebras, Illinois J. Math., **11**, 92—111 (1967).
19. — Deformations of subalgebras of Lie algebras, J. Diff. Geom., to appear.

Non-Compact Lie Group Actions

R. F. WILLIAMS*

The second half of this paper will be an outline of certain recent results of the author; the first half is an introduction to the program of "global analysis" formulated by S. SMALE, required to explain the second half. Though the basic ideas of this program go back to POINCARÉ and G. D. BIRKHOFF, and important contributions were made by PONTRJAGIN, KOLMOGOROFF, ANASOV and others, it is largely the work of SMALE that has inspired the current flowering of the subject in this country.

For a real exposition of this program and a good bibliography, see Smale's long expository paper, Differential Dynamical Systems [1]. The author would like to thank SMALE for many helpful conversations and for early copies of [1]. At the same time, the reader should be warned that the following is only my understanding of his program.

1. Smale's Program

The principal goal is the study of smooth actions of the reals $\mathbb{R}$ on a (compact) smooth manifold M. However, it turns out to be very profitable to study first the (simpler) smooth actions of the integers $\mathbb{Z}$ on M. Thus one is led to the space $\mathrm{Diff}(M)$ of C^r diffeomorphisms in the C^r-topology, (r large). Then Smale's program, in this setting, is:

To find and classify, up to a reasonable equivalence relation, a dense Baire subset of $\mathrm{Diff}(M)$. *The classification should be manageable in the sense that it is via a countable set of invariants.*

It seems quite possible that a set of diffeomorphisms – called $\mathscr{S}$ here – recently defined by SMALE will turn out to be such. In fact, $\mathscr{S}$ is even open and it is striking that this can be proved so easily and conceptually [1, p. 65]. We begin with definitions.

For $f \in \mathrm{Diff}(M)$, $x \in M$ is a *non-wandering* point provided that for each neighborhood N of x, there is an integer n such that $f^n(N) \cap N \neq \phi$. The set $\Omega(f)$ of all non-wandering points is closed and invariant; it contains all periodic points.

* Supported in part by NSF Grant GP 5591.

Example. There is a diffeomorphism $f: S^n \to S^n$ with two fixed points, p^+ and p^- such that if x is a third point, then $f^n(x) \to p^+$ or p^- according as $n \to +\infty$ or $-\infty$. Then $\Omega(f) = \{p^+, p^-\}$. Note that $\Omega(f)$ is where the "action" is.

A closed invariant subset Λ of M has a *hyperbolic structure* provided there is a splitting of the tangent bundle to M, restricted to Λ,

$$T_M | \Lambda = E^u \oplus E^s,$$

where the splitting is invariant under the differential Df, $Df|E^u$ is an expansion and $Df|E^s$ is a contraction. (There are constants $C>0$, $\lambda>1$ such that for $v \in E^u$, $\|(Df)^n v\| \geqslant C\lambda^n \|v\|$, for some metric. Similarly, for $v \in E^s$.) For $\Lambda = M$, this definition was given by Anasov.

Example. The matrix $A = \begin{pmatrix} 2 & 1 \\ 1 & 1 \end{pmatrix}$ describes a linear map of the plane onto itself which, as A is unimodular, induces an automorphism f of the two dimensional torus, T^2. The eigen vectors define two line bundles on the plane which pass down to the torus as E^u and E^s. As one eigen value is larger than 1 and the other smaller, it follows that $E^u \oplus E^s$ is a hyperbolic structure for f on all of T^2. (Such a diffeomorphism, for which the whole manifold has a hyperbolic structure, is called an ANASOV diffeomorphism. See [1, p. 20].)

Axiom A. a) *$\Omega(f)$ has a hyperbolic structure;*
b) *the periodic points of f are dense in $\Omega(f)$.*

Smale's Axiom *B* will not be reproduced here as it is more technical and does not directly concern us. It requires transversal intersections of stable and unstable manifolds [1, p. 54]. Then for our purposes, $\mathcal{S}$ could be either the set of all $f \in \mathrm{Diff}(M)$ satisfying either A or both A and B.

It is a basic and unsolved problem to decide whether these sets are dense in $\mathrm{Diff}(M)$.

A candidate for the equivalence relation for diffeomorphisms in $\mathcal{S}$ is topological conjugacy on the non-wandering set. That is, $f \sim f'$ if there is a *topological* homeomorphism $h: \Omega(f) \to \Omega(f')$ such that $hf = f'h$. (Or h could be required to be defined on a neighborhood of $\Omega(f)$.) One of the invariants of such equivalence classes is the Weil-Artin-Mazur zeta function [1, p. 31] defined for a function $g: X \to X$ by

$$\zeta(t) = \exp \sum_{i=1}^{\infty} \frac{N_i t^i}{i}$$

where N_i is the number of points of X fixed under g^i. There is the important

Conjecture (Smale). If $f \in \mathrm{Diff}(M)$ satisfies Axioms A and B and M is compact, then the zeta function of f is rational.

Smale has proved this for f satisfying Axioms A and B in case the bundle E^u is orientable over $\Omega(f)$; I have proved it [4] for the zeta function of $f|\Lambda$, where Λ is an attractor (see below), if f satisfies Axiom $A(a)$. If these two proofs could be combined, it would prove the conjecture in general.

It is a basic theorem of Smale's that if f satisfies Axiom A, then $\Omega(f)=\Omega_1 \cup \Omega_2 \cup \cdots \cup \Omega_n$, where the Ω_i's are closed, invariant and mutually disjoint and this partition cannot be further refined. The Ω_i's are called *basic sets.* An important first step in Smale's program is to characterize and then classify basic sets. An *attractor* Λ is a basic set having a (compact) neighborhood N such that $f(N) \subset N$ and $\bigcap_{n \geqslant 0} f^n(N)=\Lambda$. If, in addition, f satisfies Axiom B, there is a partial ordering $\Omega_i \leqslant \Omega_j$, which means, roughly, "some points are tending from Ω_i to Ω_j". Note that attractors of f and f^{-1} occupy the extremes in this partial ordering.

It is not hard to see that a zero-dimensional attractor with hyperbolic structure would have to be finite. Thus in a sense, the simplest non-trivial case is that of 1-dimensional attractors. In this case, the characterization problem is solved in [3]: 1-dimensional attractors are (generalized) solenoids.

2. An Example

Example. There is a diffeomorphism $f: S^3 \to S^3$ which satisfies Axiom A (and B) where $\Omega(f)$ is the union of two disjoint p-adic solenoids $\Sigma^- \cup \Sigma^+$. If $x \in S^3-(\Sigma^- \cup \Sigma^+)$, then $f^n(x) \to \Sigma^-$ or Σ^+ according as to whether $n \to -\infty$ or $+\infty$.

Let S^3 be the join of two smooth 1-spheres S_A and S_B. Then S_A and S_B have tubular neighborhoods A and B which are chosen so that $A \cup B=S^3$ and $A \cap B=T^2$, a smooth torus. Let S_1 be a smooth unknotted 1-sphere passing p times around B in its interior and let S_{-1} be situated in A, just as S_1 is in B.

Let $T_1 \subset \mathrm{Int}\, B$ be the boundary of a tubular neighborhood $v(S_1)$ of S_1 and $T_{-1} \subset \mathrm{Int}(A)$ be the boundary of a tubular neighborhood $v(S_{-1})$ of S_{-1}. Then the linking numbers $\ell(S_{-1}, S_B)=p=\ell(S_A, S_1)$. Hence there is a diffeomorphism $f: S^3 \to S^3$ taking S_{-1} to S_A and S_B to S_1. We take f to send $v(S_{-1})$ to A and B to $v(S_1)$.

f can be described intuitively as follows: first, stretch B out to p times its length, shrinking its cross-sectional diameter by about $1/4p$.

Then twisting it forms it into p loops and sends it back into B's interior. Note that any map near f in the C^1-topology does this as well.

Then $f(T_i)=T_{i+1}$ for $i=-1$ and 0; this equation defines T_i for all other $i\in\mathbb{Z}$. Similarly $f(S_i)=S_{i+1}$ and $f(v(S_i))=v(S_{i+1})$ defines S_i and $v(S_i)$ for $i\in\mathbb{Z}$. Let $\Sigma_+=\bigcap_{i>0} v(S_i)$ and $\Sigma_-=\bigcap_{i<0} v(S_i)$. Note that if x is between or on S_i and S_{i+1}, then $f(x)$ is between or on S_{i+1} and S_{i+2}. It follows that $\Omega(f)\subset\Sigma_-\cup\Sigma_+$. It follows from general principles that the periodic points of f are dense in Σ_α, $\alpha=\pm$, so that $\Omega(f)=\Sigma_+\cup\Sigma_-$. Note that for $p=2$, $f|\Sigma_\alpha$ is just the expansive map (or its inverse) described in [2], $\alpha=\pm$.

Finally, note that the directions "parallel" to S_1 in $v(S_1)$ are stretched by f; these determine a line bundle E^u over $v(S_1)$. Similarly, the directions parallel to a fiber of the tubular neighborhood $v(S_1)$ are shrunk, and determine a plane bundle E^s over $v(s_1)$. Then $E^u\oplus E^s$ is the tangent bundle of M, restricted to $v(S_1)$ so that f has a hyperbolic structure over Σ_+. Similarly, f has a hyperbolic structure over Σ_-.

3. One-dimensional Attractors

The remainder of this paper is an outline of [3]. First, smooth one-dimensional branched manifolds are defined, just as the unbranched ones, except that three kinds of neighborhood are allowed: the reals, the halfclosed reals and a copy of the letter upsilon, Y. This last is chosen so that the two branches are tangent at their juncture with the trunk. Note that branched manifolds have tangent bundles. Thus smooth mappings as well as immersions between them can be defined. The boundary is defined as usual; the branch set of a branched 1-manifold is likewise clear. Both these sets are finite, if the manifold is compact.

Let K be a branched 1-manifold and $g: K\to K$ be an immersion satisfying:

Axiom 1. *g is an expansion;*

Axiom 2. *all points of K are non-wandering under g;*

Axiom 3. *each point of K has a neighborhood whose image under g is an arc.*

Then let Σ be the inverse limit of the sequence

$$K \xleftarrow[g]{} K \xleftarrow[g]{} K \xleftarrow[g]{} \cdots.$$

For a point $a=(a_0,a_1,a_2,\ldots)\in\Sigma$, let $h(a)=(g a_0, a_0, a_1,\ldots)$. Then h is a homeomorphism. Σ is called a (generalized) *solenoid* and h the *shift map* determined by g and K.

Characterization Theorem. *If f is a diffeomorphism of a compact manifold which has a 1-dimensional attractor Λ, then Λ is a generalized solenoid and $f|\Lambda$ is a shift map. Conversely, each such solenoid and shift map occur as a 1-dimensional attractor Λ with hyperbolic structure (even for a map $f: S^4 \to S^4$).*

Structure Theorem. *a) Each point of a solenoid has a neighborhood of the form $[0,1] \times C$ where C is a Cantor set. b) The periodic points of the shift map of a solenoid are dense. c) The Weil-Artin-Mazur zeta function of a solenoid is rational.*

I have recently found with the aid of R. D. Anderson that solenoids are *not* in general homogeneous, so that a) cannot be greatly improved. c) is subsumed by the theorem of the author alluded to in Section 1.

A large problem concerning these "solenoids" remains: their classification. In particular, given $g: K \to K$ and $g': K' \to K'$ satisfying Axioms 1–3, when are the two resulting systems, $h: \Sigma \to \Sigma$ and $h': \Sigma' \to \Sigma'$ topologically conjugate? Though certain sufficient conditions are known (see [3; 3.6]) a better approach might be along the following lines. If K is orientable, so is Σ; if not, a double cover arranges a closely related Σ'. Orientable Σ's are all fiber bundles over a 1-sphere with a Cantor set as fiber, though not vice-versa.

Finally, it appears to the author that this program can be carried out as well for n-dimensional attractors.

Added in proof. Abraham and Smale have recently found an example showing that the set $\mathscr{S}$ of §1 is *not* dense. The classification problem for solenoids has essentially been solved by the author (mimeo. notes, Northwestern University).

References

1. Smale, S.: Differentiable Dynamical Systems, Mimeographed notes, Berkeley 1966—67. Bull. Amer. Math. Soc. **73**, 747—817 (1967).
2. Williams, R. F.: A note on unstable homeomorphisms, Proc. A. M. S. **6**, 308—9 (1955).
3. — One-dimensional non-wandering sets, Topology **6**, 473—487 (1967).
4. — The zeta function of an attractor, to appear in the Proceedings of the Michigan State meeting on Topology 1967. (Mimeo notes, Northwestern University).

Fixed Point Sets of Transformation Groups on Separable Infinite-Dimensional Frechet Spaces*

JAMES E. WEST

Let G be a topological group, and let X be a topological space. An action of G on X as a transformation group is a continuous function f from $G \times X$ to X with the property that the correspondence $g \to f(g, .)$ defines a homomorphism of G into the group $G(X)$ of all homeomorphisms of X onto itself. The fixed point set of an action f is defined to be the set of all x in X such that $f(g, x) = x$ for each g in G. Under certain conditions on G and X it is possible to determine something about the structure of the fixed point set of any action of G on X as a transformation group. An example of this type of theorem was established by P. A. SMITH, who proved in 1941 [8] that if p is a prime number, G is a group of order p, and X is a compact Hausdorff space (or a locally compact Hausdorff space of finite covering dimension) which is acyclic in mod p (Čech) homology, then the fixed point set of each action of G on X is non-empty and acyclic modulo p.

On the other hand, it is sometimes possible to prove that nearly every closed subset of X is the fixed point set of some action of G on X. A theorem in this direction was proven by V. L. KLEE, Jr. [6] which states that each compact subset of a separable, infinite-dimensional, Hilbert space is the fixed point set of periodic homeomorphisms of the space onto itself of all finite periods (greater than one, of course). Another theorem, proven by ANATOLE BECK in 1958 [3], states that if G is the real numbers and X is a metric space, then the existence of an action of G on X as a transformation group with an empty fixed point set is equivalent to the existence for each closed subset F of X of an action of G on X with fixed point set F.

The purpose of this paper is to state a generalization of the above theorem of KLEE which in some sense complements Beck's theorem, for

* This paper is essentially the text of a talk presented by the author to the Tulane Conference on Compact Transformation Groups in May of 1967. The material is from the author's dissertation (Louisiana State University, 1967), which was directed by Professor R. D. ANDERSON. The author was supported as a graduate student by a National Defense Education Act fellowship.

instead of restricting only the group, we restrict only the space and obtain a very similar result. More specifically, let f be an action of G on X as a transformation group. If $f(g,x)=x$ for all x in X implies that g is the identity of G, then f is said to be "effective", and if $f(g,x)=x$ for some x in X implies that g is the identity, then f is said to be "free". If F is the fixed point set of f, let f be called "semi-free" if $f(g,x)=x$ implies that x is in F or that g is the identity of G. Our principal result is as follows (Theorem 5):

If G is a topological group and X is a separable, infinite-dimensional Frechet space, then there exists an effective action of G on X if and only if for each closed subset F of X there is a semi-free action of G on X with fixed point set F.

Here, the term "Frechet space" means a metrizable, complete, locally convex, linear, topological space. This theorem immediately implies the following (Corollary 1): "Let G be a topological group. There exists a non-trivial action of G on a separable infinite-dimensional Frechet space if and only if for each closed subset F of such a space X there exists an action of G on X with fixed point set F." We also have another corollary (Corollary 2): "Let X be a separable, infinite-dimensional Frechet space, and let G be a compact topological group. There exists for each closed subset F of X a semi-free action of G on X with fixed point set F if and only if G is metrizable."

General Outline

By results of C. Bessaga and A. Pełczynski [4], of M. I. Kadec [5], and R. D. Anderson [2], the following is true:

Theorem 1. *All separable, infinite-dimensional, Frechet spaces are homeomorphic.*

Because of Theorem 1, any seperable, infinite-dimensional, Frechet space may be chosen as a model in which to set the proof. The one chosen is s, the countably infinite Cartesian product of the real line with itself. This choice is made because of the following results made possible by the structure of s as a product space:

Theorem 2. *Let s' denote the countably infinite Cartesian product of s with itself, and let T be the set of all points of s'.whose set of coordinates is dense in s. T is homeomorphic to s.*

The proof of Theorem 2 is by a geometric argument ending in an appeal to a theorem of R. D. Anderson [1, Cor. 5.4]. From Theorem 2, we obtain as a corollary:

Theorem 3. *Let G be a topological group, and suppose that there exists an effective action of G on s as a transformation group. Then there exists a free action of G on s as a transformation group.*

Theorem 3 may be proven simply from Theorem 2 as follows: Let f be an effective action of G on s as a transformation group. Construct an action of G on s' from f in a coordinate-wise fashion by the formula $(g,(x,y,z,\ldots))\to(f(g,x), f(g,y), f(g,z),\ldots)$. This new action leaves T as an invariant set, and the restriction to T defines a free action of G on T, which by Theorem 2 is homeomorphic to s.

We now state the final theorem leading to our main result.

Theorem 4. *Let G be a topological group, and suppose that there exists a free action of G on s as a transformation group. Then for each closed subset F of s there exists a semi-free action of G on s as a transformation group with fixed point set F.*

The proof of Theorem 4 consists of finding a suitable sequence of closed neighborhoods of F each of which contains its successor in its interior and the intersection of which is F. An action of G is defined on the relative complement in each set of the interior of its successor in such a manner that these actions together with the trivial action of G on F simultaneously define a semi-free action of G on s with fixed point set F. The product structure is used strongly to determine that the separate actions of G can be easily induced from a given free action of G on s.

This construction is not particularly dependent upon the structure of s as a linear space. Indeed, Theorem 4 can be obtained from a more general theorem proven in a similar manner. Let a space X be said to have the reflective isotopy property if the homeomorphism h of $X\times X$ onto itself defined by $h(x,x')=(x',x)$ is invertibly isotopic to the identity. (An invertible isotopy on a space Y is a continuous function H from $Y\times[0,1]$ to Y such that for each t in $[0,1]$, the function $y\to H(y,t)$ is in $G(Y)$ and the function F from $Y\times[0,1]$ to Y such that for each t in $[0,1]$, $F(H(y,t),t)=y$ is continuous.) A space is perfectly normal if each closed subset of it is a G_δ set.

Theorem 4′. *Let X be the countably infinite Cartesian product of a space Y with itself, and suppose that for each positive integer n the n-fold Cartesian product of Y with itself is perfectly normal. If X has the reflective isotopy property, then for a topological group G, there exists a free action of G on X if and only if for each closed subset F of X there is a semi-free action of G on X with fixed point set F.*

Theorem 4 follows from Theorem 4′ by a theorem of R. Y. T. WONG who proved in his dissertation [9] that every element of $G(s)$ is invertibly isotopic to the identity. As $s \times s$ is homeomorphic to s, the result follows.

Collecting theorems 1, 3, and 4 yields the following:

Theorem 5. *If X is a separable, infinite -dimensional, Frechet space and G is a topological group, then there exists an effective action of G on X as transformation group if and only if for each closed subset F of X there is a semi-free action of G on X as a transformation group with fixed point set F.*

From Theorem 5, Corollary 1 is immediate.

Corollary 1. *Let G be a topological group, and let X be a separable, infinite-dimensional, Frechet space. If there exists a non-trivial action of G on X as a transformation group, then for each closed subset F of X there exists an action of G on X with fixed point set F.*

Because each compact metrizable group is isomorphically embeddable in a countably infinite Cartesian product of General Linear groups of finite-dimensional real vector spaces [7, p. 100], and because for each such product group there is an effective action on s as a transformation group, and because the exsistence of a free action of a topological group as a transformation group on a metric space implies metrizability of the group, the following is true:

Corollary 2. *Let X be a separable, infinite-dimensional, Frechet space, and let G be a compact topological group. There exists for each closed subset F of X a semi-free action of G on X as a transformation group with fixed point set F if and only if G is metrizable.*

References

1. ANDERSON, R. D.: Topological properties of the Hilbert cube and the infinite product of open intervals, Trans. Amer. Math. Soc., **126,** 200—216 (1967).
2. — Hilbert space is homeomorphic to the countable infinite product of lines, Bull. Amer. Math. Soc., **72,** 515—519 (1966).
3. BECK, A.: On invariant sets, Ann. of Math., **67,** 99—103 (1958).
4. BESSAGA, C., and A. PEŁCZYNSKI: Some remarks on homeomorphisms of F-spaces, Bull. Acad. Polon. Sci. Ser. Sci. Math. Astron. Phys., **10,** 265—270 (1962).
5. KADEC, M. I.: On topological equivalence of all separable Banach spaces, (Russian) Dokl. Akad. Nauk. S. S. S.R., **167,** 23—25 (1966).

6. KLEE, V. L., Jr.: Fixed point sets of periodic homeomorphisms of Hilbert space, Ann. of Math. (2) **64,** 393—395 (1956).
7. MONTGOMERY, D., and LEO F. ZIPPIN: Topological Transformation groups, New York: Interscience Publishers, Inc. (1955).
8. SMITH, P. A.: Fixed point theorems for periodic transformations, Amer. J. Math., **63,** 1—8 (1941).
9. WONG, R. Y. T.: On homeomorphisms of certain infinite dimensional product spaces, Trans. Amer. Math. Soc., **128,** 148—154 (1967).

On a Kind of Discrete Transformation Group*

S. Kinoshita

In this note we present an attempt at a topological study of a part of the theory of discontinuous groups (see for instance [1]). The idea is to study it as an effective transformation group G on a compact metric space X, where any element of G is a homeomorphism of type 2 (see § 1.1) or the identity of X. Some results are given in § 2, which are more or less like that of discontinuous groups. However the author does not know how to characterize the discontinuity theorem topologically (see for instance [1]). In § 1 we study a special case; and results will be applied in § 2. In § 3 a theorem related to covering transformation groups is proved.

§ 1

1.1. Let X be a compact metric space which contains at least three points. An autohomeomorphism g of X is called *of type* 2, if there exist two distinct points a and b such that for each $x \in X-b$, $\lim_{n\to\infty} g^n(x)=a$ and for each $x \in X-a$, $\lim_{n\to-\infty} g^n(x)=b$ (See [2]). Both a and b are fixed points of g called the *attractive* point and *repulsive* point of g, respectively. The set $a \cup b$ is called the *fixed point set* of g. It was proved in [3] that if g is of type 2, then for each compact subset $C \subset X-b$, $\lim_{n\to\infty} g^n(C)=a$ and for each compact subset $C \subset X-a$, $\lim_{n\to-\infty} g^n(C)=b$.

1.2. Now let G be an effective transformation group on X. Suppose that there exist two distinct points a and b such that each element $g \in G$ is a homeomorphism of type 2, except for the identity, whose fixed point set is $a \cup b$. Let G_+ be the set of all g whose attractive point is a, and G_- the set of all g whose attractive point is b. Clearly $G=G_+ \cup G_- \cup e$ and these three sets are mutually disjoint. We put $O=X-a\cup b$. Further, assume that there exists a point $c \in O$ such that $\underline{\lim}_{g\in G_+} g(c)=a$ and $\overline{\lim}_{g\in G_-} g(c)=b$. Then we may say that G is *of type* 2. The purpose of this paragraph is to prove the following:

* This research was supported in part by NSF Grant GP-5458.

Theorem 1. *Let X be a compact metric space which contains at least three points and let G be a commutative transformation group of type 2. Then G is an infinite cyclic group.*

1.3. Assume that X is a compact metric space which contains at least three points and that G is a commutative transformation group of type 2 on X. Then G has no element of finite order and from this it follows that if $g_1^n = g_2^n$, then $g_1 = g_2$.

Lemma 1.1. *Let $g_0 \in G$ and $c' = g_0(c)$. Then $\overline{\lim}_{g \in G_+} g(c') = a$ and $\overline{\lim}_{g \in G_-} g(c') = b$.*

Proof. Since $\overline{\lim}_{g \in G_+} g(c') = \overline{\lim}_{g \in G_+} g(g_0(c))$ and $\overline{\lim}_{g \in G} g(c) = a \cup b$, we have $\overline{\lim}_{g \in G_+} g(c') \subset a \cup b$. Now assume on the contrary that there is a sequence $\{g_n\} (g_n \in G_+)$ such that $\lim_{n \to \infty} g_n(c') = b$. Since $\lim_{n \to \infty} g_n g_0(c) = b$, we have $b = g_0^{-1}(b) = g_0^{-1}(\lim_{n \to \infty} g_n g_0(c)) = \lim_{n \to \infty} g_0^{-1} g_n g_0(c) = \lim_{n \to \infty} g_n(c)$, which is a contradiction.

1.4. **Lemma 1.2.** *For any g_1 and g_2 of G_+ there exists g_3 of G_+ such that both g_1 and g_2 are powers of g_3.*

Proof. First assume that there are g_1 and g_2 of G_+ such that all $g_1^m g_2^m(c)$ are different from each other. There exist disjoint neighborhoods U of a and V of b such that $g_1(V) \cap U = \phi$. Since $\overline{\lim}_{g \in G} g(c) = a \cup b$, there is only a finite number of $g(c)$ contained in $X - (U \cup V)$. Then there exists $N > 0$ such that if $n > N$, all $g_1^m g_2^n(c)$ are contained in $U \cup V$. Since $\lim_{m \to \infty} g_1^m g_2^n(c) = a$ and $\lim_{m \to -\infty} g_1^m g_2^n(c) = b$, there is an integer m' such that $g_1^{m'} g_2^n(c) \in V$ and $g_1^{m'+1} g_2^n(c) \in U$. But this contradicts the assumption that $g_1(V) \cap U = \phi$. Hence we have proved that for any g_1 and g_2 of G_+ there exist m_1 and m_2 such that $g_1^{m_1} = g_2^{m_2}$.

Put $m = \text{g.c.d.}(m_1, m_2)$. Then there are integers s_1 and s_2 such that $m = s_1 m_1 + s_2 m_2$. Put $m_1 = d_1 m$ and $m_2 = d_2 m$. Then $s_1 d_1 + s_2 d_2 = 1$. Since $g_1^{m_1} = g_2^{m_2}$, $g_1^{d_1} = g_2^{d_2}$. Now put $g = g_1^{s_2} g_2^{s_1}$. Then we have $g^{d_2} = g_1^{s_2 d_2} g_2^{s_1 d_2} = g_1^{s_2 d_2} g_1^{s_1 d_1} = g_1$, and similarly $g^{d_1} = g_2$. Put $g_3 = g$ if $d_2 > 0$ and $g_3 = g^{-1}$ if $d_2 < 0$. Then clearly $g_3 \in G_+$, and hence g_3 satisfies the required condition.

1.5. **Proof of Theorem 1.** Let $g \in G_+$ and V a neighborhood of b such that $a \notin \bar{V}$. Assume on the contrary that G is not an infinite cyclic group. Then by Lemma 1.2, there exists a sequence $g = g_0, g_1, \ldots$ of G_+ such that $g_{i-1} = g_i^{n_i}$, where $n_i > 1$, for each $i = 1, 2, \ldots$ Since $\lim_{g \in G_+} g(c) = a$, there is only a finite number of g_i such that $g_i(c) \in V$. Hence there exists $N>0$ such that if $i>N$, $g_i(c) \notin V$. Therefore if $i>N$, $g_i^{1-n_1 n_2 \dots n_i}(g_0(c)) = g_i(c) \notin V$. Clearly $g_i' = g_i^{1-n_1 n_2 \dots n_i}$ is an element of G_- and $g_i' \neq g_j'$ if $i \neq j$. Hence $\overline{\lim_{g \in G-}}\, g(c') \neq b$, where $c' = g_0(c)$, which contradicts Lemma 1.1.

§ 2

2.1. Let X be a compact metric space which contains at least three points and let G be an effective transformation group on X such that each element g of G is a homeomorphism of type 2 or the identity e. We denote the attractive point of $g_\alpha(\neq e)$ by a_α and the repulsive point of g_α by b_α, respectively. The set $a_\alpha \cup b_\alpha$ is called the fixed point set of $g_\alpha(\neq e)$. Let L be the set of all attractive points of G, i. e. $a \in L$ if and only if a is the attractive point for some g of G. The following properties of G follow directly from the definition of G: G has no element of finite order; for each x of $X-L$, the correspondence of $g \in G$ to $g(x)$ is a one to one correspondence of G onto the orbit $G(x)$ of x.

2.2. **Theorem 2.1.** *For any* $g \in G, G(L) = L$.

Proof. Let $a_\alpha \in L$ be the attractive point and b_α the repulsive point of g_α, respectively. Then for any $x \in X-b$, $\lim_{n\to\infty} g_\alpha^n(x) = a_\alpha$. Therefore for any $x \in X-b$, $g(a_\alpha) = g(\lim_{n\to\infty} g_\alpha^n(x)) = \lim_{n\to\infty} g(g_\alpha^n(x)) = \lim_{n\to\infty} (g g_\alpha g^{-1})^n g(x)$ $= \lim_{n\to\infty} h^n(g(x))$, where $h = g^{-1} g_\alpha g$. Hence $g(a_\alpha)$ is the attractive point of h^n, and therefore $g(L) \subset L$. Since $g^{-1} \in G$, similarly we have $g^{-1}(L) \subset L$, and this means that $L \subset g(L)$. Therefore $g(L) = L$.

Theorem 2.2. *L is a set of two points or dense in itself.*

Proof. Suppose that there are three distinct attractive points a_1, a_2 and a_3 of g_1, g_2 and g_3, respectively. Let b_1 be the repulsive point of g_1. Then $b_1 \neq a_2$ or $b_1 \neq a_3$. Assume that $b_1 \neq a_2$. Then $\lim_{n\to\infty} g_1^n(a_2) = a_1$ and $g_1^n(a_2) \in L$ for any n. Therefore a_1 is a limit point of L.

2.3. A transformation group G is said to satisfy *condition CM*, if for any g_1, g_2 of G with the same attractive point, $g_1 g_2 = g_2 g_1$.

Theorem 2.3. *Assume that G satisfies the condition CM. Then for any $g_1, g_2(\neq e) \in G$, the fixed point sets of g_1 and g_2 are the same or disjoint.*

Proof. Let a_1 and a_2 be attractive points of g_1 and g_2, and b_1 and b_2 repulsive points, respectively.

First assume that $a_1 = a_2$. Suppose on the contrary that $b_1 \neq b_2$. Then $g_2(b_1) \neq b_1$ and $g_2(b_1) \neq a_1$. Therefore $g_1 g_2(b_1) \neq g_2(b_1)$. On the other hand since $a_1=a_2, g_1g_2=g_2g_1$. Therefore $g_1g_2(b_1)=g_2g_1(b_1)=g_2(b_1)$, which is a contradiction. Therefore $b_1=b_2$.

Next we assume that $a_1=b_2$. Then a_1 is the attractive point of g_2^{-1}. Therefore $b_1=a_2$ by the above argument.

Theorem 2.4. *Assume that G satisfies the condition CM. Then for any* $g_1, g_2(\neq e)$ *of G,* $g_1, g_2=g_2g_1$ *if and only if they have the same fixed point set.*

Proof. Let $a_1 \cup b_1$ be the fixed point set of g_1 and $a_2 \cup b_2$ that of g_2. First suppose that $a_1 \cup b_1=a_2 \cup b_2$. If $a_1=a_2$ (and $b_1=b_2$), then $g_1g_2=g_2g_1$. If $a_1=b_2$ (and $a_2=b_1$), then $g_1g_2^{-1}=g_2^{-1}g_1$. Hence we have $g_2g_1=g_1g_2$.

Next we suppose that a_1, b_1, a_2 and b_2 are four distinct points. Further assume that $g_2(a_1)\neq b_1$. Since $g_2(a_1)\neq a_1$, $g_1g_2(a_1)\neq g_2(a_1)=g_2g_1(a_1)$. Therefore $g_1g_2\neq g_2g_1$. Now assume that $g_2(a_1)=b_1$. Then $g_2(b_1)=g_2^2(a_1)\neq a_1$. Since $g_2(b_1)\neq b_1$, $g_1g_2(b_1)\neq g_2(b_1)=g_2g_1(b_1)$. Therefore $g_1g_2\neq g_2g_1$.

2.4. Let X be a topological space and G a transformation group on X. Then G is called *properly discontinuous,* if for each $x\in X$ there exists a neighborhood U of x such that $U\cap g(U)=\phi$ for any $g\in G$ $(g\neq e)$.

Lemma 2. *Let X be a Hausdorff space and G a properly discontinuous transformation group on X. Then for each* $x\in X$, $\underline{\lim}_{g\in G}\, g(x)=\phi$.

Proof. Assume on the contrary that there exist points x and y such that $y\in \underline{\lim}_{g\in G}\, g(x)$. Since G is properly discontinuous, there exists a neighborhood U of y such that $U\cap g(U)=\phi$ for any $g(\neq e)$. But U contains $g_1(x)$ and $g_2(x)$ for some g_1 and $g_2(g_1\neq g_2)$. Hence $g_2g_1^{-1}(g_1(x))\in U$ and $g_1(x)\in U$. Therefore $g_2g_1^{-1}(U)\cap U\neq\phi$, which is a contradiction.

2.5. Again we assume that X is a compact metric space which contains at least three points and that G is an effective transformation group on X such that each element g of G is a homeomorphism of type 2 or the identity. Put $O=X-\bar{L}$. Then G is said to satisfy the *condition CT,* if there exists a point $c\in O$ such that for any $a\in\bar{L}$ and for any neighborhood U of a there is a neigborhood V of a which satisfies the following condition: If $g_\alpha(c)\in V$, then the attractive point a_α of g_α is contained in U.

Theorem 2.5. *Assume that G satisfies the conditions CM and CT and that G is properly discontinuous on O. Put* $G_a=\{g\in G|g(a)=a\}$, *where* $a\in L$. *(*G_a *is the isotropy subgroup of G at a.) Then* G_a *is an infinite cyclic group.*

Proof. By Theorem 2.3 there exists the unique point $b \in L$ such that $g(b)=b$ for any $g \in G_a$. Hence any element $g(\neq e)$ of G_a is a homeomorphism of type 2 whose fixed point set is $a \cup b$. Further by Theorem 2.4, G_a is a commutative group. Therefore by Theorem 1 we have only to prove that there exists $c \in X - a \cup b$ such that

$$\overline{\lim_{g \in G_+}} g(c)=a \quad \text{and} \quad \overline{\lim_{g \in G_-}} g(c)=b,$$

where $G_+=\{g_\alpha \in G_a | a_\alpha = a\}$ and $G_-=\{g_\alpha \in G_- | a_\alpha = b\}$.

Let c be a point of O at which condition CT is satisfied. First we prove that $\underline{\lim}_{g \in G} g(c)=a \cup b$. Since G is properly discontinuous on O, $\overline{\lim}_{g \in G_a} g(c) \subset \bar{L}$ by Lemma 2. Let $y \in \overline{\lim}_{g \in G_a} g(c)$, and assume on the contrary that $y \notin a \cup b$. Then by condition CT there is a neighborhood V of y such that if $g_\alpha(c) \in V$, then $a_\alpha \in X - a \cup b$. But this contradicts our definition of G_a.

Now again by condition CT there is a point $c \in O$ and a neighborhood V of a such that if $g_\alpha(c) \in V$, then $a_\alpha \in X - b$, where $g_\alpha \in G_a$. But then $\underline{a_\alpha}$ must be a, and therefore g_α is an element of G_+. This means that $\underline{\lim}_{g \in G_-} g(c)=b$. Similarly we have $\overline{\lim}_{g \in G_+} g(c)=a$. Hence, as stated, G_a is an infinite cyclic group by Theorem 1.

§ 3

Let X be a topological space and G a transformation group on X. Then G is called *fixed point free*, if for any $g \in G$ $(g \neq e)$, g has no fixed point. If G is properly discontinuous on X, then G is fixed point free. G is said to satisfy *Sperner's condition*, if for any compact subset C of X, $C \cap g(C) = \phi$, except for a finite number of g. It is proved in [4] that if G is fixed point free and satisfies Sperner's condition, then G is properly discontinuous on X, where X is a locally compact Hausdorff space. Of course these studies are closely related to covering transformation groups (See [4]).

Now let X be a compact metric space and O an open subset of X. Let G be a transformation group on X such that $g(O)=O$ for any $g \in G$. Then G is called *regular* on O, if for any $x \in O$ and for any $\varepsilon > 0$ there exists $\delta > 0$ such that $d(g(x), g(y)) < \varepsilon$ for any $g \in G$, whenever $d(x, y) < \delta$. In this paragraph we prove the following

Theorem 3. *Let X be a compact metric space and O an open subset of X. Let G be a transformation group on X such that $g(O)=O$ for any $g \in G$. If G is properly discontinuous and regular on O, then G satisfies Sperner's condition on O.*

Proof. Since G is properly discontinuous on O, if $O=X$, then G is a finite group. Therefore the theorem is true for this case. Now suppose that O is a proper subset of X. Let C be a compact subset of O. Since $X-O$ is compact, there is 2ε-neighborhood U of $X-O$ such that $U \cap C = \phi$. Let U' be an ε-neighborhood of $X-O$. Let $x \in C$. Since G is properly discontinuous on O, $\underline{\lim}_{g\in G}\, g(x) \subset X-O$ by Lemma 2. Therefore $g(x) \in U'$ for any $g \in G$, except for a finite number of g, say $g_1, \ldots, g_n$. Since G is regular on O, there is $\delta > 0$ such that if $d(x,y) < \delta$, then $d(g(x), g(y) < \varepsilon$. Then if V is a δ-neighborhood of x, $g(V) \subset U$ for any $g \in G$, except for these $g_1, \ldots, g_n$. The set C is covered by a finite number of such kind of neighborhood V. Then it is easy to see that $g(C) \subset U$, except for a finite number of g. Therefore G satisfies Sperner's condition on O.

References

1. LEHNER, J.: Discontinuous groups and automorphic functions, Amer. Math. Soc. (1964).
2. KINOSHITA, S.: On quasi-translations in 3-space, Topology of 3-manifolds, edited by M. K. Fort, Prentice-Hall, Inc. 223—226 (1962).
3. HOMMA, T., and S. KINOSHITA: On the regularity of homeomorphisms of E^n, Jour. Math. Soc. Japan, **5**, 365—371 (1953).
4. KINOSHITA, S.: Notes on covering transformation groups, Proc. Amer. Math. Soc. **19**, 421—424 (1968).

Problems

Let G be an infinite cyclic transformation group acting on an n-dimensional Euclidean space R^n. Is there a point $x \in R^n$ such that $G(x)^- \neq R^n$? (The answer is yes for $n=2$ by Brouwer's translation theorem.) There is an infinite cyclic transformation group G acting on the plane such that $G(y)^- = R^2$ for some $y \in R^2$ (BESICOVITCH, HEDLUND). It seems to be unknown whether or not there is an infinite cyclic transformation group G acting on R^n $(n \geqslant 3)$ such that $G(y)^- = R^n$ for some $y \in R^n$. (See also KINOSHITA, Coll. Math. *6* (1958).) – S. KINOSHITA.